MCQ Based Practice Book on
Food Technology, Process & Food Engineering

NIPA® GENX ELECTRONIC RESOURCES & SOLUTIONS P. LTD.
New Delhi-110 034

About the Authors

Dr. Suresh Chandra is working as Professor in the Department of Processing and Food Engineering, College of Technology, of Sardar Vallabhbhai Patel University of Agricultural and Technology, Meerut (UP). He obtained his B.Tech. in Ag. Engg. from GBPUAT, Pantnagar; M.Tech. in Post Harvest Engineering and Technology from AMU Aligarh, and Ph.D. in Agric. Process and Food Engineering from SVPUAT, Meerut. He has more than 20 year's service experience in teaching, research and extension. Dr. Chandra published more than 100 research paper, 12 books, 8 practical manuals, 3 training bulletins and various book chapters. He has been conferred with various prestigious awards in his carrier of teaching. He is reviewers of various national and international journals. He has guided 10 M.Tech. and 05 Ph.D. scholars and many more in his guidance. He established Agro-Processing Centre and Food Processing Unit in the University. He has handled many projects funded by ICAR/RKVY/UPCAR and Deptt. of Agriculture, U.P as a PI and Co-PI. He has organized various seminar/symposia/webinar and entrepreneurship development training for farmers, women and rural unemployed persons, and delivered invited talks as resource person and guest lecturers as subject expert in external institutions. He has participated in various summer/winter schools and short courses organized by ICAR-Institutions/SAUs.

Er. Shobhi Choudhary is a Research Scholar in the Department of Processing and Food Engineering, College of Technology, SVPUA&T Meerut Uttar Pradesh, India 250110. She completed her B.Tech. in Agricultural Engineering in 2021 from CCSU, Meerut, India. She obtained his M. Tech. degree in Agricultural Engineering (Process and Food Engineering) from Department of Agricultural Engineering, College of Technology, SVPUA&T Meerut Uttar Pradesh, India in 2023 and awarded PG Scholar award in international conference and is currently pursuing a Ph.D. in the same field from the same college and university. She has published more than 5 research papers in National and International Journals and more than 10 abstracts in various national and international conferences, 1 book chapter as main and co-author.

Er. Alka Singh is a Research Scholar in the Department of Processing and Food Engineering, College of Technology, SVPUA&T Meerut Uttar Pradesh, India 250110. She completed her B.Tech. in Agricultural Engineering in 2018 from TMU, Moradabad India. She obtained his M. Tech. degree in Agricultural Engineering (Process and Food Engineering) from Department of Agricultural Engineering, College of Technology, SVPUA&T Meerut Uttar Pradesh, India in 2023 and awarded PG Scholar award in international conference and is currently pursuing a Ph.D. in the same field from the same college and university. She has published more than 5 research papers in National and International Journals and more than 10 abstracts in various national and international conferences, 1 book chapter as main and co-author.

Er. Amit Kumar is a Research Scholar in the Department of Processing and Food Engineering, College of Technology, SVPUA&T, Meerut (UP). He completed his B.Tech. in Agricultural Engineering in 2021 from Dr. A.P.J. Abdul Kalam Technical University, Lucknow, India. He obtained his M. Tech. degree in Agricultural Engineering (Process and Food Engineering) from Department of Agricultural Engineering, College of Technology, SVPUA&T Meerut Uttar Pradesh, India in 2023 and awarded PG Scholar award in an international conference and is currently pursuing a Ph.D. in the same field from the same college and university. He has published more than 5 research papers in National and International Journals, and more than 10 abstracts in various national and international conferences,

To
My Grand Parent

MCQ Based Practice Book on Food Technology, Process & Food Engineering

Suresh Chandra
Professor
Department of Processing and Food Engineering
S.V.P. University of Agriculture & Technology
Meerut, Uttar Pradesh

Shobhi Choudhary
Research Scholar
Department of Processing and Food Engineering
S.V.P. University of Agriculture & Technology
Meerut, Uttar Pradesh

Alka Singh
Research Scholar
Department of Processing and Food Engineering
S.V.P. University of Agriculture & Technology
Meerut, Uttar Pradesh

Amit Kumar
Research Scholar
Department of Processing and Food Engineering
S.V.P. University of Agriculture & Technology
Meerut, Uttar Pradesh

NIPA® GENX ELECTRONIC RESOURCES & SOLUTIONS P. LTD.
New Delhi-110 034

NIPA® GENX ELECTRONIC RESOURCES & SOLUTIONS P. LTD.

101,103, Vikas Surya Plaza, CU Block
L.S.C. Market, Pitam Pura, New Delhi-110 034
Ph : +91-11-43860225, Mob.: +91 9717133558, 9540816132
E-mail: newindiapublishingagency@gmail.com
Website: www.nipaersources.com

Print ISBN: 978-93-58874-06-8
ebook ISBN: 978-93-58875-21-8

Composed and Designed by NIPA®.

Preface

Food Safety Officers (FSOs) are responsible for analyzing food samples gathered from different places for examination. These professionals examine if there is any suspicious or hazardous material in the ingredients or the food product ensuring the safety of consumers. Many organizations/institutions conduct entrance examination for admission in various disciplines related to food science, dairy science and technology, postharvest technology, food processing technology, agricultural process and food engineering. There are also some competitive examinations such as ICAR-NET/ARS, CSIR-NET, SRF/JRF, GATE and Food Safety Officer by State Public Service Commissions. Hence there is a great need of terminological books covering a lot of important technical words related to Food Science.

Every effort has been made to ensure that the information compiles from various sources such as publications and internet sites are accurate and up to date. However, the readers are strictly advised to confirm the data in case of doubt. Mistakes and omissions if any are inadvertent and will be corrected when pointed out by the readers. We request all readers to send their valuable suggestions with regard to the improvement of this book.

Finally, we are making a plea to those who make use of this book to supply us with information or points of view that differ with those expressed in. We know that there will be errors, for which we alone are responsible, and we will appreciate the opportunity to correct those. This effort should be viewed as a first try at the comprehensive view of agricultural process and food engineering. In subsequent editions or volumes, there will be opportunity not only to correct, but to expand the perspective and thus to achieve more complete our objective of providing the serious students with a consistent and correct answer of how we make better. Feedback from readers of this book will certainly help us to improve its quality in future.

Though the book is sure to make more useful to the reader as syllabus but the authors and it is earnestly hoped that suggestions to improve the book shall be forthcoming from the readers. In last, authors are thankful to staff of NIPA, New Delhi; giving us an opportunity to publish this important book in the favor of student's keen interest.

Authors

Food Safety Officer Examination Syllabus

Uttar Pradesh

Desirable and potentially undesirable food constituents and their importance. Water, Carbohydrates, Lipids, Proteins, Vitamins, Minerals: Classes, Nomenclature, structure and their chemistry.

Enzymes: Nomenclature, Classification and specificity of enzymes, enzyme Kinetics, enzymes and their role in modification of foods. Single cell protein (SCP). Metabolic Pathways: Carbohydrates, proteins and fats, catabolism and anabolism, Digestion and absorption, Assimilation and Transport of nutrients in human beings.

General characteristics of microorganism: Classification, morphology, physiology, growth, nutrition and reproduction; Pure culture techniques and maintenance of cultures, control of microorganism. Sources of contamination. Microbiological standards of foods. Food poisoning and food borne infections, Food toxins, food plant sanitation, inspection and control, personnel hygiene, beneficial microorganisms and their utilization in food fermentation.

Balance diet, Recommended Dietary Allowances (RDA), Dietary utilization and disturbances, Functions and energy values of foods, Basal energy metabolism: BV, NPU, BMR, PER calculations, Protein quality as reference protein. Dietary allowances and standards for different age groups, Techniques for assessment of human nutritional status, Causes and preventions of malnutrition. Biochemical, clinical manifestation & preventive measures due to vitamins deficiency. Diseases caused due to food adulteration.

Basic knowledge of major Indian crops, their total production, losses in storage, storage of perishable and non-perishable foods, Storage requirements: Storage environment and its interaction with stored product; temperature and moisture migration. Storage practices (including fumigation and aeration of stored product). Storage structures for Perishable and non-perishables: bulk storage structures (bins and silos) Modified and Controlled atmosphere storage systems.

Technology of fruits and vegetable products: Juices, pulp, Puree, Paste, Ketchup, Sauce, soup, concentrates and powders, Squashes and cordials. Beverage: Still and carbonated. James, Jellies and Marmalades. Preserves, candies and crystallized fruits.

Processing of oilseeds Pretreatments of oilseeds, oil milling, solvent extraction, impurities present in oils and fats. Vegetable oil refining, hydrogenation, enter esterification, processing of de-oiled cake into protein products, textured protein. Peanut butter, Margarine and Spread. Different quality parameters: Peroxide value, Saponification value, Iodine value, acid value, RM value, Polanski value, Adulteration and its detection in oils and fats. Rancidity, auto oxidation and anti-oxidants.

Fluid Milk: Physico-chemical properties, Production and collection, cooling and transportation of milk. Clarification and filtration of milk, pasteurized and homogenized milk.: Whole, Standardized, Toned, Double toned, skim and special milks, Test for milk quality and detection of dairy adulterant. Milk products: cream, butter, Butter oil/ Ghee, Cheeses, Curd and Yoghurt, Sweet meat, Ice cream. Evaporated and Condensed milk, milk powders. Indian Dairy Products, Manufacture of casein, and Lactose. Causes and prevention of defects in dairy products.

Scientific slaughtering, Tendering and curing of meat, Beef Mutton, Pork Sausages and other meat products. Fish and poultry processing. Physical, Chemical Nutritional and Functional characteristics of Egg. Processing of Egg and egg products.

Principle underlying spoilage and preservation of foods: Low and high temperature, drying and dehydration, chemical, salt and sugar, Radiation, HHP, hurdle technology,

Parboiling and Milling of paddy, Quality characteristics, Curing and aging of rice, Processed rice products. Wheat milling into flour and semolina, Flour grades and their suitability for baking purposes, Flour quality, Macaroni products, Baked products: Bread, Biscuits and cakes, Breakfast cereals. Dry and Wet milling of corn, Starches and its conversion products, Malting of barely, beer, wine, cider, vinegar, Milling of legume-pulses.

Spices: Composition, Structure and characteristics. Preservation and processing

of spices, Processing of Tea and coffee, cocoa processing, chocolate and confectionary products.

Concept of food packaging, polymeric, glass, metal packaging materials, Aseptic Packaging. Newer trends, Package printing, Packaging Laws and Regulations, Evaluation of food packaging materials and package performance.

Quality Control/Quality Assurance, Legislation for food safety: National and International criteria, sampling, records, risk analysis and management. Microbial contamination, Nutritional imbalance. Pesticide residues, permitted food additives. Relevant Foods laws: PFA, FPO, SWMA, MPO, Ag Mark, and BIS Standards. Food Safety and Standard Act (FSSA). Testing food for its safety. The HACCP system and food Safety management systems used in Controlling microbiological hazards. Food standards and Specifications, NABL and ISO certification requirements.

Tamil Nadu, Assam and Andhra Pradesh

1. **Food Chemistry:** Carbohydrates-Structure and functional properties of mono, di & oligo- polysaccharides including starch, cellulose, pectic substances and dietary fiber; Proteins – Classification and structure of proteins in food. Lipids-Classification and structure of lipids, Rancidity of fats, Polymerization and polymorphism; Pigments-Carotenoids, chlorophylls, anthocyanins, tannins and myoglobin; Food flavours-Terpenes, esters, ketones and quinones; Enzymes-Enzymatic and non-enzymatic browning in different foods
2. **Food Microbiology, Hygiene & Sanitation:** Characteristics of microorganisms-Morphology, structure and detection of bacteria, yeast and mold in food,

Spores and vegetative cells; Microbial growth in food- Intrinsic and extrinsic factors, Growth and death kinetics, serial dilution method for quantification; Food spoilage- Contributing factors, Spoilage bacteria, Microbial spoilage of milk and milk products, meat and meat products; Food borne disease- Toxins produced by Staphylococcus, Clostridium and Aspergillus; Bacterial pathogens- Salmonella, Bacillus, Listeria, Pseudomonas, Candida, Escherichia coli, Shigella, Campylobacter; Food Hazards of natural origin – sea food toxins, biogenic amines, alkaloids, phenolic compounds, protease inhibitors and phytates. Types of hazards, biological, chemical, physical hazards, importance of safe foods. Priniciples and methods of food preservation. Hygiene and sanitation in food sector – pest control measures, Garbage and Sewage disposal, Water – Sources, purification, Hazards Analysis & Critical Control Point (HACCP), Good Manufacturing Practices (GMP), Good Hygienic Practices (GHP), Good laboratory Practices (GLP).

3. **Food Processing Technology:** Food processing- structure, composition, nutritional significance and types of processing methods for various categories of foods: Cereals (Rice-milling, parboiling, Barley- Pearling, malting, brewing, Corn- wet and dry milling, Wheat-milling, pulses (milling, germination, cooking, roasting, frying, canning and fermentation), and oil seeds (extraction and refining), Fruits & vegetables (canning, drying and dehydration, concentration, freezing, IQF, thawing and fermentation); plantation crops (primary and secondary processing of Tea, Coffee and Cocoa), Spices (Oleoresin and essential oil extraction), Meat, fish and poultry (ante mortem inspection, slaughtering and dressing, post mortem examination, canning, curing, smoking, freezing, dehydration and fermentation), eggs (quality inspection and dehydration) milk (receiving, separation, clarification, pasteurization, standardization, homogenization, sterilization, UHT and fermentation). Unit operations of food processing – grading, sorting, peeling and size reduction. Product development – Consumer trends and their impact on new product development; stages- to conceive ideas, evaluation of ideas, developing ideas into products, test marketing and commercialization; criteria for selection of raw materials, sensory evaluation, objective evaluation, standardization. Types & functions of packaging materials including smart packaging and biodegradable materials used in foods. Packaging material as a threat, impact on health and controlling measures. Surveys – types, sampling procedures for conducting surveys and for quality control.
4. **Food Laws and Organizations:** Laws and Regulations – Brief review of regulatory status in India before the advent of FSSAI: FPO, MMPO, MFPO, Prevention of Food Adulteration Act, Paradigm shift from PFA to FSSAI; Overview of Food Safety Standards Act 2006, Food Safety Standards Rules & Regulations, 2011 (Licensing and registration of food businesses, Food product standards & Additives, Packaging & labelling, Contaminants, toxins an residues, Laboratory and sample analysis, Prohibition and restriction on sales), Organizational hierarchy, Powers and duties of Food business Operator, Food

Safety Officer, Designated Officer, Food Analyst; Food recall and Traceability, Other Acts: Essential commodities Act, Legal Metrology Act, AGMARK Codex Alimentarius – development and issue of standards, Committees under Codex, role in maintaining harmony in food standards. National Organizations – Bureau of Indian Standards, ICMR, ICAR, NABL, Council for social welfare, Ministry of Health & Family Welfare – delivery Health Services in India. Export and Quality Control through Export Inspection Council (EIC), APEDA and MPEDA. International Organizations FAO (Food & Agriculture Organization), WHO (World Health Organization), ISO, WTO, APLAC, ILAC.

5. **Public & Occupational health and Nutrition:** Public Health: Definition of Public Health and Associated Terms, Current Concerns in Public Health : Global and Local, Core functions and scope of public health, History of public health and evolution of Public Health, Concept of health and disease, Natural history of disease, Levels of prevention, Concept of health and disease, Natural history of disease, Determinants of health, Infectious Disease and Germ Theory, Introduction to public health ethics, Globalization and Health, Governance in Health, International Health Regulations, Indian Health Systems. Occupational Health – health of workers in industries safety measures, occupational diseases. Nutrition – Assessment of nutritional status, Balanced diet, food sources of nutrients, essential vitamins, amino acids and fatty acids, their deficiency diseases and toxicity, PER, Recommended dietary allowances for various nutrients, Antinutrients, clinical and diet surveys. Programmes on Nutrition in India (mid-day meals at schools, anganwadi systems, ICDS, NIDDCP, NNAPP, WIFS, National Food Security Mission, SABLA, FSSAI initiatives on food fortification, FFRC, FFWP, NPPNB due to VAD, NGCP).

Maharashtra

CHEMISRTY

- Table Periodic / Concepts of qualitative Analysis.
- Bonding & structure of organic compound & their nomenclature.
- Mechanism of organic Reactions Alkanels & Cycloalkanes.
- Chemical calculations -Introduction to Molecular Spectroscopy.
- Chemical Thermodynamics.
- Industrial Organic Chemistry.
- Aromatic Hydrocarbons -Haloarenes, Phenols, Ether and Epoxides.
- Source of organic compounds chemical industry.

INTRODUCTION OF FOOD PROCESSING AND TECHNOLOGY

- F&VP, Milk, Meat, Oil, Grain milling, Tea-Coffee, Spices & Condiments processing.
- Food processing techniques (Minimal processing technologies, Photochemical processes, Pulsed electric field, Hurdle technology.)

- Food Preservation Techniques (Pickling, Drying, Smoking, Curing, Canning, Bottling, Jellying, Modified atmosphere, Pasteurization etc.)

CODEX ALIMENTARIUS COMMISSION (CODEX)

- Introduction.
- Standards, Codex of practice, Guidelines and Recommendations.
- Applying Codex Standards.
- Codex India – Role of Codex Contact point, National Codex Contact Point (NCCP)
- Core functions of NCCP-India.
- National Codex Committee of India – TOR (Terms of Reference), Functions, Shadow Committees etc.

EMERGING ISSUES IN FOOD PROCESSING

- Organic food.
- Identifying organic foods, Advantages, The Organic Certification Process, Organic food labeling.
- GM food.
- Why are GM food produced. Main issues of concern for Human Health, How are GM Food regulated Internationally, Regulation in India.
- Role of WHO to improve evaluation of GM food.
- Benefits & Controversies.
- Irradiated Food
- How is food Irradiated, Sources of radiations used.
- Potential uses of Food Irradiation.
- Labelling of Irradiated Food.
- Freeze dried food
- Definition, Principle of Freeze-drying, Process.
- The benefits of Freeze-Drying.
- Functional Foods & Nutraceuticals.
- Functional foods from plant sources, Animal sources.
- Nutraceuticals, Dietary supplements, Regulation.
- Nano-tech in food processing.
- What is Nanotechnology, use in food products and processing.
- Food Fortification & Modification.

FSSAI Act 2006

- Food Safety & Standards Act, 2006,
- Major aspects of the act.

ADVANCES IN FOOD SAFETY & QUALITY MANAGEMENT

- Pre-requisite Programme – Good Hygienic Practices (GHP)
- Objective, Scope & Use, Key aspects of Hygiene Control Systems etc.
- Pre-requisite Programme – Good Manufacturing Practices (GMP)
- What is GMP, GMP in Food Industry etc.
- Food Safety Management Systems – HACCP
- HACCP seven principles, HACCP applications
- HACCP based SOPs by USDA
- Food Safety Management Systems – ISO 22000
- Key elements of ISO 22000
- What does ISO 22000 bring to the HACCP Method
- Why to use ISO 22000
- Quality Management Systems – ISO 9001
- Introduction, Clauses of ISO 9001 : 2000
- Documentation structure of ISO 9001 : 2000
- Quality Manual, Mandatory Procedures, SOP's, Formats and Records etc.

OTHER RELATED ACTS

- B.I.S
- C.R.P.C / I.P.C.
- EC (Essential Commodities Act)
- Consumer Protection Act.

Odisha

Food Safety Standards Act Rules & Regulations (50 Marks)

- Food: Definition,
- Food additives and processing aid,
- Contaminants,
- Toxic substances
- Heavy Metals
- Pesticides: Pest control,
- Antibiotic Residues and microbiological control
- Genetically modified foods,
- Organic foods,
- Functional foods,
- Proprietary foods
- Packaging and Labelling of Foods

- Restriction on advertisement & prohibition to unfair trade practices
- Responsibilities of the Food Business Operator
- Liability of manufactures, packers, whole-sellers, distributors and sellers
- Food Recall procedures
- Licensing and Registration of Food Business
- General requirements on Hygienic and Sanitary practices by the FBOs (Good Manufacturing Practice, Good Hygienic practice, Hazard Analysis & Critical Control Point/Personal Hygiene)
- Water pollution: Portable water: sources and methods of purification,
- Food and Water
- Borne Diseases

Chemistry (50 Marks)

- Carbohydrates,
- Proteins,
- Lipids – Definition,
- Classification and Properties, significance
- Essential and Non-Essential Acids
- Vitamins: Fat-soluble and water-soluble vitamins, their sources, properties and Significance.
- Enzymes- definitions, classification properties
- Chemical components of foods-Micro and Micronutrients and their dietary sources and biochemical rules
- Nutrition deficiencies and disorders
- To estimate acid value iodine value saponification value of lipids
- Food spoilage- microorganisms conditions (Factors & Control)
- Food poisoning (intoxication)- Bacterial, fungal, algal.
- Food preservation: Storage, physical and chemical methods
- Fermentation of food items: Types of cheese, yoghurt, bread, etc.

Jharkhand

PAPER-2: Food Technology

1. **Food Chemistry:** Carbohydrates -Structure and functional properties of mono, di & oligo-polysaccharides including starch, cellulose, pectic substances and dietary fibre; Proteins – Classification and structure of proteins in food. Lipids-Classification and structure of lipids, Rancidity of fats, Polymerization and polymorphism; Pigments-Carotenoids, chlorophylls, anthocyanins, tannins and myoglobin; Food flavours-Terpenes, esters, ketones and quinones; Enzymes-Enzymatic and non-enzymatic browning in different foods.

2. **Food Microbiology:** Characteristics of microorganisms-Morphology, structure and detection of bacteria, yeast and mold in food, Spores and vegetative cells; Microbial growth in food- Intrinsic and extrinsic factors, Growth and death kinetics, serial dilution method for quantification; Food spoilage- Contributing factors, Spoilage bacteria, Microbial spoilage of milk and milk products, meat and meat products; Food borne disease-Toxins produced by Staphylococcus, Clostridium and Aspergillus; Bacterial pathogens- Salmonella, Bacillus, Listeria, Escherichia coli, Shigella, Campylobacter; Food Hazards of natural origin – sea food toxins, biogenic amines, alkaloids, phenolic compounds, protease inhibitors and phytates. Types of hazards, biological, chemical, physical hazards factors affecting safety, importance of safe foods. General methods of food preservation and food processing preservation of food spoilage.
3. **Food Processing technology:** Food processing- structure, composition, nutritional significance and types of processing methods for various categories of foods: Cereals (Rice-milling, parboiling, Barley- Pearling, malting, brewing, Corn- wet and dry milling, Wheat-milling, pulses (milling, germination, cooking, roasting, frying, canning and fermentation), and oil seeds (extraction and refining) , Fruits & vegetables (canning, drying and dehydration, concentration, freezing, IQF, thawing and plantation crops (primary and secondary processing of Tea, Coffee and Cocoa), Spices (Oleoresin and essential oil extraction), Meat, fish and poultry (ante mortem inspection, slaughtering and dressing, post mortem examination, canning, curing, smoking, freezing, dehydration), milk (receiving, separation, clarification, pasteurization, standardization, homogenization, sterilization, UHT). Unit operations of food processing – grading, sorting, peeling and size reduction. Product development – Consumer trends and their impact on new product development; stages- to conceive ideas, evaluation of ideas, developing ideas into products, test marketing and commercialization; criteria for selection of raw materials, sensory evaluation, objective evaluation, standardization. Types & functions of packaging materials used in foods. Packaging material as a threat, impact on health and controlling measures. Surveys – types, sampling procedures for conducting surveys and for quality control.
4. **Food Laws and Organizations:** Laws and Regulations – Brief review of regulatory status in India before the advent of FSSAI: FPO, MMPO, MFPO, Prevention of Food Adulteration Act, Paradigm shift from PFA to FSSAI; Overview of Food Safety Standards Act 2006, Food Safety Standards Rules & Regulations, 2011 (Licensing and registration of food businesses, Food product standards & Additives, Packaging & labelling, Contaminants, toxins and residues, Laboratory and sample analysis, Prohibition and restriction on sales), Organizational hierarchy, Powers and duties of Food business Operator, Food Safety Officer, Designated Officer, Food Analyst; Food recall and Traceability, Other Acts: Essential commodities Act, Legal Metrology Act, AGMARK Codex Alimentarius – development and issue of standards, Committees under Codex, role in maintaining harmony in food standards. National Organizations

– Bureau of Indian Standards, ICMR, ICAR, NABL, Council for social welfare, Ministry of Health & Family Welfare – delivery Health Services in India. Export and Quality Control through Export Inspection Council (EIC), APEDA and MPEDA. International Organizations FAO (Food & Agriculture Organization), WHO (World Health Organization), ISO, WTO, APLAC, ILAC.

5. **Hygiene & Sanitation:** Hygiene and sanitation in food sector – pest control measures, Garbage and Sewage disposal, Water – Sources, purification, Hazards Analysis & Critical Control Point (HACCP), Good Manufacturing Practices (GMP), Good Hygienic Practices (GHP), Good laboratory Practices (GLP).

6. **Public & Occupational Health and Nutrition:** Public Health- Definition of Public Health and Associated Terms, Current Concerns in Public Health: Global and Local, Core functions and scope of public health, History of public health and evolution of Public Health, Concept of health and disease, Natural history of disease, Levels of prevention, Concept of health and disease, Natural history of disease, Determinants of health, Infectious Disease and Germ Theory, Introduction to public health ethics, Globalization and Health, Governance in Health, International Health Regulations, Indian Health Systems. Occupational Health- health of workers in industries safety measures, occupational diseases. Nutrition- Assessment of nutritional status, Balanced diet, food sources of nutrients, essential vitamins, amino acids and fatty acids, their deficiency diseases and toxicity, PER, Recommended dietary allowances for various nutrients, Antinutrients, clinical and diet surveys. Programmes on Nutrition in India (mid-day meals at schools, anganwadi systems, ICDS, NIDDCP, NNAPP, WIFS, National Food Security Mission, SABLA, FSSAI initiatives on food fortification, FFRC)

Contents

Section-I MCQ Practice Set

Section-II Short Notes for ICAR-NET/ARS/FSO

Section-I

MCQ Practice Set

Model Paper 1

1. Enzymes are made up of
   a) Carbohydrates b) Proteins
   c) Fats d) Lipids
2. Immediate precursor of chlorophyll a and chorophyll b is
   a) Protochlorophyllide b) Haemoglobin
   c) Biluribin d) Xanthophylls
3. Water soluble flavonoid pigments are
   a) Lycopene b) Lutein
   c) Anthocyanins d) Carotene
4. Which enzyme is used for the production of maltose syrup
   a) Destrin b) Fungal alpha-amylase
   c) Ligase d) Dextrose
5. The Molecular formula of Xanthone is
   a) $C_{13}H_8O_2$ b) $C_{10}H_2O_6$
   c) $C_{15}H_8O_2$ d) $C_{11}H_5O$
6. Joining of apoenzyme and co-enzyme is called
   a) Prosthestic group b) Holoenzyme
   c) Enzyme complex d) Helical Complex
7. Which of the following is a water soluble pigment?
   a) Chlorophyll a b) Chlorophyll b
   c) Carotenoids d) Phycobilins
8. Xanthophylls differ from carotenes in having
   a) Mg b) Cl
   c) Oxygen d) Hydrogen
9. Enzyme used in bread making is
   a) Amylase E b) Cellulase
   c) Pectinase d) None of the above
10. Chloroplast does not contain
   a) Xanthopyll b) Anthocyanins
   c) Carotene d) Plastids

11. Name the pigment which impart red colour to Tomato
    a) Erytrocyanin b) Alpha- Carotene
    c) Beta-Carotene d) Lycopene
12. The Co-enzyme is
    a) A metal b) A Vitamin
    c) Protein d) Inorganic Compound
13. Pigment responsible for the process of sunlight
    a) Chlorophyll b b) Chlorophyll a
    c) Anthocyanin d) Carotenoids
14. Enzyme involved in alcholoic fermentation is
    a) Lipase b) Ligase
    c) Zymase d) Sucrase
15. For Photosynthesis green plants require
    a) Chlorophyll b) Light
    c) CO_2 and H_2O d) All of these
16. Enzyme secreted by pancreas
    a) Trypsin b) Pepsin
    c) Chymotrypsin d) Alcohol Dehydrogensae
17. Hydrolyzable tannins are also called
    a) Acetic acid b) Citric acid
    c) Gallic acid d) Stearic acid
18. Diastase enzyme take part in digestion of
    a) Lipids b) Starch
    c) Carbohydrates d) Protein
19. Father of Microbiology was...
    a) E. Jenner b) L. Pasteur
    c) L.Philips d) R. Koch
20. Monoclonal Antibody was discovered by...
    a) AV Leuveanhocek b) Fletcher
    c) Kohler & Milstein d) L. Pasteur
21. Pencillin was discovered by...
    a) Fleming b) Walkshman
    c) Robert Fleming d) J. Lister
22. Best example of primary metabolite used in large scale is...
    a) Formaline b) Oil
    c) Alcohol d) Acetone
23. Citric acid is produced by...
    a) Bacteria b) Yeast
    c) Virus d) Fungus

24. Enrichmnet culture technique was discovered by...
    a) Bacteria b) Beijerinck
    c) E. Hesse d) L. Pasteur
25. For preservation of anaerobic culture in laboratory we use...
    a) Stab b) Slants
    c) Broth d) Petri plate
26. Short-term upto 6 months preservation of microbes was done by...
    a) Broth b) Glycerol culture
    c) Slants d) Stab
27. Peptone and Tryptone are used as a source of...
    a) Carbon b) Copper
    c) Water d) Nitrogen
28. Temperature of hot air oven for sterilization is best
    a) 100C for 30 min b) 170C for 30 min
    c) 160C for 30 min d) 180C for 30 min
29. Tyndallization is the process of...
    a) Autoclaving b) Sterilization
    c) Transparenting d) Hot air oven
30. Fermentation occurs in the
    a) Presence of O b) Absence of O
    c) Presence of N d) Absence of C
31. Bacteria that are responsible for fermentation of dairy milk are
    a) Azetobacter b) Rhizobium
    c) Aspergillus d) Lactobacillus
32. Plastic syringes, rubber gloves are sterilized by...
    a) Radiation b) Autoclaving
    c) Filtration d) Spirit
33. The germicidal range of UV light is...
    a) 254nm b) 1000nm
    c) 1500nm d) 500nm
34. Temperature in pasteurization is...
    a) 50°C b) 65°C
    c) 63°C d) 100°C
35. Industrial microbiology, mainly depends on phenomenon...
    a) Fermentation b) Pasteurization
    c) Sterilization d) Solidification
36. Bacitracin is produced by...
    a) Clostridium Spp. b) Staphylococcus spp.
    c) Bacillus subtilis d) E. coli

37. Father of microscopy was...
    a) Fletcher
    b) Fleming
    c) Flavus
    d) Leeuwenhoek
38. Size reduction of fruits and vegetables are mostly performed by:
    a) Crushing
    b) Shearing
    c) Cutting
    d) Squeezing
39. Percentage of husk in paddy is about:
    a) 18-22 per cent
    b) 15-18 per cent
    c) 20-24 per cent
    d) 25-30 per cent
40. Percentage of bran in the rice is:
    a) 2-4%
    b) 6-8%
    c) 4-6%
    d) 8-10%
41. Percentage of endosperm in the paddy:
    a) 68-70%
    b) 70-72%
    c) 72-74%
    d) 74-78%
42. In modern rice milling, vibratory separator used for
    a) Dehusking
    b) De-stoning
    c) Whitening
    d) Polishing
43. Parboiling of paddy is done to:
    a) Achieve maximum recovery of head rice
    b) Minimise the broken percentage
    c) Reduce the milling losses
    d) All are correct
44. In single boiling method, paddy is soaked in water for:
    a) 24-48 hours
    b) 24-72 hours
    c) 48-72 hours
    d) 12-24 hours
45. In steaming process, a soaked paddy exposed to steam at a pressure of about:
    a) 2 kg/cm^2
    b) 3 kg/cm^2
    c) 4 kg/cm^2
    d) 5 kg/cm^2
46. In comparison to raw rice is parboiled rice rich in:
    a) Protein
    b) Vitamin
    c) Minerals
    d) All are correct
47. Compared to raw rice, parboiled rice needs:
    a) Less polishing
    b) More polishing
    c) Doesn't need polishing
    d) None of these
48. Cooking quality of rice is represented by:
    a) Time of cooking
    b) Swelling capacity and expansion ratio
    c) Gruel quality and pastiness
    d) All are correct

49. Parboiled rice is inferior in:
    a) Starch b) Fat
    c) Protein d) Mineral
50. The shelling of paddy by centrifugal dehusker is due to:
    a) Shear b) Impact
    c) Crushing d) Rubbing
51. The shelling of paddy in rubber roll sheller is due to:
    a) Rubbing b) Crushing
    c) Shearing d) All are correct
52. Percentage of bran in paddy:
    a) 2-6% b) 2-4%
    c) 6-8% d) 8-10%
53. The steps of parboiling process are:
    a) Steaming-Soaking-Drying b) Dring-Soaking-Steaming
    c) Soaking-Steaming-Drying d) None of these
54. Which statement is not correct?
    a) India is the biggest producer of vegetables in the world.
    b) India is the second largest producer of fruits in the world.
    c) Production of fruits, vegetables and spices called horticulture.
    d) Uttar Pradesh is the largest producer of the wheat in India.
55. Highest productivity of wheat is in ________ state of India
    a) West Benga b) Punjab
    c) Maharashtra d) Odisha
56. Which one of the following sequence is correct in the context of three largest wheat producing states in India?
    a) Punjab, Uttar Pradesh and Haryana
    b) Uttar Pradesh, Haryana and Punjab
    c) Uttar Pradesh, Punjab and Haryana
    d) Punjab, Odisha , Uttar Pradesh
57. Which of the following has highest fat content?
    a) Ragi b) Wheat
    c) Bajra d) Oat
58. How many major constituents are in milk?
    a) 2 b) 6
    c) 8 d) 4
59. Which is a milk process that makes milk more easily digested by those with a sensitive digestive system
    a) Homogenization b) Ionization
    c) Evaporation d) Pasteurization

60. Yogurt contains mixed lactic acid culture containing
    a) Lactobacillus bulgaricus and Leuconostoc
    b) Lactobacillus bulgaricus and Propionibacterium
    c) None of the option
    d) Lactobacillus bulgaricus and Streptococcus thermophilus
61. Phosphatase test is applicable for
    a) Pasteurized milk b) Sterilized milk
    c) Homogenized milk d) Irradiated milk
62. Boiling point of milk?
    a) 90°C b) 100.17°C
    c) 150°C d) 121°C
63. Buffalo milk contain solid not fat (SNF) is
    a) 8.5 b) 6.5
    c) 10 d) 9
64. Yellow color of milk fat is due to presence of
    a) Vitamin D b) Carotenoids
    c) Calcium d) Magnesium
65. Plasma term used for
    a) MSNF b) MF
    c) Total milk fat d) Whey
66. In acidophilus milk how much culture is incubated?
    a) 1-2% b) 7-9%
    c) 3-5% d) 9-10%
67. Enzyme responsible for starch splitting?
    a) Lipase b) Protease
    c) Phosphatase d) Diastase
68. Skim milk contain fat content less than
    a) 0.5 b) 1.5
    c) 3 d) 4.5
69. In koumiss, finished product contain alcohol upto
    a) 5% b) 7.50%
    c) 1% d) 2.50%
70. The optimum pH and temperature for the growth of *S. thermophilus* is
    a) 5.1 and 60°C b) 3.5 and 95°C
    c) 6.8 and 38°C d) 4.7 and 72°C
71. Homogenized milk is the milk in which fat globules are subdivided into
    a) 2 micron b) 5 micron
    c) 7 micron d) 0.5 micron

72. Sequence involves in production of homogenized milk according to Indian conditions
   a) Preheating, cooling, homogenization, clarification, pasteurization
   b) Preheating, homogenization, pasteurization, clarification, cooling
   c) Preheating, clarification, homogenization, pasteurization, cooling
   d) Preheating, homogenization, clarification, pasteurization, cooing
73. Nature of milk is
   a) Basic b) Acidic
   c) Alkali d) Amphoteric
74. Bacteria used for fermentation should be
   a) Gram negative b) Gram positive
   c) Spore forming d) None of the option
75. Fat globules membrane is made up of
   a) Phospholipid b) Protein
   c) Small chain of Amino acid d) Phospholipid and protein
76. Carotene is a precursor of
   a) Vitamin D b) Vitamin C
   c) Vitamin A d) Vitamin K
77. In bottle sterilization is done at temperature
   a) 120-125°C b) 105-110°C
   c) 90-100°C d) 140-150°C
78. Which of the following involves preservation with the help of microorganisms?
   a) Aseptic processing b) Fermentation
   c) Pasteurization d) None of the option
79. Which of the following is not used as a preservation in fruits and vegetables?
   a) Acetic acid b) Bavistin
   c) Citric acid d) None of the option
80. Asepsis is the process of
   a) Maintenance of a clean and hygienic environment for storage and processing
   b) Keeping milk at low temperature and clean environment
   c) Treatment with hydrogen peroxide for storage
   d) Removing water from the food material
81. The classification of commodities into various fractions depending on commercial value and usage is known as
   a) Sorting b) Scalping
   c) Grading d) Cleaning
82. Drying is a process of
   a) Heat transfer b) Mass transfer
   c) Heat and Mass transfer d) None of the option

83. Which method is used to preserve food by high temperature?
   a) Refrigeration b) Ohmic heating
   c) Ultrasound d) Pasteurization
84. Freezing refers to the storage of food at temperature between
   a) 0 to 10°C b) -5 to 0°C
   c) -10 to -5°C d) -18 to -30°C
85. Liquid nitrogen is used in
   a) Cryogenic freezing b) Plate freezing
   c) Fluidized freezing d) Blast freezing
86. The destruction of all microorganisms in food by thermal processing is known as
   a) Pasteurization b) Blanching
   c) Sterilization d) Scalping
87. Which method is a non-thermal method?
   a) Drying b) Ultrasound
   c) Dehydration d) Blanching
88. Ohmic heating is also known as
   a) Electrical resistance heating b) Irradiation
   c) Ultrasound d) HPP
89. Nitrate and Nitrite is used to preserve
   a) Fruits and vegetables b) Juices
   c) Meat d) Bakery products
90. Which one is not an antioxidant preservative?
   a) BHA b) Sugar
   c) Ascorbic acid d) Sulphite
91. Dry ice is known as
   a) Solidified nitrogen b) Solidified CO_2
   c) Solidified water d) Solidified freon 12
92. Low energy ultrasound application is performed at frequency higher than
   a) 50KHz b) 100KHz
   c) 20KHz d) 500KHz
93. Pasteurization of milk is to be carried out at minimum temperature of
   a) 125°C b) 63°C
   c) 90°C d) 71°C
94. Which food is a perishable food?
   a) Onion b) Grains
   c) Fish d) Jaggery
95. Food spoilage by microbiological is due to
   a) Bacteria b) Rodent
   c) Insect d) Parasites

96. Mesophilic bacteria grow at temperature

a) 60º C b) 100º C

c) 120º C d) 38º C

97. Microbial and enzyme inactivation in ultrasound method is due to

a) Low temperature b) Cavitation process

c) High temperature d) None of the option

98. Size reduction of grains is caused by impact in

a) Gyratory b) Hammer mill

c) Crushing roll d) Jaw crusher

99. Which law gives good results for fine grinding

a) Kick law b) Rittinger law

c) Bond law d) All the option

100. Standard screens are used to measure the grain size in the range of

a) 0 - 30 µm b) 76mm - 250mm

c) 76mm - 38µm d) More than 100mm

Answer Key

1	b	2	a	3	c	4	b	5	a	6	b	7	d
8	d	9	a	10	b	11	d	12	b	13	b	14	c
15	d	16	a	17	c	18	b	19	b	20	c	21	a
22	c	23	d	24	b	25	a	26	b	27	d	28	c
29	b	30	b	31	d	32	a	33	a	34	c	35	a
36	c	37	d	38	c	39	a	40	c	41	b	42	b
43	d	44	b	45	c	46	d	47	a	48	d	49	c
50	b	51	c	52	a	53	c	54	a	55	b	56	c
57	d	58	d	59	a	60	d	61	a	62	b	63	d
64	b	65	a	66	c	67	d	68	a	69	d	70	c
71	a	72	d	73	d	74	b	75	d	76	c	77	b
78	b	79	b	80	a	81	c	82	c	83	d	84	d
85	a	86	c	87	b	88	a	89	c	90	b	91	b
92	b	93	b	94	c	95	a	96	d	97	b	98	b
99	b	100	c										

Model Paper 2

1. Ascorbic acid is the chemical name of
   a) Vitamin K b) Vitamin C
   c) Vitamin B d) Vitamin A
2. In alkalimetric titrations, the titrant is
   a) Iodine b) NaOH
   c) EDTA d) Magnesium
3. EDTA is a
   a) Tetradentate Ligand b) Octadentate Ligand
   c) Pentadentate Ligand d) Hexadentate Ligand
4. The indicator used in the EDTA method is
   a) Erichrome black T b) Benzene
   c) Phenopthalene d) Ethylene diamine
5. Which nutrient do adult women require in greater amounts than adult men?
   a) Vitamin D b) Protein
   c) Iron d) Calcium
6. Father of Industrial Microbiology is...
   a) E. Jenner b) L. Pasteur
   c) L. Philips d) R. Koch
7. Differential staining is....
   a) Simple Stain b) Gram Stain
   c) Negative Stain d) Congo stain
8. Which of the following foods is not a good source of iron?
   a) Meat b) Milk
   c) Egg d) Liver
9. Best example of primary metabolite used in large scale is...
   a) Formalin b) Oil
   c) Acetone d) Alcohol
10. Citric acid is produced by.............
    a) Fungus b) Bacteria
    c) Yeast d) Virus

11. Rubber roll sheller consists of
    a) 5 Rolls
    b) 2 Rolls
    c) 6 Rolls
    d) 4 Rolls
12. Centrifugal sheller works on the principle of
    a) Centrifugal force
    b) Magnetic force
    c) Centripetal force
    d) Gravitational force
13. Rubber roll sheller used for the shelling of
    a) Green pea
    b) Groundnut
    c) Paddy
    d) Sugarcane
14. Bran is by product of
    a) Green pea
    b) Groundnut
    c) Rice
    d) Sugarcane
15. Destoner in paddy processing is used for removing of
    a) Stone
    b) Paddy
    c) Rice
    d) Husk
16. When resorcinol solution added in cane sugar containing milk then----colour is obtained
    a) Bluish green
    b) Violet
    c) Red
    d) Orange
17. Which stage does adulteration take place in?
    a) Producer
    b) Distributor
    c) Retailer
    d) All of the mentioned
18. The full form of PFA
    a) Prevention of Food Adulteration
    b) Protection of Food Act
    c) Prevention of Food Act
    d) None of the mentioned
19. When iodine solution is added to starch containing milk then---colour is obtained
    a) Red
    b) Blue
    c) Yellow
    d) Violet
20. Why are adulterants added
    a) To increase shelf life of product
    b) To improve flavour, colour and appearance
    c) To sell lesser quantity at the same price
    d) All of the mentioned
21. A fermentation process in which a large amount of salt is used is called
    a) Salting
    b) Pickling
    c) Boiling
    d) Preserving
22. Sugar processed fruit juice is called
    a) Jam
    b) Punch
    c) Jelly
    d) Marmalade

23. TSS of jam is
    a) 68%
    b) 90%
    c) 50%
    d) 75%
24. Which of the following ingredients is added to a jelly?
    a) Pectin
    b) Sugar
    c) Acid
    d) All of the mentioned
25. Which of the following in jelly responsible for the formation of gel?
    a) Sugar
    b) Acid
    c) Pectin
    d) Water
26. Which law gives good results for fine grinding?
    a) Kick Law
    b) Bond Law
    c) Rittinger Law
    d) All of the mentioned
27. According to Kick Law, the energy requirement is proportional to
    a) Product of length of feed and products
    b) Ratio of length of feed and products
    c) Subtraction of length of feed and products
    d) Addition of length of feed and products
28. Average size particle is calculated by
    a) 0.135(13.66)^FM
    b) 1.25(1.366)^FM
    c) 0.135(1.366)^FM
    d) 13.5(1.366)^FM
29. Size reduction increases the
    a) Particle size
    b) Weight
    c) Surface area
    d) Volume
30. Which law gives good results for coarse grinding?
    a) Kick law
    b) Rittinger Law
    c) Bond Law
    d) Newton Law
31. What type of heat transfer is used when cooking in a microwave?
    a) Conduction
    b) Radiation
    c) Convection
    d) None of these
32. Frozen foods have a higher transfer co-efficient hence they lose heat faster than other food items when exposed to the outside environment.
    a) TRUE
    b) False
    c) May be false and may be true as per situation
    d) None of these
33. will most likely occur when raw egg is placed on black pavement on a hot, summer day?
    a) The pavement will conduct the heat and cook the egg.
    b) The egg will reflect the heat and make the pavement hotter underneath.
    c) The egg will absorb the heat and leave the pavement hotter underneath.
    d) The pavement will absorb the heat from the egg, and the egg will stay raw.

34. Which best explains how food is fried on a stove burner?
    a) The food transfers heat directly to the pan, and pan transfers heat directly to the stove burner.
    b) The stove burner transfer heat directly to the pan, and the pan directly transfers heat to the food.
    c) The stove burner radiates heat to pan and the pan radiates heat to the foods.
    d) The food creates a convection current; the heat rises to the pan and then rises to the stove burner.
35. Water boiling is an example of ___
    a) Convection b) Conduction
    c) Radiation d) None of these
36. In a ball mill if 'R' is radius of mill, 'g' is acceleration due to gravity and 'r' is the radius of ball, then critical speed is given by
    a) $n=2\pi\sqrt{((R-r)/g)}$ b) $n= 1/2\ \pi\ \sqrt{(g/(R-r)}$
    c) $n= 1/2\ \pi\ \sqrt{((R-r)/g)}$ d) $n=2\ \pi\ \sqrt{(g/(R-r))}$
37. Mixing index _________ with time.
    a) Decreases b) Increases
    c) Equals d) Does not change
38. Ribbon blenders are used for
    a) Powder b) Liquid
    c) Dough and paste d) All the option
39. The arms of kneaders are of ____________ shape.
    a) Rectangular b) Sigmoid
    c) Triangular d) Circular
40. Propeller agitator are used for
    a) High viscosity b) Powder
    c) Low viscosity d) Pastes
41. Milling refers to
    a) Size reduction for converting raw commodity into edible form
    b) Separation and cleaning process
    c) None of the option
    d) Size reduction and separation process for converting raw material into edible form
42. Which of the following relation is true?
    a) $E_{hulling} = N_{hulling} \times E_{wk}$ b) $\eta_{hulling} = E_{hulling} \times E_{wk}$
    c) $Ewk = \eta_{hulling} \times E_{hulling}$ d) None of the option
43. Which of the following indicates the uniformity of grind in a reduced material?
    a) Degree of grinding b) Work index
    c) None of the option d) Fineness modulus

44. Work index is related to
 a) Bond law b) Kick law
 c) Rittinger law d) None of the option
45. The angle between two jaws in a jaw crusher is
 a) 10-20° b) 30-40°
 c) 60-70° d) 20-30°
46. Attrition mill is operated between
 a) 350 - 700 rpm b) 1000 -1500 rpm
 c) 1500 - 2000 rpm d) 100 - 300 rpm
47. Attrition mill reduced size by
 a) Impact and shear b) Compression and shear
 c) Crushing and shear d) Shear
48. Centrifugation is done when particle size is less than
 a) 10μm b) 5μm
 c) 2μm d) 12μm
49. Reitz mill is used when solid content in the range between
 a) 10-15% b) 5-10%
 c) 15-20% d) 40-80%
50. Boiling point of n-hexane
 a) 90°C b) 65.5°C
 c) 45°C d) 100°C
51. Hydraulic press used 28MPa for
 a) 5 - 10 min b) 13-15 min
 c) 20 - 25 min d) 25 - 30 min
52. Communition means
 a) Extraction b) Size reduction
 c) Leaching d) Pressing
53. Which one of the following forms of water have the highest value of thermal conductivity?
 a) Normal water b) Boiling water
 c) Solid Ice d) Steam
54. Identify the very good insulator
 a) Saw Dust b) Glass wool
 c) Asbestos d) Cork
55. Most metals are good conductor of heat because of
 a) Transport of energy
 b) Lattice defects
 c) Capacity to absorb energy
 d) Free electrons and frequent collision of atoms

56. Consider the following parameters

i) Composition ii) Density

iii) Porosity iv) Structure,

Then, thermal conductivity of glass wool varies from sample to sample because of variation is

a) i and ii b) i and iii

c) i, ii and iii d) i, ii, iii and iv

57. During a cold winter season, a person prefers to sit near a fire. Which of the following modes of heat transfer provides him the maximum heat?

a) Conduction from the fire

b) Radiation will provide quick warmth

c) If it is near the fire, convection sounds good

d) Convection and radiation together

58. Let us assume two walls of same thickness and cross-sectional area having thermal conductivities in the ratio 1/2. Let us say there is same temperature difference across the wall faces, the ratio of heat flow will be

a) 0.5 b) 4

c) 2 d) 1

59. A plane slab of thickness 60 cm is made of a material of thermal conductivity k = 17.45 W/m K. Let us assume that one side of the slab absorbs a net amount of radiant energy at the rate q = 530.5 watt/m2. If the other face of the slab is at a constant temperature t2 = 38 degree Celsius. Comment on the temperature with respect to the slab?

a) 87.5 degree Celsius b) 47.08 degree Celsius

c) 32 degree Celsius d) 32.87 degree Celsius

60. A composite wall is made of two layers of thickness δ1 and δ2 having thermal conductivities k and 2k and equal surface area normal to the direction of heat flow. The outer surface of composite wall is at 100 degree Celsius and 200 degree Celsius. The minimum surface temperature at the junction is 150 degree Celsius. What will be the ratio of wall thickness?

a) 02:03 b) 01:01

c) 01:02 d) 02:01

61. The current in the electrical system is analogous to

a) the thermal conductance in the heat transfer system

b) the heat flow in the heat transfer system

c) the temperature in the heat transfer system

d) the thermal resistance in the heat transfer system

62. What will be the simple correct formula for the rate of heat flow (q) through a wall of surface area A and thickness of t ? The two surfaces of the wall are at temperatures T1 and T2 (T1>T2). Consider thermal resistance for the heat transfer system as R_h.

a) $q = (T1 - T2) / R_h$
b) $q = R_h A ((T1 - T2) / t)$
c) $q = R_h A (T1 - T2)$
d) $q = R_h (T1 - T2)$

63. All radiations in a black body are

a) Absorbed
b) Reflected
c) Refracted
d) Transmitted

64. Which of the following is an example of forced convection?

a) Chilling effect of cold wind on a warm body
b) Cooling of billets in the atmosphere
c) Heat exchange on cold and warm pipes
d) Flow of water in condenser tubes through pump

65. The unit of overall coefficient of heat transfer is

a) W/m^2
b) W/mK
c) s
d) W/m^2K

66. Heat Transfer takes place as per -----law of thermodynamics

a) Zeroth
b) Third
c) Second
d) First

67. When heat is transferred from one particle of hot body to another by actual motion of the heated particles, it is referred to as heat transfer by

a) Conduction
b) Radiation
c) Convection
d) All three

68. Sensible heat is the heat required to

a) Change vapour into liquid
b) Change liquid into vapour
c) Convert water into steam and superheat it
d) Increase the temperature of a liquid of vapour

69. 40% of incident radiant energy on the surface of a thermally transparent body is reflected back. If the transmissivity of the body be 0.15, then the emissivity of surface is

a) 0.15
b) 0.45
c) 0.60
d) 0.55

70. A perfect black body is one which

a) Is black in colour
b) Transmits all heat radiations
c) Reflects all heat
d) Absorbs heat radiations of all wave lengths falling on it

71. The concept of overall coefficient of heat transfer is used in heat transfer problems of
   a) Conduction
   b) Conduction and Convection
   c) Convection
   d) Radiation
72. The negative sign in the Fourier heat conduction equation indicates
   a) Heat always flow is in the direction of positive temperature gradient
   b) No heat flow is there
   c) None of these
   d) Heat always flows in the direction of negative temperature gradient
73. Father of Industrial Microbiology is…
   a) E. Jenner
   b) L. Pasteur
   c) L.Philips
   d) R. Koch
74. Differential staining is….
   a) Gram Stain
   b) Simple Stain
   c) Negative Stain
   d) Congo stain
75. Aspergillus can be grown on…
   a) Blood Agar
   b) BHI
   c) SDA
   d) MLA
76. Best example of primary metabolite used in large scale is…
   a) Formaline
   b) Oil
   c) Alcohol
   d) Acetone
77. Citric acid is produced by…
   a) Fungus
   b) Bacteria
   c) Yeast
   d) Virus
78. Basic function of fermenter is...
   a) To sterilize media
   b) to purify the media
   c) To recover media
   d) Optimum condition
79. Atropine is a competitive antagonist of the actions of…
   a) Neuron
   b) Nerve
   c) Vein
   d) Acetylcholine
80. Enrichment culture technique was discovered by…
   a) Bacteria
   b) Beijerinck
   c) E. Hesse
   d) L. Pasteur
81. For preservation of anaerobic culture in laboratory we use…
   a) Stab
   b) Slants
   c) Broth
   d) Petri plate
82. Long-term preservation of microbes was done by
   a) Glycerol culture
   b) Slants
   c) Lyophilization
   d) Stab

83. Peptone and Tryptone are used as a source of...
    a) Carbon b) Nitrogen
    c) Copper d) Water

84. Temperature of autoclave is
    a) 121°C b) 56°C
    c) 131°C d) 141°C

85. Tyndallization is the process of...
    a) Autoclaving b) Sterilization
    c) Transparenting d) Hot air oven

86. Fermentation occurs in the
    a) Presence of O b) Absence of O
    c) Presence of N d) Absence of C

87. Bacteria that are responsible for fermentation of dairy milk are
    a) Azotobacter b) Rhizobium
    c) Aspergillus d) Lactobacillus

88. Plastic syringes, rubber gloves are sterilized by...
    a) Autoclaving b) Radiation
    c) Filtration d) Sprit

89. The germicidal range of UV light is...
    a) 1000nm b) 1500nm
    c) 500nm d) 254nm

90. Temperature in pasteurization is...
    a) 50°C b) 65°C
    c) 100°C d) 60°C

91. Industrial microbiology, mainly depends on phenomenon
    a) Fermentation b) Pasteurization
    c) Sterilization d) Solidification

92. Bacitracin is produced by...
    a) Clostridium Spp. b) Staphylococcus spp.
    c) Bacillus subtilis d) E. coli

93. Which nutrient do adult women require in greater amounts than adult men
    a) Calcium b) Iron
    c) Protein d) Vitamin D

94. The highest levels of vitamin C is present in
    a) Black Currants b) Parsley
    c) Broccoli d) Orange juice

95. Which one of the following is not a monosaccharide sugar
    a) Glucose b) Sucrose
    c) Fructose d) Galactose

96. Acetic acid gives
   a) Spicy flavor
   b) Vinegar sour taste
   c) Cardamom flavour
   d) Fruity flavour
97. Curcumin pigment is
   a) Insoluble in oil
   b) Sparingly soluble in oil
   c) Soluble in oil
   d) None
98. Maillard Browning Reaction is a
   a) Enzymatic reaction
   b) Non –Enzymatic reaction
   c) Acidic reaction
   d) Basic Reaction
99. Enzymatic Browning occurs due to
   a) Carmelans
   b) Carotenoids
   c) anthocyanins
   d) polyphenoloxidase
100. Flavonoids show their antioxidant property through
   a) Peroxyl radical scavenging activity
   b) Transition metal chelation
   c) None of the above
   d) All the above

Answer Key

1	b	2	b	3	b	4	a	5	c	6	b	7	b
8	b	9	d	10	a	11	b	12	a	13	c	14	c
15	a	16	c	17	d	18	a	19	b	20	d	21	b
22	c	23	a	24	d	25	c	26	c	27	b	28	c
29	c	30	a	31	a	32	a	33	b	34	a	35	a
36	b	37	b	38	a	39	b	40	c	41	d	42	b
43	d	44	a	45	d	46	a	47	c	48	b	49	d
50	b	51	a	52	b	53	c	54	b	55	d	56	d
57	b	58	a	59	b	60	c	61	b	62	a	63	a
64	d	65	d	66	c	67	a	68	d	69	b	70	d
71	b	72	d	73	b	74	a	75	c	76	c	77	a
78	b	79	d	80	b	81	a	82	c	83	b	84	a
85	b	86	b	87	d	88	b	89	d	90	d	91	a
92	c	93	b	94	a	95	b	96	b	97	c	98	b
99	d	100	d										

Model Paper 3

1. Which law gives good results for fine grinding?
   a) Kick Law
   b) Bond Law
   c) Rittinger Law
   d) All of the mentioned
2. According to Kick Law, the energy requirement is proportional to
   a) Product of length of feed and products
   b) Ratio of length of feed and products
   c) Subtraction of length of feed and products
   d) Addition of length of feed and products
3. Average size particle is calculated by
   a) 0.135(1.366)^FM
   b) 0.135(13.66)^FM
   c) 1.25(1.366)^FM
   d) 13.5(1.366)^FM
4. Size reduction increases the
   a) Particle size
   b) Surface area
   c) Weight
   d) Volume
5. Which law gives good results for coarse grinding?
   a) Rittinger Law
   b) Kick law
   c) Bond Law
   d) Newton Law
6. Grinding is a process of
   a) Drying
   b) Size reduction
   c) Polishing
   d) Packaging
7. According to which law, the energy required for size reduction to proportional to the change in dimension (or to the ratio of initial size to the final size)
   a) Rittinger Law
   b) Kick law
   c) Bond Law
   d) None of these
8. Attrition mill is operated at speed
   a) 350 to 700 rpm
   b) 700 to 1400 rpm
   c) 1400 to 2000 rpm
   d) 100 to 350 rpm
9. Rotational speed of ball mill is kept at __________ of critical speed
   a) 10 to 30 %
   b) 30 to 50 %
   c) 65 to 80 %
   d) 50 to 65 %

10. Reitz mill able to grind material below
   a) 5μm b) 20μm
   c) 25μm d) 15μm
11. Hydraulic press uses 5MPa for
   a) 5 to 10 min b) 10 to 15 min
   c) 15 to 20 min d) 01 to 05 min
12. Double cone mixer is used for
   a) Liquid b) Powder
   c) Dough and Paste d) all of these
13. Ribbon blender are used for
   a) Powder b) Liquid
   c) Dough and Paste d) all of these
14. Full form of HEPA
   a) High effective particles air filter
   b) High efficiency particulate air filter
   c) Heavy efficiency particulate air filter
   d) High effective particulate air filter
15. Rate of filtration depends on
   a) Volume b) Pressure
   c) Temperature d) Density
16. Which equation defines rates of filtration
   a) Darcy equation b) Henderson equation
   c) Smith equation d) Harkins Jura equation
17. How much pressure is applied in Ultrafiltration ?
   a) less than 1 bar b) 20 - 40 bar
   c) 30 to 60 bar d) 1 to 10 bar
18. How much pressure is applied in Reverse Osmosis ?
   a) less than 1 bar b) 30 to 60 bar
   c) 20 - 40 bar d) 1 - 10 bar
19. Substance that prevents cake to become compact and increase the rate of filtration
   a) Slurry b) Filtrate
   c) Filter medium d) Filter aids
20. Woven wire cloth used in
   a) Depth filtration b) Surface filtration
   c) Membrane filtration d) All of these
21. Which one of the following forms of water have the highest value of thermal conductivity?
   a) Normal water b) Solid Ice
   c) Boiling water d) Steam

22. The extended surface used for the enhancement of heat dissipation is

a) Fin b) Fourier number

c) Convective Coefficient d) No finned surface

23. What is the unit of diffusion coefficient?

a) m.s b) g/m-s

c) m/s d) m^2/s

24. Ratio of inertia force to viscous force is known as

a) Grashof number b) Reynolds number

c) Stanton numbe d) Prabdl Number

25. During a cold winter season, a person prefers to sit near a fire. Which of the following modes of heat transfer provides him the maximum heat?

a) Conduction from the fire

b) Radiation will provide quick warmth

c) If it is near the fire, convection sounds good

d) Convection and radiation together

26. Let us assume two walls of same thickness and cross-sectional area having thermal conductivities in the ratio 1/2. Let us say there is same temperature difference across the wall faces, the ratio of heat flow will be

a) 4 b) 2

c) 1 d) 0.5

27. A plane slab of thickness 60 cm is made of a material of thermal conductivity k = 17.45 W/m K. Let us assume that one side of the slab absorbs a net amount of radiant energy at the rate q = 530.5 watt/m2. If the other face of the slab is at a constant temperature t2 = 38 degree Celsius. Comment on the temperature with respect to the slab?

a) 87.5 degree Celsius b) 47.08 degree Celsius

c) 32 degree Celsius d) 32.87 degree Celsius

28. Iodine value of ghee ranges in between

a) 20-25 b) 26-38

c) 40-45 d) 45-49

29. Moisture content in a unit volume of air is called

a) Relative humidity b) Absolute humidity

c) Humidity d) Specific humidity

30. Milk is deficient in

a) Iron b) Iron and copper

c) Copper d) Non

31. All radiations in a black body are

a) Reflected b) Refracted

c) Absorbed d) Transmitted

32. What is effectiveness of fin?
   a) The heat which would be transferred if entire fin area was at base temperature to the ratio of actual heat transferred from fin area
   b) The heat which would be transferred if entire fin area was at minimum temperature to the ratio of actual heat transferred from fin area
   c) The ratio of actual heat transferred from fin area to the heat which would be transferred if entire fin area was at base temperature
   d) The ratio of actual heat transferred from fin area to the heat which would be transferred if entire fin area was at minimum temperature

33. The unit of overall coefficient of heat transfer is
   a) W/m^2K b) W/m^2
   c) W/mK d) s

34. Heat Transfer takes place as per -----law of thermodynamics
   a) Zeroth b) First
   c) Second d) Third

35. When heat is transferred from one particle of hot body to another by actual motion of the heated particles, it is referred to as heat transfer by
   a) Radiation b) Convection
   c) Conduction d) All three

36. The ratio of heat transfer coefficient to the flow of heat per unit temperature rise due to the velocity of the fluid is known as
   a) Fourier number b) Stanton number
   c) Weber Number d) Nusselt Number

37. 40% of incident radiant energy on the surface of a thermally transparent body is reflected back. If the transmissivity of the body be 0.15, then the emissivity of surface is
   a) 0.15 b) 0.6
   c) 0.55 d) 0.45

38. Perfect black body is one which
   a) Absorbs heat radiations of all wave lengths falling on it
   b) Is black in colour
   c) Transmits all heat radiations
   d) Reflects all heat

39. The concept of overall coefficient of heat transfer is used in heat transfer problems of
   a) Conduction and Convection b) Conduction
   c) Convection d) Radiation

40. Radiation dryers use _____
   a) Ultraviolet rays b) Infrared radiation
   c) Visible range d) None of these

41. A fermentation process in which a large amount of salt is used is called
   a) Salting b) Pickling
   c) Boiling d) Preserving
42. Sugar processed fruit juice is called
   a) Jam b) Jelly
   c) Punch d) Marmalade
43. TSS of jam is
   a) 90% b) 50%
   c) 68% d) 75%
44. Which of the following ingredients is added to a jelly?
   a) Pectin b) Sugar
   c) Acid d) All of the mentioned
45. Which of the following in jelly responsible for the formation of gel?
   a) Sugar b) Acid
   c) Pectin d) Water
46. Low energy ultrasound application are performed at frequency higher than
   a) 10KHz b) 100KHz
   c) 250KHz d) 500KHz
47. Sulphur dioxide and sulphites is used to preserve
   a) Fruits b) Meat
   c) Fish d) Wine
48. Sterilization is done at temperature
   a) 73°C b) 63°C
   c) 121°C d) 221°C
49. In spray drier particle are separated by
   a) Hydrocyclone b) Centrifuge
   c) Cyclone d) None of these
50. In drying of food materials in a mechanical dryer _________ is taking place
   a) Natural convection b) Forced convection
   c) Radiation d) Conduction
51. The chilling of fruits and vegetables involves bringing down the temperature of commodities to
   a) -1 to 8°C b) -4 to 4°C
   c) 4 to 12°C d) 10 to 18°C
52. The process of exposing the food of either electromagnetic or ionizing radiation to destroy the microorganism is known as
   a) Irradiation b) Dehydration
   c) Sterilization d) Pasteurization

53. Which of the following is not a fermented product?
   a) Fruit wine
   b) Apple Cider
   c) Vinegar
   d) None of these
54. Vinegar is also known as
   a) Citric acid
   b) Acetic acid
   c) Forming acid
   d) None of these
55. Beer, fruit juices and aerated drinks are preserved by
   a) Sterilization
   b) Pasteurization
   c) Blanching
   d) None of these
56. HPP usually applied at pressure
   a) 200MPa
   b) 400MPa
   c) 500MPa
   d) 600MPa
57. PEF is applied on intensity
   a) 160 - 120 Kv/cm
   b) 10 - 80Kv/cm
   c) 80 - 100 Kv/cm
   d) 100 - 120 Kv/cm
58. Which of the following is a non thermal method
   a) Ultrasound
   b) Ohmic heating
   c) HPP
   d) All of these
59. Cold sterilization refers to the preservation of food by
   a) Refrigeration
   b) Radiation
   c) Dehydration
   d) Hyophilisation
60. Type of moisture that can be removed by common drying techniques is
   a) Equilibrium moisture
   b) Bound moisture
   c) Total moisture
   d) Free moisture
61. When resorcinol solution added in cane sugar containing milk then----colour is obtained
   a) Bluish green
   b) Red
   c) Violet
   d) Orange
62. Which stage does adulteration take place in?
   a) Producer
   b) Distributor
   c) Retailer
   d) All of the mentioned
63. The full form of PFA
   a) Protection of Food Act
   b) Prevention of Food Act
   c) Prevention of Food Adulteration
   d) None of the mentioned
64. When iodine solution is added to starch containing milk then---colour is obtained
   a) Red
   b) Blue
   c) Yellow
   d) Violet

65. Why are adulterants added
    a) To increase shelf life of product
    b) To improve flavor, color and appearance
    c) To sell lesser quantity at the same price
    d) All of the mentioned

66. The acidity and pH of fresh milk vary with
    a) Species, breed
    b) Individuality
    c) Stage of lactation, health of animal
    d) All of these

67. Which of the following statements is correct?
    a) Platform tests are those tests carried to check only microbiological quality of the milk
    b) Platform tests include the test performed to check the quality of the incoming milk on the receiving platform
    c) Platform tests are those tests which measure the fat content and total solids present in the milk
    d) None of these

68. Casein is present in milk in
    a) Dispersed form
    b) Emulsion form
    c) Colloidal form
    d) All of these

69. Which of the following enzymes is not present in milk
    a) Analase
    b) Lipase
    c) Peroxide
    d) None of these

70. Which of the following is a lactic acid - alcohol fermented milk, originated in Russia
    a) Kumiss
    b) Kefir
    c) Yoghurt
    d) Bulgarin buttermilk

71. Which one of the following is a self carbonated beverage
    a) Kumiss
    b) Yoghurt
    c) Dahi
    d) None of these

72. According to PFA, cream contains
    a) Not less than 15% milk fat
    b) Not less than 25% milk fat
    c) Not less than 10% milk fat
    d) Not less than 20% milk fat

73. Whipping cream contains
    a) 65 - 85% milk fat
    b) 20 - 25 % milk fat
    c) 30 - 40 % milk fat
    d) 10 - 15 % milk fat

74. Full form of DVS
    a) Directly Vat Set
    b) Direct Verified Set
    c) Direct Vat Solution
    d) Direct Vat Set

75. Full form of COB
    a) Clot on beverage
    b) Clothing on boiling
    c) None of these
    d) Clot on boiling

76. Analase is responsible for the hydrolysis of
    a) Phosphate   b) Protein
    c) Starch   d) Lipid
77. pH of fresh and normal buffalo milk is
    a) 3.8   b) 6.7
    c) 4.5   d) 5.9
78. Milk fat is
    a) Oil in water types emulsion   b) Water in oil type emulsion
    c) Colloidal state   d) None of these
79. Dye reduction test (MBR) is carried out to
    a) To determine heat stability
    b) To identify the type of microorganisms present in milk
    c) To detect adulteration of milk with water
    d) To determine the extent of bacterial contamination and growth in milk
80. Phosphatase test is not applicable for
    a) Pasteurized milk   b) Sterilized milk
    c) Both (a &b)   d) None of these
81. Flaked rice is made from
    a) Raw rice   b) Brown rice
    c) Parboiled rice   d) Bulgur
82. Malting is a basic operation employed in
    a) Barley   b) Wheat
    c) Rice   d) Pulse
83. In milling, cleaned oats are subjected to kiln drying to inactivate lipase. This process is known as
    a) Drying   b) Dehydration
    c) Sterilization   d) Stabilization
84. All cereals are rich in
    a) Carbohydrates   b) Fats
    c) Protein   d) None
85. Milling are consists
    a) Cleaning   b) Grading
    c) Separating   d) All
86. Rice bran contains about-----percent fat
    a) 17-19%   b) 10-12%
    c) 12-15%   d) 15-17%
87. The purpose of tempering of wheat through tempering bin is employed to
    a) Equalize the moisture in whole grain   b) Raise the moisture
    c) Reduce the moisture   d) Raise the temperature

88. The removal of colour of rice is called
   a) Blanching b) Bleaching
   c) Parboiling d) Polishing
89. The percentage of endosperm in mature corn is about
   a) 65% b) 82%
   c) 70% d) 85%
90. Percentage of bran in wheat milling is
   a) 12% b) 5%
   c) 10% d) 15%
91. Barley has indigestible fibre percentage
   a) 6 b) 7
   c) 8 d) 9
92. Brewing and malting is related to
   a) Juice b) Beer
   c) Milk d) Sausages
93. The steeping of corn is carried out at 50℃ for a period of
   a) 10-18 hrs b) 20-25 hrs
   c) 40-50 hrs d) 28-48 hrs
94. Paddy, corn, wheat, buck wheat seeds are specially rich source of
   a) Carbohydrates b) Vitamin
   c) Fat d) Oil
95. The endosperm of cereals grain contains the highest amount of carbohydrates in the form of
   a) Pectin b) Dextrin
   c) Starch d) Sugar
96. Debranned, polished and cracked parboiled wheat is known as
   a) Bulgur b) Sela
   c) Pasta d) Milled wheat
97. The husk of grain is removed by the action of
   a) Compression and shear b) Abrasion and friction
   c) Impact and friction d) All of them
98. Physico-thermal properties of grain is classified into
   a) Strength or hardness b) Density and hardness
   c) Thermal properties d) All of them
99. In the refining process of bran oil, neutralization process is used for
   a) Removal of hard wax b) Removal of FFA
   c) Removal of odours d) Removal of colouring matter

100. Paddy is normally stored at
   a) 12 % moisture content on wet basis
   b) 15 % moisture content on wet basis
   c) 15 % moisture content on dry basis
   d) 12 % moisture content on dry basis

Answer Key

1	c	2	b	3	a	4	b	5	b	6	b	7	b
8	a	9	c	10	d	11	c	12	b	13	a	14	b
15	b	16	a	17	d	18	b	19	d	20	b	21	b
22	a	23	d	24	b	25	b	26	d	27	b	28	b
29	b	30	b	31	c	32	c	33	a	34	c	35	c
36	b	37	d	38	a	39	a	40	b	41	b	42	b
43	c	44	d	45	c	46	b	47	a	48	c	49	c
50	b	51	a	52	a	53	d	54	b	55	b	56	d
57	b	58	d	59	b	60	d	61	b	62	d	63	c
64	b	65	d	66	d	67	b	68	c	69	d	70	a
71	d	72	b	73	c	74	d	75	d	76	c	77	b
78	a	79	d	80	b	81	c	82	a	83	d	84	a
85	d	86	a	87	a	88	b	89	b	90	a	91	b
92	b	93	d	94	a	95	c	96	a	97	d	98	d
99	b	100	a										

Model Paper 4

1. Which is not a consequence of vitamin A deficiency
   a) Corneal ulcers b) Bitot's spots
   c) Night blindness d) Optic neuropathy
2. Which of the following is the best source for lutein?
   a) Tomatoes b) Mandarin oranges
   c) Spinach d) Cantaloupe
3. The inducer of invertase enzyme is
   a) Starch b) Sucrose
   c) Galactosidase d) Fatty acids
4. Discoloration may be done by using
   a) Anthocyanase b) Sulphloydryl oxidase
   c) Proteases d) All of these
5. Vitamin helps in blood clotting
   a) Vitamin K b) Vitamin A
   c) Vitamin C d) Vitamin D
6. Which enzyme is used in Liquefaction of starch to dextrin?
   a) Cellulase b) Alpha-amylase
   c) Pectin d) All of these
7. Which of the following vitamins is also known as cobalamin?
   a) Vitamin B11 b) Vitamin B2
   c) Vitamin B12 d) Vitamin B6
8. Milk digestion occurs due the presence of
   a) Beta-amylase b) None of these
   c) Reductase d) Lactase
9. Vitamin B deficiency disease beriberi occurs due to deficiency of which enzyme
   a) Thiaminase b) Ascorbic Acid
   c) Lipogenase d) None of these
10. Which type of additive is required in foods containing fats and oils to prevent rancidity
    a) Preservatives b) Emulsifiers
    c) Anti-oxidants d) Solvents

11. Vitamin functions as both, hormone and visual pigment
    a) Thiamine b) Retinal
    c) Riboflavin d) Folic Acid
12. Which of the following foods is not a good source of iron?
    a) Meat b) Egg
    c) Milk d) Liver
13. A unit operation which converts liquid into a vapour and leaving behind a concentrated solution or a higher boiling point liquid is called
    a) Dehydration b) Freezing
    c) Condensation d) Evaporation
14. Which evaporator is used for concentrating highly heat-sensitive materials such as orange juice, food materials and require short residence times?
    a) Agitated film b) Falling film
    c) Forced circulation d) Rising film
15. Overall material balance for single evaporator is
    a) $m_s = m_p + m_v$ b) $m_f = m_p + m_v$
    c) $m_f = m_s + m_v$ d) $m_f = m_p + m_s$
16. Solute balance in single effect evaporator is
    a) $m_v x_v = m_p x_p$ b) $m_f x_f = m_s x_s$
    c) $m_f x_f = m_p x_p$ d) $m_f x_f = m_s x_v$
17. Single effect evaporator consists of
    a) Heating source b) Vapour chamber
    c) Condenser d) All the above
18. heat supplied by condensing steam in heat exchanger load in single effect evaporator is
    a) $m_S \lambda_S = (m_P H_P + m_V H_V) - m_f H_f$ b) $m_S \lambda_S = m_f C_{Pf} (T - T_f) + m_V \lambda_V$
    c) $m_f = m_p + m_v$ d) $m_f x_f = m_P x_P$
19. Boiling point elevation or rise is calculated by
    a) $\Delta Tb = 0.91$ m b) $\Delta Tb = 0.51$ m
    c) $\Delta Tb = 0.71$ m d) $\Delta Tb = 0.31$ m
20. Dühring rules is employed in
    a) Leaching b) Distillation
    c) Evaporation d) Drying
21. The steam economy of evaporator is calculated by
    a) m_v/m_s b) m_s/m_f
    c) m_f/m_s d) m_s/m_v
22. Boiling point elevation or boiling point rise of a solution is defined as the increase in boiling point over the boiling point of pure water at a particular__________.
    a) Enthalpy b) Temperature
    c) Pressure d) Mass

23. To achieve freezing of a food product, the product must be exposed to a________________ medium for sufficient time to remove sensible heat and latent heat of fusion from the product.

    a) High pressure b) Low-temperature
    c) Low volume d) High temperature

24. Removal of the sensible and latent heat results in a reduction in the product temperature as well as a conversion of the water from liquid to solid state _____________.

    a) Smoke b) Ice
    c) Vapour d) Water

25. The use of low-temperature air at high speeds in direct contact with small product objects is a form of _____________________.

    a) IQF b) COP
    c) PQF d) EMC

26. Air Blast is method of

    a) Separation b) Drying
    c) Freezing d) Cooling

27. IQF means

    a) Individual quick freezing b) Internal quick freezing
    c) Internal quick flow d) Intelligent quick feeding

28. The density of solid water (ice) is ______ than the density of liquid water.

    a) Unknown b) Less
    c) Equal d) More

29. The thermal conductivity of ice is approximately a factor of_____ larger than the thermal conductivity of liquid water.

    a) Zero b) Six
    c) Two d) Four

30. The first and most popular equation for predicting freezing time was proposed by _____________.

    a) Fourier b) Stefan
    c) Heldmann d) Plank

31. Plank's Equation is used for calculating the

    a) Freezing time b) Blanching time
    c) Roasting time d) Drying time

32. Ice cream is example of

    a) Chilling system b) Freezing system
    c) Baking system d) Blanching system

33. The queen of spices is

    a) Cardamom b) Pepper
    c) Ginger d) Chilly

34. Which of the following parts of a plant are spices NOT made from
    a) Bark b) Cell
    c) Leaf d) Root
35. The aromatic volatile components of spices are called
    a) Spice fat b) Spice oil
    c) Spice gel d) Spice paste
36. From which part of turmeric plant, the turmeric powder is obtained
    a) Dried rhizome b) Seed
    c) Dried root d) Dried fruit
37. Ginger is a
    a) Rhizome b) Tuber
    c) Bulb d) Seed
38. The optimum moisture content of black pepper is
    a) 5-7% b) 15-20%
    c) 8-10% d) 20-22%
39. Ginger belongs to the family
    a) Zingiberaceae b) Solanaceae
    c) Papilonaceae d) Piperaceae
40. Which of the following spices known as the king of spices
    a) Aniseed b) Black pepper
    c) Chilly d) Ginger
41. Commonly used spice clove obtained from the
    a) Root b) Stem
    c) Flower bud d) Fruit
42. Biological name of black pepper is
    a) Mangifera indica b) Centella asiatica
    c) Pisum sativam d) Piper nigrum
43. Black pepper is which type of fruits
    a) Dried and unripe fruit b) Ripe and dried fruit
    c) Dried, Unripe and Drupe fruit d) It is a drupe fruit
44. Concentration of piperine in black pepper ?
    a) 15-18% b) 6-9%
    c) 4-5% d) 10-11%
45. How long do all spices last?
    a) 6 months b) 4 weeks
    c) 1-2 years d) 5 years
46. Cinnamon is obtained from which part of the plant
    a) Roots b) Bark
    c) Stem d) Fruits

47. Chillies are pungent due to the presence of the compound?
    a) Peperine   b) Eugienol
    c) Capsaicin   d) Curcumin
48. Leading State in black pepper production in India ?
    a) Karnataka   b) Maharashtra
    c) None   d) Kerala
49. Origin of ginger is
    a) India   b) South East Aisa
    c) Australia   d) Mediterranean region
50. Black pepper belongs to the family
    a) Solanaceae   b) Piperaceae
    c) Papilonaceae   d) Zingiberaceae
51. Oleoresin means
    a) Extracted nutrients   b) Oil of spices
    c) Carbohydrate of spices   d) None
52. In India, Fenugreek is commonly called
    a) Palau   b) Methi
    c) Mint   d) None
53. Which of the following is not a legume?
    a) Corn   b) Peas
    c) Lentils   d) Peanuts
54. All are examples of legumes except
    a) Kidney beans   b) Quinoa
    c) Chickpeas   d) Soybeans
55. Which of the following crop is the major source of protein in a vegetarian diet?
    a) Oilseeds   b) Wheat
    c) Rice   d) Pulses
56. Which of the following is the leading producer of soybean in India?
    a) Madhya Pradesh   b) Chhattisgarh
    c) Maharashtra   d) Uttar Pradesh
57. Which of the crop gives the highest protein equals to egg, milk and meat?
    a) Black gram   b) Soybeans
    c) Lentil   d) Pegion pea
58. Cooking of pulses
    a) Destroy enzyme inhibitors   b) Improves nutritional quality
    c) Improve palatability   d) All of these option
59. Fermentation of pulses leads to the reduction of _________ present in it
    a) Proteins   b) Anti nutritional factors
    c) Vitamins   d) Minerals

60. Which of the following is untrue about solvent extraction of oil?
   a) The most common solvent used is Hexane
   b) This process is difficult/complex for small scale operators
   c) This suitable for materials containing low percentage of oil
   d) None of these option
61. Maximum edible oil in India is contributed by
   a) Sunflower b) Mustard
   c) Ground nut d) Castor
62. Anti nutrients of Legumes are
   a) Primary metabolites b) Secondary metabolites
   c) Tertiary metabolites d) None of these option
63. Heat stable anti nutrients is
   a) Goitrogen b) Saponins
   c) Lectins d) None of these option
64. Which of the following physical method used to inactivate anti nutrients ?
   a) Reducing agent b) Germination
   c) Extrusion d) Fermentation
65. Which temperature is used to inactivate trypsin inhibitor?
   a) 50-80°C b) 105-110° C
   c) 180-200°C d) Above 200°C
66. Trypsin inhibitor is
   a) Proteins b) Glycosides
   c) Phenols d) Others
67. Ultrasound method used to inactivate anti-nutrients at frequency
   a) 30KHz b) 90KHz
   c) 15KHz d) 100KHz
68. Principle of ultrasound based on
   a) Le Chatelier principle b) Resistance heating
   c) Cavitation process d) None of these option
69. Tannin is
   a) Glycosides b) Proteins
   c) Phenols d) Others
70. Which one of the following is not anti-nutrient compound?
   a) Saponin b) Phenols
   c) Tannin d) Trypsin
71. Tannin is
   a) Heat stable b) Heat liable
   c) Heat unstable d) None of these option

72. Protein content in legumes is
  a) 1-3% b) 5-10%
  c) 10-12% d) 20-30%
73. Glucose and mannose are epimers around carbon
  a) C1 b) C6
  c) C2 d) C4
74. During equilibrium optical rotation of Glucose is
  a) +98.7° b) +112.7°
  c) +52.7° d) +18.7°
75. On hydrolysing Starch, it is changed to
  a) Amylose b) Amylopectin
  c) Galactose d) Glucose
76. Glycosidic bond present in the milk sugar is
  a) α(1-6) b) α(1-4)
  c) β(1-4) d) β(1-6)
77. Unbranched homopolysaccharide is
  a) Starch b) Amylopectin
  c) Cellulose d) Glycogen
78. Which of the following amino sugar are present in the bacterial cell wall
  a) Sialic acid b) Aminoglycoside
  c) N-acetylmuramic acid d) Azide
79. Carbohydrate generally exist in biological system
  a) L-Form b) D-form
  c) Racemic Mixture d) Cis-Form
80. Which of the following amino acid does not have chiral centre
  a) Phenylalanine b) Glycine
  c) Tyrosine d) Isoleucine
81. D-isomers of amino acid are found in
  a) Virus b) Plant
  c) Bacteria d) All of the above
82. The tertiary structure of protein represents
  a) Linear sequence of amino acids b) 3-D structure of protein
  c) Spatial arrangement of subunits d) Amino acids in free form
83. Esters of fatty acids with higher molecular weight with monohydric alcohols are known as
  a) Fats b) Wax
  c) TAG d) All of the above

84. Which of the following amino acid have indole ring in its structure?
 a) Phenylalanine b) Tyrosine
 c) Histidine d) Tryptophane
85. Cystine is a
 a) Amino sugar b) Dimer
 c) Carbohydrate d) Monomer
86. Secondary structure of protein is stabilized by
 a) Covalent bonds b) Hydrophobic bonds
 c) Ionic bonds d) Hydrogen bonds
87. Which of the following protein need not to acquire the quaternary structure?
 a) Monomeric protein b) Multimeric protein
 c) Dimeric protein d) None
88. Ramachandran plot is used to explain the protein
 a) Primary structure b) Secondary structure
 c) Quaternary structure d) Tertiary structure
89. Isomers differ from each other in the configuration only around C1 known as
 a) Anomers b) Polymers
 c) Enantiomers d) Epimers
90. In connective tissues, extracellular fluids that lubricate joints is a
 a) Amino acid b) Protein
 c) Heteropolysaccharides d) Homopolysaccharide
91. The simplest form of lipids is
 a) Ghee b) Fatty acids
 c) Oil d) Wax
92. Fatty acids which contain only single C–C bonds are known as
 a) Lipids b) Wax
 c) Saturated fatty acids d) Unsaturated Fatty acids
93. Subcooling after condensation increases the refrigeration effect.
 a) FALSE b) TRUE
 c) depends on other factors d) nothing can be said
94. The C.O.P of a refrigeration cycle with increase in evaporator temperature, keeping condenser temperature constant, will
 a) Decrease b) Increase
 c) Remain unaffected d) None of these
95. Which of the following is not an advantage of vapor compression refrigeration system over air refrigeration system?
 a) Smaller size for a given capacity of refrigeration
 b) It has less running cost
 c) It can be employed over wide range of temperature
 d) There is no problem of leakage in vapor compression refrigeration system

96. What are the desirable characteristics of a refrigerant?
   a) It should not be non-corrosive
   b) It should have minimum enthalpy of vaporization
   c) It should not be toxic
   d) All of the above

97. In refrigeration system refrigerant gains heat at
   a) Throttling device b) Compressor
   c) Condenser d) Evaporator

98. The number of hydrogen atoms in the refrigerant R-12 is
   a) 1 b) 2
   c) 3 d) Nil

99. One Ton of Refrigeration is equivalent to
   a) 50 k Cal/min b) 1000 kg
   c) 210 KJ/hr d) 210 KJ/s

100. Compressor used in domestic refrigerator is a
   a) Hermetic compressors b) Centrifugal
   c) Depends on manufactures d) Semi-hermetic Compressor

Answer Key

1	d	2	c	3	b	4	a	5	a	6	b	7	c
8	d	9	a	10	c	11	b	12	c	13	d	14	b
15	b	16	c	17	d	18	b	19	b	20	c	21	a
22	c	23	b	24	b	25	a	26	c	27	a	28	b
29	d	30	d	31	a	32	b	33	a	34	b	35	b
36	a	37	a	38	c	39	a	40	b	41	c	42	d
43	c	44	b	45	c	46	a	47	c	48	d	49	b
50	b	51	b	52	b	53	a	54	b	55	d	56	a
57	b	58	d	59	b	60	d	61	c	62	b	63	b
64	c	65	b	66	a	67	a	68	c	69	c	70	b
71	a	72	d	73	c	74	c	75	d	76	c	77	c
78	c	79	b	80	b	81	c	82	b	83	b	84	d
85	b	86	d	87	a	88	b	89	a	90	a	91	b
92	c	93	b	94	b	95	d	96	c	97	d	98	d
99	a	100	a										

Model Paper 5

1. In a domestic air-conditioner unit, the __________ acts as secondary refrigerant.
   a) Air
   b) Brine
   c) Water
   d) None of these
2. At what temperature Fahrenheit and Celsius gives the same readings (degree) of temperature
   a) -40
   b) 40
   c) 100
   d) None of these
3. Specific volume is a extensive property
   a) True
   b) False
   c) Depends on other factors
   d) None of these
4. Following is the example of path function
   a) Entropy
   b) Work done
   c) Internal Energy
   d) Enthalpy
5. Calculate the dryness fraction of steam which has 1.25 kg of water in the suspension with 40kg of steam.
   a) 0.13
   b) 0.97
   c) It has no dryness fraction
   d) None
6. A cyclic heat engine operates between a heat source temperature of 900 degree C and heat sink temperature of 30°C. Find the least rate of heat rejection per kW of net power output produced by the engine.
   a) 1.35kW
   b) 742W
   c) 350W
   d) 258W
7. COP of a refrigerator is 4, what would be the efficiency of corresponding heat engine.
   a) 40%
   b) 33%
   c) 25%
   d) 20.00%
8. The exergy of a fluid at a higher temperature is ___ than that at a lower temperature and ___ as the
   a) More, decreases
   b) Less, increases
   c) More, increases
   d) Less, decreases

9. The work done by a closed system in a reversible process is always ___ that done in an irreversible process.
   a) Less than
   b) Less than or more than
   c) More than
   d) Equal to

10. Which of the following is true?
   a) The first law states that the energy is always conserved quantity-wise
   b) The second law states that the energy always degrades quality-wise
   c) Nothing can be said precisely
   d) Both of the mentioned

11. Which of the following statements is true? X is the dryness fraction.
   a) The value of x varies between 0 and 1
   b) For saturated water, x=0
   c) For saturated vapour, x=1
   d) All mentioned

12. Kelvin-Planck's and Clausius' statements of second law of thermodynamics are
   a) Equivalent to each other
   b) Not Connected to each other
   c) Violation of one doesn't violate the other
   d) None of these

13. The correct sequence of the processes taking place in a Carnot cycle is
   a) Adiabatic →adiabatic→isothermal→isothermal
   b) Isothermal→isothermal→adiabatic→adiabatic
   c) Adiabatic→isothermal→isothermal→adiabetic
   d) Isothermal→adiabatic→isothermal→adiabatic

14. According to Carnot's theorem, all heat engines operating between a given constant temperature source and sink, none has a higher efficiency than a reversible engine.
   a) False
   b) True
   c) Depends on various other factors like working fluid, etc
   d) None of these

15. The limitation of the first law of thermodynamics is
   a) Indicates the direction of any spontaneous process
   b) It assigns a quality to different forms of energy
   c) None of mentioned
   d) Does not indicate the possibility of a spontaneous process proceeding in a definite direction

16. Triple point for water is
   a) 0.01 K
   b) 100^0C
   c) 0.01^0C
   d) No such point exists

17. A cold storage is maintained at -5 degree C while the surroundings are at 35 degree C, the heat leakage from the surroundings into the cold storage to be 29kW. What would be the minimum power required to drive the system ideally.
    a) 4.33kW b) 13kW
    c) 29kW d) data insufficient
18. The value of Mach number for a compressible fluid
    a) >0.3 b) <0.2
    c) >0.2 d) <0.3
19. If a person studies about a fluid which is at rest, what will you call his domain of study?
    a) Fluid Mechanics b) Fluid Kinematics
    c) Fluid Statics d) Fluid Dynamics
20. Specific gravity of oil is
    a) 0.2 b) 0.8
    c) 0.4 d) 0.5
21. The value of the Bulk Modulus of an ideal fluid is
    a) Zero b) Unity
    c) Less than that of a real fluid d) Infinity
22. The principle of floatation of bodies is based on the premise of
    a) Newtons law of viscosity b) Newtons first law
    c) Metacentre d) None of the mentioned
23. The value of the surface tension of an ideal fluid is
    a) Zero b) More than that of a real fluid
    c) Infinity d) Unity
24. The specific gravity of a liquid has
    a) No unit
    b) The same unit as that of mass density
    c) The same unit as that of weight density
    d) The same unit as that of specific volume
25. The specific volume of a liquid is the reciprocal of
    a) Mass density b) Specific volume
    c) Specific weight d) Weight density
26. Which one of the following is the unit of specific weight?
    a) N/m b) N/m^2
    c) N/ms d) N/m^3
27. Which one of the following is the unit of mass density?
    a) Kg/m^3 b) Kg/ms
    c) Kg/m^2 d) Kg/m

28. For an incompressible fluid does density vary with temperature and pressure?
    a) It varies only for higher values of temperature and pressure
    b) It varies only for lower values of temperature and pressure
    c) It varies for all temperature and pressure range
    d) It remains constant
29. Specific gravity is what kind of property?
    a) Extensive
    b) Intensive
    c) It depends on external conditions
    d) None of the mentioned
30. Which one of the following is not a unit of dynamic viscosity?
    a) Pa-s b) N-s/m^2
    c) Stokes d) Poise
31. Which one of the following is the CGS unit of dynamic viscosity?
    a) Stokes b) Pa-s
    c) m^2/s d) Poise
32. Which of the following is a unit of kinematic viscosity?
    a) Stokes b) Pa/sec
    c) m^2 /s d) Poise
33. What happens to viscosity in the case of incompressible fluids as temperature is increased?
    a) It increases b) It decreases
    c) It remains constant d) None of the mentioned
34. Which of the following contribute to the reason behind the origin of surface tension?
    a) Only cohesive forces
    b) Only adhesive forces
    c) Neither cohesive forces nor adhesive forces
    d) Both cohesive forces and adhesive forces
35. Which of the following is the correct relation between centroid (G) and the centre of pressure (P) of a plane submerged in a liquid?
    a) G is always below P b) P is always below G
    c) P is either at G or below it d) G is either at P or below it.
36. For liquid fluids will capillarity rise (or fall) increase or decrease with rise in temperature.
    a) Increase b) Decrease
    c) First decrease then increase d) Remain constant
37. Which principle is used for calculating the centre of pressure?
    a) Principle of conservation of energy b) Principle of momentum
    c) Principle of balancing of momentum d) None of the mentioned

38. Carbohydrate which cannot be hydrolyzed further known as
    a) Glycans b) Polysaccharides
    c) Disaccharides d) Monosaccharides
39. Invert sugar is also known as
    a) Sucrose b) Fructose
    c) Dextrose d) Glucose
40. The major storage form of carbohydrates in animals is
    a) Starch b) Chitin
    c) Glycogen d) Cellulose
41. Polysaccharides are also known as
    a) Hetroglycans b) Simple sugars
    c) Homoglycans d) Glycans
42. Which of the following is the example of oligosaccharide?
    a) Disaccharides b) Gum
    c) Trioses d) Starch
43. Digestion of lipids occur in
    a) Stomach b) Mouth
    c) Small Intestine d) All of these
44. Naturally occurring fats are
    a) D-type b) L-type
    c) Symmetrical d) Asymmetrical
45. Which one is unsaturated fatty acid?
    a) Linoleic acid b) Lauric acid
    c) Palmitic acid d) Myristic acid
46. The reagent which is used in Saponification
    a) NaOH / KOH b) Ammonia
    c) Acetic acid d) Butanone
47. Which of the following is a non-reducing Disaccharide?
    a) Sucrose b) Galactose
    c) Maltose d) Glucose
48. Name the sugar present in Honey is
    a) Maltose b) Fructose
    c) Ribulose d) Lactose
49. Which one of the following is not a monosaccharide sugar?
    a) Glucose b) Fructose
    c) Galactose d) Sucrose

50. Which type of additive is required in foods containing fats and oils to prevent rancidity
   a) Antioxidants b) Preservatives
   c) Emulsifiers d) Solvents
51. The degree of unsaturation of lipid can be measured as
   a) Saponification value b) Iodine value
   c) Acetyl value d) Ester value
52. Natural lipids are
   a) Oil b) Mercury
   c) Water d) None of these
53. Which sugar is also known as fruit sugar
   a) Fructose b) Galactose
   c) Sucrose d) Maltose
54. Table sugar is a combination of
   a) Galactose & Glucose b) Sucrose & Galactose
   c) Maltose & Fructose d) Glucose & Fructose
55. Essential fatty acid is
   a) Palmitic acid b) Stearic acid
   c) Linolenic acid d) Oleic acid
56. Which is Omega-6 fatty acid
   a) Linoleic acid b) Acetic acid
   c) Glycerol d) Linolenic acid
57. Which one is not a lipid?
   a) Fats b) Oils
   c) Waxes d) Proteins
58. Angle of Repose is a
   a) Chemical properties b) Frictional properties
   c) Thermal properties d) Physical properties
59. The amount of heat required to increase the temperature of a unit mass of the substance through one degree
   a) Specific heat b) Heat enthalpy
   c) Sensible heat d) Latent heat
60. Air oven method is used for determination of
   a) Pressure b) Temperature
   c) Moisture content d) Humidity
61. $CaCl_2$ is used to measure the grain moisture content. What is CaC_2?
   a) Calcium carbide b) Calcium chloride
   c) Calcium dicarbon d) Calcium carbon

62. What is full form of EMC?
a) Equivalent moisture content
b) Equal moisture content
c) Evaporated moisture content
d) Equilibrium moisture content

63. The removal of moisture from grains to a predetermined level is called
a) Drying
b) Steeping
c) Hydration
d) Dehydration

64. The value of sphericity ranges from
a) 0 to 1
b) 0-10
c) 0-30
d) unknown

65. Porosity is the percentage of total volume occupied by the
a) Ice
b) Blood
c) Milk
d) Air

66. Shape, size and volume is
a) Chemical properties
b) Physical properties
c) Thermal properties
d) Frictional properties

67. Thermal conductivity of grain is a
a) Physical properties
b) Frictional properties
c) Chemical properties
d) Thermal properties

68. Reynold's number is the ratio of the inertia force to the
a) Surface tension force
b) Viscous force
c) Gravity force
d) Elastic force

69. The unit of kinematic viscosity in S. I. units is
a) N-m/s
b) N-s/m^2
c) N-m
d) m^2/s

70. The product of mass and acceleration of flowing liquid is called
a) Inertia force
b) Viscous force
c) Gravity force
d) Pressure force

71. A differential manometer is used to measure
a) Difference of pressures between two points in a pipe
b) Atmospheric pressure
c) Pressure in pipes and channels
d) Pressure in venturi meter

72. One poise is equal to
a) 0.1 N-s/m^2
b) 1 N-s/m^2
c) 10 N-s/m^2
d) 100 N-s/m^2

73. The pressure at a point 4 m below the free surface of water is
a) 19.24 kPa
b) 29.24 kPa
c) 49.24 kPa
d) 39.24 kPa

74. The specific gravity of water is taken as
   a) 1 b) 10
   c) 0.1 d) 0.001
75. Free surface of a liquid tends to contract to the smallest possible area due to force of
   a) Viscosity b) Friction
   c) Surface tension d) Cohesion
76. Property of a fluid by which molecules of different kinds of fluids are attracted to each other is called
   a) Cohesion b) Adhesion
   c) Compressibility d) Viscosity
77. The property of fluid by virtue of which it offers resistance to shear is called
   a) Viscosity b) Surface tension force
   c) Adhesion d) Cohesion
78. The surface tension of mercury at normal temperature compared to that of water is
   a) Less
   b) Same
   c) More
   d) More or less depending on the size of glass tube
79. A pressure of 25 m of head of water is equal to
   a) 245 kN/ m^2 b) 25 kN/ m^2
   c) 2.5 kN/ m^2 d) 2500 kN/m^2
80. If the surface of liquid is convex, men
   a) Cohesion pressure is negligible b) Cohesion pressure is decreased
   c) There is no cohesion pressure d) Cohesion pressure is increased
81. Dynamic viscosity of most of the gases with rise in temperature
   a) Decreases b) Increases
   c) Remain unaffected d) Unpredictable
82. Newton's law of viscosity is a relationship between
   a) Pressure, velocity and temperature
   b) Rate of shear strain and temperature
   c) Shear stress and rate of shear strain
   d) Shear stress and velocity
83. Stoke is the unit of
   a) Dynamic viscosity in S. I. units
   b) Kinematic viscosity in C. G. S. units
   c) Dynamic viscosity in M. K. S. units
   d) Kinematic viscosity in M. K. S. units

84. Mercury is often used in barometer because
  a) It is the best liquid
  b) The height of barometer will be less
  c) Its vapour pressure is so low that it may be neglected
  d) Both (B) and (C)
85. The viscosity of water at 20°C is
  a) One stoke b) One centipoise
  c) One centistoke d) One poise
86. Bernoulli equation deals with the law of conservation of
  a) Energy b) Mass
  c) Momentum d) Work
87. The kinematic viscosity of an oil (in stokes) whose specific gravity is 0.95 and viscosity 0.011 poise, is
  a) 0.116 stoke b) 0.0611 stoke
  c) 0.611 stoke d) 0.0116 stoke
88. The unit of thermal conductivity is
  a) m/wK b) W/m K
  c) K/mW d) Wm/K
89. A process of heat transfer from hot body to cold body in straight line is called
  a) Radiation b) Diffusion
  c) Distillation d) Evaporation
90. A thermo-physical and physico-chemical operation by which the excess moisture from a food products is called
  a) Diffusion b) Distillation
  c) Evaporation d) Drying
91. EMC is related to
  a) Roasting b) Drying
  c) Frying d) Baking
92. Henderson equation is related to
  a) IMC b) EMC
  c) CMC d) FMC
93. Terminal velocity is applied in to
  a) Tray dryer b) LSU dryer
  c) Fluidized bed dryer d) Vacuum dryer
94. A curve between sorption and adsorption is related to
  a) IMC b) CMC
  c) FMC d) EMC
95. Evaporation process is only applied to remove moisture from
  a) Grain b) Milk
  c) Rice d) Pickles

96. Distilled water is prepared by
   a) Diffusion
   b) Distillation
   c) Radiation
   d) Dehydration

97. Roult's law is related to
   a) Diffusion
   b) Radiation
   c) Dehydration
   d) Distillation

98. Blanching is a
   a) Thermal process
   b) Distillation process
   c) Drying process
   d) Freezing process

99. Total mass balance in evaporation is calculated as
   a) F + V + P = 0
   b) V = P +F
   c) F = P - V
   d) F = P + V

100. The amount of heat required to increase the temperature of a unit mass of the substance through one degree
   a) Specific heat
   b) Heat enthalpy
   c) Sensible heat
   d) Latent heat

Answer Key

1	a	2	a	3	b	4	b	5	b	6	c	7	d
8	a	9	c	10	d	11	d	12	a	13	d	14	b
15	d	16	c	17	a	18	a	19	c	20	b	21	d
22	c	23	a	24	a	25	a	26	d	27	a	28	d
29	b	30	c	31	d	32	a	33	b	34	d	35	c
36	b	37	c	38	d	39	a	40	c	41	d	42	a
43	d	44	b	45	a	46	a	47	a	48	b	49	d
50	a	51	b	52	d	53	a	54	d	55	c	56	a
57	d	58	b	59	a	60	c	61	b	62	d	63	a
64	a	65	d	66	b	67	d	68	b	69	d	70	a
71	a	72	a	73	d	74	a	75	c	76	b	77	a
78	c	79	a	80	d	81	b	82	c	83	b	84	d
85	b	86	a	87	d	88	b	89	a	90	d	91	b
92	b	93	c	94	d	95	b	96	b	97	d	98	a
99	d	100											

Model Paper 6

1. The science of rheology is only about years of age.
   a) 70 b) 76
   c) 72 d) 80
2. Textural characteristics are an important factor in the overall quality of many
   a) Food products b) Mobil oil
   c) Petrol d) Vegetables
3. Rheological parameters are also useful in defining the quality attribute of
   a) Linseed oil b) Oil
   c) Vegetable d) Food products
4. Study of rheology helps to select proper method of harvesting and sorting
   a) Shape b) Seed
   c) Raw materials d) Size
5. Rheological studies also help to evaluate quality of cheese and applicability of cheese for various applications like suitability for
   a) Cheese topping b) Pizza topping
   c) Capsicum topping d) Red Chilly topping
6. Place sample between molar teeth and bite down evenly, evaluating the force required to compress the food.
   a) Hardness b) Cohesiveness
   c) Roughness d) Adhesiveness
7. Place sample between molar teeth, compress and evaluate the amount of deformation before rupture.
   a) Hardness b) Viscosity
   c) Cohesiveness d) Gumminess
8. Place sample between molar teeth and bite down evenly until the food crumbles, cracks or shatters, evaluating the force with which the food moved away from the teeth
   a) Cohesiveness b) Fracturability
   c) Hardness d) Roughness
9. Place sample in the mouth and manipulate with the tongue against the palate, evaluating the amount of manipulation necessary before the food disintegrates.
   a) Flow ability b) Viscosity
   c) gumminess d) Cohesiveness

10. Place spoon with sample directly in front of mouth and draw liquid from spoon over tongue by slurping, evaluating the force required to draw liquid over tongue at a steady rate.
    a) Viscosity
    b) Gumminess
    c) Cohesiveness
    d) Adhesiveness
11. SEM refers to
    a) Scanning Electron Milk
    b) Steel Electron Mass
    c) Steel Electron Molecule
    d) Scanning Electron Microscopy
12. TEM refers to
    a) Transmission Electron Mass
    b) Transmission Electron Microbiology
    c) Transmission Electron Microscopy
    d) Transmission Electron Minimum
13. Cooking of Chhana in sugar syrup for
    a) 15 minute
    b) 10 minute
    c) 25 minute
    d) 30 minute
14. Hardness, mN for Cow milk chhana is
    a) 11.6
    b) 13
    c) 15
    d) 17
15. Cohesiveness for Cow milk chhana is
    a) 1
    b) 3
    c) 6
    d) 0.59
16. Springiness, mm for Buffalo milk chhana is
    a) 3
    b) 5
    c) 1
    d) 6
17. Chewiness, mN. Mmfor Buffalo milk chhana is
    a) 50
    b) 55
    c) 65.32
    d) 60
18. Adhesiveness mN for Cow milk chhana is
    a) 0.35
    b) 0.25
    c) 0.45
    d) 0.55
19. Relaxation curve (stress relaxation) – It is the curve obtained when stress is applied as a function of time at
    a) Constant speed
    b) Constant strain
    c) Constant stress
    d) Flexibility
20. The tests conducted under conditions of static/quasi-static loading are known as
    a) ANOVA test
    b) Quasi-static tests
    c) T-Test
    d) Determination test

21. Fourier's Law of heat conduction is represented

a) qx/Q= -PdT/dx
b) qx/R= -CdT/dx
c) qx/A= -KdT/dx
d) qx/C= -MdT/dx

22. Unit of heat flux in S.I. system is

a) W/mK
b) W/pK
c) W/°C K
d) W/m^2

23. At very low pressure (vacuum), the thermal conductivity approaches as

a) Zero
b) <1
c) >1
d) >1.5

24. Ice has a thermal conductivity much greater than

a) Vapour
b) Less
c) Water
d) Rate of generation of heat

25. Conduction through a hollow sphere, the temperature varies hyperbolically with the

a) Length
b) Radius
c) Diameter
d) Resistance

26. The total heat lost from the two forces at steady state is equal to the

a) Thermal conductivity of medium
b) Total heat generated
c) Volumetric rate of heat generation
d) Gross sectional area of the plate

27. Interface resistance is also known as

a) Contact resistance
b) Heat flux
c) Heat resistance
d) Series resistance

28. If 10 w/m^2 heat flux is conducted across a wall of 2cm thickness experiencing a temperature gradient of 2°C m the thermal conductivity of the wall is

a) $1x10^{-1}$ w/mk
b) $1x10^{-2}$ w/mk
c) $1x10^{-3}$ w/mk
d) $1x10^{3}$ w/mk

29. A cark slab of 100mm thickness has one face as t-12°c and the other face at 21°c, If the mean thermal conductivity (K) of the cork is 0.042J/m/s/k, the rate of heat transfer (J/s) through one mm^2 of the wall will be

a) 2.5
b) 5
c) 10
d) 13.9

30. Conduction heat transfer is quantified by

a) Laplace law
b) Blake-kognry
c) Fourier's law
d) Burke-Plummer equation

31. Convective mass transfer coefficient of water vapour in air depends on

a) Velocity of air
b) Density of water vapour
c) Viscosity of water vapour
d) None of the above

32. Flux of the property being transferred per unit time per unit cross-sectional area perpendicular to the

a) X-direction of flow
b) Y-direction of flow
c) Z-direction of flow
d) None of the above

33. For a primary gas mixture of A and B, the diffusivity coefficient D for B diffusing into

a) A
b) B
c) Both A & B
d) None of the above

34. Thermal conductivity of the same material varies with

a) The thickness of the material
b) The temperature of the material
c) Thermal conductivity of the same material does not vary at all
d) Both C & D

35. As the radius increases in hollow cylinder the temperature

a) Also increases
b) Decreases
c) Unpredictable
d) It depends upon the direction of heat flow

36. What is the formula for the thermal resistance (R) of a hollow cylinder of internal radius R_i outer radius R_o and the length l, thermal conductivity of the cylinder material in k. The heat flows from inner side to outer side

a) $R= \ln(R_o/R_i)/2\ \pi kl$
b) $R= \ln(R_i /R_o)/2\pi kl$
c) $R= (R_i +R_o)/2\ \pi kl$
d) $R= (R_i\ R_o)/2\pi kl$

37. At the critical radius of insulation of a hollow sphere in the heat transfer will be

a) Minimum
b) Maximum
c) Does not change
d) Un predictable

38. What is the formula for thermal resistance for convection at the outer surface of hollow cylinder?

a) Rc = 1/hrl
b) Rc = 2π/hrl
c) Rc = 1/2π/hrl
d) None of the above

39. What is the purpose of using fins in a particular heat transfer system

a) To increase rate of heat transfer
b) To decrease rate of heat transfer
c) To maintain rate of heat transfer at a constant rate
d) Can not say

40. Fluidized bed dryer is used for drying

a) Tea Leaves
b) Vegetables
c) Milk
d) Fruit Juices

41. The average particle size in spray dryer is

a) 100pm
b) 50pm
c) 10pm
d) 0.01pm

42. In forced air-drying method the moisture moves from:

a) Centre to grain surface
b) Grain to air
c) Air to grain
d) All are correct

43. In deep bed drying, the depth of grain layer is:
   a) 10 cm b) 20 cm
   c) Less than 10 cm d) More than 20cm
44. LSU dryer was developed in
   a) Japan b) England
   c) USA d) None of these
45. Thermal conductivity of single grain ranges from 1-5 is:
   a) 0.4-0.8 Kcal/mhr/°C b) 0.3-0.6 Kcal/mhr/°C
   c) 0.20-0.25 Kcal/mhr/°C d) 0.25-0.30 Kcal/mhr/°C
46. In sack drying, a maximum air temperature should be:
   a) 43.3°C b) 41
   c) 42°C d) 45°C
47. Heat sensitive or easily oxidizable materials are dried by:
   a) Fluidized bed dryer b) Flash dryer
   c) Rotary dryer d) Drum dryer
48. Which dryer used radiation for drying?
   a) Microwave dryer b) Spray dryer
   c) Drum dryer d) Flash dryer
49. Which dryer is used to dry atomized droplets of a feed?
   a) Fluidized bed dryer b) Spray dryer
   c) Drum dryer d) Flash dryer
50. During drying, only __________ can be evaporated, which depends upon the vapour concentration in the gas.
   a) Bound moisture b) Fixed moisture
   c) Free moisture d) Total Moisture
51. ____________ involves the sublimation of water from ice under high vacuum at temperatures well below 0°C.
   a) Flash drying b) Freeze-drying
   c) Furnace drying d) Ice drying
52. Radiation dryers usually use ___________waves.
   a) Low frequency b) High frequency
   c) Yellow light d) Bright red visible frequency
53. Which temperature can be measured by an instrument called psychrometer?
   a) Dry bulb temperature b) Wet bulb temperature
   c) Both a and b d) None of these
54. Sling psychrometer is used to measure
   a) Dry bulb temperature b) Wet bulb temperature
   c) Relative humidity d) Dry and Wet bulb temperature

55. For biological materials, the relationship between EMC and RH has been given by:
a) Rankine b) Janssen
c) Henderson d) Chung fast

56. The capacity of thermal dryer depends on the:
a) Rate of mass transfer b) Rate of heat transfer
c) Rate of heat and mass transfer d) None of these

57. A critical moisture content varies with:
a) Rate of drying b) Thickness of material
c) Both a and b d) None of these

58. The moisture content is determined by
a) Direct method b) Indirect method
c) Both a and b d) None of these

59. The amount of water mechanically entrapped in the void spaces of the system called
a) Free moisture b) Unbound moisture
c) Bound moisture d) All of these

60. Area required to dry one ton of grain in the sun is about:
a) 8 sq. m b) 10 sq. m
c) 15 sq. m d) 20 sq. m

61. Temperature at which a given unsaturated air-vapor mixture became saturated known as
a) Dew point b) Bound moisture
c) Equilibrium moisture d) Critical moisture

62. The amount of water evaporated by using 1 kg of steam is called:
a) Steam ratio b) Steam economy
c) Evaporator efficiency d) Boiling efficiency

63. When you concentrate orange juice by boiling off the excess water, the unit operation in the process is known as:
a) Drying b) Distillation
c) Evaporation d) Crystallization

64. The most effective evaporator type is:
a) Plate b) Rising film
c) Shell and tube d) Falling film

65. The term "Cold Sterilization" is also known as:
a) Irradiation b) Cold storage
c) Refrigeration d) Ultrafiltration

66. A point where solid, liquid and vapor phase of substance exist is called:
a) Triple point b) Melting point
c) Critical point d) Boiling point

67. The amount of heat required to change the state of a substance is called:
    a) Specific heat b) Sensible heat
    c) Latent heat d) Total heat
68. Time of freezing of food is longer than that for thawing. The reason for this is:
    a) Water has lower thermal conductivity than ice
    b) Water has higher thermal diffusivity than ice
    c) Water has higher density than ice
    d) Water has higher thermal conductivity than ice
69. The density of food after freezing:
    a) Increases b) Decreases
    c) Unchanged d) First increase then decreases
70. The removal of moisture from food products to very low levels usually to bone dry condition is called:
    a) Drying b) Screening
    c) Grading d) Dehydration
71. Heat transfer in liquids and gases is essentially due to:
    a) Conduction
    b) Convection
    c) Radiation
    d) Conduction and Radiation put together
72. Separation of dissolved solids by application of pressure is:
    a) Pressing b) Filtration
    c) Vacuum filtration d) Membrane separation
73. ------------------------------is the separation of solids from liquid by passing the flow of mixture through fine pores:
    a) Grading b) Sieving
    c) Filtration d) Sedimentation
74. Which of the following can be used to separate two liquid components with different densities?
    a) Centrifugation b) Homogenization
    c) Aspiration d) Specific gravity separation
75. Freeze drying time is directly proportional to the .... of the material being dried:
    a) Thickness b) Cube of thickness
    c) Square of the thickness d) Fourth power of thickness
76. The full form of KGY is:
    a) Kilo Grant b) Kilo Gray Yield
    c) All of these d) Kilo Gray
77. The intensity of pulsed electric field is:
    a) 10 – 20Kg/cm b) 20 – 80Kv/cm
    c) 80 – 100Kv/cm d) 100 – 150Kv/cm

78. Ohmic heating works on the principle of:
a) Electrical resistance
b) Thermal resistance
c) Mechanical resistance
d) Electrical conductance

79. Microwave oven works on the .... Properties of material
a) Physical
b) Mechanical
c) Dielectric
d) Thermal

80. HHPP is stands for:
a) Hydrostatic high pressure processing
b) High hydrostatic power processing
c) High hydrostatic pressure processing
d) High hydrodynamic power processing

81. In India, the frequency of microwaves used in commercial appliances is:
a) 896 MHz
b) 1650 MHz
c) 915 MHz
d) 2450 MHz

82. BIOS stands for
a) Binary Input output system
b) Basic Input Off system
c) Basic Input Output system
d) all the above

83. ALU and Control Unit jointly known as
a) PC
b) CPU
c) RAM
d) ROM

84. Microorganism multiply everywhere, except....
a) Nutrient Agar
b) Atmosphere
c) Nutrient Broth
d) Slants

85. Who had seen minute living worms in putrid meat and milk worms?
a) Athanasius Kircher
b) Robert Koch
c) Louis pasteur
d) H. Kircher

86. Toxin of Clostridium Botulinum is...
a) Thermo stable
b) Thermo labile
c) Stable
d) E. coli

87. Food intoxication is the ingestion of....
a) Mineral toxicity
b) Salt toxicity
c) toxin formed by food
d) Toxin produced by microorganisms

88. Bacteria able to survive in high salt concentration are known as...
a) Halotolerant
b) Psychrophiles
c) Mesophiles
d) Thermophiles

89. Example of Osmophilic Bacteria is….
 a) Streptococcus thermophilus b) Lactobacillus
 c) E.coli d) Salmonella
90. In sewage the most contaminating microorganism is …
 a) Staphylococcus spp. b) E. coli
 c) Viruses d) Corona virus
91. Bacteria that are able to grow in both the presence and absence of oxygen…
 a) Obligate anearobes b) Aerobes
 c) Coliform d) Facultative Anaerobes
92. Majority of microbe grow around …………… and molds around ………………
 a) 30°C and 35°C b) 35°C and 25°C
 c) 37°C and 25°C d) 45°C and 15°C
93. Temperature of hot air oven is…
 a) 160°C for 60 min. b) 100°C for 30 min.
 c) 90°C for 30 min. d) 115°C for 45 min.
94. In spore forming bacteria maximum resistance occurs at pH…
 a) 4 b) 5
 c) 7 d) 9
95. Food spoilage is caused by many microbes, except
 a) Viruses b) Bacteria
 c) Fungi d) Parasites
96. Any change that renders food unfit for human consumption is called…
 a) Spoilage b) Processing
 c) Deterioration d) Preservation
97. Who showed that wine fermentation from grapes and souring of wine was caused by microorganisms?
 a) Louis Phillips b) Louis Pasteur
 c) Robert koch d) Metchnikoff
98. Microbes that grow at refrigerated temperature
 a) Osmotrophes b) Psychrophiles
 c) Psychrotrophs d) Mesotrophs
99. Microorganisms that survive at the temperatures of pasteurization are…
 a) Staphylococcus spp. b) Thermodurics
 c) Xerotolerant d) E.coli
100. Which type of fermentation process is used by E. coli
 a) Alcoholic b) Mixed
 c) Butyric d) Acidic

Answer Key

1	b	2	a	3	d	4	c	5	b	6	a	7	c
8	b	9	c	10	a	11	d	12	b	13	a	14	a
15	d	16	b	17	c	18	a	19	b	20	b	21	c
22	d	23	a	24	c	25	b	26	b	27	a	28	a
29	d	30	c	31	a	32	c	33	a	34	b	35	d
36	a	37	b	38	c	39	a	40	a	41	a	42	b
43	d	44	c	45	b	46	a	47	b	48	a	49	b
50	c	51	b	52	b	53	c	54	d	55	c	56	c
57	c	58	c	59	a	60	c	61	a	62	b	63	c
64	d	65	a	66	a	67	c	68	a	69	b	70	d
71	b	72	d	73	c	74	a	75	c	76	d	77	b
78	a	79	c	80	c	81	d	82	c	83	b	84	b
85	a	86	b	87	d	88	a	89	b	90	b	91	d
92	c	93	a	94	c	95	a	96	a	97	b	98	c
99	b	100	b										

Model Paper 7

1. …………….. regulates that flow of substances in and out of the cell
   a) Cytoplasm b) Cellwall
   c) Ribosome d) Cell membrane
2. Bacterial cell lack in which organelle?
   a) True nucleus b) Mitochondria
   c) Chloroplasts d) All
3. Bacterial cell wall is made of
   a) Lipid b) Protein
   c) Phospholipids d) Peptidoglycan
4. Fungi cell wall is made of
   a) Chitin b) Protein
   c) Phospholipids d) Peptidoglycan
5. Bacteria reproduce through
   a) Asexual reproduction b) Binary fission
   c) Sexual reproduction d) Both a & b
6. The extracellular infectious virus particle is called
   a) Spike b) Virion
   c) Plasmid d) Both a & b
7. Protein coat of virus which pretects its nucleic acid is called
   a) Virion b) Plasmid
   c) Capsid d) Spike
8. Envelop is known as ………… of virus
   a) Cell wall b) Capsid
   c) Spike d) Outer covering
9. Confirmation of chemical, antigenic and biological properties of viruses is done by
   a) Envelope b) Capsid
   c) Spike d) Cell Wall
10. Projecting spikes of virus are found on
    a) Flagellum b) Envelope
    c) Capsid d) Cell Wall

11. …….. is plant-like because they contain the green pigment chlorophyll
    a) Bacteria b) Virus
    c) Algae d) Fungi
12. Which microorganism forms characteristic hyphae called mycelium
    a) Bacteria b) Virus
    c) Algae d) Fungi
13. Fruiting structure of fungi called
    a) Conidia b) Exospores
    c) Endospores d) All
14. Which mold is use full in preparation of antibiotic drugs?
    a) Penicillium b) Roquefort
    c) Saccharomyces cerviceae d) Mushroom
15. Toxins produced by fungi is known as
    a) Bacteriocin b) Saccharomyces cerviceae
    c) Nisin d) Mushroom
16. Yeasts grow best in environment which is
    a) Neutral or slightly acidic pH b) Acidiacidic
    c) Basic d) Neutral or slightly basic
17. Reproduction of yeast may occur by
    a) Sexual reproducing b) Asexual reproducing
    c) Both a and b d) None
18. Which microorganism is used as biological raising, or leavening, agent?
    a) Bacteria b) Yeast
    c) Both a and b d) Virus
19. Alcoholic beverages and vinegar is prepared by
    a) Bacteria b) Yeast
    c) Fungi d) Virus
20. Pertussis is also known as
    a) Dipthria b) Whooping cough
    c) Common flu d) Cough and cold
21. ………. is a serious bacterial infection usually affecting the mucous membranes of your nose and throat
    a) Diphtheria b) Whooping cough
    c) Both a and b d) Swine flu
22. Characteristic thick, gray membrane covering throat and tonsil is a sign of
    a) Tonsillitis b) Diphtheria
    c) Whooping cough d) Allergy
23. Antibiotic drugs are effective on
    a) Virus b) Fungi
    c) Mold d) Bacteria

24. Chronic cough with blood-tinged sputum is a sing of
    a) Whooping cough b) Tuberculosis
    c) Tonsillitis d) Diphtheria
25. ………… is used to examine objects in two dimensions
    a) Simple microscope b) Compound microscope
    c) Dissecting microscope d) TEM
26. In which microscope electron beam passes through the specimen?
    a) Dissecting microscope b) TEM
    c) SEM d) All
27. Which microorganism is a very common cause of diarrhea in developing countries?
    a) E. Coli b) Staphylococcus aureus
    c) Clostridium Botulinum d) Shigella
28. Which microorganism is the most common cause of bacillary dysentery in India
    a) E. Coli b) Staphylococcus aureus
    c) Clostridium Botulinum d) Shigella
29. The most abundant bacteria found in fecal matter are
    a) E. Coli b) Staphylococcus aureus
    c) Clostridium Botulinum d) Shigella
30. Which virus is involved in food borne diseases?
    a) Hepatitis A b) Norovirus
    c) Vibrio cholera d) All
31. Skin ring worm infection is caused due to
    a) Worm b) Fungus
    c) Virus d) Bacteria
32. Fungal ring worm infection is caused by ……………….. organism
    a) Ring worm b) Tinea
    c) Helminthes d) Both a and c
33. A packaged food that can be prepared quickly & easily is called
    a) Convenience Food b) Health food
    c) Fast food d) Junk food
34. Convenience foods developed at the time of
    a) World war I b) World war II
    c) World war III d) Recently
35. Convenience foods are convenient in terms of
    a) Preparing b) Eating
    c) Serving d) All of above
36. Extrusion is defined as
    a) Preserved foods b) Shaping by force
    c) Convenience foods d) Light, fluffy food

37. Example of extruded food is
    a) Pasta b) Noodles
    c) Cracked cereals d) All of above
38. Extruded foods considered as
    a) Low cost high productivity b) Low cost low productivity
    c) High cost low productivity d) High cost high productivity
39. A typical food extruder consist
    a) Feeding zone b) Kneading zone
    c) Cooling zone d) All
40. Pasta is originally cuisine of
    a) Japan b) Italy
    c) Germany d) India
41. FOSU terms for
    a) Food for specific Use b) Food for Social Utilization
    c) Food of Special Use d) Food of Systematic Utilization
42. Dispersion of one substance into on ether substance is called
    a) Solution b) Mixture
    c) Colloid d) Gel
43. Tyndall effect can be seen in
    a) Solution b) Gel
    c) Colloid d) Suspension
44. Example of emulsion is
    a) Milk b) Mayonnaise
    c) Both a and b d) Salad oil dressing
45. For a good jelly formation, which is/are important
    a) Sugar b) Pectin
    c) Acid d) All of above
46. Pectin is used as
    a) Oxidative agent b) Thickening agent
    c) Reducing agent d) Emulsifying agent
47. Decomposition of fats to glycerol and free fatty acids is called
    a) Pyrolysis b) Rancidity
    c) Hydrolysis d) Both a and c
48. Thermal breakdown of fat molecules is known as
    a) Pyrolyisis b) Oxidation
    c) Smoking point d) None
49. Butyric & Caproc acids are responsible for
    a) Rancid flavor b) Gelatinization
    c) Coagulation d) Leaving

50. Use of baking soda will decrease
    a) Thiamine b) Vitamin A
    c) Protein d) Calcium
51. Iodine number of fat indicates
    a) Unsaturated fatty acids b) Saturated fatty acids
    c) Essential fatty acids d) All
52. Enzymes responsible of r browning are
    a) Amylase b) Lipase
    c) Protease d) Phenolase
53. Astringency occurs due to precipitation of ...... in saliva
    a) Protein b) Enzyme
    c) Hormone d) Starch
54. Percolation & filtration is a method of making of
    a) Tea b) Fruits juice
    c) Alcohol d) Coffee
55. Brownian movement can be seen in
    a) Solution b) Emulsion
    c) Suspension d) Colloid
56. True solution is always
    a) Turbid b) Semi-transparent
    c) Transparent d) Non transparent
57. Which food pigment blocks UV light radiation of sun?
    a) Chlorophyll b) Betalin
    c) Anthocyanin d) Flavonoids
58. End product of browning reaction is
    a) Orthoquinones b) Melanins
    c) Flavanoids d) All
59. Wheat protein & corn protein are used as
    a) Protein Source b) Edible coating
    c) Filler material d) None of above
60. Pasteurization is the heat treatment designed primarily to kill...
    a) Only spores
    b) Vegetable forms of microorganisms
    c) Spore forming bacteria
    d) All form of microorganisms
61. Microorganisms are responsible for causing food borne disease, except...
    a) Bacteria b) Viruses
    c) Mold d) Yeast

62. Clostridium botulinum mainly result in spoilage of -------------- foods
   a) High acid Food b) Low acid Food
   c) Acidic Food d) Medium acid Food

63. Long, whip-like protrusion that aids in cellular locomotion in prokaryotic cell is called.
   a) Nucleus b) Cytoplasm
   c) Flagellum d) Nucleoid

64. Unhygienic and unauthorized killing of food animal is known as :
   a) Ritual slaughter b) Clandestine slaughter
   c) Human slaughter d) Sacrifice

65. An animal showing excessive laceration may be placed in the category of :
   a) Emergency slaughter b) Slaughter under special condition
   c) Delayed slaughter d) Casualty slaughter

66. An incision to be given without any pause, pressure, tearing, stabbing, slanting is the thumb rule of :
   a) Halal method b) Clandestine slaughter
   c) Humane method d) Jewish method

67. Lymph nodes are absent in :
   a) Poultry b) Pig
   c) Horse d) Cattle

68. 'Warmed – over flavor ' in meat is caused by :
   a) Volatile compound b) Steroid
   c) Phospholipase d) H_2O_2

69. An imaginary zone around the animal, which it always tries to maintain by moving away while being handled is called :
   a) Fight zone b) Flight zone
   c) Fright zone d) Prohibited zone

70. The condition seen in tetanus affected animal on antemortem inspection is :
   a) Locked jaw b) Dropped jaw
   c) Lumpy jaw d) Open jaw

71. The organ of choice for determination of antibiotic residues in meat is:
   a) Liver b) Kidney
   c) Both a and b d) Spleen

72. The change that is not associated with incomplete bleeding is ;
   a) Blood in left ventricle b) Engorged S/ C blood vessel
   c) Dark flabby flesh d) Cloudy swelling

73. The test used for detection of icterus in meat is :
   a) Rimmington and Fowrie test b) Rothera's test
   c) Malachite green test d) Casoni test

74. Boiling test in meat should be performed for detection of ;
 a) Abnormal odour b) Freshness
 c) Abnormal taste d) Staleness

75. The condition not found in an emaciated animal is :
 a) Wrinkled and dry skin b) Pormnent bones
 c) Rough hair coat d) Prominent eyes

76. The change not associated with porcine stress syndrome is :
 a) Decrease in temperature b) Arrhythmia
 c) Skin blanching d) Dyspnoea

77. As per ICMSF standards , the count of Salmonella in good quality meat should be :
 a) 0 b) Not more than 1/ g
 c) 1/10 g d) 10/10 g

78. The color of colonies of Statpylococci on Baird Parker medium is
 a) Yellow b) Jet black
 c) Violet d) Pink

79. The color of colonies of Statpylococci on Brilliant green agar medium is:
 a) Pale yellow b) Green
 c) Violet d) Pink

80. Spleen is congested and contains tiny foci of necrosis in bovine salmonellosis. This speenomegaly is known as :
 a) Burst spleen b) Board spleen
 c) Slaughter spleen d) Balloon spleen

81. The extract release volume of good quality meat should be ;
 a) More than 30 ml b) More than 20 ml
 c) More than 25 ml d) Less than 20 ml

82. The device use for gaseous stunning is known as :
 a) Oval tunnel b) Ferris wheel
 c) Dip lift d) All of above

83. The method of gaseous stunning , having largest thorough put is :
 a) Oval tunnel b) Ferris wheel
 c) Dip lift d) Pilot wheel

84. On ante mortem inspection, if uncontrolled haemorrahages are seen, the judgment should be :
 a) Casualty slaughter b) Delayed slaughter
 c) Emergency slaughter d) Unfit for slaughter

85. The judgment of emergency slaughter is given in case of :
 a) Postpartum paraplegia b) Uterine prolapsed
 c) Lactation tetany d) Recumbency

86. The animal should be declared 'unfit for slaughter' after antemortem examination if it is suffering from :
   a) Rabies b) Haemorrhagic septicemia
   c) Tetanus d) All of above

87. It the animal stock is found very dirty on ante mortem inspection, the judgment should be :
   a) Casualty slaughter b) Slaughter at last
   c) Delayed slaughter d) Unfit for slaughter

88. The term 'bob-veal' is used to denote the meat of :
   a) Day old calf b) 10 days old calf
   c) 7 days old calf d) 30 days old calf

89. The major by-product obtained from animal fat is /are :
   a) Candy b) Soap and detergent
   c) Chewingum d) All of above

90. The possible source of contamination to establish critical control points in fresh meat production in buffalo abattoir is/are :
   a) Evisceration b) Cutting
   c) Chilling d) All of above

91. The humane method of slaughter by designing of a gaseous chamber to induce anesthesia in animal to kill them, was developed for the first time by :
   a) Benjamin Ward Richordson b) James Steel
   c) Martin Kaplan d) John Snow

92. The blue print of slaughter house was developed by :
   a) Benjamin Ward Richordson b) James Steel
   c) Martin Kaplan d) John Snow

93. The main purpose of stunning is :
   a) To induce unconsciousness b) To improve meat quality
   c) To enhance bleeding efficiency d) To relive pain to animal

94. The person who assist the cutter in Jewish method is called :
   a) Schochet b) Kosher
   c) Shomer d) Chalef

95. Trimethylamine level more than 7 mg per cent in fish meat indicates :
   a) Fresh fish b) Onset of spoilage
   c) Spoiled fish d) Very fish

96. The first bacterium to invade the carcass from bowl after death is :
   a) E. coli b) Klebsiella
   c) Cl. Perfringens d) Pseudomonas

97. Use of antibiotics should be withheld at least ….. prior to slaughter :
   a) 3 days b) 5 days
   c) 15 days d) 20 days

98. Kidney fat is abundant in (animal species) :
   a) Sheep b) Cattle
   c) Goat d) Pig

99. Amonical odour is the characteristic of :
   a) Pork b) Mutton
   c) Chevon d) Beef

100. A thick subcutaneous fatty layer is a characteristic of :
   a) Pork b) Mutton
   c) Chevon d) Beef

Answer Key

1	d	2	d	3	c	4	a	5	d	6	b	7	c
8	d	9	a	10	b	11	c	12	d	13	d	14	a
15	b	16	a	17	c	18	c	19	b	20	b	21	c
22	b	23	d	24	b	25	b	26	b	27	a	28	d
29	a	30	d	31	b	32	b	33	a	34	b	35	d
36	b	37	d	38	a	39	d	40	b	41	a	42	b
43	c	44	c	45	d	46	b	47	d	48	a	49	a
50	a	51	a	52	d	53	a	54	d	55	d	56	c
57	d	58	b	59	b	60	b	61	d	62	b	63	c
64	b	65	a	66	d	67	a	68	c	69	c	70	a
71	a	72	d	73	a	74	a	75	d	76	a	77	a
78	b	79	a	80	a	81	c	82	d	83	a	84	b
85	b	86	d	87	c	88	c	89	d	90	d	91	a
92	a	93	d	94	c	95	c	96	a	97	b	98	c
99	b	100	a										

Model Paper 8

1. The process preservation of fruits and vegetable with common salt and vinegar is called
   a) Fermentation b) Pickling
   c) Steeping d) Asepsis
2. The by product of apple is
   a) Rags b) Peel
   c) Pomace d) Stones
3. Fruit for making jelly should be rich in
   a) Pectin and acids b) Acids and protein
   c) Protein and sugars d) Sugars and acids
4. Sweetest sugar in fruit is
   a) Sucrose b) Fructose
   c) Glucose d) Galactose
5. Guava is a rich source of
   a) Vitamin C b) Carbohydrates
   c) Vitamin A d) ß- carotene
6. Ripening fruits exist which gas is
   a) NH_3 b) Methane
   c) CO_2 d) Ethylene
7. Why does blanching in vegetables?
   a) Microorganism removal b) Inactivate enzymes
   c) For softening the vegetables d) All
8. Vegetables are called as protective food because they are rich source of
   a) Fat b) Viamins and minerals
   c) Proteins d) Sugars
9. Daily requirement of vegetable per capita/day is
   a) 150 g b) 285 g
   c) 200 g d) 400 g
10. The process of removal of moisture to bone dry condition is called
    a) Drying b) Dehydration
    c) Freeze drying d) Osmotic drying

11. Specific gravity of milk can be increased by adding
    a) Fat
    b) Enzyme
    c) SNF
    d) water
12. Controlled atmosphere storage is used for storage of
    a) Fresh fruits & vegetables
    b) Jams
    c) Juices
    d) Dried products
13. Most fruits and vegetable are stored at:
    a) Low temperature and low humidity
    b) High temperature and low humidity
    c) High temperature and high humidity
    d) Low temperature and high humidity
14. Fruits and vegetables stored in a cold chamber enhance storage life because
    a) There is an increase in humidity
    b) CO_2 concentration in the environment is increased
    c) Rate of respiration is decreased
    d) Exposure to sum light is prevented
15. Standard unit used in expressing refrigeration capacity is
    a) Calorie
    b) Ton
    c) Ampere
    d) Joule
16. Which gas is helpful for freezing of foods at-196°C?
    a) Nitrogen
    b) Oxygen
    c) Hydrogen
    d) Ozone
17. The process of removal of moisture to bone dry condition is called
    a) Drying
    b) Dehydration
    c) Freeze drying
    d) Osmotic drying
18. Microwave oven is as dryer is an example of
    a) Vacuum drying
    b) Radiation
    c) Di electric drying
    d) Freeze drying
19. Method of drying used for heat sensitive produce is
    a) Spray drying
    b) Tray drying
    c) Flow drying
    d) Freeze drying
20. Cryo-preservation is associated with
    a) Liquid oxygen
    b) Liquid nitrogen
    c) Liquid ammonia
    d) Liquid carbon dioxide
21. In foam mat drying, foaming agent help to
    a) Decrease surface area
    b) Stabilizes
    c) Neither increase nor decrease surface area
    d) Increase surface area

22. The removal of water from a food material by sublimation from the frozen state to the vapor state is known as
    a) Freeze drying b) Cryogenic Freezing
    c) Freeze concentration d) Heat pump drying
23. Freeze drying time is directly proportional to the
    a) Square of the thickness of material being drying
    b) Thickness of material being drying
    c) Cube of the thickness of material being drying
    d) Fourth power of the thickness of material being drying
24. During dehydration of vegetable, the temperature of dehydration should be
    a) 40 to 50°C b) 60 to 70°C
    c) 80 to 90°C d) 100 to 115°C
25. Drying of fruit pulp can be accomplished by a
    a) Tray dryer b) Fluidized bed dryer
    c) Drum dryer d) Spray dryer
26. Fruits and vegetables are preferably dried in a
    a) Cabinet dryer b) Spray dryer
    c) Freeze dryer d) Fluidized bed dryer
27. Cool storage temperature means storing below a temperature of
    a) 25°C b) 16°C
    c) 2.2°C d) 4°C
28. Controlled atmosphere storage applied to
    a) Low Oxygen b) High CO_2 atmosphere
    c) Burner gas d) All the above
29. Ideal storage temperature for potato is
    a) 0°C b) 2°C
    c) 4°C d) 8°C
30. Hypobaric storage is also known as:
    a) MA storage b) CA storage
    c) All the them d) Low pressure storage
31. The ideal concentration of sodium benzoate for preserving fruit juice
    a) 0.06 - 0.10 % b) 0.01-0.05 %
    c) 1.1-1.5 % d) 1.6 - 2.0 %
32. Canned vegetable is dipped in brine solution is made up from
    a) Black salt b) Common salt
    c) Sugar d) Salt and sugar

33. Fruits and vegetables stored in a cold chamber enhance storage life because
    a) There is an increase in humidity
    b) CO_2 concentration in the environment is increased
    c) Exposure to sum light is prevented
    d) Rate of respiration is decreased

34. The lye peeling of fruits and vegetables is done at the temperature of
    a) 93°C b) 75°C
    c) 80°C d) 100°C

35. The composition of following gases is kept control in CA storage
    a) O2 & N_2 b) O2 & CO_2
    c) O2 & H d) O2, N_2 & CO_2

36. Good quality of pectin is found when guava is at
    a) Green stage b) Dark yellow in colour
    c) Ripe but green stage d) Over ripe

37. In freeze drying the liquid phase comes at the pressure above
    a) 4.7 mm b) 1.7 mm
    c) 2.7 mm d) 3.7 mm

38. Irradiation can be delay of sprouting for
    a) Potato b) Onion
    c) Garlic d) All

39. Controlled atmosphere storage is used for storage of
    a) Jams b) Juices
    c) Fresh fruits & vegetables d) Dried products

40. Whipping cream contains fat content in the range of
    a) 10-20% b) 20-25%
    c) 30-40% d) 65-85 %

41. Vitamin B deficiency disease beriberi occurs due to deficiency of which enzyme
    a) Ascorbic Acid b) Lipogenase
    c) Thiaminase d) None of these

42. Flavonoids show their antioxidant property through
    a) Peroxyl radical scavenging activity b) Transition metal chelation
    c) None of the above d) All the above

43. Which type of additive is required in foods containing fats and oils to prevent rancidity
    a) Preservatives b) Anti-oxidants
    c) Emulsifiers d) Solvents

44. Milk digestion occurs due the presence of
    a) Reductase b) Beta-amylase
    c) Lactase d) None of the above

45. Vitamin helps in blood clotting
 a) Vitamin K b) Vitamin A
 c) Vitamin C d) Vitamin D
46. Which vitamins is also known as Cobalamin
 a) Vitamin B11 b) Vitamin B2
 c) Vitamin B6 d) Vitamin B12
47. Discoloration may be done by using
 a) Sulphloydryl oxidase b) Proteases
 c) Anthocyanase d) All of these
48. Hydrolysable tannins are also called
 a) Acetic acid b) Gallic acid
 c) Citric Acid d) Stearic Acid
49. Chloroplast does not contain
 a) Xanthopyll b) Plastids
 c) Anthocyanins d) Carotene
50. Alura-red comes under which category
 a) Flavours b) Natural colour
 c) Artificial colour d) None of these
51. which colour pigment is present in Marigold flower
 a) Lutein b) Annatto
 c) Beta carotene d) None
52. Joining of apoenzyme and co-enzyme is called
 a) Prothestic group b) Enzyme complex
 c) Holoenzyme d) Helical Complex
53. Pigment responsible for the process of sunlight
 a) Chlorophyll a b) Chlorophyll b
 c) Anthocyanin d) Carotenoids
54. Diastase enzyme take part in digestion of
 a) Lipids b) Starch
 c) Carbohydrates d) Protein
55. The tail attached to the Pyrrole ring is called
 a) Porphyrin b) Tail
 c) Phytol d) None of these
56. The Molecular formula of Xanthone is
 a) $C_{10}H_2O_6$ b) $C_{15}H_8O_2$
 c) $C_{11}H_5O$ d) $C_{13}H_8O_2$
57. Immediate precursor of chlorophyll a and chorophyll b is
 a) Protochlorophyllide b) Haemoglobin
 c) Biluribin d) Xanthophylls

58. Xanthophylls differ from carotenes in having
    a) Mg b) Oxygen
    c) Cl d) Hydrogen
59. The photosynthetic pigments are located in
    a) Chloroplast b) Grana
    c) Thylakoid d) Stroma
60. BHA and BHT both are
    a) Natural antioxidants b) Synthetic antioxidants
    c) Food colours d) Synthetic sweeteners
61. Refrigeration usually operate at
    a) 0 to 5°C b) 10 to 20°C
    c) 4 to 7°C d) 10 to 15°C
62. Which types of microorganisms grow at low temperature
    a) Psychrophilic b) Mesophilic
    c) Thermopilic d) All of above
63. Bread should be stored at
    a) Freezing temperature b) High temperature
    c) Room temperature d) None of above
64. Slow freezing allows formation of
    a) Large in size ice crystals b) Small in size ice crystals
    c) Medium in size ice crystals d) None of above
65. Quick freezing allows formation of
    a) Large in size ice crystals b) Medium in size ice crystals
    c) Small in size ice crystals d) None of above
66. Frozen food, in which large crystals are formed is of
    a) Inferior quality b) Good quality
    c) Both a and b d) None of above
67. In which of the following, Maillard reaction takes place?
    a) Prolonged cooking of carbohydrate with fatty acids
    b) Prolonged cooking of carbohydrate with salt
    c) Prolonged cooking of carbohydrate with amino acids
    d) Prolonged cooking of carbohydrate with water
68. In old refrigerator gas used is
    a) Freon b) Nitrogen
    c) Carbon dioxide d) None of above
69. In modern refrigerator gas used is
    a) Freon b) Nitrogen
    c) Tetrafluoroethane d) None of above

70. Pasteurization by High temperature and short time use temperature
   a) Above 100°C b) Above 120°C
   c) Above 50°C d) Above 70°C

71. Pasteurization by Low temperature and higher time by holding method, use temperature
   a) 100°C to 150°C b) 100°C to 200°C
   c) 60°C to 70°C d) None of above

72. Irradiation is also termed as
   a) Hot sterilization b) Cold sterilization
   c) Natural sterilization d) None of above

73. Slow/ sharp freezing is carried out at temperature between
   a) -18 to -35°C b) -10 to -15°C
   c) -0 to -7°C d) -15 to -23°C

74. Quick freezing is carried out at temperature between
   a) -40to -45°C b) -18 to -45°C
   c) -15to -23°C d) -10 to -15°C

75. In food preservation, salt water is termed as
   a) Syrup b) Brine
   c) Saline d) None of above

76. In food preservation sugar water is termed as
   a) Syrup b) Brine
   c) Saline d) None of above

77. Which is an inert gas, filed in food packets to prevent oxidation of fats ?
   a) Nitrogen b) Oxygen
   c) Ammonia d) All of above

78. Food on which pasteurization is applied, should be stored in
   a) Room temperature b) Medium temperature
   c) Refrigeration temperature d) None of above

79. In irradiation process radiation of wavelength should be
   a) Less than 200nm b) More than 200nm
   c) Less than 400nm d) More than 200nm

80. Irradiation method causes
   a) Destruction of natural antioxidants
   b) Promote oxidation
   c) Decreases thiamine, pyridoxine, B12, C, D, E and K
   d) All of above

81. Examples of perishable foods are
   a) Pulses b) Cereals
   c) Oils and fats d) Milk, fruits, vegetables

82. Non-perishable food is
    a) Milk
    b) Fruits
    c) Pulse and cereals
    d) Vegetables
83. Processing methods applied on various food
    a) Increases nutrients contents
    b) Reduce the nutrient contents
    c) Do not change nutrient contents
    d) None of above
84. Pigmented food should be packed or stored in
    a) Transparent packaging materials
    b) Thin plastic bags
    c) Opaque packaging materials
    d) None of above
85. For Fruit jam preparation
    a) Food colour is required
    b) Pectin is required
    c) Lecithin is required
    d) All of above
86. To check the TSS of Jam
    a) Refractometer is used
    b) Spectrometer is used
    c) pH meter is used
    d) all of above
87. Fortification means
    a) To add additives in processed food products
    b) To add nutrients in processed food products
    c) To aid Oil and fit in food products
    d) All of above
88. Convenience food are
    a) Easy to cook
    b) Easy to store
    c) Economical
    d) All of above
89. Sodium glutamate is also known as
    a) Hing
    b) Black salt
    c) Ajino moto
    d) None of above
90. Sugar and salt are used in food preservation
    a) For good taste
    b) For prevention of spoilage
    c) For prevention of microbial growth
    d) All of above
91. Oil and fats are used in food products
    a) For good taste
    b) For prevention of spoilage
    c) For prevention of microbial growth
    d) All of above
92. PFA act is applied
    a) To prevent adulteration
    b) To prevent misbranding
    c) To prevent economic fraud
    d) All of above
93. AGMARK is applicable on
    a) Milk and milk products
    b) Fruits and fruit products
    c) Cereals, pulsed, oil. Flour
    d) All of above

94. Spices in food products are used
   a) For good taste b) For prevention of spoilage
   c) For prevention of microbial growth d) All of above
95. Example of GRAS substances are
   a) Spices, oil, sodium bi-carbonate b) Food additives
   c) Food preservatives d) All of above
96. In canned food, there is the chances of growing
   a) Streptococcus b) Lacto bacillus
   c) Clostridium Botulinum d) All of above
97. Isoelectric point at which maximum coagulation of milk protein takes place
   a) 4.2 pH b) 4.6 pH
   c) 5.4 pH d) 6.6 pH
98. Beta-carotene in the body changes into Retinol in
   a) Liver b) Pancreases
   c) Gall balder d) Intestine
99. GRAS stands for
   a) General Requirement of Amino acids b) Generally Recognized as Safe
   c) General Requirement of Acetic acid d) None of above
100. The fortificant used in iodized salt is
   a) Potassium Iodide and Potassium Iodate b) Sodium Iodide
   c) Both a and b d) None of above

Answer Key

1	b	2	c	3	a	4	b	5	a	6	d	7	d
8	b	9	b	10	b	11	c	12	a	13	d	14	c
15	b	16	a	17	b	18	c	19	d	20	b	21	d
22	a	23	a	24	b	25	c	26	a	27	b	28	d
29	c	30	d	31	a	32	b	33	d	34	a	35	b
36	c	37	a	38	d	39	c	40	c	41	c	42	d
43	b	44	c	45	a	46	d	47	c	48	b	49	c
50	c	51	a	52	c	53	a	54	b	55	c	56	d
57	a	58	b	59	c	60	b	61	c	62	a	63	c
64	a	65	c	66	a	67	c	68	a	69	c	70	d
71	c	72	b	73	d	74	b	75	b	76	a	77	a
78	c	79	a	80	d	81	d	82	c	83	b	84	c
85	b	86	a	87	b	88	d	89	c	90	d	91	d
92	d	93	b	94	d	95	a	96	c	97	d	98	a
99	b	100	c										

Model Paper 9

1. Defects in fresh egg include
   a) Bloom
   b) Meat spots
   c) Cracks
   d) All of them
2. Green rots in eggs is caused by
   a) P. fluorescens
   b) P. graveolens
   c) Achromobacterperolens
   d) Paracolobactrum
3. Proteus causes spoilage of eggs by producing
   a) Red rots
   b) Colorless rot
   c) Black rots
   d) Pink rot
4. Hay odor in fishes is caused by
   a) Sporotrichum
   b) Enterobacter cloacae
   c) P. fluorescens
   d) Cladosporium
5. The final stage of spoilage of eggs by molds is
   a) Black rot
   b) Pin-spot molding
   c) Fungal rotting
   d) Superficial fungal spoilage
6. Untreated eggs lose moisture during storage and hence
   a) Lose weight
   b) Become heavy
   c) Get discolored
   d) None of the above
7. Dry packing of eggs are done by using
   a) Salt and sand
   b) Lime and sawdust
   c) Oiling and waxing
   d) Both a and b
8. Sodium pentachloro phenate inhibits
   a) Molds
   b) Yeasts
   c) Bacteria
   d) Virus
9. Thermostabilization is a method of dipping eggs into hot water to
   a) Wash the eggs
   b) Increase the water content
   c) Reduce evaporation of moisture
   d) Reduce contamination of eggs
10. Eggs are selected for storage by
    a) Waxing
    b) Oiling
    c) Candling
    d) None of them

11. Salt fish are spoiled by
    a) Acidophilic bacteria
    b) Halophilic bacteria
    c) Thermophilic bacteria
    d) Psychrophilic bacteria
12. Pseudomonas fluorescens causes spoilage of fish indicated by
    a) Yellow to greenish yellow color
    b) Greenish blue color
    c) Reddish tinge
    d) No color change
13. The kind and rate of spoilage of fish vary with
    a) The kind of fish
    b) Temperature
    c) The condition of the fish when caught
    d) All of the above
14. What is germicidal ice?
    a) Ice with chemical preservatives
    b) Ice which is crushed
    c) Ice with antiseptic solution
    d) None of the above
15. The pH of the flesh after the death of fishes is related to the amount of
    a) Oxygen available at death
    b) Protein content of the flesh
    c) Bacteria in the flesh
    d) Glycogen available at death
16. Fish can be preserved by
    a) Freezing
    b) Canning
    c) Drying
    d) All of the above
17. Spoilage of smoked fish can be delayed using
    a) Acetic acid
    b) Sorbic acid
    c) Benzoic acid
    d) Boric acid
18. Antioxidants to preserve fishes are used as
    a) Glazes
    b) Gases
    c) Dips
    d) All of the above
19. Deterioration of fatty fish produces appreciable amounts of 'stale fishy', which is
    a) Trimethylamine
    b) Chloramines
    c) Ammonia
    d) Unsaturated fatty acids
20. Chocolate-brown discoloration in fish is caused by
    a) Serratia
    b) Bacillus
    c) Proteus
    d) Asporogenous yeast
21. Increase in concentrations of carbon dioxide in the atmosphere of stored chicken inhibits the growth of
    a) Psychrotrophs
    b) Mesotrophs
    c) Thermophiles
    d) Alkalophiles
22. To reduce the number of organisms in a chill tank
    a) Calcium is added
    b) Carbon dioxide is added
    c) Mercury is added
    d) Chlorine is added
23. Most poultry are preserved by
    a) Chilling
    b) Freezing
    c) Heat treatment
    d) Both a and b

24. An 'outside cut' is
    a) If the trachea is cut from inside
    b) If the trachea is partially cut
    c) If the trachea is left intact
    d) None of the above
25. Soaking cut-up poultry in solutions of organic acids is to
    a) Lengthen shelf life
    b) Shorten shelf life
    c) Increase the bacterial count
    d) Reduce the bacterial count
26. The pigment pyoverdine in meat is produced by
    a) Serratia
    b) Pseudomonas
    c) Bacillus
    d) Vibrio
27. Microaerophillic bacteria are seen in
    a) chicken wrapped in oxygen-impermeable films
    b) Vacuum-packed chicken
    c) White meat
    d) Dark meat
28. Vacuum-packed chicken contains spoilage organisms like
    a) Enterobacter
    b) Pseudomonas
    c) Clostridium
    d) Bacillus
29. Pseudomonas, the chief spoilage organism of cut-up poultry, develops a slime accompanied by
    a) Taint
    b) Acidic smell
    c) Sour taste
    d) All the above
30. Chicken carcasses are irradiated to effectively destroy
    a) Salmonella
    b) Enterococci
    c) Clostridium
    d) Pseudomonas
31. 'Rabbito' is also called
    a) Flavor enhancer
    b) Sweaty feet
    c) Fishiness
    d) Surface taint
32. Black smudge in salted butter is caused by
    a) Phoma
    b) Pseudomonas nigrificans
    c) P. fluorescens
    d) Penicillium
33. Aeromonas hydrophila causes fishiness in
    a) Milk
    b) Cheese
    c) Butter
    d) Cream
34. Pseudomonas syncyanea causes
    a) Yellow milk
    b) Red milk
    c) Brown milk
    d) Blue milk
35. Inhibitory substances present in milk are
    a) Lactoperoxidase and agglutunins
    b) Renin and casein
    c) Lactose and fats
    d) None of them

36. Stormy fermentation of milk results in
    a) Foam at the top if the milk is liquid
    b) Gas bubbles caught in the curd
    c) Floating curd containing gas bubbles
    d) All of the above
37. Sweet curdling occurs at an early stage of
    a) Proteolysis
    b) Souring
    c) Lipolysis
    d) Curdling
38. Evaporated milk is made by removing
    a) 50% of water from whole milk
    b) 40% of water from whole milk
    c) 60% of water from whole milk
    d) 80% of water from whole milk
39. Rapid heating of cream, accompanied by injecting steam is known as
    a) Pasteurization
    b) Vacreation
    c) Sterilization
    d) Cooking
40. The centrifugal procedure used for removing bacteria from milk is
    a) Centrifugation
    b) Bactofugation
    c) Separation
    d) None of them
41. Soft swell has
    a) Both ends of the can bulged
    b) Distorted ends of the can
    c) Flat ends of the can
    d) Dented seams of the can
42. Hydrogen swells is favored by
    a) Increased acidities in foods
    b) Increasing temperature of storage
    c) A poor exhaust
    d) All of the above
43. Bacillus coagulans causes
    a) Flat sour spoilage
    b) TA spoilage
    c) Sulfide spoilage
    d) None of them
44. TA spoilage is caused by
    a) C. perfringens
    b) C. botulinum
    c) C. nigrificans
    d) C. thermosaccharolyticum
45. Molds found growing in jellies and candied fruits can grow in sugar concentrations up to
    a) 67.5%
    b) 60%
    c) 67%
    d) 65%
46. Black beets caused by Bacillus betanigrificans occurs in the presence of
    a) Low salt content
    b) High salt content
    c) High content of soluble iron
    d) Low content of soluble iron
47. 'Sulphide stinker' is caused by
    a) C. thermosaccharolyticum
    b) B. coagulans
    c) C. butyricum
    d) Desulfotomaculumnigrificans
48. Canned hams often show the presence of
    a) S. faecalis
    b) Lactobacillus
    c) Leuconostoc
    d) Bacillus

49. Byssochlamys fulva in canned fruits produces
    a) Sclerotia
    b) Ascospores
    c) Chlamydospores
    d) Oospores
50. Putrid swells of meat is caused by
    a) C. sporogenes
    b) C. pasteurianum
    c) C. putrefaciens
    d) C. butyricum
51. Yellow pigmented Micrococci and Bacilli cause stamping ink which is
    a) Discoloration of bones
    b) Discoloration of meat fat
    c) Discoloration of hooves
    d) Discoloration of hide
52. Spoilage of soft drinks is evidenced by
    a) Rancidity and sourness
    b) Offensive odor and taste
    c) Cloudiness and ropiness
    d) Discoloration and clarity
53. Carbonation is inhibitory and
    a) Germicidal
    b) Antiseptic
    c) Bactericidal
    d) None of these
54. Musty odor and taste in root beer is caused by
    a) Clostridium
    b) Achromobacter
    c) Bacillus
    d) Aspergillus
55. Solar salt is obtained from the
    a) Evaporation of surface salt water
    b) Sun drying of salt
    c) Filtration of water
    d) Heating in furnaces
56. Halobacterium salinarium is present in
    a) Wet salt
    b) Purified salt
    c) Welled salt
    d) Solar salt
57. Undesirable flavor in fats or oils is a result of
    a) Souring
    b) Cloudiness
    c) Rancidity
    d) Contamination
58. As a result of oxidation, butter fat and meat fat become
    a) Tallowy
    b) Discolored
    c) Supple
    d) Dehydrated
59. The content of microorganisms in spices can be reduced with a treatment of
    a) Alcohol
    b) Glutaraldehyde
    c) Propylene oxide
    d) Benzoic acid
60. Growth of Clostridium botulinum is seen in
    a) Imitation cheese
    b) Fresh mushroom
    c) Both a and b
    d) None
61. The microorganisms added to Swiss cheese to improve flavor and assist eye formation is
    a) Lactobacillus cremoris
    b) Lactobacillus bulgaricus
    c) Streptococcus thermophilus
    d) Propionibacterium freudenreichii

62. Moistened sterile crackers support the growth of
    a) P. camemberti b) A. niger
    c) R. stolonifer d) P. chrysogenum
63. Distiller's yeast is a high-alcohol yielding strain of
    a) Candida b) Torula
    c) S. cerivisiae d) Torulopsis
64. Soil stocks are preserved by
    a) Freeze drying b) Glycerol stocks
    c) Drying d) Heat fixing
65. Dried yeast contains
    a) 8% moisture b) 5% moisture
    c) 10% moisture d) 0% moisture
66. Baker's yeast can also be prepared from
    a) Grain mashes b) Waste sulfite liquor
    c) Wood hydrolysate d) All the above
67. Impure mixed cultures are required for the production of
    a) Citric acid b) Lactic acid
    c) Vinegar d) Alcohol
68. Bacterial cultures have been preserved for months at room temperature on agar slants which contain
    a) 5% NaCl b) 1% NaCl
    c) 2% NaCl d) 4% NaCl
69. The activity of a culture is judged by its
    a) Rate of growth b) Production of products
    c) Both a and b d) None of the above
70. Deterioration of cultures may result from
    a) Improper handling
    b) Cultivation
    c) Frequent transfer over long periods in an inadequate culture medium
    d) All the above
71. 'Wort' is a
    a) Dense solution b) Clear solution
    c) Solution with suspended particles d) None of them
72. Boiling of the wort with hops is to
    a) Inactivate enzymes b) Sterilize it
    c) To caramelize sugar slightly d) All of the above
73. 'Sarcinasickness' of beer is caused by
    a) Saccharomyces cerevisiae b) Pediococcus cerevisiae
    c) Lactobacillus pastorianus d) Hansenulaanomala

74. Silky turbidity in beer is caused by
   a) Zymomonas anaerobium
   b) Acetobacter pasteurianus
   c) Gluconobacteroxydans
   d) Lactobacillus diastaticus
75. Sake a yellow rice beer has an alcohol content of
   a) 5-10%
   b) 14-17%
   c) 20-25%
   d) 2-5%
76. An undesirable process called aceti- ficaion involving the oxidation of alcohol is caused by
   a) Streptococcus mucilaginosus
   b) Zymomonasanaeobium
   c) Gluconobacteroxydans
   d) Lactobacillus diastaticus
77. Fermentation of glycerol in wine results in
   a) Pousse
   b) Amertume
   c) Mousy flavor
   d) Tourne
78. 4% commercially available acetic acid is
   a) Brine
   b) Vinegar
   c) Tartar
   d) Salt
79. 'Fish Eye' spoilage of ripe olives is caused by
   a) Bacillus subtilis
   b) Bacillus pumilus
   c) Bacillus polymyxa
   d) only a and b
80. Slimy or ropy kraut is caused by
   a) Lactobacillus plantarum
   b) Erwinia herbicola
   c) Leuconostocmesentroides
   d) Lactobacillus brevis
81. SCP can be obtained from
   a) Yeasts
   b) Fungi
   c) Algae
   d) All of the above
82. Dextran can be obtained from
   a) Molasses
   b) Sugar beet
   c) Parafin
   d) Ethanol
83. A. niger is the principal mold used in the production of
   a) Oxalic acid
   b) Acetic acid
   c) Amino acids
   d) Citric acid
84. The most important organism in the production of amylase is
   a) Bacillus subtilis
   b) Candida pulcherima
   c) Geotrichumcandidum
   d) Saccharomyces cerevisiae
85. Invertase is used in the manufacture of
   a) Confectionery
   b) Fruit juices
   c) Jams
   d) Jellies
86. Pectinases are used in food industries to
   a) Enhance flavor
   b) Clarify fruit juices
   c) Enhance color of juices
   d) Retain freshness of juices

87. Addition of which acid makes milk more digestible to infants?
    a) Citric acid b) Gluconic acid
    c) Amino acid d) Lactic acid
88. The substrate used in the production of SCP is
    a) Molasses b) Acid hydrolysate of wood
    c) Fruit wastes d) All the above
89. The limitations on the use of bacteria as SCP is
    a) Poor public acceptance b) Small size
    c) High content of nucleic acids d) All the above
90. S. maxima is grown commercially in
    a) Lake Texcoco b) Lake Victoria
    c) Lake Missisipi d) none of the above
91. L. monocytogenes can be isolated from
    a) Milk b) Sewage
    c) Soil d) Both a and b
92. The diarrheal syndrome and the emetic syndrome are characteristic of
    a) Staphylococcal food poisoning b) Salmonellosis
    c) Perfringens poisoning d) Bacillus cereus food poisoning
93. Sea foods and sea water may contain
    a) Vibrio vulnificus b) Streptococcus faecalis
    c) Aeromonas hydrophilia d) Vibrio parahaemolyticus
94. A characteristic cysteine loop in the structure of the enterotoxin is seen in the toxin produced by
    a) Salmonella b) Staphylococcus
    c) Yersinia d) Bacillus
95. Kauffman-White scheme identifies isolates of
    a) Salmonella b) Clostridium
    c) Staphylococcus d) none of the above
96. Food poisoning caused by a gram negative, curved motile rod identified in Japan is
    a) V. parahaemolyticus b) Salmonella typhimurium
    c) Escherichia coli d) Shigella sonnei
97. C. jejuni has been isolated from
    a) Chicken carcasses b) Turkeys
    c) Pork sausages d) All of the above
98. Aeromonas hydrophilia may be pathogenic to
    a) Frogs b) Ducks
    c) Rabbits d) None of the above
99. Traveler'sdiarrhea is caused due to the toxins produced by
    a) Clostridium b) Bacillus
    c) Escherichia d) Yersinia

100. Scarlet fever and septic sore throat are diseases caused by
   a) Bacillus cereus
   b) Streptococcus pyogenes
   c) Arizona hinshawii
   d) Shigella boydii

Answer Key

1	d	2	a	3	c	4	b	5	c	6	a	7	d
8	a	9	c	10	c	11	b	12	a	13	d	14	a
15	d	16	d	17	b	18	d	19	a	20	d	21	a
22	d	23	d	24	c	25	a	26	b	27	a	28	a
29	d	30	a	31	d	32	b	33	c	34	d	35	a
36	d	37	a	38	c	39	b	40	b	41	a	42	d
43	a	44	d	45	a	46	c	47	d	48	a	49	b
50	a	51	b	52	c	53	a	54	b	55	a	56	d
57	c	58	a	59	c	60	c	61	d	62	a	63	c
64	a	65	a	66	d	67	c	68	b	69	c	70	d
71	b	72	d	73	b	74	a	75	b	76	c	77	b
78	b	79	d	80	a	81	d	82	a	83	d	84	a
85	a	86	b	87	d	88	d	89	a	90	a	91	d
92	d	93	a	94	b	95	a	96	a	97	d	98	a
99	c	100	b										

Model Paper 10

1. The syndrome resulting from the ingestion of toxin in a mold contaminated food is
   a) Aspergillosis
   b) Enterotoxicosis
   c) Mycotoxicosis
   d) Neurotoxicosis
2. Patulin is toxic to
   a) Mice
   b) Fishes
   c) Cattle
   d) Human beings
3. Alimentary toxic aleukia isolated from grain is produced by
   a) Alternaria
   b) Cladosporium
   c) Fusarium
   d) All of the above
4. Mold nephrosis, a disease in pigs has been associated with
   a) Aflatoxin
   b) Patulin
   c) Citrinin
   d) Ochratoxin
5. Scombroid fish poisoning has been associated with
   a) Proteus morganii
   b) Klebsiella pneumonia
   c) Lyngbyamajuscula
   d) Both a and b
6. Lyngbya majuscule causes
   a) Ciguatera poisoning
   b) Scombroid poisoning
   c) Shellfish poisoning
   d) None of the above
7. A toxic substance detected in blue cheese is
   a) Luteoskyrin
   b) Roquefortine
   c) Patulin
   d) Ochratoxin
8. Penicillium islandicum produces
   a) Citrinin
   b) Sterigmatocystin
   c) Luteoskyrin
   d) Patulin
9. Aflatoxins are produced by
   a) A. flavus
   b) A. parasiticus
   c) A. niger
   d) Both a and b
10. A chlorinated isocoumarin derivative is
    a) Ochratoxin
    b) Aflatoxin
    c) Patulin
    d) Citirinin

11. V. parahaemolyticus has an incubation period of
    a) 15-24 hours
    b) 30 min-8 hours
    c) 12-50 hours
    d) 6-48 hours
12. Soft drinks are tested for chemicals when they are contained in a
    a) Glass bottle
    b) Tetrapack
    c) Metal container
    d) None of them
13. The toxic substance of the 'Chinese restaurant' syndrome is
    a) Acetic acid
    b) MSG
    c) Vinegar
    d) Chilli sauce
14. Corynebacterium is detected by the analysis of
    a) Throat swab
    b) Urine sample
    c) Blood sample
    d) Skin lesions
15. Samples of suspected perishable foods must be kept cold
    a) During collection of the sample
    b) During analysis
    c) During transit to laboratory
    d) There is no need to keep the samples cold
16. Cultures from nose, throat or skin lesions are used to test for
    a) Salmomnella
    b) Bacillus
    c) Clostridium
    d) Staphylococcus
17. Mucoid diarrhea is the predominant symptom in the detection of
    a) Giardia lamblia
    b) Enteric viruses
    c) Entamoeba histolytica
    d) Taenia solium
18. Neurological symptoms caused by puffer-fish poisoning is because of
    a) Shellfish toxin
    b) Tetraodon toxin
    c) Both a and b
    d) None of them
19. The predominant symptoms in the lower gastro-intestinal tract infection is
    a) Fever
    b) Abdominal cramps and diarrhea
    c) Chills
    d) Malaise
20. Cyanosis occurs due to
    a) Phosphate poisoning
    b) Loss of iron
    c) Nitrite poisoning
    d) None of the above
21. A system to control the safety of a manufactured product is determined by
    a) HACCP
    b) ISO 9000
    c) CCP
    d) FAO

22. HACCP is designed to
    a) Prevent problems before they occur
    b) Correct deviations as soon as detected
    c) Both a and b
    d) None of the above
23. National academy of sciences is an authority on
    a) Food additives b) Food safety
    c) Food preservation d) Food processing
24. The Codex guidelines for HACCP has
    a) Five principles b) Six principles
    c) Seven principles d) None of the above
25. CIP refers to
    a) Clean in place systems b) Cleanliness in place systems
    c) Clean in places systems d) None of them
26. During transportation, perishable foods should be kept at
    a) 3.3–4.4°C b) 66°C
    c) 0°C d) Both a and b
27. An example of a wetting agent is
    a) Polyether alcohol b) Chlorine
    c) Trisodium phosphate d) Citric acid
28. CCP aims at preventing or reducing
    a) Physical hazards b) Chemical hazards
    c) Biological hazards d) All the above
29. Validation ensures that the industry
    a) Complies with the required plan b) Has random sampling
    c) Prevents deviations d) Lists the significant hazards
30. The simplest record-keeping system to ensure effectiveness is by
    a) Establishing procedures for verification
    b) Establishing documentation
    c) Establishing monitoring systems
    d) Establishing critical control points
31. The principle behind the controlled and modified atmosphere technologies is
    a) To reduce the rate of respiration,
    b) Reduce microbial growth, and
    c) Retard enzymatic spoilage by changing the gaseous environment surrounding the food product
    d) All the above

32. The CA implies a ____________ degree of control than MA in maintaining specific levels of O_2, CO_2, and other gases.
    a) Lower b) Higher
    c) Same d) Unknown
33. In CA storage facilities, both temperature and gas composition of the storage atmosphere are regulated or controlled. The gas concentration ranges encountered in CA storages are 1 to 10% O_2, 0 to 3% CO_2, and the balance is
    a) Nitrogen (N_2) b) Ammonia (NH_3)
    c) H_2S d) All the above
34. Food grains get oxygen from the air and burn food from its endosperm. This process liberates heat, water vapours and carbon dioxide. This process in grains is called
    a) Fermentation b) Respiration
    c) Preservation d) Evaporation
35. Respiration is a process of slow combustion of ______________ in presence of oxygen in living system to produce the energy.
    a) Protein b) Water
    c) Vitamins d) Carbohydrates
36. Winter season over and summer season starts, the atmospheric temperature is higher. The grain stored in winter is ____________ than the atmospheric temperature.
    a) Cooler b) Warmer
    c) Same d) None
37. The moist air condenses due to cool grain and moisture accumulation at bottom. Because of increased grain moisture, spoilage takes place at ____________ of the bin.
    a) Top b) Centre
    c) Bottom d) Sides
38. Decreasing the concentration of O_2 and increasing concentration of CO_2 in stored grains, the rate of respiration is
    a) Decreased b) Increased
    c) No changed d) None
39. The total amount of heat required to be removed from the space in order to bring it at the desired temperature by the air conditioning and refrigeration equipment is known as
    a) Refrigeration load b) Heating load
    c) Cooling load d) Heat stability
40. The heat that must be removed from the refrigerated products in order to reduce the temperature of the product to the desired level is called
    a) Product load b) Air change load
    c) Respiration load d) Wall gain load
41. Fruits, vegetables, juices, raw meat, fish and milk belong to the
    a) High moisture foods b) Intermediate moisture foods
    c) Low moisture foods d) None

42. Bread, hard cheeses and sausages are examples of

a) High moisture foods a) Low moisture foods

b) Intermediate moisture foods c) None

43. The dehydrated vegetables, grains, milk powder and dry soup mixtures categorised in

a) High moisture foods b) Intermediate moisture foods

c) Low moisture foods d) None

44. Water activity, a_w, is defined as the ratio of the water vapor pressure of the food to the vapor pressure of pure water at the same ___________

a) Moisture b) Temperature

c) Viscosity d) Porosity

45. The function representing the relationship between water content (e.g. as grams of water per gram of dry matter) and water activity at constant temperature is called

a) Water vapor sorption isotherm b) Moisture sorption isotherm

c) Both (a & b) d) None

46. Bacterial growth does not occur at water activity levels below

a) 0.05 b) 0.9

c) 0.1 d) 0.6

47. The water activity limit for the growth of molds and yeasts is between

a) 0.01-0.1 b) 0.1-0.2

c) 0.8 -0.9 d) 0.9-1.0

48. Most enzymatic reactions require water activity levels of

a) 0.01-0.1 b) 0.1-0.2

c) 0.85 or higher d) Above 0.99

49. Differential scanning calorimetry (DSC) measures and records the heat capacity (i.e. the amount of heat necessary to increase the temperature by 1 degree Celsius) of a sample and of a reference as a function of

a) Temperature b) Pressure

c) Humidity d) Viscosity

50. Newton's law states that the shearing force F_x required to maintain the upper plate in movement is proportional to the area of the plate A and to the ____________ gradient (dv_x /dz).

a) Velocity b) Temperature

c) Momentum d) density

51. Which one of the following is not the example of disaccharide?

a) Maltose b) Glucose

c) Sucrose d) Hactose

52. Which one of the following is the disaccharide?

a) Sucrose b) Lactose

c) Maltose d) All of the above

53. Which one of the following is the trisaccharide?
    a) Raffinose b) Stachyose
    c) Verbascose d) None
54. Polysaccharides are the polymer of
    a) Disaccharides b) Trisaccharides
    c) Monosaccharide d) Oligosaccharides
55. Polysaccharides are usually tasteless and form colloids with
    a) Alcohol b) Water
    c) Acid d) Formaldehyde
56. Which one of the following is the homopolysaccharide
    a) Starch b) Glycogen
    c) Cellulose d) All of the above
57. Insulin is the example of
    a) Monosaccharide b) Disaccharides
    c) Oligosaccharides d) Polysaccharides
58. D-ribose and D-deoxyribose is example of
    a) Triose sugar b) Tetrose sugar
    c) Pentose sugar d) Hexoses sugar
59. The constituents of cane sugar and beet sugar is
    a) Hactose b) Maltose
    c) Sucrose d) Fructose
60. The product of starch hydrolysis and occurs in germinating seed is
    a) Sucrose b) Hactose
    c) Maltose d) Fructose
61. Which of the following contains nitrogen?
    a) Fats b) Proteins
    c) Carbohydrates d) None
62. On heating with conc. HNO_3, proteins give yellow colour. This test is called
    a) Oxidizing test b) Xanthoproteic test
    c) Hoppe's test d) Acid base test
63. Enzymes are
    a) Proteins b) Minerals
    c) Oils d) Fatty acids
64. Which is an essential constituent of a diet?
    a) Starch b) Glucose
    c) Carbohydrates d) Protein
65. Which of the following foods stuffs contain nitrogen?
    a) Carbohydrates b) Fats
    c) Protein's d) None

66. Protein is a polymer of
   a) Glucose
   b) Terephthalic acid
   c) Amino acid
   d) Glycol
67. Oleic, satiric, palmitic acids are
   a) Nucleic acids
   b) Amino acid
   c) Fatty acid
   d) None
68. Which amino acid has no asymmetric carbon atom?
   a) Histidine
   b) Glycine
   c) α – alanine
   d) Threonine
69. Which of the following is a test for protein?
   a) Beilstein test
   b) Biuret test
   c) Benedict's test
   d) Molisch's test
70. Which of the following is not a function of proteins?
   a) Nail formation
   b) Skin formation
   c) Muscle formation
   d) Providing energy for metabolism
71. Match the food items and their principal flavoring agents given in the two columns below

Column I	Column II
P. Butter	Menthol
Q. Orange	Limonene
R. Cloves	Eugenol
S. Mint	Diacetyl

   a) P – 3, Q – 2, R – 4, S – 1
   b) P – 2, Q – 3, R – 1, S – 4
   c) P – 4, Q – 1, R – 3, S – 2
   d) P – 4, Q – 2, R – 3, S – 1
72. Isolation of food flavor is accomplished by
   a) Head space method
   b) Destination method
   c) Solvent extraction
   d) HPLC method
   e) All the above
73. Concentration of dilute organic and aqueous flavor isolates is done by
   a) Evaporation
   b) Freeze concentration
   c) Adsorption
   d) All the above
74. The method used for solvent extraction of food flavor is determined by
   a) The physical state of food (i.e., solid vs liquid)
   b) Quantity of food (small batch process or large continuous operation)
   c) Extracting solvent (heavier or lighter than water)
   d) All the above

75. The simplest method of solvent extraction is batch extraction using
   a) Beaker and keep
   b) Conical flask
   c) Pipette and burette
   d) Separatory funnel
76. Chlorophyllides in green vegetables are _______ stable than chlorophylls
   a) More
   b) Less
   c) Unstable
   d) None
77. Chlorophylls are converted to chlorophyllides by means of the natural occurring
   a) Catalase
   b) PPO
   c) Chlorophyllase
   d) Phylase
78. The colour of fresh muscle or meat is due predominantly to
   a) Chlorophylls
   b) Hemoglobin
   c) Anthocyanins
   d) Myoglobin
79. Muscle (meat) pigments includes
   a) Myoglobin
   b) Cytochromes
   c) Hemoglobin and vitamin B – 12
   d) Yellow coenzymes associated with cytochromes
   e) All the above
80. Hemoglobin is the red pigment of
   a) Tomato
   b) Carrot
   c) Blood
   d) Beet sugar
81. Enzymes in living systems to
   a) Provide energy
   b) Provide immunity
   c) Transport oxygen
   d) Catalyze biochemical processes
82. The enzymes which converts glucose into ethyl alcohol is
   a) Diastase
   b) Invertase
   c) Maltase
   d) Zymase
83. Enzymes belongs to the class of compounds
   a) Polysaccharides
   b) Polypeptides
   c) Polynitrogen heterocyclic compounds
   d) Hydrocarbon
84. Enzymes are made up of
   a) Edible protein
   b) Protein with specific structure
   c) Nitrogen containing carbohydrates
   d) Carbohydrates
85. The conversion of maltose to glucose is possible by the enzyme
   a) Zymase
   b) Lactase
   c) Maltase
   d) Diastase
86. The enzyme rennin is used for the manufacturing of
   a) Cheese
   b) Curd
   c) Khoa
   d) Sauce

87. Lecithin is an example of
    a) Carbohydrates b) Hormones
    c) Vitamins d) Phospholipids
88. Which of the following acts as storehouse of energy in body?
    a) Hormone b) Vitamins
    c) Fats d) All of these
89. Saponification of coconut oil yields glycerol and
    a) Palmitic acid b) Sodium palmitate
    c) Oleic acid d) Stearic acid
90. Which of the following is an unsaturated acid?
    a) Linoleic acid b) Stearic acid
    c) Myristic acid d) Lauric acid
91. Foods rich in protein, vitamins and minerals are called
    a) Energy giving foods b) Growth inducing foods
    c) Protective foods d) All of above
92. Rancidity of oil can be prevented with the use of
    a) Opaque packaging b) Air tight packaging
    c) Antioxidants d) All of above
93. Nutritional requirement of boy and girls is same up to the age of
    a) 9 years b) 12 years
    c) 15 years d) 17 years
94. Yellowish figment found in papaya and mangoes
    a) β- Carotene b) Lycopene
    c) Anthocynin d) All of above
95. Folic acid requirement of pregnant women
    a) 200mcg/day b) 400 mcg/day
    c) 600 mcg/day d) 800 mcg/day
96. Protein requirement of lactating mother during first six months
    a) 50g/day b) 65 g/day
    c) 75 g/day d) 45 g/day
97. Vitamin C requirement of adolescent
    a) 40mg/day b) 50 mg/day
    c) 25 mg/day d) 36 mg/day
98. Period between childhood and adulthood
    a) Late childhood b) Adolescence
    c) Early adulthood d) None of above
99. Each kidney in the human contain about ……. of nephrone
    a) 2-4 millions b) 3-8 millions
    c) 1-1 millions d) 2-6 millions

100. Renal system of human includes

a) Kidneys
b) Uterus
c) Urinary bladder
d) All of above

Answer Key

1	c	2	a	3	d	4	c	5	d	6	a	7	b
8	c	9	d	10	a	11	a	12	c	13	b	14	a
15	c	16	d	17	a	18	b	19	b	20	c	21	a
22	c	23	b	24	c	25	a	26	d	27	a	28	d
29	a	30	b	31	d	32	b	33	a	34	b	35	d
36	a	37	c	38	a	39	c	40	a	41	a	42	b
43	c	44	b	45	c	46	b	47	c	48	c	49	a
50	a	51	b	52	d	53	a	54	c	55	b	56	d
57	d	58	c	59	c	60	c	61	b	62	b	63	a
64	b	65	c	66	c	67	c	68	b	69	b	70	d
71	d	72	e	73	d	74	d	75	d	76	a	77	c
78	d	79	e	80	c	81	d	82	d	83	b	84	b
85	c	86	a	87	d	88	c	89	b	90	a	91	c
92	d	93	a	94	a	95	b	96	c	97	a	98	b
99	c	100	d										

Model Paper 11

1. Maize is deficient in
   a) Methionine b) Lysine
   c) Tryptophan d) None
2. Winterization is accomplished to remove.
   a) Waxes b) Cloudiness of Oils
   c) SFA d) All the above
3. What do you mean from 'Dhal' ?
   a) Whole dehisced pulse b) Whole cleaned pulse
   c) Dehusked and splitted pulse d) All the above
4. The separation of the bran from the endosperm of which cereal grain in more difficult.
   a) Rye b) Wheat
   c) Triticale d) None of above
5. Stabilization of enzymes in required before milling of which cereal grain.
   a) Maize b) Oat
   c) Rice d) Barley
6. The purpose of steeping of grain in wet milling is to
   a) Help in hull separation b) Soften kernel
   c) Easy separation d) All of above
7. In the CFTRI process of rice parboiling, the steeping water temperature in increased as high as.
   a) 100°C b) 60°C
   c) 90°C d) 70-75°C
8. The term 'Protein shift' used in the air classification of a wheat flour in for.
   a) Protein removal b) Protein grinding
   c) Protein concentration d) None of the above
9. Lipid deterioration in stored cereals occurs due to.
   a) Hydrolysis b) Oxidation
   c) Both a and b d) None of the above
10. Which is not the grain storage fungus?
    a) *Fusarium* b) *Alternaria*
    c) *Penicillium* d) *Aspergillum*

11. Natural air drying is not preferred when the RH of the air exceed to.

a) 60% b) 70%

c) 80% d) 90%

12. The relationship between moisture content and water activity of foods is given by.

a) Fourier's equation b) Stefan's law

c) Plank's equation d) BET equation

13. Crude fiber mainly composed of.

a) Glucose + Cellulose b) Lignin + Pectin

c) Cellulose + hemi cellulose d) All carbohydrate

14. The saccharifying enzyme is

a) β - amylase b) α - amylase

c) Lactase d) Both a and b

15. The smallest food grain is

a) Milled rice b) Rye

c) Fox taint millet d) Teff millet

16. The property of a fluid that resists the force tending to cause the fluid to flow is termed as

a) Osmotic pressure b) Viscosity

c) Vapor pressure d) Water activity

17. The amount of air incorporated in the products like whipped cream, egg-white foam and cake better is due to.

a) Specific gravity b) Specific heat

c) Surface tension d) Viscosity

18. A colloid system where in the colloid particles are dispersed a water is called

a) Hydro Colloid b) flocculation

c) Peptization d) Tyndall effect

19. Lactose cleaved to glucose and galactose in the intestine by an enzyme activity is called.

a) Catalase b) Amylase

c) Lactase d) Lipase

20. The brown colour of scorched milk is due to the reaction of

a) Enzymatic browning b) Non- Enzymatic browning

c) Both a and b d) None

21. A milk coagulated product in which all the milk proteins by action of heat coagulate into a uniform mass is termed as.

a) Ice- cream b) Chhana

c) Khoa d) Fermented milk

22. A milk product in which proteins are coagulated from milk by heat and acid treatment by the addition of Lime juice, acetic acid etc. to heat milk is called

a) Khoa b) Chhana

c) Whey d) Curd

23. Which enzyme causes the splitting of cheese proteins into peptones and peptides in cheese processing?
    a) Lipase b) Amylase
    c) Lactose d) Rennin
24. The process transforms the cheese from a tasteless hard compact mass to the unique flavored product called cheese, which is ready for consumption?
    a) Ripening b) Ageing
    c) Homogenization d) Salting
25. Rancidity of milk fat takes place due to the acceleration of lipase activity by
    a) Cooling b) Separation
    c) Pasteurization d) Homogenization
26. Milk is an Oil in water ...., with the fat globules dispersed in a continuous skim milk phase.
    a) Emulsion b) Colloid
    c) Gel d) Sol
27. A mechanical treatment in which the fat globules in milk brought about by passing milk under high pressure through a tiny aperture.
    a) Filtration b) Heat exchanger
    c) Homogenization d) Curdling
28. The stabilization process in extensively employed in the manufacturing of
    a) Chhana b) Ice- cream
    c) Curd d) Whey protein
29. The prestratification is the process used in
    a) Curd making b) Khoa making
    c) Ghee making d) Cheese making
30. Butter milk is a byproduct of the
    a) Milk industry b) Juice industry
    c) Rice industry d) Butter industry
31. Water forms a major portion of fruits as
    a) 20-30% b) 30-40%
    c) 40-60% d) 75-90%
32. Normally fruits are poor source of protein and fat except
    a) Guava b) Banana
    c) Avocado d) Custard apple
33. Avocado has a fat content about
    a) 8% b) 18%
    c) 28% d) 38%
34. Pumpkin and melons are very good source of
    a) Iron b) Calcium
    c) Sodium d) Iodine

35. Pear, guava, apple, banana, litchi etc. exhibit a pinkish brown discoloration when cut. This has been found to be caused by
    a) Chlorophyll b) Carotenoid
    c) Lylopene d) Leucoanthocyanidin
36. Flavones naringin and neohesperidin are flavonoids compounds which are responsible for astringent and bitter taste of
    a) Citrus fruits b) Apple
    c) Guava d) Custard apple
37. The optimum temperature range for the best activity of the enzyme is
    a) 20-25 b) 43-50
    c) 30-40 d) 40-45
38. The enzyme acts best at a pH between
    a) 3-4 b) 4-5
    c) 5-6 d) 6-7
39. Betalains, a water soluble pigment are normally found in.
    a) Carrot b) Sweet potato
    c) Beet d) Brinjal
40. Carotenoids a yellow – orange red coloured pigments are precursor of
    a) Vitamin A b) Vitamin E
    c) Vitamin D d) Both (a) and (c)
41. The empty space between the white and shall at the large and of the egg is called
    a) Cuticle b) Air cell
    c) Egg white d) Yolk
42. A dispersion consisting of solid part called granules and a liquid part, the plasma is called
    a) Egg white b) Air cell
    c) Yolk d) Shell
43. A yellow or greenish tint in egg raw white may indicate the presence of
    a) Ovaflavin and riboflavin b) Gluten and globulin
    c) Globulin and riboflavin d) zein and pectin
44. The yolk of a large egg contains about
    a) 10 calories b) 29 calories
    c) 59 calories d) 75 calories
45. Omega -3 enhanced eggs are from hens fed a diet contain 10-20%
    a) Rice husk b) Sesame hull
    c) Flax seed d) Soybean meal
46. Omega- 3 enhanced eggs contain .... times omega-3 fatty acids than the classic eggs.
    a) 3 b) 7
    c) 6 d) 10

47. The optimum storage temperature for eggs should be between
   a) 4 - 5°C b) 10- 16°C
   c) 15-20°C d) Any temp.

48. *Salmonella enteritidis* in eggs will not grow at temperature below 4 and is killed at
   a) 45°C b) 71°C
   c) 96°C d) 121°C

49. Biological value of the eggs is
   a) 84.5% b) 96%
   c) 76% d) 74.3%

50. Eggs yolks are one of the few foods that naturally contain
   a) Vitamin A b) Vitamin C
   c) Vitamin K d) Vitamin D

51. The commodity is used as a reference against which other food products are compared for their protein content
   a) Egg b) Milk
   c) Barley d) Pulse

52. Meat is muscle tissues of
   a) Cereals b) Milk
   c) Animal d) Egg

53. Glycogen and glucose are the two carbohydrates provided by
   a) Fruits b) Meat
   c) Fish d) Eggs

54. Muscles tissues of meat have a function to convert chemical energy (ATP) into
   a) Electrical energy (current) b) Frictional energy (sliding)
   c) Mechanical energy (movement) d) Thermal energy (heat)

55. Braising is a moist heat cooking method recommended for less tender cuts of
   a) Meat b) Fruits
   c) Bamboo shoots d) Vegetables

56. All the carbohydrates are made up of one or more molecules of simple
   a) Fats b) Sugars
   c) Amino acids d) Pigments

57. Inversion or chemical break down of sucrose results
   a) Fine sugar b) Invert sugar
   c) Reducing sugar d) Non reducing sugar

58. The first stage of processing the raw sugar is to soften and then remove the layer of mother liquor surrounding the crystals with a process is called.
   a) Carbonation b) Boiling
   c) Flocculation d) Affination

59. A syrup made from maize, composed mainly of fructose is known as
    a) Maple syrup b) High fructose syrup
    c) Corn syrup d) Corn sweetener
60. A natural sweetener made by the concentration of sugar cane juice in pan is called
    a) Jaggery b) Sugar
    c) Maple syrup d) Khand sari
61. Invert sugar contains a mixture of fructose (levulose) and dextrose (D-glucose) obtained through acid or enzymatic hydrolysis of
    a) Sugar b) Fruits juice
    c) Corn syrup d) Honey
62. Fuller's earth is now used in conjunction with various activated carbon for
    a) Bleaching of oil b) Hydrogenation of oil
    c) Deodorization of oil d) Sedimentation of juice
63. Hydrogenation converts unsaturated triglycerides into saturated and it changes oils into
    a) Emulsion b) Colloids
    c) Butter d) Solid fats
64. Margarine belongs to
    a) Fat b) Fruit
    c) Meat d) Pickles
65. Which of the following fat has higher shortening power?
    a) Lard b) Hydrogenated fat
    c) Butter d) All the above
66. Soy bean and cotton seed oil, refined and partially hydrogenated to the desired consistency are extensively used to produce
    a) Butter b) Shortening
    c) Lard d) Margarine
67. A disadvantage of the UHT is to destroy the heat sensitive vitamins such as
    a) Vitamin K b) Vitamin D
    c) Vitamin E d) Vitamin C
68. Food irradiation is another sterilizing technique in which foods are bombarded by high energy rays called
    a) X-rays b) UV rays
    c) Gamma rays d) Cosmic rays
69. Minimum temperature and time for milk pasteurization are based on the thermal death time studies for the most heat resistance pathogen found in milk is
    a) *Bacillus cereus* b) *Coxielliae Brunetti*
    c) *Lactobacillus sps.* d) *Listeria*
70. The symbol employed an irradiated food in called.
    a) Futura b) Radura
    c) Fostoc d) Irrada

71. The percentage of the test material in the mixture of the standard product is called
    a) Threshold limit b) Composite score
    c) Dilution number d) Hedonic score
72. Which part of the saffron used as a flavoring agent?
    a) Leaf b) Floral
    c) Stigma d) Stem
73. A substitute of milk paneer made from soybean is called
    a) Tofu b) Tempeh
    c) Koumiss d) Natto
74. Air compression pycnometer is used to measure.
    a) Wight b) Volume
    c) Temperature d) Sp. Gravity
75. Infra-red spectroscopy is used for
    a) Dynamic measurement b) Quality control
    c) Monitoring application d) All the above
76. Which of the following is not a thermodynamic property?
    a) Temperature b) Heat
    c) Pressure d) Specific volume
77. Pigeon pea is considered as hard to mill due to
    a) High protein content b) High gum content
    c) Low husk content d) None of these
78. Time independent non-Newtonian fluids are called
    a) Thixotropic b) Rheopectic
    c) Shear thinning d) None of above
79. In an HTST pasteurizer, booster pump is located just before
    a) Heating b) Regeneration
    c) Cooling d) Holding
80. Which of the fouling is not a type of condenser?
    a) Air cooled b) Water cooked
    c) Evaporative d) Freezer
81. The increase in boiling point of a solution over boiling point of water is known as
    a) Boiling point elevation b) Boiling point depression
    c) Boiling point constant d) None of the above
82. A package is considered to provide protection to the food from
    a) Light b) Oxygen
    c) Contamination d) All the above
83. Jenssen's equation is related to
    a) Storage silo design b) Size reduction of particles
    c) Grain transportation system d) Size separation of grains

84. A chromatograph is used for
   a) Measuring rain fall
   b) Measuring temperature of a gas
   c) Analyzing the composition of a gas
   d) Measuring the pressure of a gas
85. Food safety refer to
   a) That it does not cause harm when consumed any way.
   b) That it does not cause harm when consumed as per intended use.
   c) That it causes harm when consumed
   d) All the above
86. What does symbol 'D' imply in work study?
   a) Inspection
   b) Delay /temporary storage
   c) Transport
   d) Permanent storage
87. ISO 22000 represents
   a) Quality system
   b) Safety system
   c) Quality and safety
   d) Environment management system
88. Codex Alimentarius Commission (ACA) was established in
   a) 1960
   b) 1963
   c) 1966
   d) 1969
89. The FSS act was introduced in the year
   a) 2002
   b) 2006
   c) 2008
   d) 2011
90. Which of the following is a food safety standard?
   a) ISO 9001
   b) ISO 22000
   c) ISO 14000
   d) All of above
91. Thermo couples are used to measure
   a) Volume
   b) RH
   c) Temperature
   d) Moisture content
92. Recommended air temperature for a grain drying is nearly.
   a) 45
   b) 60
   c) 90
   d) 100
93. Unit for measurement of vacuum is
   a) Kg_f/cm^2
   b) Torr
   c) mm of Hg
   d) None
94. Hot water treatment of fruits and vegetables is mainly used for controlling
   a) Insect
   b) Fungi
   c) Bacteria
   d) Viruses
95. The basis for measuring temperature is given by
   a) Zeroth law of thermodynamics
   b) First law of thermodynamics
   c) Second law of thermodynamics
   d) Newton's law of cooling

96. The dimension of energy is

a) $ML^2 T^{-2}$ b) MLT^{-2}

c) $ML^{-2} T^{-1}$ d) $ML^2 T^{-1}$

97. In addition to permeability, the other mass transport phenomenon in package system is

a) Sorption b) Migration

c) Sorption and migration d) None of the above

98. Maximum lateral and vertical pressures in a silo exist at

a) Top of silo b) Bottom of silo

c) Centre of silo d) Varies from silo to silo

99. Repeatability of the instrument with respect to a given fixed input is

a) Sensitivity b) Resolution

c) Accuracy d) Precision

100. Process layout is employed for

a) Batch production b) Effective utilization of m/c

c) Continuous type of product d) All the above

Answer Key

1	b	2	d	3	c	4	c	5	b	6	d	7	d
8	c	9	c	10	b	11	b	12	d	13	c	14	a
15	d	16	b	17	a	18	a	19	c	20	b	21	c
22	b	23	d	24	a	25	d	26	a	27	c	28	b
29	c	30	d	31	d	32	c	33	c	34	a	35	b
36	a	37	b	38	d	39	c	40	a	41	b	42	c
43	a	44	c	45	c	46	d	47	b	48	b	49	b
50	d	51	a	52	c	53	b	54	c	55	a	56	b
57	b	58	d	59	c	60	a	61	a	62	a	63	d
64	a	65	a	66	d	67	d	68	c	69	b	70	b
71	c	72	c	73	a	74	b	75	d	76	b	77	b
78	a	79	b	80	d	81	a	82	b	83	a	84	c
85	a	86	b	87	a	88	b	89	d	90	b	91	c
92	a	93	c	94	b	95	a	96	a	97	c	98	b
99	d	100	a										

Model Paper 12

1. The food qualities such as taste, odour, colour etc. are measured in term of
   a) Quick index
   b) Organoleptic scale
   c) Score card
   d) Hedonic scale
2. In UV spectroscopy samples are typically placed in a.
   a) Transparent cell
   b) Cuvette
   c) Silicon glass
   d) Both a and b
3. A definite area or a space where some thermodynamic process takes place is known as.
   a) Thermodynamic law
   b) Thermodynamic process
   c) Thermodynamic system
   d) Thermodynamic cycle
4. LMTD in case of parallel flow compared to counter flow would be.
   a) More
   b) Less
   c) Same
   d) Negative
5. The difference between wet bulb temp and dry bulb temperature.
   a) Dry bulb depression
   b) Wet bulb depression
   c) Dew point depression
   d) Degree saturation
6. Reynolds number for pipe flow in given by.
   a) PVD/m
   b) m/PVD
   c) VD/m
   d) m/VD
7. Angle of repose (θ) for a give food materials is equal to.
   a) $(\tan^{-1}h)/D$
   b) $(\tan^{-1}D)/h$
   c) $(\tan^{-1}h)/r$
   d) $(\tan^{-1}r)/h$
8. The height value of angle of internal friction indicates the materiel is
   a) Cohesive
   b) Easy flowing
   c) Normal flowing
   d) No indication of flow
9. During sensible heating or cooling, there is no change in the values of
   a) RH
   b) Humid ratio
   c) Specific humidity
   d) Absolute humidity
10. Cow milk is a protein in
    a) Albumin
    b) Zein
    c) Casein
    d) Lactalbumin

11. In autoclaved sausages which are stored unrefrigerated, the primary hurdles is
    a) pH b) Deter activity
    c) Sub lethal damage to spores d) Eh
12. Which is defined as the ratio of reaction rate at T+10 to that at T i.e. the increase in reaction rate caused by an increase in temperature of 10°C
    a) Q_{10} b) F
    c) D d) Z
13. The expansion joint is mostly use for pipes which carry steam at .... pressure.
    a) Low b) High
    c) Constant d) Very high
14. Belt conveyer discharge pattern follows?
    a) Belt shape b) Non linear
    c) Parabolic d) Sigmoid
15. Most important general requirements of food packaging are
    a) Be non-toxic b) Acts as a barrier
    c) Protects against contamination d) All the above
16. Kanaj is a type of
    a) Grain storage structure b) Food products
    c) Drying methods d) Grain separator
17. Prime cost is also called as.
    a) Fixed cost b) Variable cost
    c) Total coast d) Depreciation cost
18. The two principal proteins in the myofibril fraction of the protein are ....
    a) Actin and myosin b) Collagen and elastin
    c) Chymosin and actin d) Chitin and elastin
19. Spore forming bacteria do not grow at pH values
    a) Less than 7 b) More than 7
    c) Less than 3.7 d) More than 5
20. Consumers are now demanding to changes in food packaging materials because of their interest in
    a) Nutrition b) The environment
    c) Saving money d) Leisure activities
21. A water closet is usually made of
    a) Cement concrete b) Porcelain
    c) Glazed berth ware d) Stone ware
22. Pure lime contain ....% or more Calcium Oxide.
    a) 78 b) 85
    c) 90 d) 95

23. The desired input to the instrument may be constant or varying slowly with respect to time
   a) Static character b) Dynamic character
   c) Both a and b d) ISO static character
24. Meat product order was established in the year
   a) 1963 b) 1973
   c) 1986 d) 1996
25. Nisin is an antibiotic produced by
   a) *Streptococcus lactis* b) *Staphylococcus*
   c) *Bacillus subtilis* d) *Clostridium*
26. Macaroni wheat is
   a) T. aestivum b) T. compactum
   c) T. monococcum d) T. durum
27. Wheat Bran is rich in nutrients such as
   a) Carbohydrates b) Protein
   c) Vitamin and minerals d) None of the above
28. The amount of silica in the ash of rice hull is
   a) 95% b) 50%
   c) 20% d) 80%
29. Maize germ oil is rich source of essential fatty acids, about half of which is
   a) Linolenic acid b) Archadonic acid
   c) Oleic acid d) Linoleic acid
30. The soluble protein which acts as storage protein in rice and barley is
   a) Albumin b) Globulin
   c) Prolamines d) Glutelin
31. In barley and oat, the cell wall of starchy endosperm is made up of
   a) - D- glucans b) Cellulose
   c) Arabinoxylan d) None
32. The protein 'Kefrin' is present in.
   a) Corn b) Wheat
   c) Sorghum d) Millet
33. Gelatinization temperature of rye starch granules is about
   a) 65-70ºC b) 55-70ºC
   c) 65-75ºC d) 50-60ºC
34. The cereal kernel having largest amount of germ is
   a) Sorghum b) Corn
   c) Rye d) Pearl millet
35. The principal protein in buck wheat is
   a) Glutelin b) Protamina
   c) Albumin d) Globulin

36. As compared to True cereals such as wheat, the pseudo cereals (Buck wheat) are rich in
   a) Soluble sugar
   b) Protein
   c) Crude fiber
   d) All of them
37. Highest amount of erucic acid is present in which oil
   a) Cocoa butter
   b) Groundnut Oil
   c) Rape seed Oil
   d) Sesame Oil
38. Among the fats/oil, the highest amount of linoleic acid is present in
   a) Corn oil
   b) Cotton seed oil
   c) Safflower oil
   d) Sunflower oil
39. The trans form of oleic acid is
   a) Palmtoleic acid
   b) Trans-oleic acid
   c) Eladiac acid
   d) Behanic acid
40. Coconut oil traditionally is extracted from
   a) Whole coconut
   b) Copra
   c) Coconut meal
   d) Coconut water
41. The sugar present in the cereal grain kernel germ/bran are composed mainly of
   a) Sucrose + Raffinose
   b) Fructose + Glucose
   c) Sucrose + Glucose
   d) Fructose + Raffinose
42. The part of crude fiber in cereal grains is made up of
   a) Lignin
   b) Hemi-cellulose
   c) Cellulose
   d) Starch
43. Cellulose is hydrolyzed by
   a) Amylase
   b) Iso-amylase
   c) Pullulanase
   d) Cellulase
44. Gelatinization is the property of.
   a) Starch
   b) Protein
   c) Cellulose
   d) Hemi-cellulose
45. Resistant starches are type
   a) Indigestible
   b) Fiber like
   c) Both(a) and (b)
   d) Digestible
46. Caramel is
   a) Synthetic colour
   b) Non- synthetic colour
   c) Pigment produced from algae
   d) None of above
47. The largest tissue of cereal kernel is
   a) Bran
   b) Husk
   c) Endosperm
   d) Germ

48. The bleaching of the yellowish pigment of wheat flour during storage is also carried out by the enzyme

a) Gluco-amylase b) Lipase

c) Lipoxidase d) Gluco- oxidase

49. HDL stands for

a) High dietary lipids b) Highly dense lipids

c) High dietary levels d) High density lipoproteins

50. Anti-foaming agents used in fermentation process is

a) Sulphurous acid b) Acetic acid

c) Silicon dioxide d) Titanium oxide

51. Size of an irregular shape object express in term of

a) Sphere city b) Roundness

c) Equivalent diameter d) None

52. Diffusion of sugar in fruit jam is example of

a) Heat transfer b) Mass transfer

c) Momentum transfer d) Both(a) and (b)

53. Which one is widely used for cleaning and separating of seeds?

a) Grader b) Screens

c) Driers d) None

54. Pulsed electric field is suitable for

a) Liquids b) Semi solids

c) Both a and b d) Solids

55. Which of the following is used for transporting the large quantities of materials over a very large distance at a low cost?

a) Chain conveyor b) Screw conveyor

c) Belt conveyor d) Pneumatic conveyor

56. The length of the screen should be between ....times its width.

a) 1-2 b) 2-3

c) 1-3 d) 2-4

57. The chemical used to kill the rats and mice is called.

a) Fungicide b) Rodenticide

c) Both a and b d) None

58. Functional layout is also called

a) Process layout b) Product layout

c) Plant layout d) All of these

59. Hydrometer works on the principle of

a) Measuring the level of floating in the liquid in reference of water

b) Measuring the level of measuring of weight of known volume

c) Finding the volume of the water displaced when known weight is dipped.

d) Vibration is sent and resonant frequency is found out which is dependent on the density of liquid.

60. How many acts are repealed by Food safety and standard act 2006.

a) 4 b) 5

c) 7 d) 8

61. Which of the following shows shear thinking behavior

a) Dilatant fluid b) Pseudo plastic fluid

c) Rheopectic fluid d) Thixotropic fluid

62. The fractional coefficient for food grains depends on

a) Grain shape b) Surface characteristics

c) Moisture content d) All the above

63. Wheat flours of desired characteristic can be obtained by

a) Blending different varieties of wheat b) Mixing of impurities

c) Removing impurities d) Washing

64. Which of the following is not a thermodynamic property?

a) Temperature b) Heat

c) Pressure d) Specific volume

65. For Newtonian fluids, the slope of the shear stress versus shear rate graph in always

a) Curve b) Non linear

c) Constant d) Equal to zero

66. The most common fumigant used for storage of cereals is

a) Aluminum phosphide b) DDT

c) Zinc phosphide d) Ethylene dibromide

67. If 'L' and 'H' are width and depth of storage structure then the structure can only be called deep when

a) $H < L \tan[(90+Ø)/2]$ b) $H = L \tan[(90+Ø)/2]$

c) $H < 1.732 L$ d) None of these

68. Voigt model (Kelvin model) consists of

a) Spring and dashpot in series b) Spring absent

c) Dashpot absent d) Spring and dashpot in parallel

69. Porosity is a property that

a) Changes as a result of change in colour and surface of the product

b) Changes as a result of change in shape and particle density

c) Changes as a result of change in the angle of repose.

d) Remains constant

70. Parboiling of paddy is a

a) Chemical treatment b) Hydro treatment

c) Thermal treatment d) Hydro thermal treatment

71. Thermal conductivity of materials depends on

a) Specific gravity b) Mass density

c) Weight density d) All the above

72. The most appropriate enzyme preparation used commercially to increase juice yields in fruit processing is

a) Amylases b) Lipases

c) Proteases d) Pectinases

73. At STP, one gram of hydrogen occupies a volume of

a) 9.4 liter b) 10.5 liter

c) 11.2 liter d) None

74. The dimension less number relating buoyant and viscous forces in natural convection is

a) Nusselt number b) Reynold number

c) Grashoff number d) Prandtl number

75. LDPE contain the chemical elements

a) Carbon and hydrogen

b) Carbon and hydrogen and oxygen

c) Carbon and hydrogen and nitrogen

d) Carbon and hydrogen and oxygen and nitrogen

76. Fumigation can be done in a sealed grain storage structure using

a) Methyl bromide/ethyl ditromide (EDB) and phosphine

b) DDT and BHC powder

c) Chloroform and sulphur

d) Aluminum phosphate and zinc phosphate

77. Mycotoxins are toxic substances produced by

a) Rats b) Bacteria

c) Fungi d) Insects

78. Load cell related to

a) Force b) Time

c) Torque d) Displacement

79. Environmental protection act was established in the year

a) 1955 b) 1976

c) 1986 d) 2006

80. The process layout in best suited where

a) Specialization exists

b) Machines are arranged according to sequence of operation

c) Few numbers of non-standardized units are to be produced

d) Mass production is envisaged

81. Volume of large objects such as fruits and vegetable can be measured by

a) Platform scale method b) Air compressor pycnometer

c) Specific gravity tube d) None of these

82. Warming of water in a beaker is an example for .... Mode of heat transfer.
    a) Radiation
    b) Conduction
    c) Convection
    d) All of these
83. The common ways of screen agitation is
    a) Revolving a cylindrical screen about a horizontal axis.
    b) Revolving a cylindrical screen about a vertical axis.
    c) By using stirrers.
    d) None of these.
84. The ability of screen is closely separating the feed into over flow and under flow is called
    a) Effectiveness
    b) Efficiency
    c) Both a and b
    d) None of these
85. Which of the following accessories are not used for improving the efficiency of screen
    a) Screen brusher
    b) Oil cloth cover
    c) Rubber balls
    d) Eccentric unit
86. Pneumatic separation is based on the difference in
    a) Mechanical properties
    b) Electrical
    c) Rheological properties
    d) Aerodynamic properties
87. Which of the following is used to measure dry and wet bulb temperature of air
    a) Hygrometer
    b) Rotameter
    c) Sling psychrometer
    d) All the above
88. During sensible cooling of air, the enthalpy
    a) Increases
    b) Decreases
    c) Remains constant
    d) None of these
89. Which of the following chemical used for measuring moisture content of grain?
    a) H_2SO_4
    b) NaCl
    c) HCl
    d) $CaCl_2$
90. The process of losing moisture and attains EMC with surrounding, the known as
    a) Absorption EMC
    b) Desorption EMC
    c) Both a and b
    d) None of these
91. The full form of LSU dryer is
    a) Ludhiana state university dryer
    b) Louisiana state university dryer
    c) Louisiana state urban dryer
    d) None of these
92. The evaporative cooling process takes place on constant
    a) Enthalpy line
    b) Humidity ratio line
    c) RH line
    d) None of the above
93. Which of the following is a conduction type dryer
    a) Tunnel dryer
    b) Belt convection dryer
    c) Vacuum shelf dryer
    d) None of the above

94. In convection drying of agricultural materials, rate of drying can be controlled by
   a) Air velocity b) Air temperature
   c) Humidity d) All the above

95. The main limitation of pulsed light technology
   a) Penetration depth b) Side effects
   c) Heavy cost d) None of above

96. Which of the following equation is used to calculate dry basis moisture content (Mwb) in term of wet basis moisture content (Mwb)
   a) $Wdb = \frac{Mwb}{1+Mwb}$ b) $Wdb = \frac{Mwb}{1-Mwb}$
   c) $Wds = \frac{Mwb+1}{Mwb}$ d) $Wds = \frac{1-Mwb}{Mwb}$

97. Which of the following is not a component of belt conveyor
   a) Idles b) Blowers
   c) pulley d) both (a) and (c)

98. The storage capacity of muda type grain storage structure is about
   a) 1-3 tonnes b) 3-18 tonnes
   c) 9-35 tonnes d) 35-50 tonnes

99. Quantity and value of every movable item is mentioned in
   a) Flow pattern b) Plant layout
   c) Product layout d) Inventory

100. The error in the experience of observer is known as
   a) Tanning error b) Mistakes
   c) Personal error d) None

Answer Key

1	d	2	d	3	c	4	b	5	b	6	a	7	c
8	a	9	b	10	c	11	c	12	a	13	a	14	c
15	d	16	a	17	b	18	a	19	c	20	d	21	c
22	d	23	a	24	d	25	a	26	d	27	c	28	a
29	d	30	b	31	a	32	c	33	a	34	d	35	d
36	d	37	c	38	c	39	c	40	b	41	a	42	b
43	d	44	a	45	c	46	b	47	c	48	c	49	d
50	c	51	c	52	b	53	b	54	c	55	c	56	b
57	b	58	a	59	d	60	d	61	a	62	d	63	a
64	d	65	c	66	a	67	d	68	d	69	b	70	d
71	b	72	d	73	d	74	c	75	a	76	a	77	c
78	a	79	c	80	c	81	a	82	c	83	a	84	c
85	d	86	d	87	c	88	a	89	d	90	b	91	b
92	a	93	c	94	d	95	a	96	b	97	b	98	a
99	d	100	c										

Model Paper 13

1. Which the increase in temperature, the tensile strength and elastic modulus of material
   a) Increase
   b) Decrease
   c) Remains constant
   d) None
2. Screens are used for mechanical reparation of
   a) Solid liquid mixture
   b) Solid gas mixture
   c) Solid solid mixture
   d) Liquid liquid mixture
3. Primary treatment of waste water in a plant will cause a drop of about .... in the BOD.
   a) 60%
   b) 50%
   c) 30%
   d) 40%
4. Kicks assumed that the energy required to reduce the material in size was ______________ proportional to the size reduction ratio.
   a) Directly
   b) Inversely
   c) Equally
   d) None
5. The grain of water vapour transmitted from 1m^2 of film are in 24 h is known as
   a) Water vapor permeability
   b) Water vapor permeance
   c) Water vapor transmission rate
   d) Moisture transfer coefficient
6. The evaporative cooled storage structures work on the principle of...................... caused by evaporation of water made to drip over the bricks or coder pads.
   a) Sensible cooling
   b) Sensible heating
   c) Adiabatic heating
   d) Adiabatic cooling
7. Which of the following statement is wrong?
   a) The ISO 9000 series is an applicable only in case of food industries
   b) The TQM process includes quality control, quality insurance and quality improvement.
   c) Monitoring of CCP comes under the purview of HACCP
   d) Statistical analysis is an integral part of quality of food products.
8. Electrical resistance of a transmitter.
   a) Increase with temperature increase
   b) Decrease with temperature increase
   c) Remain unaffected with the change in temperature.
   d) None of these

9. Consumer protection act was passed in the year
   a) 1946 b) 1966
   c) 1986 d) None of these
10. The specific heat of dry grain in Kcal / kg°C ranges from
   a) 0.25 - 0.35 b) 0.35 - 0.45
   c) 0.40 - 0.60 d) 0.80 - 1.00
11. Infrared spectroscopy is used for
   a) Quality control b) Dynamic measurement
   c) Monitoring application d) All the above
12. The ratio of specific heat at constant pressure and specific heat constant volume is
   a) Less than 1.0 b) More than 1.0
   c) Equal to 1.0 d) Higher than 1.0
13. Heat exchange between the wall and fluid is due to
   a) Conduction b) Convection
   c) Mutual exchange d) Radiation
14. Standard barometric pressure is
   a) 1.0332 kg/cm^2 b) 760 mm of hg
   c) 1013.25 mbar d) All the above
15. Bernoulli equation represents conservation of
   a) Force b) Mass
   c) Energy d) Momentum
16. Electrolux refrigerator is operated in the principle of.
   a) Vapor compression system
   b) Vapor absorption system
   c) Air compression system
   d) Vapor compression and absorption
17. Centrifugal dehusker removes husk from paddy based on
   a) Impact b) Friction
   c) Shear d) Abrasion
18. In single effect evaporator, the economy is
   a) Equal to 1.0 b) More than 1.0
   c) Less than 1.0 d) More than 1.0
19. Density of milk is
   a) 1030 kg/m^3 b) 1035 kg/m^3
   c) 930 kg/m^3 d) 950 kg/m^3
20. A Maillard reaction is an example of the browning of
   a) Protein b) Starch
   c) Sugar d) Protein and sugar

21. In microware heating, water heats much faster than ice due to the high .... factor of water.
    a) Dielectric constant b) Dielectric loss
    c) Both a and b d) Frequency
22. Drying constant is determined by
    a) Graphical method b) Half life period method
    c) Both a and b d) Statistical method
23. In elevating of grain, the discharge from bucket elevator is a combination of
    a) Centrifugal b) Gravitation
    c) Both (a) and(b) d) Frictional
24. The degree of moisture protection of a package is measured in term of
    a) Water vapor transpiring rate b) Water vapor turning rate
    c) Water vapor transmission rate d) Water vapor diffusion rate
25. Aeration can provide major benefits in the storage of grain if
    a) It cools the grain and slow down inset activity
    b) Prolongs the effectiveness of pesticides by cooling
    c) It can provide an appreciable drying function
    d) All the above
26. The common flow patterns are
    a) L b) Line
    c) Circular d) All of the above
27. Meat preservative agents are
    a) Nitrite and nitrate b) Sugar
    c) Acetic acid d) Benzoic acid
28. Spore forming bacteria do not grow at pH values
    a) <7 b) >7
    c) <3.7 d) >5
29. Potatoes have a higher satiety value if they are
    a) Fried b) Baked
    c) Boiled d) Roasted
30. The unit slug is used for expressing
    a) pound force b) pound mass
    c) gram mass d) gram
31. Absolute alcohol without trace of water is known as
    a) Dehydrated alcohol b) Dry alcohol
    c) Pure alcohol d) Power alcohol

32. The head loss of an orifice meter is
    a) Less than that of the venturimeter
    b) Less than that of the nozzle flow meter
    c) Greater than that of the flow meter
    d) Greater than that of the venturimeter
33. Who is responsible for implementation of HACCP?
    a) Farmers b) Processor
    c) Government d) Both (b) and (c)
34. Starch is made up of
    a) Amylose + Amylopectin b) Glucose + Sucrose
    c) Fructose + Amylose d) Glucose + Amylose
35. The elastic behavior of dough is due to
    a) Lipid b) Cellulose
    c) Gluten d) Moisture
36. Isomerization of glucose to fructose is done with
    a) Glucose isomerase b) Galacto isomerase
    c) Mannose isomerase d) None
37. Sorbitol is found in
    a) Plums b) Apples
    c) Pears d) All the above
38. Isomalt is also known as
    a) Hydrogenated Isomaltose b) Hydrogenated Palatinose
    c) Both a and b d) None
39. Starch is found as storage carbohydrate in plants in the form of
    a) Grains b) Granules
    c) Crystal d) None
40. Waxy starch contains
    a) Amylase b) Amylopectin
    c) Both a and b d) None
41. Ropiness in beer in caused by
    a) Acetobacter b) Lactobacillus
    c) Pediococcus d) All
42. Coffee beans are fermented by
    a) Erwinia b) Saccharomyces
    c) Both a and b d) All
43. Miso is a fermented product of
    a) Soybean b) Moong dal
    c) Cucumber d) Rice

44. Which of the following is a type of microbial spoilage of sauer kraut
   a) Soft kraut b) Sling kraut
   c) Rotted kraut d) All the above
45. Cottage cheese undergoes spoilage by
   a) Bacteria b) Yeasts
   c) Molds d) All the above
46. Tempeh is a fermented product of
   a) Soybean b) Rice
   c) Wheat d) Corn
47. Sake is a fermented product from
   a) Japan b) Thailand
   c) Mexico d) All
48. Which of the following yeast is used for Single Cell Protein?
   a) Candida b) Rhodotorula
   c) Saccharomyces d) All the above
49. Which of the following has antimicrobial activity
   a) Lactoferrin b) Ozone
   c) Sodium chlorite d) All the above
50. Which of the following antibiotic is widely used in food
   a) Subtilin b) Tylosine
   c) Nisin d) All the above
51. Phase diagrams are important in studying of
   a) Crystallization b) Distillation
   c) Precipitation d) All
52. Shear thickening fluids are also known as
   a) Elastic b) Plastic
   c) Dilatant d) None
53. Mechanism of microwave heating is based on
   a) Ionic polarization b) Dipole rotation
   c) Both a and b d) None
54. Which of the following is dimensionless
   a) Shape factor b) Friction factor
   c) Relative dielectric d) All the above
55. Which of the following has m^2/s unit
   a) Mass diffusivity b) Kinematic viscosity
   c) Thermal diffusivity d) All the above
56. Pectins are polymer of
   a) Sugar and protein b) Protein and fat
   c) Sugar and acid derivatives d) Fat and acid derivatives

57. The pigments and their precursors found in fruits and vegetables occurs in
    a) Chloroplast b) Chromoplast
    c) Both a and b d) None
58. Important characteristics of tannin is
    a) Astringency b) Pungency
    c) Sweetness d) Sourness
59. Flavors of a food product is a combination of
    a) Taste and smell b) Taste and appearance
    c) Smell and appearance d) None
60. The target organism for canning is
    a) Bacillus cereus b) *E. coli*
    c) Pediococcus botulinium d) All the above
61. The heat stability of milk is judged by
    a) MBRT b) COB
    c) Lactometer d) Freezing point
62. The detection of adulteration of milk with water is done by
    a) T.A. b) Ethanol test
    c) Lactometer d) Alcohol alizarin test
63. Bactofugation is a process to remove bacteria by
    a) Sedimentation b) Centrifugation
    c) Precipitation d) Homogenization
64. Bitter taste in milk is due to enzymatic break down of
    a) Fats b) Carbohydrates
    c) Proteins d) Calcium
65. As the size of fat globules increases, the rate at which cream rises/ separates
    a) Decreases b) Increases
    c) No effect d) None
66. Which of the following operation leads to increase in moisture control of the wheat grain
    a) Tempering b) Conditioning
    c) Milling d) Both a and b
67. Conditioning and tempering operation............................. the yield of flour from wheat grain.
    a) Increase b) Decrease
    c) No effect d) None
68. The chemical used for bleaching of wheat flour is
    a) Chlorine b) Benzoyl peroxide
    c) Acetone peroxide d) All

69. Flour low in gluten and protein content is called
   a) Strong flour b) Weak flour
   c) Soft flour d) Hard flour
70. Week flour is recommended for
   a) Biscuit b) Cookies
   c) Cakes d) All
71. Strong flour is recommended for
   a) Bread b) Cookie
   c) Cakes d) Biscuits
72. Saki, an alcoholic beverage is prepared from
   a) Wheat b) Barley
   c) Rice d) Maize
73. The flint, dent, sweet, waxy are variety of
   a) Corn b) Wheat
   c) Rice d) Barley
74. Malting of barley involves
   a) Soaking b) Germination
   c) Drying d) All the above
75. The kernel of oat is called
   a) Caryopsis b) Groat
   c) Grit d) None
76. Farina is another name of
   a) Suji b) Chunk
   c) Nugget d) None
77. Which of the following require strong gluten characteristics?
   a) Bread b) Cookies
   c) Biscuit d) Cakes
78. The chemical leaving agent perform by producing
   a) CO_2 b) O_2
   c) NO_2 d) All
79. Commonly chapatti is prepared from
   a) Whole wheat flour b) Refined flour
   c) Semolina d) None
80. Which of the following pulse has protein value similar to animal protein.
   a) Red grain b) Lentil
   c) Soybean d) Black grain
81. Cluster bean is used as
   a) As dal b) Thickness
   c) Leavener d) All

82. The function of shortening in dough is
   a) Reduce toughness of dough
   b) High machinability
   c) Enhance flavor
   d) All the above
83. PDR stands for
   a) Peak dough result
   b) Peak dough resistance
   c) Protein dough resistance
   d) None
84. The major source of dietary protein in India is
   a) Cereals
   b) Pulses
   c) Milk
   d) Meat
85. Which of the following is not a legume?
   a) Horse gram
   b) Lentil
   c) Rice bean
   d) Maize bean
86. Gelatin is obtained from
   a) Collagen
   b) Elastin
   c) Actin
   d) Myosin
87. The cut from the belly portion of hog carcass is called
   a) Mutton
   b) Bacon
   c) Veal
   d) Ham
88. The musty or earthy flavor of meat in due to
   a) Actinomycetes
   b) Flavobacterium
   c) Chromo bacterium
   d) None
89. The major lipid of egg yolk is
   a) Triglycerides
   b) Manoglycerides
   c) Diglycerides
   d) None
90. The process for evaluating the quality of egg in
   a) Candling
   b) Handling
   c) Scalding
   d) Stewing
91. Fish protein are.................. digestible.
   a) Highly
   b) Poorly
   c) Non
   d) None of these
92. Single cell protein (SCP) may contain protein (%) up to
   a) 10%
   b) 80%
   c) 30%
   d) 20%
93. Lag phase in growth curve in also known as
   a) Death phase
   b) Period of adaptation
   c) Telophase
   d) None
94. Compressed yeast contains water content around
   a) 27%
   b) 40%
   c) 70%
   d) 90%

95. Production of ethanol from glucose occur by
   a) EM pathway b) TCA cycle
   c) Glycolysis d) None
96. Hops are obtained from
   a) Cyamopsis tetragonolobus b) Humulus lupulus
   c) Cicer arietinum d) None
97. Which of the following in a distilled beverage ?
   a) Brandy b) Rum
   c) Whisky d) All
98. A process that changes the shape of protein molecule without breaking in covalent bonds is
   a) Denaturation b) Coagulation
   c) Agglutination d) Saturation
99. Energy lost when water molecules from ice crystals is
   a) Specific heat b) Latent heat
   c) Heat of fusion d) Heat of vaporization
100. Which of the following is natural colour?
   a) Betalains b) Riboflavin
   c) Chlorophyll d) All

Answer Key

1	b	2	c	3	c	4	a	5	c	6	d	7	a
8	b	9	c	10	b	11	d	12	b	13	b	14	d
15	c	16	b	17	c	18	c	19	c	20	d	21	b
22	c	23	c	24	c	25	d	26	d	27	a	28	c
29	a	30	b	31	d	32	c	33	d	34	a	35	c
36	a	37	d	38	c	39	b	40	b	41	d	42	c
43	a	44	d	45	d	46	a	47	a	48	d	49	d
50	c	51	d	52	c	53	c	54	d	55	d	56	c
57	c	58	a	59	a	60	c	61	b	62	c	63	b
64	c	65	b	66	d	67	a	68	d	69	b	70	d
71	a	72	c	73	a	74	d	75	b	76	a	77	a
78	a	79	a	80	c	81	b	82	d	83	b	84	b
85	d	86	a	87	b	88	a	89	a	90	a	91	a
92	b	93	b	94	c	95	a	96	b	97	d	98	a
99	b	100	d										

Model Paper 14

1. Which acid is found in carbonated non alcoholic beverages?
   a) Propionic acid　　b) Acetic acid
   c) Citric acid　　d) Phosphoric acid
2. Ripening of cheese in presence of fungus penicillium roqueforti results change in
   a) Texture　　b) Flavor
   c) Colour　　d) Both a and b
3. Microorganism responsible for discoloration in fish during spoilage.
   a) Alcaligenes　　b) Flavobacterium
   c) Pseudomonas　　d) Rhodotorula
4. Common microbial contaminant is poultry (freshly dressed, eviscerated) food product in
   a) Coliforms　　b) Pediococcus sp.
   c) Bacillus subtilis　　d) Pseudomonas
5. Sarcina sickness is the defect of
   a) Wine　　b) Sauerkraut
   c) Beer　　d) Bread
6. CIPHET, a premier institute is related to
   a) Fiber　　b) Fluid
   c) Feed　　d) Food
7. Which of the following is not a texture character?
   a) Firmness　　b) Fibrousness
   c) Smoothness　　d) Cohesiveness
8. Porosity of food grains is in the rang e of
   a) Above 75%　　b) 30-40%
   c) 10-20%　　d) 50-70%
9. Specific gravity of grains is determined by
   a) Pycrometer　　b) Toluene displacement method
   c) Refractometer　　d) None of above
10. To calculate the protein in percentage, percent nitrogen is multiplied by
    a) 2.65　　b) 5.62
    c) 2.56　　d) 6.25

11. The wave lengths for UV lights which range from
    a) 10 nm to 400nm
    b) 400nm to 780nm
    c) 780nm to 920nm
    d) 920nm
12. The purpose of having number of tubes in shell and tube exchanger is
    a) To transfer heat
    b) To increase surface area for higher heat transfer rate
    c) To decrease the rate
    d) None of the above
13. The product of Grashoff and Prandtl number is called
    a) Euler number
    b) Weber number
    c) Cauchy number
    d) Rayleigh number
14. UV-Visible spectroscopy is also known as
    a) Electric spectroscopy
    b) NIR spectroscopy
    c) NMR spectroscopy
    d) CT spectroscopy
15. A falling film evaporator is designed to concentrate
    a) Viscous feed
    b) Framing feed
    c) Scaling feed
    d) Feed which is heat sensitive
16. The temperature for gelatinization is about
    a) 65°C
    b) 70°C
    c) 75°C
    d) 80°C
17. The value $\frac{C_p}{C_v}$ for air is
    a) 1.2
    b) 1.4
    c) 1.8
    d) 2.3
18. Triple point temperature of water in
    a) 0.1°C
    b) 0.01°C
    c) 0.4°C
    d) 0°C
19. Yoghurt is similar to
    a) Dahi
    b) Paneer
    c) Condensed milk
    d) lassi
20. The AGMARK ghee grading scheme was initiated in
    a) 1937
    b) 1938
    c) 1954
    d) 1964
21. Lassi is a by-product of .... making
    a) Butter
    b) Desi-butter
    c) Ice cream
    d) Dahi
22. The pH of natural milk is around
    a) 5.8
    b) 6.8
    c) 8.8
    d) 1.2

23. Lactoferrin inhibits bacteria by removal of
    a) Calcium b) Copper
    c) Iron d) None
24. Protein efficiency ratio of milk protein in
    a) 2.5 b) 3.3
    c) 1.0 d) 10
25. Butter is a ....material
    a) Elastic b) Plastic
    c) Viscoelastic d) None
26. Rennet is used in processing of
    a) Ice cream b) Butter
    c) Cheese d) All
27. Bloat defect in condensed milk is due to
    a) Bacteria b) Yeast
    c) Virus d) Mold
28. Instantization refers to
    a) Making powder soluble b) Making powder insoluble
    c) Restricting powder soluble d) None
29. Unit of enthalpy is
    a) kw/kgm b) kj/kg.cm
    c) kw d) kJ/kg
30. Canning usually employs temperature .... to kill spoilage organism and to inactivate enzyme action.
    a) Below 80°C b) Above 80°C
    c) Below 100°C d) Above 100°C
31. Rate of mass transfer depends on
    a) Viscosity b) Thermal conductivity
    c) Heat input d) Thermal diffusivity
32. Low acids foods such as meat, poultry and fish have to be processed at a temperature of
    a) Equal to 100°C b) Higher than 100°C
    c) Less than 100°C d) Higher than 88°C
33. Agar is extracted from
    a) Algae b) Fungi
    c) Bacteria d) Crops
34. Fruits are generally deficient in
    a) Protein b) Carbohydrates
    c) Vitamins d) Water

35. Saffron is obtained from
    a) Dried stigma b) Dried ovary
    c) Dried pistel d) Dried carpel
36. The flavor of asafoetida is due to the presence of compound
    a) Sulphur b) iron
    c) magnesium d) Calcium
37. Green tea is .... Tea
    a) Fermented b) Unfermented
    c) Semi-fermented d) None
38. The caffeine content of tea is about
    a) 1.3% b) 2.3%
    c) 3.3% d) 4.3%
39. Saver kraut is .... Acid fermentation of cabbage.
    a) Acetic acid b) Citric
    c) Lactic d) Glucose
40. Very round wheat sample will have falling number
    a) 60 b) 250
    c) 400 d) None
41. The evaporator in the vapour compression refrigeration system is normally installed in
    a) High pressure side b) Low pressure side
    c) Neutral side d) No pressure side
42. The coefficient of performance is always .... one
    a) Less than b) More than
    c) Slightly greater than d) Equal to
43. Centrifugal force developed in a centrifuge is larger than gravity force by a factor.
    a) $V^2/g.r$ b) $V^2/g.r^2$
    c) $V/g.r$ d) $V^2/g^2.r$
44. The pressure head of centrifugal pump varies.
    a) as square of speed b) as cube of speed
    c) as fourth power of speed d) directly as speed
45. Retrogradation is a property of
    a) Fat b) Starch
    c) Protein d) All
46. NIR stands for
    a) Near Infraction Reflection b) Near Infrared Refraction
    c) Near Induction Refraction d) None
47. Air oven method and Karl fisher method is used to evaluate
    a) Moisture b) Ash
    c) Fat d) Protein

48. Fat marbling makes the muscle
   a) Non-juicy b) Tough
   c) Tender d) None
49. Increased toughening of meat refers to
   a) Ageing of meat b) Rigor mortis
   c) Slough tearing d) All
50. DFD stands for
   a) Dark firm and dry b) Dry firm and dark
   c) Dry fat and dark d) Dark fat and dry
51. PSE stands for
   a) Poly soft and exudate b) Pale soft and exudate
   c) Protein soft and exudate d) None
52. The equation of continuity of flow in
   a) Mass b) Momentum
   c) Energy d) All the above
53. The equation of continuity of flow in applicable whom
   a) Flow is compressible
   b) Flow is one dimensional
   c) Velocity is uniform over the cross section
   d) All of the above
54. Which conveyers are used as metering and feeding device
   a) Belt conveyor b) Chain conveyor
   c) Pneumatic conveyor d) Screw conveyor
55. Which conveyor were on the principle of inertia
   a) Belt b) Chain
   c) Vibratory d) Screw
56. Trans Form of fatty acid is
   a) Bent form b) Straight form
   c) Mixture of bent and straight form d) Circular form
57. The conversion of glucose into glycogen in human body is called
   a) Glucogenesis b) Glucolysis
   c) Glyconeogensis d) Lipolysis
58. The conversion of glucose into fat in body is called
   a) Glucogenesis b) Glucolysis
   c) Glyconeogensis d) Lipogenesis
59. Break down of glucose releasing energy is called
   a) Glucogenesis b) Glucolysis
   c) Glyconeogensis d) Lipolysis

60. The biological value of Egg proteins is
  a) 96 b) 90
  c) 74 d) 80

61. The biological value of milk proteins is
  a) 96 b) 90
  c) 74 d) 80

62. The science of fermentation is known as
  a) Fermentology b) Zymology
  c) Biotechnology d) phycology

63. Penicillin was discovered in
  a) 1901 b) 1935
  c) 1928 d) 1941

64. Downstream processing in bioprocesses is related to
  a) Fermentation b) Culture
  c) Substrate d) Product recovery

65. Penicillin may cure
  a) Pneumonia b) Scarlet fever
  c) Tetanus d) All

66. One ton of refrigeration is the amount of heat required to melt one ton of ice at 0 in
  a) 1 hr b) 24 hrs
  c) 6h d) 12h

67. Regeneration is economical when product is
  a) Heat b) Cooled
  c) Heated and cooled d) Sub cooled

68. Janssen equation is related to
  a) Storage silo design b) Grain transportation
  c) Grain marketing system d) Grain packaging storage

69. If 'L' and 'H' are with and depth of storage structure then the structure can only be called deep when.
  a) H<L tan[(90+)/2] b) H=L tan[(90+)/2]
  c) H<1.732L d) None

70. In crystal growth of sucrose, mass transfer of sucrose molecules to surface of crystal is
  a) First order process b) Second order process
  c) Zero order process d) None

71. Guar gum is also known as
  a) Cluster bean gum b) Carob gum
  c) Gum Arabic d) None

72. Which of the following is not a bacteria ?
   a) Acenetobacter b) Aeromonas
   c) Alcaligenes d) Alternaria
73. Stiffening of muscles upon slaughter is due to formation of
   a) Actin b) Myosin
   c) Actomosin d) None
74. Blue mold rot generally occurs in
   a) Apples b) Pepper
   c) Cucumber d) Onion
75. Koji is a fermented product made from
   a) Wheat b) Rice
   c) Soybean d) All
76. Which of the following is particularly associated with starch analysis?
   a) Viscometer b) Rheometer
   c) RVA d) None
77. Carcass is a
   a) Refrigerant b) Coolant
   c) Type of beverage d) Slaughter animal body
78. Lactometer is used to measure ....of milk.
   a) Protein b) Moisture
   c) Density d) Pressure
79. Which of the following is generally used for determining elastic modulus of material
   a) Viscometer b) Rheometer
   c) Texture analyzer d) All the above
80. Rheo fermentometer is used for the analysis of
   a) Cookies b) Bread
   c) Yoghurt d) All
81. An activity in project walk is represented by
   a) A straight line b) A curve
   c) An arrow d) A circle
82. Slack is also known as
   a) Activity b) Event
   c) Float d) Time
83. Which of the following is a food safety, standard?
   a) ISO 1901 b) ISO 22000
   c) ISO 14000 d) All the above
84. Dummy activity indicates
   a) Activity slack b) Critical path
   c) Interfering float d) Zero stack

85. FSSAI is governed by
   a) Ministry of agriculture
   b) Ministry of health and family welfare
   c) Ministry of law and justice
   d) Ministry of food processing Industries
86. The reference electrode in pH measurement is
   a) Gloss electrode b) Hydrogen electrode
   c) Antimony electrode d) Hg-calomel electrode
87. In PID controller, D starch for
   a) Direct b) Derivatives
   c) Double d) None
88. Fuzzy logic process is related to
   a) Mathematical model b) Artificial model
   c) Senior model d) Senior network model
89. Transducer are
   a) Detective element b) Signal processor
   c) Biological material d) None
90. The least count of any instrument is known as
   a) Dead band b) Resolution
   c) Accuracy d) Frequency response
91. Basic flow chart shape for decision is
   a) j b) ◇
   c) ⬠ d) ∧
92. A process flow chart in a plant
   a) Gives a clear understanding of entire process
   b) Does not show relation between major parts of process
   c) Does not show relation between minor parts of process
   d) Accommodates all the above characteristics
93. Moisture migration in stored grains results from
   a) Moisture changes b) Temperature changes
   c) Vapour pressure d) Volume changes
94. Which one is a liquid type of fumigant?
   a) Methyl bromide b) Ethylene bromide
   c) Aluminum phosphide d) Hydrogen phosphide
95. Critical thickness of insulation for sphere is given by
   a) k/h b) 2k/h
   c) h/2k d) 4k/h
96. The number of holes per square inch in a 20 mm screen will be
   a) 20 b) 200
   c) 400 d) 500

97. An ideal storage temperature for potato is

a) 0°C b) 4°C

c) 10°C d) 16°C

98. Packaging film which is used for better MAP in

a) LDPE b) HDPE

c) Polypropylene d) PVC

99. The radiation which possesses least penetrating power are

a) Gamma rays b) Microwaves

c) Ultraviolet rays d) Infrared rays

100. Asepsis refers to prevention from

a) Air b) Light

c) Moisture d) Microorganism

Answer Key

1	a	2	b	3	c	4	d	5	c	6	d	7	d
8	b	9	a	10	d	11	a	12	b	13	d	14	a
15	c	16	b	17	b	18	b	19	a	20	b	21	b
22	b	23	c	24	b	25	b	26	c	27	b	28	a
29	c	30	d	31	c	32	b	33	a	34	a	35	a
36	a	37	b	38	d	39	c	40	c	41	a	42	a
43	a	44	a	45	b	46	b	47	a	48	c	49	b
50	a	51	b	52	a	53	d	54	d	55	c	56	b
57	a	58	d	59	b	60	a	61	b	62	b	63	c
64	d	65	d	66	b	67	c	68	a	69	d	70	a
71	a	72	d	73	c	74	a	75	c	76	c	77	d
78	c	79	b	80	b	81	c	82	c	83	b	84	d
85	b	86	a	87	b	88	a	89	a	90	b	91	b
92	a	93	b	94	a	95	b	96	c	97	b	98	c
99	c	100	d										

Model Paper 15

1. Texture property of paneer can be best represented by
   a) Hardness
   b) Gumminess
   c) Springiness
   d) Chewiness
2. Electromagnetic radiation wave length for UV, VIS, NIR radiation should be
   a) 300-400 nm, 400-765 nm, 765-3200 nm
   b) 765-3200 nm, 400-765 nm, 300-400 nm
   c) 400-765 nm, 300-400 nm, 765-3200 nm
   d) 300-400 nm, 765-3200 nm, 400-765 nm
3. Voigt model (Kelvin model) consists of
   a) Spring and dashpot in series
   b) Spring absent
   c) Dashpot absent
   d) Spring and dashpot in parallel
4. Thermal conductivity of a single grain is following times greater than that of the bulk grain.
   a) 1-2
   b) 2-3
   c) 3-4
   d) None
5. Hagen- poiseuille equation is useful for measuring
   a) Viscosity
   b) Density
   c) Heat capacity of fluid
   d) Reynolds number
6. Rheology is the science of
   a) Deformation in the metals
   b) Stress and strain behavior of metals
   c) Deformation and flow in the viscoelastic materials
   d) Flow of viscous products.
7. The relationship between the true demty (TD) and bulk demty (BD) is
   a) $P = 1 - \frac{BD}{TD}$
   b) $P = 1 + \frac{TD}{BD}$
   c) $P = 1 + \frac{BD}{TD}$
   d) $P = 1 - \frac{TD}{BD}$
8. The quantity of wax required to coat on fruits is estimated based on
   a) Surface area
   b) Projected area
   c) Roundness
   d) All the above

9. Advantages of dielectric heating are
   a) Uniformity of heating b) High rate of temperature
   c) Both a and b d) None of the above
10. The farinograph gives information about .... Properties of flour.
   a) Drying b) Mixing
   c) Size reduction d) Absorption
11. Size of the irregular shaped food grain is represented by
   a) Total diameter b) Average diameter
   c) Perimeter d) Equivalent diameter
12. The type of can is used for acidic foods
   a) SR cans b) AR cans
   c) Tin containers d) Glass containers
13. Gluten free diet (GFD) is gives to patients with
   a) Crohn's disease b) Colic disease
   c) Irritable bowel disease (IBD) d) Liver disease
14. The SI unit of energy in
   a) Kilo joule b) Kilogram
   c) Kilo calorie d) Kilo power
15. The enzyme presents in pineapple is
   a) Papain b) Pepsin
   c) Ficin d) Bromelin
16. Starch is made up of
   a) Glucose + Sucrose b) Amylase + Amylopectin
   c) Fructose + Amylose d) Glucose + Amylose
17. The existence of more than one compound with the same molecular formula is called
   a) Isomerism b) Carbonization
   c) Levomerism d) Reductomerism
18. Process of adding vitamins and nutrients to food and drinks is called
   a) Standardization b) Enrichment
   c) Nitrification d) Food fortification
19. The activity of .... in fresh rice is probably responsible of it sticky consistencies of tea cooking.
   a) α- amylase b) Protease
   c) Lipase d) Oxidase
20. Husk separators are used in
   a) Wheat processing b) Barley processing
   c) Paddy processing d) Corn processing

21. Baking powder is used in baking as
    a) Antimicrobial agent
    b) Leavening agent
    c) Anticaking agent
    d) Curing agent
22. Yeast used in bakery product is
    a) Cake
    b) Bread
    c) Cookies
    d) Biscuit
23. Most important ingredient in the spongy cake is
    a) Sugar
    b) Fat
    c) Egg
    d) Milk
24. Non crystalline candies are called as
    a) Soft candy
    b) Hard candy
    c) Crystalline
    d) Caramelized candy
25. The disease occur in bread due to
    a) Bacteria
    b) Yeast
    c) Virus
    d) Fungi
26. Sweetness which do not provide energy is
    a) Sucrose
    b) Glucose
    c) Maltose
    d) Asparatama
27. World Food Day is celebrated on
    a) October 16
    b) July 16
    c) November 16
    d) August 16
28. Type of yeast used commercially in baking industry is
    a) Compressed yeast
    b) Dry yeast
    c) Powder yeast
    d) Dry and powder
29. Which of the following is a water soluble pigments in foods
    a) Anthocyamin , anthoxanthine and flavonoids
    b) Chlorophyll
    c) Betalains
    d) Both (a) and (c)
30. Pungency of chilly is expressed by
    a) Scoville value
    b) Hedonic score value
    c) L-value
    d) Colour A value
31. To depress the water activity of food to a given level
    a) Nacl will be required in large amounts compared to sugar
    b) Nacl will be required in smaller amounts compared to sugar
    c) Nacl and sugar will be required in same amounts
    d) None of the above
32. Heart transfer coefficient in natural convection increases with increasing
    a) Biot number
    b) Musselt number
    c) Grashoff number
    d) None

33. One ton of refrigeration in terms of kcal/min is
    a) 30 b) 50
    c) 3000 d) 600
34. Which of the following is the primary refrigerant?
    a) Nacl b) H_2O
    c) NH_3 d) $Calcl_2$
35. Diffusion of water molecules within solid food follows
    a) Kick's law b) Fick's law
    c) Fourier's law d) Bernoulli's law
36. The part of a vapour compression refrigeration system, which absorbs heat from the refrigerated space is
    a) Compressor b) Condenser
    c) Evaporator d) None
37. A mixture of air and water vapour is adiabatically cooled. The lowest temperature of the mixture this achieved is equal to
    a) Dry bulb temperature b) Wet bulb temperature
    c) Dew point temperature d) Situation temperature
38. Chose the correct statement....
    a) The thermal conductivity of gases decreases with temperature
    b) The thermal conductivity of insulating solids increase with temperature
    c) The thermal conductivity of good electrical conductors is generally low
    d) The thermal conductivity variation is of low percentage in gages as compared to solids
39. In the given fin configuration increase in conductivity will
    a) Decrease the total heat flow
    b) Will affect only the temperature gradient
    c) Increase the total heat flow
    d) Heat flow is influenced only by the base temperature and sectional area
40. In forced convection, molecular diffusion causes
    a) Momentum flow in turbulent region
    b) Momentum flow in laminar region
    c) Heat flow in turbulent region
    d) Diffusion has no part in energy transfer
41. The traditional batch pasteurization is accomplished using a holding time of 30 min at
    a) 63°C b) 65°C
    c) 72°C d) 45°C
42. ....flow, in which the fluid on the outside flow in a perpendicular direction to the pipe axes.
    a) Parallel b) Counter
    c) Cross d) Newtonian

43. A gray surface is one for which

a) Reflectivity equals emissivity
b) Emissivity equals transmissivity
c) Emissivity is constant
d) Absorptivity equals reflectivity

44. For a given shape, partial pressure and temperature, the emissivity of

a) O_2 is higher than that of N_2
b) N_2 is higher than that of O_2
c) O_2 is higher than that of CO_2
d) CO_2 is higher than that of O_2

45. Insulating material have ....emissivity at angle near horizontal

a) High
b) Low
c) Same
d) Zero

46. With increase in temperature, thermal conductivity of good conductors will

a) Increase
b) Decrease
c) Initially increase and then decreases
d) Does not change

47. Rayleigh number in the product of

a) Reynolds number and prandtl number
b) Prandtl number and Nusselt number
c) Grashoff and Prandtl number
d) Reynold number and nusselt number

48. Triple point temperature and pressure of water are

a) 0.1°c and 1 atm
b) 0.01°c and 0.006028 atm
c) 0.1°c and 0.06028 atm
d) 0°c and 1 atm

49. Which of the following is an extensive property of a thermodynamic system?

a) Volume
b) Pressure
c) Density
d) Temperature

50. Viscosity is a property that manifests.

a) At fluid & did boundary only
b) Between two adjacent fluid layers in relative motion
c) In uniform incompressible flours
d) Only in turbulent flours

51. Conduction of heat within a solid body in governed by

a) Fourier's law
b) Newtori's law
c) Fick's law
d) Lambert's law

52. Drying process involves

a) Gas transfer
b) Mass transfer
c) Heat and mass transfer
d) Heat transfer

53. Constant rate of drying of food material is independent of

a) Air velocity
b) Thickness of bed
c) Air humidity
d) Air temperature

54. In refining of crude oil to edible oil, winterization refer to

a) Removal of FFA
b) Removal of odorous matters
c) Removal of colouring matters
d) Removal of soft wax

55. The moisture content at which the drying rate ceases to be constant is known as
   a) Equilibrium m.c.
   b) Critical m.c.
   c) Saturation m.c
   d) Average m.c.
56. Develop on expression to compute thousand kernel weights at any moisture content dry basis
   a) $W_{1000} = H_d(1+W_d)g$
   b) $W_{1000} = W_m(1+M_c)g$
   c) $W_{1000} = M_d(W_d +100)g$
   d) $W_{1000} = W_d(1+M_d)g$
57. Standard screens are used to measure the particle size range between
   a) 70mm to 30m
   b) 76mm to 38m
   c) 78mm to 38m
   d) 80mm to 40m
58. Which of the following parameters affect the dry rate most, during drying process?
   a) Drying air temperature
   b) Air humidity
   c) Air velocity
   d) All the above
59. When is a body is at thermal equilibrium with its soundings, its absorvity equals emissivity is given by
   a) Kirchhoff's law
   b) Plank's law
   c) Steffen- Boltzmann law
   d) Maxwell law
60. Neutralization process, in refining of bran oil, is carried out mainly to remove
   a) Gums
   b) Free fatty acids
   c) Colour
   d) Wax
61. In a cyclone separator, the centrifugal force which acts upon the particle is given as
   a) wv^2
   b) W^2v/gr
   c) wv^2/gr
   d) w^2/r
62. Cereal grain is dried under ....of drying.
   a) Constant rate drying
   b) Falling rate period
   c) Microwave
   d) None
63. Which one of the following is true with regard to moisture sorption behavior of foods
   a) Equilibrium air moisture content for the desired moisture content with in the food will be higher drying rehydration than while drying.
   b) Equilibrium air moisture content for the desired moisture content with in the food will be lower drying rehydration than while drying.
   c) Equilibrium air moisture content for the desired moisture content with in the food drying rehydration and drying will be equal.
   d) None of the above.
64. Suitable moisture content for safe storage of paddy is in the range of.
   a) 4 – 8%
   b) 10-12%
   c) 16-18%
   d) 22-24%
65. The unit operation not included in the agro processing
   a) Drying & dehydration
   b) Milling
   c) Harvesting and threshing
   d) Cleaning and grading

66. 100 tonnes of paddy were harvested at 22% (wb) moisture content and it has to be brought to a moisture content of 15% (wb). What is the estimated Weight of paddy after drying.

a) 90.4t b) 91.8t
c) 93.6t d) 94.2t

67. A mixture of solid-liquid food can be separated by

a) Filtration b) Screening
c) Expression d) All the above

68. The emissive power of a body depended on

a) Temperature of the body b) Physical nature
c) Nature of the body d) All the above

69. The purpose of pitting in the pulse milling is

a) Splitting of pulse
b) Pearling of the pulse
c) Polishing
d) To aid the process of oil penetration for loosing of husk.

70. LSUdying is most suitable for drying of.

a) Wheat b) Corn
c) Paddy d) Barley

71. In milling of pulses, the whole dehusked grain is called.

a) Gota b) Bhusi
c) Chuni d) Grade-I dhal

72. The freezing point of a solution is affected by the concentration level of.

a) Salts b) Sugars
c) Salts and sugars d) None of the above

73. An isochoric process occurs at

a) Constant pressure b) Constant volume
c) Constant temperature d) Constant entropy

74. Purification in wheat milling refer to

a) Separation of pure endosperm b) Breaking of grain
c) Tempering of grain d) Degerming of grain

75. The height of the conical portion of a cyclone separator is ....times the height of the cylindrical portion.

a) 1-2 b) 2-3
c) 3-4 d) 4-5

76. Soxhlet apparatus in mainly used for

a) Expression b) Extraction
c) Leaching d) Infiltration

77. Which reduces the loss of flavoring components?

a) Freeze drying b) Cryogenic grinding
c) Sub atmospheric grinding d) Vacuum drying

78. Which of the following law is related to cooling of a substance?
   a) Newton's law b) Fourier's law
   c) Nain's law d) Kirchhoff's law
79. Ohmic heating is .... method of processing foods.
   a) Thermal b) Non- thermal
   c) Both a and b d) None
80. The area under the lethal rate curve is called
   a) Log phase b) Lethality
   c) ºF d) None of these
81. High intensity pulsed electric field can be used to
   a) Inactivation of bacteria
   b) Inactivation of vegetative bacteria
   c) Inactivation of yeast
   d) Inactivation of bacteria and yeasts
82. Cans treated with acid resistance lacquer are called as
   a) R-enameled can b) C- enameled can
   c) SR- enameled can d) Plain can
83. Detrimental effect of texture damages is more during freezing preservation in
   a) Fruits b) Vegetable
   c) Meat d) Both a and b
84. In high pressure processing, which of the following may be used as pressure transmitting medium
   a) Air b) Water
   c) Vapor d) Metallic can
85. Pasteurization of milk is necessary for
   a) Long shelf life at room temperature
   b) Destruction of all microorganism
   c) Destruction of pathogenic microorganism
   d) Long shelf life under refrigeration
86. Microwave heating of food correspond to wave length of
   a) 12 or 34 cm b) 0 to 12 cm
   c) 0 to 5 cm d) None of above
87. Maturity inducer for fruits and vegetables includes
   a) Shape b) Skin colour
   c) Abscission d) All the above
88. Thermal processing of food material should be based on the heat sensitivity of the
   a) Spores of fungi b) Vegetative cells of fungi
   c) Spores of bacteria d) Vegetative of cells of bacteria

89. In canning of fruits and vegetables, flat sour refers to

a) Bulging of cans b) Overfilling of cam

c) High acid formation d) None of above

90. In case of ohmic heating, food acts as

a) Hot heat transfer surface b) Conductor of electricity

c) Electrical resistance d) None of the above

91. The process of dewaxing of oils is performed to remove

a) Wax b) Organic impurities

c) Moisture d) All the above

92. Escherichia coli enters in food chain mainly through

a) Air b) Water

c) Dust d) Field contamination

93. Moisture content is dried vegetable in

a) 2% b) 3%

c) 5% d) 6%

94. Which of the following method of heat treatment gives the best texture and flavor of milk?

a) LTLT pasteurization b) HTST pasteurization

c) UHT pasteurization d) None of the above

95. High hydrostatic pressure system of food processing is based on the principle of

a) Appertization b) Thermosonication

c) Nucleation d) Pascalization

96. When two or more fans are connected in series in a grain drying system, then

a) Volume of air through each fan will be same

b) Total pressure drop across each fan will be same

c) Total volume of air will be sum of individual fan volume

d) Total pressure will be not be the sum of individual fan total pressure

97. On a pressure system, safety, value is remained open and blow until the pressure is reduced to.

a) Set pressure is exceeded b) Set pressure is decreased

c) Below the atmospheric d) Below the pop off pressure

98. A ball bearing is provided on shaft for the purpose of

a) Reducing friction b) Reducing energy requirement

c) Reducing wearing of shaft d) All of the above

99. Coefficient of performance of dryer is

a) 1 - HUF b) 1+ HUF

c) (1 – HUF)/(1+ HUF) d) (1+ HUF)/ (1 – HUF)

100. The power requirement to reduce the size of particles given by Rittinger's law is ....than kick's law.

a) Equal
b) Less
c) Greater
d) All the above

Answer Key

1	c	2	a	3	d	4	c	5	a	6	c	7	a
8	a	9	c	10	b	11	d	12	b	13	b	14	a
15	d	16	b	17	a	18	d	19	a	20	c	21	b
22	b	23	c	24	a	25	a	26	d	27	a	28	a
29	d	30	a	31	b	32	c	33	b	34	d	35	b
36	c	37	b	38	b	39	c	40	b	41	a	42	c
43	c	44	d	45	b	46	b	47	c	48	b	49	a
50	b	51	a	52	c	53	b	54	d	55	d	56	d
57	b	58	a	59	a	60	b	61	c	62	b	63	a
64	b	65	c	66	c	67	a	68	d	69	d	70	c
71	a	72	c	73	b	74	a	75	b	76	b	77	b
78	a	79	a	80	b	81	d	82	d	83	d	84	b
85	c	86	a	87	d	88	c	89	c	90	c	91	d
92	b	93	b	94	c	95	d	96	a	97	d	98	a
99	a	100	c										

Model Paper 16

1. Speed of bucket elevator ranges from
   a) 2.5 to 2.8 m/s   b) 2.5 to 4.0 m/s
   c) 3.5 to 6.0 m/s   d) 5.0 to 10.0 m/s
2. The fundamental principle of pressure nozzle is the conversion of pressure energy into
   a) Mechanical energy   b) Rotational energy
   c) Kinetic energy   d) Static energy
3. Which conveyer is used for transporting the large quantifier of materials over a very large distance at a low cost
   a) Belt conveyer   b) Chain conveyer
   c) Pneumatic conveyer   d) Screw conveyer
4. ……………………..conveyer is used to convey sticky materials.
   a) Belt conveyer   b) Chain conveyer
   c) Pneumatic conveyer   d) Screw conveyer
5. Selection of material handling system should be taken according to
   a) Conveying material   b) Capacity and speed of conveyer
   c) Stability of conveyer   d) All the above
6. The package containing number of secondary packages is called
   a) Primary   b) Secondary
   c) Tertiary   d) Quaternary
7. Most widely used biodegradable packaging material is
   a) Wheat gluten   b) Paper
   c) Corn zein   d) Starch
8. Packaging technique where in package and food are made sterile separately and packaged thereafter under sterile condition is
   a) Coning   b) MAP
   c) Aseptic packaging   d) CAP
9. Packaging a food serve the following
   a) Protects the food   b) Contain the food
   c) Enhance marketability   d) All the above
10. Which type of package is used for retailing a single unit of food product.
   a) Tertiary   b) Primary
   c) Secondary   d) None of above

11. A screw conveyer can be used up to a maximum of .... Inclination from horizontal.
    a) 20º b) 15º
    c) 10º d) 25º
12. The clearance between screw flight edges and through wall of a screw conveyer.
    a) Decreases with the length of screw.
    b) Increases with diameter of screw.
    c) Increases with length of screw.
    d) Remains same through the length of screw.
13. In belt conveyer, trough angle for paddy grain is
    a) 10º b) 20º
    c) 25º d) 30º
14. Major disadvantages of the bagged stored system is
    a) Flexibility of storage b) Law rodent loss potential
    c) Re- adsorption of moisture d) Rapid handling
15. Which gas would be produced when calcium carbide will be added to grain having some moisture?
    a) CO_2 b) Ethylene
    c) Acetylene d) O_2
16. Grain is spoiled in bin during warm season takes place due to moisture accumulation at
    a) Top b) Bottom
    c) Center d) Near wall of bin
17. Grain is spoiled in bin due to accommodation of
    a) Temperature b) Heat
    c) RH d) Moisture
18. The rate of respiration of paddy increases with increase in
    a) Temperature b) Moisture content
    c) Mass d) Both a and b
19. Zero energy cool chambers works on the principle of
    a) Latent heat of fusion b) Latent heat of evaporation
    c) Sensible heat of evaporation d) Both a and b
20. During storage, the lipase present in the rice bran hydrolysis into
    a) Free fatty acid b) Glycerol
    c) Amino acid d) Both a and b
21. By pulling a slight vacuum and replacing the package atmosphere with a desired mixture of CO_2, O_2 and N_2 is called as
    a) Passive MAP b) Active MAP
    c) Equilibrium MAP d) Both a and b

22. The grain are placed inside an airtight container which stops oxygen and wall movement between the outside atmosphere and the stored grain is called as
    a) MAP
    b) CAS
    c) Hermetic storage
    d) Hypobaric storage
23. Air tight storage is not recommended for grain with a moisture content of
    a) Less than 10%
    b) About 12%
    c) In between 10-13%
    d) Above 13%
24. In MAP of fruits and vegetables
    a) CO_2 and O_2 level increase.
    b) CO_2 level increase and O_2 level decrease.
    c) CO_2 level decrease and O_2 level increase.
    d) CO_2 and O_2 level remain constant.
25. Other factors remaining some, lowering storage temperature by 5 should
    a) Reduce the storage period by half
    b) Increase the storage period by 50 %
    c) Increase the storage period by a factor of two
    d) Not affect the storage period significantly
26. During winter reason, moisture accumulation and spoilage of grain take place at the
    a) Top of the bin
    b) Centre of bin
    c) Bottom of the bin
    d) None of these
27. Layout sheet for layout diagram is a diagrammatic representation of
    a) Intermediate occurrence
    b) Placer of occurrence
    c) Series of occurrence
    d) Processing occurrence
28. The design of building layout in a plant should provided for
    a) Adequate ventilation
    b) Flexibility of operation
    c) Adequate drainage of water
    d) All the above
29. Quality assurance in relation to food manufacturing ensures
    a) The food produced is nutritious
    b) Quality raw material is used
    c) Food meets specification and standards
    d) The food is produced at affordable cost
30. Which of the following statement is wrong?
    a) The ISO-9000 series is applicable only in case of food industries.
    b) The TQM process includes quality control, quality insurance and quality improvement.
    c) Monitoring of ccp. Comes under the purview of HACCP.
    d) Statistical analysis is an integral part of quality analysis of food products.
31. While laying out a food processing plant, one must consider
    a) Geographical location of site
    b) Provision for expression of plant
    c) Proximity of terminal market
    d) All the above

32. A hot air anemometer is used to measure.
    a) Air velocity
    b) Air temperature
    c) Air humidity
    d) All the above
33. The tube of the rotameter
    a) Tapers upwards
    b) Is of uniform cross section
    c) Tapers downwards
    d) Tapers in the horizontal direction
34. Selective radiation pyrometer is a temperature measuring device works on the principle of
    a) Planks law
    b) Newton law
    c) Coulombs law
    d) Kirchhoff's law
35. Fuzzy logic process is related to
    a) Mathematic model
    b) Linear model
    c) Artificial computer neural work
    d) Non linear model
36. In infrared pyrometer is used for measuring.
    a) Pressure
    b) Humidity
    c) Temperature
    d) None of above
37. An iron- content thermo couple is recommended to be used in the following temperature range
    a) 63 to 1473°C
    b) 223 to 2033°C
    c) 3 to 673°C
    d) 3 to 1273°C
38. Rotameter is a
    a) Dray flow meter
    b) Variable area flow meter
    c) Variable head flow meter
    d) Rotating propeller type flow meter
39. Diode in an electronic device allows the current to flow in
    a) Single direction
    b) Two direction
    c) Three direction
    d) Multiple direction
40. An element which sense and converts a desirable input into a usable form to be handled by the measurement system.
    a) Transducer system
    b) Data presentation element
    c) Signal conditioning system
    d) Auxiliary element
41. HACCP stands for
    a) Hazardous analysis of critical control point
    b) Hazardous analysis and critical control point
    c) Hazardous analysis critical control point
    d) Hazardous analysis by critical control point
42. ISO 9000 family of standard in related to
    a) Production system
    b) Quality management system
    c) Management system
    d) None of these

43. Break even chart are often used for determining the
    a) Impact of the rate of change in sales b) Revenue cost
    c) Cost volume profit d) A & C
44. The term for activity slack time is
    a) Free float b) Independent float
    c) Total float d) All the above
45. The shape of normal distribution curve is
    a) Rectangular b) Bell shaped
    c) Parabolic d) Hyperbolic
46. If any value in XB column of final simples table is negative, the solution is ....
    a) Infeasible b) Unbounded
    c) Bounded d) Non solution
47. The objective of network analysis in to
    a) Minimize total project duration
    b) Minimize total project cost
    c) Minimize total product delays, interception and conflicts
    d) Maximize total project duration
48. An assignment problem can be solved by
    a) Simples method b) Transportation method
    c) Dual simplex method d) Simplex & transportation method
49. In the network only one activity may can meet any ....number
    a) 1 b) 2
    c) 3 d) 4
50. If the constraint of on linear programming problem has an equation of less than or equal to type, the variable to be added are ....
    a) Slack b) Surplus
    c) Artificial d) Decision
51. If one or more variable vanish than a basic solution to the system in called
    a) Non feasible region b) Feasible region
    c) Degenerative solution d) Basic solution
52. When the total demand in equal to supply than the transportation problem is said to be
    a) Balanced b) Unbalanced
    c) Maximization d) Minimization
53. The coefficient of an artificial variable in the objective function of penalty method are always assured to be
    a) 0 b) 1
    c) -m d) m

54. A project consists of a number of tasks which are called ....
    a) Activities
    b) Events
    c) Dummy activity
    d) Successor
55. Which is employed in construction and business problem
    a) Network
    b) Critical
    c) Pert
    d) CPM
56. The role of artificial variable in simplex method is
    a) To aid in finding initial base feasible solution
    b) To start phase of simplex method
    c) To find shadow prices from final simplex method
    d) None of these
57. Paddy is also called
    a) Rough rice
    b) Broun rice
    c) Polished rice
    d) Rice
58. Among the cereal, the most ancient grain is
    a) Maize
    b) Rice
    c) Wheat
    d) Barley
59. In the maize, the toxin producing fungus is
    a) Aspergillums
    b) Fusarium
    c) Both a and b
    d) None of them
60. The largest among the cereal seeds is the
    a) Sweet corn
    b) Dent corn
    c) Jowar
    d) Paddy
61. Ergot is a kind of
    a) Bacteria
    b) Protozoa
    c) Fungus
    d) Algae
62. Ergot affects which parts of the kernel
    a) Head
    b) Endosperm
    c) Germ
    d) Ear
63. Triticale is hybrid cross of
    a) Wheat and Rye
    b) Sorghum and pearl millet
    c) Corn and wheat
    d) Rye and sorghum
64. Wheat bug affect which constituent of the kernel
    a) Starch
    b) Protein
    c) Germ
    d) Vitamin
65. The per capita availability of oil/fat in kg as compared to recommended 18 kg per year by WHO to India people is
    a) 10
    b) 5
    c) 15
    d) 20

66. Grain legumes are
    a) Oilseeds b) Pseudo-cereals
    c) Pulses d) Legumes
67. The important pseudo-cereals crops are that of
    a) Foxtail millet b) Barnyard millet
    c) Buck wheat d) Sorghum
68. Finger millet is also called
    a) Bajra b) Jawar
    c) Ragi d) Teff
69. Indian millet is also called
    a) Pearl millet b) Finger millet
    c) Proso millet d) Kodo millet
70. Which cereal crop in grouped under chenopod grains
    a) Grain amarnets b) Quinoa
    c) Canihua d) All of them
71. Size of an irregular shape object expression term of
    a) Sphere city b) Roundness
    c) Equivalent diameter d) None of these
72. Diffusion of sugar in jam is an example of
    a) Heat transfer b) Mass transfer
    c) Momentum transfer d) Both a and b
73. Which are more widely used device for cleaning and separation of seeds
    a) Graders b) Screens
    c) Driers d) None of these
74. Pulsed electric field is suitable for
    a) Liquids b) Semi-solids
    c) Solids d) Both a and b
75. Which of the following is used for transporting the large quantitier of materials over a very large distance at a low cost.........
    a) Chain conveyor b) Screw conveyor
    c) Belt conveyor d) Pneumatic conveyor
76. The chemical used to kill the rats and mice in called
    a) Fungicide b) Rodenticide
    c) Both a and b d) None
77. Functional layout is also called
    a) Process layout b) Product layout
    c) Plant layout d) All of these

78. Hydrometer works on the principle of
    a) Measuring the level of floating in the liquid in reference to water.
    b) Measuring the level of measuring of weight of known volume.
    c) Finding the volume of water displayed when known weight is lipped.
    d) Vibration is sent and resonant frequency is find out which is dependent on the density of liquid

79. How many acts are repealed by food safety and standard act-2006.
    a) 4 b) 5
    c) 7 d) 8

80. Which of the following is strongest emulsifiers
    a) Mono-glycerides b) Diglycerides
    c) Triglycerides d) Glycerol

81. The husk content of whole coconut is about
    a) 50% b) 60%
    c) 40% d) 70%

82. The fat level of oil seeds is in the range of
    a) 20-50% b) 10-20%
    c) 50-80% d) 20-30%

83. The problem associated with chenopod grain that it contains a poisonous glycosides called
    a) Tannins b) Flavanoids
    c) Apigenum d) Saponins

84. Protein content of buck wheat is in the range of
    a) 10-12% b) 6-8%
    c) 14-15% d) None

85. Grain shriveling is a major problem in
    a) Sorghum b) Barley
    c) Wheat d) Triticale

86. The common ways of screen agitation is
    a) Revolving a cylindrical screen about a horizontal axis.
    b) Revolving a cylindrical screen about a vertical axis.
    c) By using stirrers
    d) None of these

87. Screen are generally constructed by perforated.
    a) Strain less steel b) Sheet metal
    c) Aluminum sheet d) Plastic sheet

88. The white wheat flour is deficient in
    a) Protein b) Starch
    c) Vitamins and minerals d) None

89. The of African people whose staple diet is pearl millet have problem of which disease
   a) Goiter   b) Anemia
   c) Scurvy   d) Flatulence
90. The important famine food is
   a) Maize   b) Wheat
   c) Sorghum   d) Finger millet
91. Brain is a excellent source of
   a) Thermino   b) Vitamin A
   c) Vitamin E   d) Both a and b
92. The vitamins and mineral generally used for enrichment of rice grain are
   a) B1 and iron   b) B2 and niacin
   c) Ca and Fe   d) All of them
93. The rancidity of parboiled rice increase during storage as compared to raw rice due to
   a) Increase in fee fatty acids
   b) Destruction of natural anti oxidant
   c) Increase in unsaturation
   d) None of these
94. The disadvantage of parboiled rice is
   a) Dark colour   b) Different flavor
   c) Prime to rancidity   d) All of them
95. After enrichment with vitamin, the rice is dusted with
   a) Ferric pyrophosphate   b) Talk
   c) Calcium   d) Both a and b
96. Flour of which cereals is used as antioxidants in food system
   a) Rye   b) Oat
   c) Sorghum   d) Millets
97. Cholesterol is absent in
   a) Lard   b) Tallow
   c) Plant oil   d) Fish oil
98. Toxins present in khesari dhal which causes paralysis can be removed by
   a) Pulse drying   b) Pulse milling
   c) Soaking in hot water   d) All of them
99. The group of anti-nutritional, sulphur- containing glucosinolates is present in which oil.
   a) Groundnut   b) Soybean oil
   c) Coconut   d) Rape mustard

100. The toxin gossypol is present in which oil bearing seeds
   a) Coconut b) Cottonseed
   c) Sunflower d) Castor oil

Answer Key

1	b	2	b	3	a	4	d	5	d	6	c	7	b
8	c	9	c	10	b	11	a	12	a	13	b	14	c
15	c	16	b	17	d	18	d	19	b	20	d	21	b
22	c	23	d	24	b	25	d	26	a	27	b	28	d
29	b	30	a	31	c	32	a	33	c	34	a	35	c
36	c	37	a	38	b	39	a	40	a	41	b	42	b
43	d	44	d	45	b	46	a	47	a	48	d	49	b
50	a	51	c	52	a	53	c	54	a	55	c	56	a
57	a	58	d	59	c	60	b	61	c	62	a	63	a
64	b	65	b	66	c	67	c	68	c	69	d	70	d
71	c	72	c	73	b	74	b	75	c	76	b	77	a
78	d	79	d	80	a	81	a	82	a	83	d	84	a
85	d	86	a	87	b	88	c	89	a	90	d	91	a
92	d	93	b	94	d	95	d	96	b	97	c	98	c
99	d	100	b										

Model Paper 17

1. Euricic acid containing group mainly contains which oil
   a) Cocoa butter
   b) Groundnut oil
   c) Rapeseed /mustard oil
   d) Sesame oil
2. Other name of lecithin is
   a) Phosphatidyl choline
   b) Phosphatidic acid
   c) Phosphatidyl ethyl amine
   d) Phosphatidyl alcohol
3. Denaturation of proteins means
   a) Loss of primary structure
   b) Loss of three dimensional structures
   c) De-polymerization
   d) Copulation
4. Caramelization takes place due to
   a) Burning of sugars
   b) Burning of starch
   c) Enzymatic browning
   d) Burning of proteins
5. In the hydrolysis of starch, the percentage of glycosidic linkages that are broken down is termed as
   a) Degree of hydrolysis
   b) Dextrose equivalent
   c) Degree of de-branching
   d) Percentage break down
6. Maillard reactions includes
   a) Reaction of amino acids with reducing sugars
   b) Reaction of sugars and organic acids
   c) Organic acid themselves
   d) All of them
7. In chemically modified starch, the abbreviation 'DS' is termed as
   a) Digestibility standard
   b) Degree of substitution
   c) Degree of solubility
   d) Dissolved starch
8. Hemi- cell houses are also termed as
   a) Pentosans
   b) Lignans
   c) β - glucans
   d) D - xylans
9. The storage proteins of cereals are
   a) Albumins and globulins
   b) Globulins and glutelins
   c) Glutelins and polyamines
   d) Prolamines and albumins

10. Diastase include
    a) ∝ - amylase
    b) β- amylase
    c) Both a and b
    d) ISO- amylase
11. Which of the following in an example for prolate shape.
    a) Lemon
    b) Carrots
    c) Grape fruits
    d) All the above
12. The device which is highly sensitive and specially designed for measuring spectral transmittance character of agricultural material is.
    a) Radiometer
    b) Rotameter
    c) Spectrophotometer
    d) All the above
13. Which of the following method is used to measure density of silage without removing sample
    a) Pycnometer
    b) Radiation method
    c) Displacement method
    d) All the above
14. The ratio of change in length to the original length is referred as
    a) Stress
    b) Strain
    c) Surface tension
    d) Viscosity
15. The portion of stress-strain curve in which the material is capable of sustaining without any permanent strain remaining upon complete release of the stress is called
    a) Proportional limit
    b) Modules of rigidity
    c) Elastic limit
    d) None of these
16. The stress corresponding to the rupture point is knows as
    a) Ultimate strength
    b) Tensile strength
    c) Compressive strength
    d) None
17. Deformation with tine when the material is suddenly subjected to a dead load, it is known as.
    a) Creep
    b) Viscosity
    c) Rigidity
    d) None of these
18. Three fundamental properties of material which describes the rheological behavior are.
    a) Elasticity, plasticity, viscosity
    b) Deformation, flow, viscosity
    c) Elasticity, time, stress
    d) Stress, strain, tine
19. Generalized Maxwell model usually represents
    a) Stress retardation
    b) Stress relaxation
    c) Creep
    d) None of these
20. The porosity of paddy varies from
    a) 40-45%
    b) 50-55%
    c) 48-50%
    d) 65-70%

21. In which of the following mode of heat transfer, the heart transfer cause by moment of individual molecular

a) Radiation
b) Coeducation
c) Convection
d) All of these

22. The resistance to heat transfer in a flat stab can be calculated

a) $R = \frac{T1 - T2}{\left(\frac{x2 - x1}{KA}\right)}$
b) $R = \frac{KA}{\Delta X}$
c) $R = \frac{\Delta X}{KA}$
d) $R = \frac{x2 - x1}{KA}$

23. The fraction of incident radiation absorbed by material is called.

a) Reflectivity
b) Absorptivity
c) Transmissivity
d) None

24. The value of steffen- Boltzmann constant is

a) $5.67x10^{-8}$ J/m^2k^4
b) $5.67x10^{-10}$ w/m^2k^4
c) $5.67x10^{-10}$ J/m^2k^4
d) $5.67x10^{-8}$ w/m^2k^4

25. Expand term 'NTU' in heat exchanger is

a) Number of temperature units
b) Number of transfer units
c) Number of thousand units
d) None of these

26. The minimum clear space between the edges of the opening in the screen surface is

a) Aperture
b) Open area
c) Mesh
d) None of these

27. Which of the following is not a type of opening is case of wire mesh

a) Square
b) Rectangle
c) Triangle
d) None of these

28. The higher percentage of open area in screen increase

a) Capacity
b) Efficiency
c) Effectiveness
d) Both a and b

29. In the disc separator, on the surface of the disc .... are present.

a) Indentation
b) Cups
c) Buckets
d) None of these

30. Which of the following separator is used for the cleaning of lighter seeds like cabbage, radish, lettuce, carrots

a) Pneumatic separator
b) Fluidized bed separator
c) Velvet roll separator
d) All of these

31. The ratio of absolute air humidity with respect to the absolute air humidity at saturation point is called.

a) Absolute humidity
b) Percentage humidity
c) Relative humidity
d) Both (a) and (c)

32. During sensible cooling of air, the enthalpy

a) Increase
b) Decrease
c) Remains constant
d) None

33. Universal moisture meter gives moisture content on .... basis.
    a) % db
    b) % wb
    c) Both (a)and (b)
    d) None of these
34. Which of the following represent Henderson equation
    a) 1-RH = exp (-CTMe^n)
    b) 1-Me = exp (-CTMe)
    c) 1-RH = - CTMe^n
    d) None
35. In which of the following drier, the heat energy can be supplied to wet products by electromagnetic waver
    a) Radiation drying
    b) Fluidized bed drying
    c) Desiccated air drying
    d) All the above
36. PHTC dryer developed at
    a) IIT Kharagpur
    b) IARI, New Delhi
    c) CIAE Bhopal
    d) CFTRI, Mysore
37. Who explained that hysteresis effect during moisture sorption in grains is due to molecular shrinkages and development of cracks and fissures.
    a) Kraemer and taylor
    b) Chong and pfost
    c) Henderson
    d) Young and nelson
38. Which of the following is not a commercial grain dryer?
    a) Bin dryer
    b) Tunnel dryer
    c) Tray dryer
    d) LSU dryer
39. Microwave oven as dryer is an example of
    a) Vacuum drying
    b) Dielectric drying
    c) Radiation drying
    d) Freeze drying
40. The drying time in falling rate period is .... wet bulb depression.
    a) Directly proportional to
    b) Inversely proportional to
    c) Not effected by
    d) Either (a) or (b)
41. Energy required for removal of bound water from the material is .... free moisture.
    a) More than
    b) Half of the
    c) Equal to
    d) 10 times higher than
42. When moisture is added to air constant dry bulb temperature, the process is known as
    a) Dehumidification
    b) Humidification
    c) Sensible cooling
    d) None
43. Rotary dryer is normally inclined at an angle with horizontal is.
    a) 5°
    b) 15°
    c) 25°
    d) 30°
44. The blade of turbine agitators are generally .... than paddle agitators.
    a) Smaller
    b) Larger
    c) Both (a) and (b)
    d) None of above

45. The membrane material used in micro filtration is.
   a) Poly sulfone b) Poly propylene
   c) Poly amide d) Poly acrylonite
46. A cyclone separator is used for separating
   a) Particles from liquids b) Liquid droplets from gases
   c) Fine particles from solids d) All of the above
47. Which of the following is also known as deep bed filters ?
   a) Clarifying filter b) Cake filter
   c) Cross flow filter d) None of above
48. If the size of ground material is 2mm, in which range it will fall?
   a) Dimension range b) Sieve range
   c) Microscopic range
d) Non of above
49. Which law gives good results for fine grinding?
   a) Kick's law b) Ratinger's law
   c) Bond's law d) All the above
50. All pulses are rich in
   a) Protein b) Carbohydrates
   c) Fats d) All the above
51. Ball mills are used for
   a) Abrasive materials b) Spices
   c) Fruits and vegetable d) All of above
52. Separation of oils from seeds by applying pressure is called
   a) Extraction b) Leaching
   c) Expression d) All of these
53. Rice milling industries regulation act 1959, has banned ....for shelling paddy.
   a) Engle berg huller b) Under runner disc huller
   c) Centrifugal huller d) Rubber roll Sheller
54. In rice milling process, the milled whole rice of 4/8and more of actual kernel size is called.
   a) Head rice b) Big broken
   c) Small broken d) Total rice
55. After pretreatment with oil, pulses are kept on floors to diffuse the oil for about.
   a) 6 h b) 9 h
   c) 12 h d) 15 h
56. The shape of steel flanks in compartment type paddy separators is
   a) S - shaped b) Z - shaped
   c) D – shaped d) All of these

57. The protein efficiency ratio of pulses are
   a) 0.1 - 0.7 b) 0.7 - 2.1
   c) 2.1 - 3.6 d) 3.6 - 4.8

58. The higher degree of gelatinization mainly reflects on
   a) Hardness b) pH
   c) sugar content d) none of these

59. Differential speed of break roll (serrated rolls) in wheat milling is
   a) 1:2.5 b) 1:1.25
   c) 1:1.35 d) 1:1.5

60. Which of the following is based for expelling cotton seed oil
   a) n- Haxane b) Hydraulic acid
   c) Leaching equipment d) All of these

61. During the freezing process, the water molecules tend to form group before the change of phase which is known as
   a) Super cooling b) Nucleation
   c) Plank's behavior d) Thawing

62. In multiple effect evaporator system, the basic objective is to improve
   a) Product quality b) Through capacity
   c) Economy of operation d) None of these

63. Viscosity of the milk at normal temperature is
   a) 0 - 1 Cp b) 1 - 1.5 Cp
   c) 1.5 - 2 Cp d) 72 Cp

64. Disc bowl centrifuge is used for
   a) Homogenization b) Cram separation
   c) Pasteurization d) Expression

65. The transformation of product during extrusion is
   a) Irreversible b) Reversible
   c) Both a and b d) None of these

66. The classification of cleaned product into various quality fractions depending upon commercial values.
   a) Sorting b) Grading
   c) Scalping d) Cleaning

67. The medium dose of irradiation is as follows
   a) < 1kgy b) 1-10kgy
   c) > 10kgy d) None of these

68. Vegetative microbial cells deactivated in HPP by
   a) Breaking non-covalent bonds b) Causing damage to cell membrane
   c) Both a and b d) None of these

69. The technique used for measuring the dose known as
   a) Dosimetry b) Radiometry
   c) Refractometry d) Spectrophotometry

70. Microwave cooking oven saves up to ....of energy as compared to conventional electric stove.

a) 50% b) 75%

c) 10% d) 25%

71. In India the frequency of microwave oven used in commercial appliances is

a) 896 MHz b) 915 MHz

c) 1650 MHz d) 2450 MHz

72. D value depends on

a) Number of microorganism present in food

b) Time of heating

c) Temperature of heating

d) None of above

73. Adequate heating of milk is indicated by

a) Phosphates test b) Boudoir test

c) Hansa test d) None

74. One I.U. of vitamin A is equal to

a) 0.3 mg of vitamin A acetate b) 0.6 mg of vitamin A acetate

c) 0.9 mg of vitamin A acetate d) None of these

75. The B.R. reading of animal body fat, range from

a) 40 to 50 b) 50 to 60

c) 44 to 51 d) None

76. Zero point is

a) Freezing point of milk b) Freezing point of water

c) Freezing point of ice cream d) None of these

77. Which of the following is known as spoon eating milk?

a) Toned milk b) Standard milk

c) Tattoo milk d) Recombined milk

78. Fat in separated milk is not more than

a) 0.1% b) 0.5%

c) 1.0% d) 2.0%

79. Plate evaporator is not suitable for

a) Milk b) Fruit juice

c) Viscous fluid d) Dilute

80. Sanitization is a process by which

a) The microbial load on objects is reduced

b) Objects are made sterile with chemicals

c) Utensils are scrubbed

d) Skin is described

81. Fruit juice is spoiled by
    a) Yeast and molds
    b) Clostridia
    c) Enterobacteria
    d) Enterococci
82. The composition of capsule of bacteria is of
    a) Fatty acid
    b) Pectin
    c) Chitin
    d) Cellulose
83. The colour of spores is Wirtz method is
    a) Red
    b) Green
    c) Pink
    d) Blue
84. Z value is denoted in the form of
    a) Minute
    b) Number
    c) Degree Celsius
    d) Log number
85. Total plate count method is being expensed is
    a) c.f.u./ml
    b) Number ell/ ml
    c) Optical demit
    d) All the above
86. Ropiness is bread is caused by
    a) *Bacillus subtilis*
    b) *Bacillus licheniformis*
    c) *Aspergillus niger*
    d) *Saccharomyces cerevisiae*
87. Softness of pickles is due to
    a) Fusarium
    b) Bacillus
    c) Penicillium
    d) Pseudomonas
88. The fishy flavor in fish in due to
    a) Diacetyl
    b) Methyl are
    c) EDTA
    d) Trimethyl amine
89. Canned meat is generally spoiled by
    a) Clostridiums
    b) Escherichia
    c) Pseudomonas
    d) Salmonella
90. *Coxiella brunetti* is responsible for
    a) Tuberculosis
    b) fever
    c) Cerebral malaria
    d) All the above
91. Sweet curdling in milk is caused by
    a) Alcaligens viscolactis
    b) Enterobacter aerogens
    c) Callus cereus
    d) Pseudomonas fluorescens
92. Rusty spot in cheese in due to
    a) L. Bulgarian
    b) L. plantarum
    c) B. subtilis
    d) Leuconotoc mesanteroides
93. Action of food acids on the can result in a
    a) Oxygen swell
    b) Hydrogen swell
    c) Soft swell
    d) Sulphur swell

94. Oil that contains linolenic acid are subjected to flavor
   a) Putrefaction   b) Enhancement
   c) Reversion   d) Suppression

95. What is ale?
   a) Type of beer   b) Fermented corn
   c) Fermented cabbage   d) Fermented carrot

96. Sarcina sickness is the defect of
   a) Wine   b) Sauerkraut
   c) Biscuit   d) Beer

97. Hopi are used in the making of
   a) Wine   b) Beer
   c) Jam   d) Pickles

98. What is obtained during the manufacturing of
   a) Jelly   b) Beer
   c) Meso   d) Tempeh

99. Presence of sterol in animal is generally known as
   a) Cholesterol   b) Phytosterol
   c) Ergo sterol   d) None

100. Milli moles of free fatty acids per 100g fat is known as
   a) Acid value   b) Polenske value
   c) RM value   d) BR reading

Answer Key

1	c	2	a	3	b	4	a	5	b	6	d	7	b
8	a	9	c	10	c	11	c	12	c	13	b	14	b
15	c	16	a	17	a	18	a	19	b	20	c	21	c
22	c	23	b	24	d	25	b	26	a	27	c	28	d
29	a	30	b	31	b	32	a	33	b	34	a	35	a
36	a	37	b	38	c	39	b	40	b	41	a	42	b
43	a	44	a	45	b	46	c	47	a	48	b	49	b
50	a	51	a	52	c	53	a	54	b	55	c	56	b
57	b	58	a	59	a	60	b	61	b	62	c	63	c
64	b	65	a	66	b	67	b	68	c	69	a	70	b
71	d	72	c	73	a	74	a	75	c	76	b	77	c
78	b	79	c	80	a	81	a	82	d	83	b	84	c
85	a	86	b	87	c	88	d	89	a	90	b	91	c
92	b	93	b	94	c	95	a	96	d	97	b	98	b
99	a	100	a										

Model Paper 18

1. In tea, fermentation, the micro organisms involved in
   a) Fusarium b) Aspergillus
   c) Alcaligens d) None
2. The content of CO_2 in beer is about
   a) 0.02 – 0.06 % b) 1.78 – 2.01 %
   c) 0.45 – 0.52 % d) 0.60 – 0.92 %
3. Meat is irradiated to present the outbreak of
   a) Shigella b) Salmonellosis
   c) T.B. d) Fever
4. Egg quality is manually checked by seeing it through a temperature it is called
   a) Lightening b) Lamping
   c) Sorting d) Candling
5. Which of the following spoilage is seen is meat by under anaerobic condition
   a) Stickiness b) Whiskers
   c) Surface slime d) Putrefaction
6. Green mold growth in fruits is due to
   a) Cladosporium b) A. niger
   c) Botrytis cineraria d) None
7. Undesirable odor in sauerkraut is due to the action of which of the following microorganisms?
   a) Pediococcus cerevisae b) Sacchromyce cerevisae
   c) A. Niger d) All the above
8. Molds are most common causes of spoilage of fruits and vegetables because
   a) They are somewhat acidic b) Fairly dry at surface
   c) Deficient in vitamin d) All the above
9. Generation time of E. coli (in min) in milk and at 37 is
   a) 6 b) 12
   c) 24 d) 36
10. D value is indicated in
    a) Minute b) Log number
    c) Degree Fahrenheit d) Degree centigrade

11. Which of the following is revolving screen

a) Grizzly | b) Shaking
c) Trammel | d) All

12. Use of filter aid is to

a) Increase filtering efficiency | b) Decrease the filtering efficiency
c) To give body to the filtrate | d) To increased the mass of cake

13. Which of the following is power number?

a) $ND^2_a\ \rho/m$ | b) N^2D_a/ρ
c) $Pg_c/N^3D^5_a\rho$ | d) NDP_2/ρ

14. Tomato ketchup is a good example of

a) Newtonian fluid | b) Non- Newtonian fluid
c) Pseudo plastic | d) Both (b) and (c)

15. One atmosphere is equal to

a) $1kg/m^2$ | b) $1kg/cm^2$
c) $1gm/m^2$ | d) $1g/cm^2$

16. Measurement of energy value of food is called

a) Calorimetry | b) Joulimetry
c) Energymetry | d) None

17. Which of the following protein is of high biological value?

a) Soy protein | b) Legume protein
c) Cereal protein | d) Meat protein

18. Legumes are deficient in

a) Methionine | b) Phenylalanine
c) Valine | d) Lysine

19. Chromium is required for

a) Protein metabolism | b) glucose metabolism
c) Fatty acid metabolism | d) Blood clotting metabolism

20. Product of glycolysis is

a) ATP | b) ADP
c) CO_2 | d) NADH

21. Elisa test in used to

a) Separate viral RNA | b) Purity
c) Protein | d) Isolated DNA

22. Aflatoxin M is found in?

a) Groundnut | b) Wheat
c) Milk | d) Soybean

23. Bitterness of coffee is attributed to

a) Caffeine | b) Chlorogenic acid
c) Insulin | d) None

24. The off flavor developed in refined oil before onset of rancidity is called as
    a) Hydrogenation b) Pre-rancidity
    c) Trans-etherification d) All the above
25. Phospho protein in
    a) Zein b) Lacto albumin
    c) Casein d) Lacto globulin
26. Class I preservative is
    a) Common salt b) Benzoic acid
    c) Sodium nitrate d) Calcium propionate
27. The end product of maillard browning is
    a) Lysine b) Caramel
    c) Furfural d) Melanoidins
28. Staling defect in bread is caused due to
    a) Caramelization b) Thermization
    c) Retro gradation phenomenon d) Fermentation
29. In gas choromatography (Ge), the basis for separation of the components of the volatile material in the deference in
    a) Partition coefficients b) Conductivity
    c) Molecular weight d) Morality
30. Karl fisher titration method is used for .... analysis in liquid milk
    a) Protein b) Fat
    c) Moisture d) Lactose
31. The most common column used in HPLC
    a) C15 b) C16
    c) C17 d) C18
32. A value at which separate acceptability from unacceptability
    a) Corrective action b) Hazard
    c) Deviation d) Critical limit
33. Failure to meet critical limit
    a) Corrective action b) Hazard
    c) Deviation d) Critical limit
34. Relating to like or dislike
    a) Bias b) Hedonic
    c) Preference d) Discrimination
35. Systematic error which may be negative or positive ....
    a) Bias b) Hedonic
    c) Preference d) Discrimination
36. Lactometer works on principle of
    a) Newton's law b) Charley low
    c) Archimedes principle d) Boyle's low

37. The milk product which exhibited the apparent viscosity is
    a) Ice cream b) Skim milk
    c) Whole milk d) Cheese
38. Detector used in HPLS is
    a) UV absorbance detector b) Electrochemical detector
    c) Refractive index detector d) All the above
39. In GLC, capillary columns are made of
    a) Glass b) Aluminum
    c) Both a and b d) Fused silica
40. Methyl orange works is pH range
    a) 3.1 - 6.3 b) 4.2 - 6.4
    c) 3.1 - 4.4 d) 4.2 - 8.0
41. Instant starches are of the type
    a) Pre- gelatinized b) Hydrolyzed
    c) Enzyme treated d) All of these
42. Crude fibers includes
    a) Cellulose b) Lignin
    c) Hemi-cellulose d) All of them
43. The phosphorous present in cereal starches is in the form of
    a) Phosphorus b) Phosphates
    c) Phospholipids d) Phosphides
44. Gelatinization of starch is due to
    a) Loss of birefringence b) Swelling of grounds
    c) De-polymerization d) Crystallization
45. LDL stands for
    a) Low dietary lipids b) Low density lipoprotein
    c) Least dense lipids d) Low dietary levels
46. BHT an antioxidant stands for
    a) Butylated hydroxyl tri-chloride b) Butylated hydroxyl toluene
    c) Benzene hydroxyl tri-chloride d) Benzene hydroxyl toluene
47. An omega six (omega-6) fatty acid is
    a) Linoleic b) Oleic
    c) Linolenic d) Iso-oleic
48. The polymer largely soluble in water is
    a) Amylose b) Amylopectin
    c) Phyto glycogen d) Starch
49. Clathrates are the complex formed by the polymer
    a) Amylose b) Amylopectin
    c) Cellulose d) β– glucans

50. Winterization is
 a) Removing long chain saturated fatty acids from oils
 b) Plasticizing solids fats
 c) Ageing of oils
 d) Removing volatiles from oil
51. The coefficient of friction between the grain depends upon its
 a) Shape b) Surface characteristics
 c) Moisture content d) All the above
52. At equilibrium moisture content (EMC), the gain or loss of moisture is
 a) High b) Low
 c) Zero d) Normal
53. Drying mode commonly used in all types of cereal grain is
 a) Radiation b) Conduction
 c) Convection d) Vacuum
54. When the unheated air supplied by the nature is utilized for the drying of the wet material, the phenomenon is known as
 a) Supplemented heat drying b) Ordinary drying
 c) Natural air drying d) Sprouted bed drying
55. Movement of moisture along the direction of heat flow is called
 a) Thermal- moisture conductivity b) Thermal conductivity
 c) Moisture evaporation d) Convection
56. The impurities first to remove from the cereal grain during cleaning is
 a) Metals b) Stones
 c) Unwanted species other than cereals d) Low density impurities
57. The impurities of lowest terminal velocity is of the type
 a) Spherical b) Cubical
 c) Diffused flaked d) None of them
58. Cleaning of cereal by aspiration is based on
 a) Aerodynamic properties b) Hydrodynamic properties
 c) Magnetic properties d) Thermal properties
59. The hazards, affecting cereal grains during storage include
 a) Biochemical deterioration b) Mechanical damage
 c) Both a and b d) None of them
60. The grain to be stored are passed through into water to get freedom from
 a) Inseets b) Rodents
 c) Moisture d) Pathogens
61. Terminal velocity of a particle is directly proportional to
 a) Diameter b) Diameter square
 c) Square root of diameter d) Diameter cube

62. Which property is important in development of milling machinery?
    a) Physical b) Mechanical
    c) Aerodynamic d) Thermal
63. Colour of the horticultural crop in the following characteristics
    a) Quantitative b) Hidden
    c) Sensory d) Specific gravity
64. Fluidity is the reciprocal of ....
    a) Density b) Porosity
    c) Viscosity d) Specific gravity
65. The young's modules (E), the shear modules (G), and the poison's ratio (M) are related by
    a) E= 2G (1-M) b) E= 2G (1+2M)
    c) E= 2G (1+M) d) None
66. Strain in a direction at right angle to the direction of applied force is known as
    a) Lateral strain b) Volumetric strain
    c) Shear stain d) None
67. Simple stress is also called as
    a) Direct stress b) Transverse stress
    c) Total stress d) Any of these
68. The combined effect of external forces acting on a body is called
    a) Stress b) Strain
    c) Load d) None
69. An empirical equation to determine surface of the food is as follows
    a) $S= k/w^3$ b) $S= kw^m$
    c) $S= kw^{1/3}$ d) $S= k/w$
70. Firmness of fruits relates
    a) Force and depth of penetration b) Force and time of application
    c) Depth of penetration and time d) All the above
71. The product of Reynolds number and prandtl number is known as
    a) Stanton number b) Peclet number
    c) Biot number d) Grashoff number
72. Kelvin – plank's law deals with
    a) Conversion of work into heat b) Conservation of work
    c) Conversion of heat to work d) Conservation of heat
73. Heat is closely related with
    a) Entropy b) Temperature
    c) Liquids d) Energy
74. Thermal diffusivity is a
    a) Dimensionless parameter b) Function of temperature
    c) Physical property of a substance d) All of the above

75. Fick's law is used in
   a) Heat transfer b) Mass transfer
   c) Momentum transfer d) None of these
76. Separation efficiency of cyclone separates is
   a) Proportional to length of cyclone b) Proportional to velocity
   c) Inversely proportion to R d) All of these
77. Ribbon blenders are used
   a) Liquid b) Powder
   c) Dough and paste d) All the above
78. Mixers for high viscosity liquids and pates are
   a) Z blade (or sigma blade) mixer b) Planetary mixer
   c) Anchor and gate agitators d) All the above
79. Fluidized bed dryer is used for drying
   a) Milk b) Tea leaver
   c) Fruit juices d) Vegetables
80. Evaporators as compared to dryers are
   a) Less efficient b) More efficient
   c) Effect is same d) None of these
81. The process of exposing food to gamma or x-rays is called
   a) HPP b) Irradiation
   c) UV Technology d) PEF
82. The main limitation of pulsed light technology
   a) Penetration depth b) Side effects
   c) Heavy cost d) None
83. Ohmic heating works on the principle of
   a) Mechanical resistance b) Electrical resistance
   c) Electrical conductance d) Thermal resistance
84. Pulsed electrical field is applied for the pasteurization of
   a) Solid foods b) Pumpable liquid
   c) Jellies d) All the above
85. NIR spectroscopy works on the principle of
   a) Beer lambert law b) Fick's law
   c) Plank's law d) Faraday's law
86. Floor area required to dry one tone of grains under sun is
   a) 5 m^2 b) 10 m^2
   c) 15 m^2 d) 20 m^2
87. Missing index (MI) is gives by
   a) $(S_0 - S)^2 / (S - S_r)^2$ b) $(S_0^2 - S^2) / (S_0^2 - S_r^2)$
   c) $(S^2 - S_0^2) / (S_0^2 - S_r^2)$ d) $(S_0^2 - S_r^2) / (S^2 - S_0^2)$

88. Fineness modulus can be estimated by
   a) $Dp= 0.135(Fm)^{1.366}$
   b) $Dp= 1.36(Fm)^{0.135}$
   c) $Dp= 0.135(1.36)^{Fm}$
   d) $Dp= 0.135(1.366)^{Fm}$
89. The idler spacing in belt conveyor should not exceeds
   a) 0.6m
   b) 1.2m
   c) 2.4m
   d) 1.8m
90. The residence time in a long tube evaporator is unusually
   a) 1 – 2 sec
   b) 5 – 30 sec
   c) 2 – 5 min
   d) 5 – 10 min
91. The enzyme playing central role in black tea processing in
   a) Protease
   b) Polyphenal oxidase
   c) Papain
   d) Phosphatidase
92. Vanilla flavor is naturally obtained by curing of which part of the plant?
   a) Seed
   b) Pod
   c) Leaves
   d) Buds
93. Asafoetida is
   a) Root extrudate
   b) Leaf
   c) Seed
   d) Petal
94. The chemicals used in aseptic packaging of food are
   a) Sulphur, per acetic acid and ethylene oxide
   b) H_2O_2, per acetic acid, ethylene oxide
   c) H_2O_2, per acetic acid, and sulphur
   d) H_2O_2, sulphur and ethylene oxide
95. Mode in expeller (oil) operate at pressure of
   a) 16 atm
   b) 160 atm
   c) 1600 atm
   d) 16000 atm
96. Extraction of aromatic compound into fat is called
   a) Expression
   b) Super critical extraction
   c) Enfluerage
   d) Soxhlet extraction
97. Poor quality egg floats in the water due to
   a) Microbial spoilage
   b) Increase in air cell
   c) Decrease in air cell
   d) All the above
98. The superior method of slaughter of meat animal as far as efficacy of bleeding in considered is
   a) Jhatka method
   b) Hold method
   c) Kosher killing
   d) None
99. Choose the pair from following which is not analogous
   a) Paneer: Cheese
   b) Ice-cream: Kulfi
   c) Butter oil: Ghee
   d) Yoghurt : Lassi

100. Maximum solubility of sugar at 20ºC in equal to

a) 43.2% b) 67.2%

c) 75.6% d) 82%

Answer Key

1	d	2	c	3	b	4	d	5	d	6	a	7	a
8	d	9	b	10	a	11	c	12	a	13	c	14	d
15	b	16	a	17	d	18	a	19	c	20	a	21	d
22	c	23	a	24	d	25	c	26	a	27	d	28	c
29	a	30	c	31	d	32	d	33	c	34	b	35	a
36	c	37	a	38	d	39	d	40	c	41	a	42	d
43	c	44	a	45	b	46	b	47	a	48	c	49	a
50	a	51	d	52	c	53	c	54	c	55	a	56	b
57	c	58	a	59	c	60	c	61	b	62	b	63	c
64	c	65	c	66	a	67	a	68	c	69	b	70	a
71	b	72	c	73	b	74	c	75	b	76	d	77	b
78	d	79	b	80	b	81	b	82	a	83	b	84	b
85	a	86	c	87	b	88	d	89	b	90	b	91	b
92	b	93	a	94	b	95	c	96	c	97	b	98	b
99	d	100	b										

Model Paper 19

1. Pressurized packed foods are called
   a) Baro foods b) Aerosols
   c) Aceituno d) Barges
2. Fo value is F value when Z is equal to
   a) 10 b) 12
   c) 16 d) 18
3. Dole process is the example of
   a) Heat- cool- fill b) Bold sterilization
   c) Batch pasteurization d) Fast canning
4. Mustard oil is effective against
   a) *Saccharomyes cerevisae* b) *Penicillium notatum*
   c) *Clostridium botulinum* d) All the above
5. Storage of food under reduced pressure is called
   a) Aseptic packaging b) Hyperbaric storage
   c) Hypobaric storage d) Gas packaging
6. Which preservative is used in tomato sauce to extend the shelf life
   a) Sodium benzoate b) Calcium propionate
   c) Tocopherol d) All the above
7. Sorbic acid is mostly used a preservative in which of the following food products?
   a) Meat b) Milk
   c) Baked product d) All the above
8. Which salt of saccharine is mainly used as sweetener?
   a) Calcium b) Potassium
   c) Magnesium d) Sodium
9. Vanilla can be persevered by using
   a) Sugar b) Alcohol
   c) Salt d) None
10. Sugar more than .... acts as preservatives
    a) 10% b) 40%
    c) 70% d) 95%

11. Wheat bran contains which of the following anti-vitamin compound
    a) Avidin   b) Hypoglycin
    c) Linatin   d) Niacytin
12. Which of the following process is responsible for the staling of bread?
    a) Gelatinization   b) Retro gradation
    c) Hydrolysis of starch   d) All the above
13. Semolina is obtained form
    a) Hard wheat   b) Soft wheat
    c) Denim wheat   d) Any of the above
14. Whole wheat flour is also known as
    a) Perfect flour   b) Straight flour
    c) Graham flour   d) Complete flour
15. Textured vegetable protein is prepared from which type of extruder?
    a) Collect extruder   b) Pasta press
    c) High shear cooking extruder   d) Low shear cooking extruder
16. The white protein which is seen after slicing of brad is called
    a) Crumb   b) Crust
    c) Cells   d) None
17. Which of the following can be used as the anti-staling in bread?
    a) GMS   b) GSM
    c) MSG   d) SGM
18. Common fumigant used for food grain is
    a) Phosgene   b) Chloroform
    c) Ethylene bromide   d) Carbon man oxide
19. Damaged starch in bread flour should be
    a) 2%   b) 9%
    c) 17%   d) 23%
20. Shortenings
    a) Impart colour of brad   b) Provide soft texture to bread
    c) Aerate dough   d) Increase the strength of bread
21. Food processing wastes contain
    a) High bio oxygen chemical content   b) Low BOD
    c) Low suspended solids   d) High BOD
22. Break even analysis is the
    a) Point of maximum profit   b) Point of minimum profit
    c) Point of zero profit   d) Point of medium profit
23. Iodophors are used is
    a) Disinfectant   b) Water conditioner
    c) Thermal disinfectant   d) Chelating agent

24. Temporary hardness is due to
    a) Sulphates b) Carbonates
    c) Nitrates d) Sulphites
25. Trickling bed is terminology used in
    a) Dehydration b) Food waste treatment
    c) Evaporation d) Distillation
26. What should be the TSS of a tomato sauce?
    a) 15% b) 20%
    c) 10% d) 5%
27. The purpose of sauce is to
    a) Enhance food's flavor b) Enhance food's texture
    c) Enhance food's appearance d) All of the above
28. The sparkling, clear sweetened fruit juice is
    a) RTS b) Squash
    c) Puree d) Cordial
29. Papain is
    a) An antibiotic b) An artificial sweetener
    c) A proteolytic enzyme d) An anticaking agent
30. Chipsona Ist is preferred for processing because of its
    a) High starch content b) Low starch content
    c) High temperature tolerance d) High glucose content
31. Ratio of Rankine and Kelvin temperature scale is
    a) 0.5 b) 1.0
    c) 1.5 d) 1.8
32. Tao's chart can be used for composition of
    a) Freezing time b) Drying time
    c) Boiling d) None of these
33. The value that defined the difference between acceptable and unacceptable is defined as
    a) Critical limit b) CCP
    c) HACCP d) Hazard
34. Heat for cooking of food materials in exterior is achieved by
    a) Shear and friction
    b) Crushing and shearing
    c) Pressure and friction
    d) Compression and external heat source
35. Which of the following method of pasteurization is done with plate heat exchanger?
    a) LTLT b) HTST
    c) UHT d) All the above

36. Marsh mallow is a
   a) Confectionary product   b) Dairy product
   c) Bakery product   d) Fruit product
37. Silver coating in confectionary is also called as
   a) Panning   b) Gilding
   c) Polishing   d) None of above
38. Quality of sugar is determined by the process specified by
   a) ICMUSA   b) IUCMSA
   c) ICUMSA   d) ISUCMA
39. Confectioner's glucose is the old name of
   a) Doctor's syrup   b) Cocoa liquor
   c) Glucose syrup   d) None of above
40. "Enrober" is a device used for
   a) Coating sweet centre with chocolate.
   b) Mixing sugar syrup with cocoa liquor.
   c) Forming boiler sugar syrup into thin rope.
   d) Cooling/ heating chocolate to form stable fat crystals.
41. The granular material behave as composite mass having the characteristics of
   a) Solid and gas   b) Semi solid and semi liquid
   c) Liquid and gel   d) Liquid and solid
42. Maximum moisture content (wb) for safe storage of wheat is
   a) 13%   b) 28%
   c) 22%   d) 18%
43. Which of the following is a unit of volume?
   a) Pascal   b) kg
   c) Bushel   d) None of these
44. Hectoliter weight is a measure of
   a) Grain true density   b) Grain bulk density
   c) Grain specific gravity   d) None
45. C A storage is used for storage of
   a) Jam   b) Juice
   c) Dried products   d) Fresh fruits and vegetables
46. Bureau of Indian standards Act introduced in the year
   a) 1986   b) 1952
   c) 1947   d) 2006
47. ISO 9000 standards was first released in
   a) 1995   b) 1987
   c) 1994   d) 2005

48. Which of the following is a food safety standard?
 a) ISO 9001 b) ISO 22000
 c) ISO 14000 d) All the above
49. ISO 22000 represents
 a) Quality system
 b) Safety system
 c) Quality and safety
 d) Environment management system
50. HACCP is a systematic approach to assure
 a) Food safety b) Food quality
 c) Both a and b d) Food supply
51. Blue cheese is also called as
 a) Roquefort cheese b) Cottage cheese
 c) Camembert cheese d) Soft cheese
52. Casein is present in milk in
 a) Dispersed form b) Colloidal form
 c) Emulsion form d) All the above
53. Green colour of the whey is due to
 a) Chlorophyll b) Riboflavin
 c) Chorophyceae d) All the above
54. Alcohol-alizarin test is done to
 a) Determine the heat stability and pH of milk.
 b) Determine the acidity of milk.
 c) Determine the extend of the microbial contamination in milk.
 d) Detect the adulteration of milk.
55. Whipping cream contains .... % fat.
 a) 10% b) 20%
 c) 40% d) 60%
56. Aim of neutralization of cream is
 a) To avoid excessive fat loss in butter milk.
 b) To present the production of an undesirable off-flavor.
 c) To improve the keeping quality of milk
 d) All the above
57. Which of the following test is used to detect sesame oil in milk
 a) Halphen test b) Baudouin test
 c) Kreis test d) Polanski value
58. Which one of the following is a self-carbonated beverage?
 a) Kumiss b) Yoghurt
 c) Dahi d) None

59. The purpose of salting of in butter manufacturing is
    a) To improve the keeping quality b) To enhance taste
    c) To increase overrun d) All the above
60. The head office of NDDB is located at
    a) Delhi b) Anand
    c) Bangalore d) Mumbai
61. Enzymes in food cause
    a) Oxidative rancidity b) Enzymatic browning
    c) Ripening and hydrolysis d) All of above
62. Class I preservatives are
    1. Sugar 2. Salt
    3. Benzoic acid 4. Vinegar
    a) 1 and 2 b) 2,3 and 4
    c) 1, 2 and 4 d) 1 and 4
63. Parabens used as
    a) Flavoring agent b) Coloring agent
    c) Preservatives d) Antioxidant
64. Gluten free diet (GFD) is given to patients with
    a) Liver disease b) Cohn's disease
    c) Celiac disease d) Irritable bowel disease
65. Which of the following terms is associated with food industrial sanitation?
    a) J I T b) HACCP
    c) BARS d) GHP
66. Total heat energy or content level of the material is called
    a) Enthalpy b) Entropy
    c) Thermal conductivity d) Heat transfer coefficient
67. A ton of refrigeration is equivalent to
    a) 50 Kcal/min b) 100 Kcal/min
    c) 500 Kcal/min d) 1000 Kcal/min
68. A fruit powder is classified an instant, if it has the following set of properties
    a) Wet ability, sink ability and solubility
    b) Wet ability, sink ability, dispensability and volubility
    c) Dispensability, and solubility
    d) Wet ability, Dispensability
69. Higher wet bulb depression will result in
    a) Larger drying time b) Shorter drying time
    c) No effect on drying time d) None of the above
70. Ideal fluid is
    a) Whose density is zero b) Whose viscosity is zero
    c) Whose conductivity is zero d) Whose velocity is zero

71. Net protein utilization is
   a) Biological value × Digestibility
   b) Biological value + Digestibility
   c) Biological value — Digestibility
   d) Biological value ÷ Digestibility
72. PER is
   a) Increase in weight ÷ g of protein eaten
   b) Increase in weight × g of protein eaten
   c) Increase in weight + g of protein eaten
   d) Increase in weight - g of protein eaten
73. Calorific value of fat is
   a) 6.1 kcal/g
   b) 7.1 kcal/g
   c) 8.1 kcal/g
   d) 9.1 kcal/g
74. Which enzyme contains iron molecule
   a) Catalase
   b) Alkaline phosphates
   c) Peroxidase
   d) Xanthine oxidase
75. Heat stability test for concentrated milk is done at
   a) 110°C
   b) 120°C
   c) 140°C
   d) 150°C
76. Carbohydrates free human diet leads to
   a) Addition's disease
   b) Hyper adrenalism
   c) Hypothyroidisms
   d) Ketosis
77. Nutrients are added in cereal grain for enrichment by using technique of
   a) Mixing in solution
   b) Syrup coating
   c) Spraying
   d) All above
78. The enrichment of rice is done in mixture of aqueous solution containing
   a) Stearic acid
   b) Zein
   c) Biotic- acid
   d) All the above
79. The antioxidant property of oat flour is due to
   a) Tocopherol
   b) Tocotrionols
   c) Phytic acid
   d) All the above
80. The group of anti nutritional, sulpher containing glucosinolates is present in which oil
   a) Groundnut
   b) Soy bean oil
   c) Coconut
   d) Rape mustard
81. Cured meats are called
   a) Beef
   b) Bacon
   c) Hors
   d) Marinader

82. Which attachment is wrong?
    a) Bromelin from pineapple can be used for the tenderization of meat.
    b) Electrical stimulation of carcass after the slaughtering of animals can cause tenderization of meat.
    c) Ficin is a proteolytic enzyme, obtained from fababean can be used for the tenderization of meat.
    d) None of these
83. A good substitute for gelatin obtained from fish in
    a) Fish silage b) Fish caviar
    c) I singles d) None
84. Glazing of fish is done to protect fish from
    a) Microbial spoilage b) Freezer burn
    c) Oxidation and freezer burn d) Chemical spoilage
85. Chicken is classified on the basis of
    a) Fat content b) Protein content
    c) Age of the bird d) None
86. Semi-viscous product obtained by the solvent extraction of essential oils is
    a) Oleoresin b) Isolate
    c) Concrete d) Absolute
87. Citral is the flavoring compound which can be obtained from
    a) Pineapple b) Apple
    c) Banana d) None
88. Jasmine have
    a) Dinethylanthranillate b) Indole
    c) Euqenol d) Heliotropin
89. When two isoprenes are joined, it is called
    a) Monoterpene b) Deterpene
    c) Triterpene d) None
90. Limonene is present in
    a) Orange b) Mandarin
    c) Lime d) Lemon
91. FSSAI stands for
    a) Food Safety Satisfy All India
    b) Food Safety Storage Authority of India
    c) Food Storage Standards Authority of India
    d) Food Safety and Standards Authority of India
92. Weight and measurement act introduced in the year
    a) 1976 b) 1952
    c) 1947 d) 2006

93. Which of the following is related to milk?
   a) FPO b) MFPO
   c) MMPO d) Milk standard act
94. FPO was introduced in
   a) 1947 b) 1963
   c) 2006 d) 1955
95. CAC is abbreviated as
   a) Critical allowable clearance
   b) CodexAlimentarius commission
   c) Central Association of commerce
   d) Consortium of applied chemist
96. The least count of any instrument is taken as the
   a) Accuracy b) Resolution
   c) Precision d) Linearity
97. Soxhlet apparatus is used for
   a) Liquid solid extract b) Protein analysis
   c) Carbohydrate analysis d) Solid gas extraction
98. The HPLC is device used for
   a) Studying rheological characteristics of foods
   b) Identifying colour of liquid foods
   c) Determining true density of foods
   d) Identifying individual components of a mixture
99. The DSC is an instrument used to measure
   a) Viscosity b) Specific heat
   c) Thermal resistance d) Electrical capacitance
100. The smallest measurable input above the zero value is called
   a) Linearity b) Threshold
   c) Accuracy d) Error

Answer Key

1	b	2	d	3	a	4	a	5	c	6	a	7	c
8	d	9	b	10	c	11	d	12	b	13	c	14	c
15	c	16	a	17	a	18	c	19	b	20	b	21	a
22	c	23	a	24	b	25	b	26	a	27	d	28	d
29	c	30	a	31	d	32	a	33	a	34	a	35	b
36	a	37	b	38	c	39	c	40	a	41	b	42	a
43	c	44	b	45	d	46	a	47	b	48	b	49	b
50	a	51	a	52	b	53	b	54	a	55	c	56	d
57	b	58	d	59	d	60	b	61	d	62	a	63	c
64	c	65	b	66	a	67	a	68	b	69	b	70	d
71	a	72	b	73	d	74	a	75	b	76	d	77	d
78	d	79	d	80	d	81	b	82	c	83	c	84	c
85	c	86	a	87	d	88	b	89	a	90	a	91	d
92	a	93	c	94	d	95	b	96	b	97	a	98	d
99	b	100	b										

Model Paper 20

1. All microorganisms need
   a) Nutrients b) Water
   c) Specific temperature d) All of above
2. Antioxidant in foods
   a) Enhance shelf life b) Prevent oxidation reaction
   c) Decrease deterioration d) All the above
3. Thermophilic bacteria grow at
   a) Low temperature b) High temperature
   c) Medium temperature d) All of the above
4. Preservation method which involves removal of water
   a) Drying b) Canning
   c) Sterilization d) Pasteurization
5. MMPO is applicable on
   a) Meat and Meat products b) Milk and Milk products
   c) Oil and Ghee d) Infant foods
6. Parabens used as
   a) Flavoring agent b) Colorant
   c) Preservative d) Antioxidant
7. The primary reason for parboiling is
   a) Enhance colour of rice b) Enhance cooking time of rice
   c) Conserve nutrients of rice d) To favor gelatinization
8. Canning process involver
   a) High temperature b) Low temperature
   c) Medium temperature d) None
9. Humectants helps to
   a) Control viscosity b) Maintain texture
   c) Reduce water activity d) All the above
10. The effective level of sorbets in foods is in the range of
    a) 0.05 - 0.30% b) 0.50 - 0.90%
    c) 1.00 - 2.00% d) None

11. Methods of membrane separation
    a) Ultra filtration
    b) Reverse osmosis
    c) Both a and b
    d) None of the above
12. BHA and BHT both are
    a) Natural and oxidant
    b) Synthetic antioxidant
    c) Food colors
    d) Synthetic sweeteners
13. Quick freezing allows formation of
    a) Large in size ice crystals
    b) Small in size ice crystal
    c) Medium in size ice crystals
    d) None of the above
14. Pasteurization by high temperature and short time used temperature
    a) Above 70°C
    b) Above 50°C
    c) Above 100°C
    d) Above 120°C
15. In food preservation, salt water is termed as
    a) Syrup
    b) Brine
    c) Saline
    d) Lye
16. Irradiation method cause
    a) Destruction of natural antioxidant
    b) Promote oxidation
    c) Decrease thiamine, pyridoxine, B-12, C,D,E and K
    d) All the above
17. For Jam preparation from fruits
    a) Food colour is required
    b) Pectin is required
    c) Lecithin is required
    d) All the above
18. Sugar and salt are used in food preservation
    a) For good taste
    b) For prevention of spoilage
    c) For prevention of microbial growth
    d) All the above
19. Examples of GRAS substances are
    a) Spices, oil, sodium bicarbonate
    b) Food additives
    c) Food preservation
    d) All the above
20. The fortificant used in iodized salt is
    a) Potassium iodide and potassium iodate
    b) Sodium iodide and sodium iodate
    c) Both a and b
    d) None of above
21. Germination can be apply on
    a) Dough and batter
    b) Milk and milk products
    c) Pulses and grains
    d) Oil and ghee
22. From which grain, gluten is derived
    a) Soy bean
    b) Wheat
    c) Rice
    d) Born

23. The record of food store inventory is kept in
    - a) Stock register
    - b) Cash Book
    - c) Note Book
    - d) Diary
24. Thiamine is also known as
    - a) Vitamin B1
    - b) Vitamin B2
    - c) Vitamin B12
    - d) none
25. Protein is known as complete protein
    - a) Milk protein
    - b) Cereal protein
    - c) Egg protein
    - d) All the above
26. Cobalt -60 is commonly used
    - a) ISO tope
    - b) Radio ISO tope
    - c) Mineral
    - d) Vitamin
27. Lower temperature at which all cells in a culture are killed in 10 min
    - a) Thermal death rate
    - b) Thermal death point
    - c) Thermal death time
    - d) None
28. A nutrient depletion by product build up stops the cells from metabolizing or reproducing the cell numbers remain static.
    - a) Lag phase
    - b) Log phase
    - c) Depth phase
    - d) Stationery phase
29. Which of the following methods of cooking food use moist heat?
    - a) Poaching
    - b) Broiling
    - c) Stewing
    - d) All the above
30. AOCS stands for
    - a) Australian oil chemists society
    - b) American oil chemist's society
    - c) American oil and cereal society
    - d) American oil chemist's survey
31. Glucose units in starch is held together by
    - a) Pectic bond
    - b) Carboxylic bond
    - c) 1,4 and 1,6 Glycosidic bond
    - d) Hydrogen bond
32. The objective of wheat milling is to get
    - a) Brain
    - b) Flour
    - c) Germ
    - d) Oil
33. Flour obtained by mixing the rational proportions of break, scratch and reduction system is called
    - a) Patent flour
    - b) Straight run grade
    - c) Whole meal flour
    - d) White flour
34. The temperature suitable for conditioning of grain is as high as
    - a) 55ºC
    - b) 45ºC
    - c) 60ºC
    - d) 30ºC

35. Hard wheat is characterized by the presence of
    a) More damaged starch b) Higher protein
    c) Less damaged starch d) Both a and b
36. The successive grinding stages in the modern flour milling are
    a) Scratch, break , reduction b) Reduction, break, scratch
    c) Break, scratch, reduction d) Break, reduction, scratch
37. The 100% extraction rate wheat flour is known as
    a) Atta b) Whole meal
    c) Whole wheat d) All of them
38. Semolina is produced by milling of durum wheat in how many stages
    a) One b) Two
    c) Three d) Four
39. The number of parts of flour by weight produced per 100 parts of wheat milled is known as
    a) Extraction rate b) Flour yield
    c) Both a and b d) None
40. The 'Dyox' used in flour bleaching is also known as
    a) Nitrogen peroxide b) Chlorine trioxide
    c) Benzoyl peroxide d) Chlorine dioxide
41. Improvers improve the colour of flour by the principle of
    a) Reduction b) Oxidation
    c) O-R reaction d) Ionization
42. Policed Rice is also called
    a) Glazed rice b) High luster rice
    c) Policed rice d) All of them
43. Milling yield of rice is affected by
    a) Carbohydrate content b) Equipment used
    c) Location of cultivation d) Protein content
44. The protein starch association in rice is weekend by socking in
    a) Sodium bisulphate b) Dilute NaOH
    c) Dilute HCL d) Sodium bicarbonate
45. The mild preservative (s) added in the water tank during wet milling of corn include
    a) Acid and SO_2 b) SO_2
    c) Potassium meta bisulphate d) Sodium hypochlorite
46. During wet milling of corn, the acid produced in the steep water is
    a) Citric acid b) Lactic acid
    c) Malic acid d) Ascorbic acid
47. The "Quern" is a type of
    a) Stone grinder b) Pounder
    c) Cone grinder d) None

48. Pearling of barley is done to remove
    a) Hull
    b) Per cap
    c) Seed coat and alluring
    d) All of them
49. Pearling of barley is based on the principle of
    a) Impact force
    b) Shear force
    c) Attrition force
    d) None
50. Barley flour is the product obtained from
    a) Pearled barley
    b) Blocked barley
    c) Hull less barley
    d) All of them
51. Separation of barn from endosperm is difficult in milling of
    a) Wheat
    b) Rye
    c) Oat
    d) Triticale
52. The yield of reasonably pure flour from rye is approximately between
    a) 64-65%
    b) 58-60%
    c) 65-70%
    d) 75%
53. De-germing of which grain is most difficult in wet milling operation
    a) Maize
    b) Millets
    c) Sorghum
    d) Wheat
54. Among the cereals which grain has good storage characteristics
    a) Maize
    b) Wheat
    c) Sorghum
    d) Millets
55. Separation of starch from protein most difficult in wet milling process of
    a) Maize
    b) Sorghum
    c) Millets
    d) Wheat
56. The average yield of flour from triticale having 13% protein and 0.61% ash is
    a) 75%
    b) 63%
    c) 70%
    d) 50%
57. Turmeric, spice, salt coating and powered basin application following by packaging of dhal comes under which pulse process
    a) Primary
    b) Secondary
    c) Tertiary
    d) Quaternary
58. The nature of freshly drawn milk is
    a) Acidic
    b) Basic
    c) Amphoteric
    d) Neutral
59. The best temperature of homogenization is in the range of
    a) 30 – 40°C
    b) 60 – 70°C
    c) 40 – 50°C
    d) 40 –70°C
60. Fallowness in milk occurs due to oxidation of
    a) Vitamin
    b) Carbohydrate
    c) Fat
    d) Protein

61. Which of the following is not a part of the solids not fat of milk?
   a) Protein
   b) Lactose
   c) Minerals
   d) Butter fat
62. Natural acidity of milk is due to
   a) Casein
   b) Acid phosphates
   c) Citrates
   d) All the above
63. Freshness of the milk can be detected by
   a) Alcohol test
   b) COB test
   c) Both a and b
   d) Phosphates test
64. If milk is adulterated with water then freezing point of the milk
   a) Increase
   b) Decrease
   c) Neither increases nor decreases
   d) None of these
65. Sediment test is carried out to check the
   a) Visible foreign matter contained in milk
   b) Bacterial count
   c) Specific gravity of milk
   d) Destroyed substances in milk
66. When fat is exposed to light then it given
   a) Musty flavor
   b) Rancid flavor
   c) Tallowy flavor
   d) Brownish colour
67. Which test is mainly done for the stability of milk?
   a) Alcohol
   b) Protein
   c) Lactose
   d) Fat
68. The unit of mass diffusivity is
   a) m/s
   b) kg/ms
   c) m^2/s
   d) m/s^2
69. Liquid level in an autoclave is measured by
   a) Single float
   b) Deferential float type monometer
   c) Glass gauge
   d) None of these
70. Iran adiabatic process
   a) Heat transfer is zero
   b) Temperature change is zero
   c) Work done is a path function
   d) Enthalpy remains constant
71. Entropy for an isolated system
   a) Never increases
   b) Never decreases
   c) Never changes
   d) None of above

72. Which empirical rule is commonly used to obtain boiling point elevation of solution?
 a) Pearson's rule b) Duhring rule
 c) Stephan's rule d) Mosley rule
73. Bernoulli's equation is derived making assumption that the flow is
 a) Uniform and incompressible
 b) Non- viscous, uniform and stead
 c) Steady, non- viscous in compressible and irrotational
 d) None of these
74. Milk having viscosity 'μ' is flowing through a circular tube of diameter (D) has a Reynolds number DG/μ the term 'G' denotes as
 a) Flow rate b) Mass velocity
 c) Volumetric flow rate d) Linear velocity
75. Stoke's law is used to determine
 a) Terminal velocity b) Drag coefficient
 c) Surface tension d) All the above
76. Stoke's law is generally used for
 a) Stream line flow b) Laminar flow
 c) Turbulent flow d) None
77. Mechanical separation is applicable to
 a) Heterogeneous mixture b) Homogeneous
 c) Colloidal mixture d) None of these
78. A two fluid nozzle is used in a
 a) Spray dryer b) Tray dryer
 c) Drum dryer d) Freeze dryer
79. What is the driving force in ultra filtration and reverse osmosis?
 a) Hydrostatic pressure b) Osmotic pressure
 c) Temperature d) None of these
80. Ultra filtration process occurs at a pressure of
 a) 1 - 7 bar b) 5-7 bar
 c) 10-12 bar d) 2-3 bar
81. What is the general range of mixing index.
 a) 1- 20 b) 10-150
 c) 50-150 d) 1-5
82. The reference condition of the steam tables is
 a) Triple point of water b) Freezing point of water
 c) Boiling point of water d) None of these
83. The quality of steam is equal to
 a) Dryness fraction of steam b) Vapor fraction of steam
 c) Liquid fraction of steam d) None

84. The specific heat for food material can be calculated by
    a) Knudsen equation
    b) Diettus boelter equation
    c) Siebel equation
    d) Fourier's equation
85. Specific heat of saturated weter at 100
    a) Negative
    b) Positive
    c) 1.0
    d) Zero
86. The flow of heat per hour is proportional to change in
    a) Mass
    b) Vapor pressure
    c) Temperature
    d) Area
87. Unsteady heat flow oceans
    a) Through the walls of furnace
    b) Through lagged pipes carrying steam
    c) Through the wall of a refrigerator
    d) During annealing of casting
88. Heat exchangers are used to
    a) Pasteurize milk only
    b) Sterilize the juice only
    c) Heat or cool the product
    d) Maintain constant temperature
89. Stassanization is the method of
    a) Pasteurization
    b) Homogenization
    c) Sterilization
    d) None
90. Spacing between the plates of plate heat exchanges is nearly
    a) 1mm
    b) 3mm
    c) 5mm
    d) 7.5mm
91. The temperature range of UHT sterilization of milk is
    a) 90 to 100°C
    b) 100 to 115°C
    c) 135 to 150°C
    d) 180 to 210°C
92. Which of the following product should have 9% TSS?
    a) Jam
    b) Tomato puree
    c) Tomato ketch up
    d) Fruit drink
93. What is TSS of fruit syrup?
    a) 20%
    b) 45%
    c) 65%
    d) 80%
94. Preservation used in squash is
    a) BTH
    b) SO_2
    c) Calcium propriate
    d) Sodium nitrate
95. Sauer kraut is the
    a) Fermented product
    b) Dairy product
    c) Fruit juice
    d) Meat product

96. Myoglobin is
   a) Protein
   b) Lipid
   c) Carbohydrate
   d) Vitamin
97. Freeze drying takes place at
   a) The triple point
   b) Below the triple point
   c) Above the triple point
   d) At zero degree Celsius
98. Use of several methods like temperature (high or low), control of acidity, control of water activity for preservation of food is known as
   a) Hurdle technology
   b) Mixed technology
   c) Stumbling technology
   d) Multiple technology
99. The head office of the Food Safety and Standards Authority of India is at
   a) Mumbai
   b) Delhi
   c) Calcutta
   d) Kochi
100. Cold Sterilization means perseveration of food by
   a) Sterilizing at –179°C
   b) Irradiation
   c) Sterilizing at 0°C
   d) Lyophilization

Answer Key

1	d	2	d	3	b	4	a	5	b	6	c	7	c
8	a	9	c	10	a	11	c	12	b	13	b	14	a
15	b	16	d	17	b	18	d	19	a	20	c	21	c
22	b	23	a	24	a	25	c	26	b	27	d	28	d
29	d	30	b	31	c	32	b	33	b	34	b	35	a
36	c	37	d	38	c	39	c	40	d	41	b	42	d
43	b	44	b	45	a	46	a	47	a	48	d	49	c
50	d	51	c	52	a	53	b	54	d	55	c	56	b
57	b	58	c	59	b	60	c	61	d	62	d	63	b
64	b	65	a	66	c	67	a	68	c	69	b	70	a
71	b	72	b	73	c	74	b	75	a	76	a	77	a
78	a	79	a	80	a	81	b	82	a	83	b	84	c
85	a	86	c	87	b	88	c	89	a	90	b	91	c
92	b	93	c	94	b	95	a	96	a	97	b	98	a
99	b	100	b										

Section-II

Short Notes for ICAR-NET/ARS/FSO

A

Abattoirs: It is the types of slaughterhouses where animals are slaughtered for meat and offal. Abattoirs usually include lairage (a holding area for live animals), a slaughtering line and cold stores. It has facilities for processing of by-products (blood, intestines, skins, fat, bristle, unusable waste products), and treatment of waste water and air are often included.

Absorbents: Materials or substances that is capable of absorption. Uses of absorbents include incorporation within food packaging (to absorb oxygen as a preservation technique, to control humidity, and to manage aroma and flavour problems in packaged foods) and for purification of foods and beverages, such as drinking water and liquid foods. Silica gel, Alumina, activated carbon or charcoal, zeolites, adsorption chillers used with refrigerants and biomaterials that adsorb proteins.

Absorption: It is a process involving molecules of one substance being taken directly into another substance. Absorption may be either a physical or a chemical process, physical absorption involving such factors as solubility and vapour-pressure relationships and chemical absorption involving chemical reactions between the absorbed substance and the absorbing medium. Absorption includes such processes as the passage of nutrients and other substances from the gastrointestinal tract into the blood and lymph, and also the uptake of water, fats and other substances into foods. It may also gain of water by dry foods.

Accelerated freeze-drying: Freeze-drying in which the rate of temperature drops between 0 and –5°C is very rapid, thus preventing the growth of ice crystals which would disrupt the cells and destroy the texture of the food, and in which heat is supplied to the food whilst under high vacuum and at a low temperature to remove the free and combined water.

Acceptability: The degree to which the quality of a food is regarded as satisfactory.

Acceptable daily intake: A safety level for substances used as food additives. The acceptable daily intake (ADI) is usually calculated as 1/100th of the maximum dose of the substance that causes no adverse effects in humans.

Acceptance: The willingness to regard the quality of a food as satisfactory.

Acesulfame K: Non-nutritive or artificial sweetener, approximately 150-200 times as sweet as sucrose. Potassium salt derived from acetoacetic acid, with good heat stability and a synergistic effect in sweetener blends. It is used in a variety of food applications, including yoghurt, table-top sweeteners, soft drinks, candy, canned foods and other confectionery. It is also called as Acesulfame potassium.

Acetic acid bacteria: Any Gram negative, aerobic, rod-shaped bacteria, e.g. Acetobacter species and Gluconobacter species, capable of oxidizing ethanol to acetic acid. It occurs on the surface of fruits, vegetables and flowers, and in soil. Used industrially in the manufacture of vinegar and may cause spoilage of beer and wines. The acetic acid bacteria are mesophilic, with optimum growth temperatures about 25°C to 30°C. These are most important in wine spoilage. Acetic acid bacteria (AAB) belong to the family Acetobacteraceae that includes several genera and species. Currently, they are classified into nineteen genera, including *Acetobacter, Acidomonas, Ameyamaea, Asaia, Bombella, Commensalibacter, Endobacter, Gluconacetobacter, Gluconobacter, Granulibacter, Komagataeibacter, Kozakia, Neoasaia, Neokomagataea, Nguyenibacter Saccharibacter, Swaminathania, Swingsia* and *Tanticharoenia*. The main species responsible for the production of vinegar belong to the genera *Acetobacter, Gluconacetobacter, Gluconobacter* and *Komagataeibacter* because of their high capacity to oxidise ethanol to acetic acid and high resistance to acetic acid released into the fermentative medium.

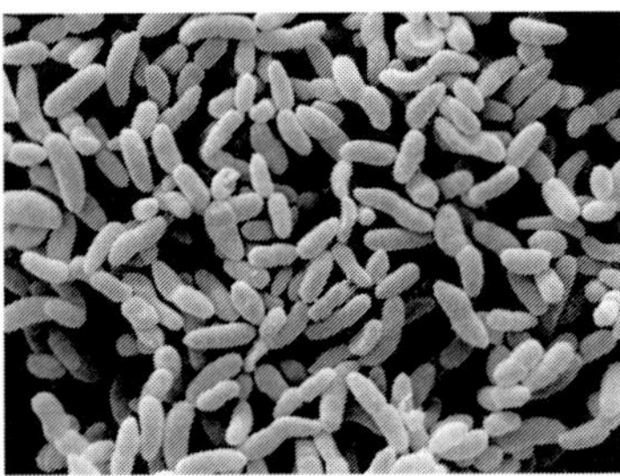

Acetic acid: It is a member of the short chain fatty acids group, which occurs in a range of foods and beverages. It may be one of the flavour compounds, or cause taints, depending on food or beverage type and the concentration at which it is present. Acetic acid (CH_3COOH) is the main constituent of vinegar. It may be used for preservation or flavouring of foods. In the food industry, acetic acid is controlled by the food additive as an acidity regulator and as a condiment. Acetic acid has 349 kcal per 100 g. Vinegar is typically 4–18% acetic acid by mass. Vinegar is used directly as a condiment, and in the pickling of vegetables and other foods. Table vinegar tends to be more diluted (4% to 8% acetic acid), while commercial food pickling employs solutions that are more concentrated. Acetic acid contained in vinegar which at 3.5% concentration will ensure the stability of pickles, etc. It is obtained by biological oxidation of alcohol in wine, ale and other fermented beverages. Very cheap vinegars are made from a chemically produced acetic acid.

Acetic Acid

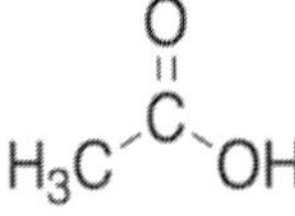

- CAS registration number:64-19-7
- Differential Formula:CH_3COOH
- Molecular weight:60.05
- Melting Point:16.7℃
- Boiling Point:118℃
- Density:1.049g/mL
- Appearance:Colorless liquid

Acetic fermentation: The process by which certain microorganisms (e.g. *Acetobacter* and *Acetomonas* spp., and *Gluconobacter oxydans*) metabolize an alcoholic substrate to form acetic acid, the main constituent of vinegar. Alcoholic substrates can be obtained from a variety of sources, such as fruits, vegetables and grain.

Acetobacter: Any of a group of microorganisms associated with rot and browning in apples, but more importantly used for the production of vinegar from wines and ales by oxidation of alcohol to acetic acid. Acetobacter species growing on the surface of alcoholic beverages, oxidize ethanol to acetic acid to produce either spoilage or vinegar.

Acetoin: Flavour compound found commonly in dairy products and wines. Synonyms include 3-hydroxy-2-butanone and acetylmethylcarbinol. Acetoin, along with diacetyl, is one of the compounds giving butter its characteristic flavor.

Acid casein: Casein produced by acid precipitation from milk at its isoelectric point, pH 4.7. Acidification can be achieved by direct addition of an acid or through the action of lactic acid bacteria.

Acid curd cheese: A cheese produced by microbial ripening of quarg (soft creamy cheese), ripening proceeding from the outside of the cheese. Cultures used include bacteria, fungi and yeasts, their selection depending on the type of cheese being made.

Acid foods/basic foods: These terms refer to the residue of the metabolism of foods. The minerals sodium, potassium, magnesium and calcium are base-forming, while phosphorus, sulphur and chlorine are acid-forming. Which of these predominates in foods determines whether the residue is acidic or basic (alkaline); meat, cheese, eggs and cereals leave an acidic residue, while milk, vegetables and some fruits leave a basic residue. Fats and sugars have no mineral content and so leave a neutral residue. Although fruits have an acid taste caused by organic acids and their salts, the acids are completely oxidized and the sodium and potassium salts yield an alkaline residue.

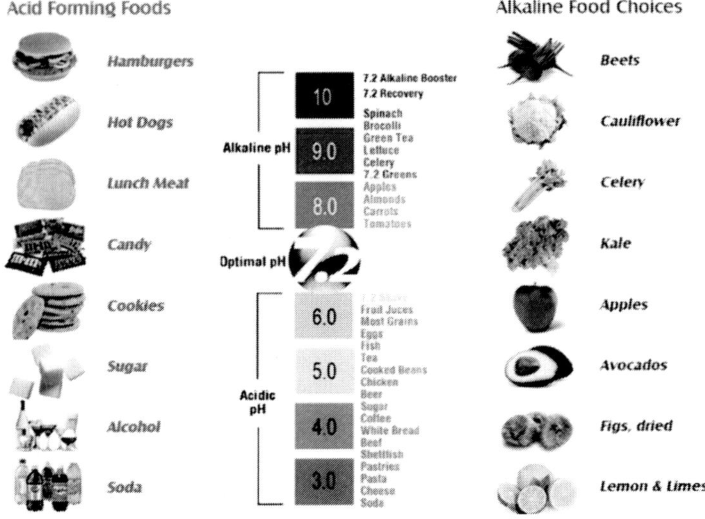

Acid number/ acid value: The level of free fatty acids presents in lipids. The acid value, also known as the acid number, is determined by measuring the amount of KOH in milligrammes that neutralizes 1 g of the lipid. Acid values of fresh edible fats tend to be low and increase with storage as the glycerides present in the lipids break down to generate free fatty acids. It is also a measure of rancidity due to hydrolysis of fat releasing free fatty acids from the triacylglycerol of the fat; serves as an index of the efficiency of refining since the fatty acids are removed during refining and increase with deterioration during storage.

$$\text{Acid Value} = \frac{\text{Volume (mL) of 0.1 mol / L KOH consumed} \times 5.611}{\text{Weight (g) of the sample}}$$

Acid whey: Whey produced by acid coagulation of milk during cheese preparation.

Acid: A substance containing an excess of hydrogen ions when compared with pure water, which contains equal amounts of hydrogen ions *(acid forming)* and hydroxyl ions *(alkali forming).* This excess causes the characteristic sour taste response.

Acidification: The process of adding acids to, or causing acids to be produced in, foods for preservation – as in pickles, sauerkraut, soused herrings, mayonnaise, etc. The usual acids are acetic, citric and lactic. Benzoic acid is effective at low concentrations. The pH of a substance is decreased to below 7 making it acidic.

Acidity: It is a measure of the strength of an acid on the scale of pH which goes from 1 (very acid) through 2 (acidity of lemon juice), 3 (apple juice) to 7 (neither acid nor alkaline), the pH of pure water. The pH values greater than 7 refer to alkalis. Very few food-spoiling organisms will grow at a pH equal to or less than 2. The degree to which a substance or solution is acidic, being dependent upon the concentration of hydrogen ions. Level of acidity is expressed using pH.

Acidophilus milk: Fermented milk produced by fermentation of milk with mixture of lactic acid bacteria, including *Lactobacillus acidophilus*, and kefir grains at 37°C to give a lactic acid content of 0.6 to 0.8%. The consumption of acidophilus milk has beneficial effects on the intestine.

Acidulants: Organic acids used in foods to control pH and fulfill a variety of functions. Applications include preservation of meat products, flavour enhancement, prevention of discoloration in sliced fruits, and prevention of development of rancidity in oils and fats. Commonly used acidulants in the food industry include citric acid, acetic acid, propionic acid and lactic acid.

Aconitic acid: One of the organic acids found in sugar cane. It is used in flavourings and acidulants for the food industry and also in the manufacture of emulsifying agents, plastics and detergents.

Actins: A family of multifunctional intracellular proteins, best known as a myofibrillar component of striated muscle fibers. They constitute about 13% of muscle proteins and are the major components of the I-band or thin filament of the sarcomere. Actins contain high levels of the amino acid proline. Imino-groups within proline contribute to the folding of actin molecules and result in formation of G-actin (globular actin). G-actin, a spherical molecule approximately 5.5 nm in diameter, constitutes the monomeric form of actin. In the presence of potassium chloride and ATP, G- actin polymerizes into long fibers of F-actin. Most vertebrate genomes contain numerous actin genes with high sequence homology in protein coding regions, but considerable variability in intron size and number. This genetic diversity can be utilized for livestock speciation and meat authenticity tests. Determination of actin content has been proposed as a means of calculating the meat content of meat products.

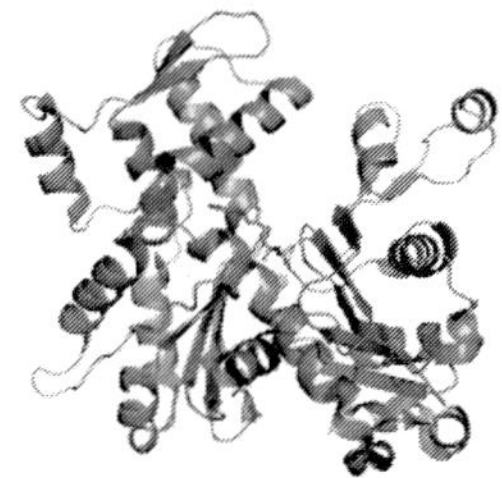

Activated carbon: Amorphous forms of elemental carbon, particularly charcoal, which have been treated, e.g. by acid or heat, to improve their powers of absorption. It is used for a variety of food and industrial applications, including drinking water purification, de- coloration of sugar solutions and sorption of residues of pesticides from wines. Activated carbon is carbon produced from carbonaceous source materials such as bamboo, coconut husk, willow peat, wood, coir, lignite, coal, and petroleum pitch.

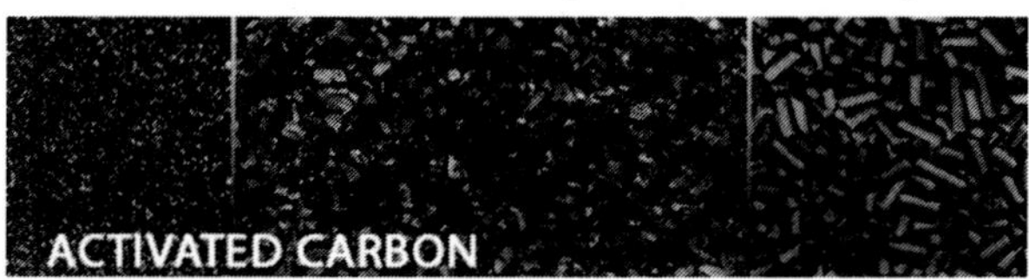

Activated dough development: A method of speeding up the primary proving of dough by adding small quantities of chemicals such as ascorbic acid, cysteine, bromates, etc., which modify the properties of the gluten.

Activation energy Activation energy is defined as the minimum amount of extra energy required for a chemical reaction to proceed; the difference in energy between that of the reactants and that at the transition state of the reaction. Activation energy determines the way in which the rate of a reaction varies with temperature. It can also be described as the minimum amount of energy needed to activate or energise molecules or atoms so that they can undergo a chemical reaction or transformation. Activation energy is denoted by E_a. It is usually measured in joules (J) and or kilojoules per mole (kJ/mol) or kilocalories per mole (kcal/mol). The Arrhenius equations relates the rate of a chemical reaction to the magnitude of the activation energy:

$$k = Ae^{\frac{-Ea}{RT}}$$

Where,

k is the reaction rate coefficient or constant

A is the frequency factor of the reaction. It is determined experimentally.

R is the Universal Gas constant

T is the temperature in Kelvin

Active packaging: Packaging materials which have functions additional to their basic barrier action. It is used for packaging a wide range of foods and beverages. Types of active packaging include: packs which adsorb ethylene to control ripening of fruits; packs which regulate moisture levels; packs which contain oxygen scavengers; packs which contain CO_2 scavengers or generators; packs which release or absorb flavours or aromas; antimicrobial packaging (e.g. packs which release ethanol to control the growth of fungi); packs with special microwave heating properties; and packaging with monitoring systems (time/temp. exposure indicators or temperature control).

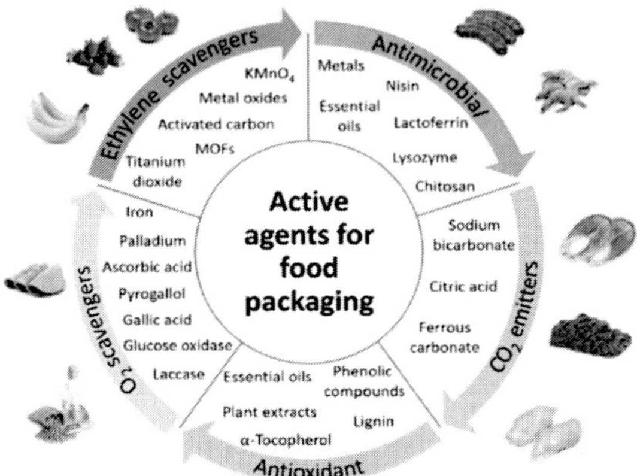

Actomyosin: A complex of the two major muscle proteins, actins and myosin. Actomyosin is formed during muscle contraction with simultaneous hydrolysis of ATP to ADP. Within myofibrils during contraction, each myosin head region on a thick myofilament attaches to a G-actin molecule within a thin myofilament. This interaction leads to formation of cross-bridges between actin and myosin and to formation of the actomyosin complex. Formation of actomyosin results in rigidity and lack of extensibility in muscles. In the presence of ATP, as in living animals, the actomyosin complex dissociates rapidly; however, *post mortem*, actomyosin is the dominant form of myofibrillar protein and it plays a major role in the development of *rigor mortis*. During *post mortem* storage, tenderness of meat is affected by modification of the actin-myosin interaction. Thermal denaturation of actomyosin occurs at temperature between 30 and 50°C.

Additive: A substance added to food to improve its properties such as keeping quality, health value, flavour, texture, colour, acidity, stability, tendency to oxidize or dry out, sweetness, cooking properties, viscosity, stickiness and the like. These include prevention of chemical and microbial spoilage, enhancement of flavour or colour, improvement of nutritional values or as an aid to processing. The most common types of additives include preservatives, colorants, sweeteners, flavourings, emulsifiers, thickeners and stabilizers etc.

Food additives				
Additive	**Examples**	**Food**	**Benefit**	**Health hazard**
Anti-oxidants	Ascorbic acid (vitamin C)	Fruit, meat	Stop food reacting with O_2 (which spoil taste, change color)	
Colourings			Improve appearance of food	
	Sunset yellow	Drinks, sweets	Give yellow/orange colour	-Cause hyperactivity in children -Trigger asthma
	Caramel	Sweet, drinks, soups	Give brown color	
Flavouring	Monosodium glutamate	Processed food, Chinese food	Enhance taste of food	
	vanillin	Desert, chocolate	Give vanilla taste	
Preserv-atives			Give food longer life	
	Sulfur dioxide	Fruit juice, dried fruit	-Kill bacteria -Preserve vitamin C	Destroy vitamin B_1
	Sodium nitrate	Meat products (sausages)	Stops growth of harmful bacteria	May cause cancer
Emulsifier	Lecithin	Powdered milk	Stop oil and H_2O separating out into different layers	

Adjunct cultures: Non-starter cultures used in addition to starters, mainly in cheese-making, to produce a specific benefit, e.g. smoother texture, improved flavour or accelerated ripening of cheese. In production of yoghurt, adjunct cultures have been used to manufacture products with increased levels of nutrients such as folates.

Adsorbents: The substances that is capable of adsorption. They are used widely in the food and biotechnology industries. Its use includes removal of unwanted materials in foods and beverages that affect either food safety or food quality. Examples include removal of proteins from white wines, pathogens from drinking water sources, radio-elements from foods, oxidation products from frying oils allowing oil recovery and reuse, and bitter compounds from fruit juices. Amongst other applications are: isolation of compounds with potential for use in foods; immobilization of enzymes; as agents in analytical techniques such as gas analysis and chromatography; and removal of unwanted aroma and flavour in packaged foods.

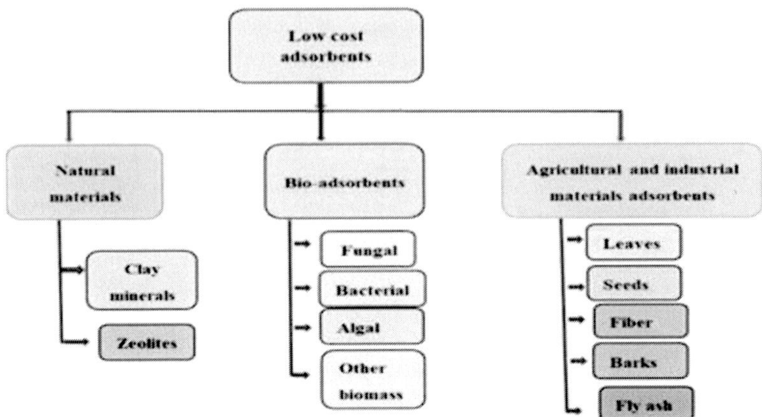

Adulterant: A cheaper substance added to food in order to increase its weight without changing its appearance and thus increase the profit margin. Some are legal, e.g. Water and phosphates in chickens, or extenders added to chilli powder.

Food & drink items	Adulterant/ extraneous substances	Purpose	Type of adulteration
Ghee	Vanaspati, anatta, & oleomargarine	To make more yellow	Deliberate
Milk	Water, skim milk	To increase volume	Deliberate
Condensed milk	Paneer, khoya	To give rich tcxlur	Deliberate
Butter	Vegetable oil, anatta, banana, oleomargarine	To increase volume & make yellowish	Deliberate
Ice cream	Slurch, rice powder or wheat Hour	To thicken cream	De liberate
Tea leaves	Black/Bengal gram dal husk wilh color	To add color	Deliberate
Red wine	Juice of bilberries	To attract/produce deep blue precipitate with lead acetate	Deliberate
Mustard oil	Papaya seed	To add hulk and weight	Deliberate
Black pepper	Papaya seed	To add bulk	Deliberate
Green chillies & peas	Malachite green	To give bright glowing green color	Deliberate
Chillies powder	Brick powder	To increase weight	Deliberate
Sugar	Chalk powder	To increase amount	Deliberate
Oils	Rancid oil	To increase volume	Deliberate
Coriander powder	Cow dung powder	To increase amount	Deliberate
Pulses	Lathyrus sativus	To increase weight	Deliberate

Common salt	White powdered xlone. chalk	To increase amount	Deliberate
Coffee	Chicory, roasted barley powder, tamarind seeds	To add bulk and color	Deliberate
Honey	Molasses, cane sugar	To increase volume	Deliberate
Wheat	Ergot (poisonous fungus)	To increase weight	Deliberate
Jaggery powder	Chalk powder	To increase amount	Deliberate
Others like preservatives, etc.	Formalin, etc.	To increases shelf life	Unintentional

Adulteration: Addition of substances to foods, or substitution of food ingredients with inferior substances, with the intent of lowering the quality of and costs of producing the food and defrauding the purchaser, e.g. addition of starch to spices, and of water to milk or beer. In order to protect the health of the consumer, the Government of India promulgated the Prevention of Food adulteration Act (PFA) in 1954.

Adverse reactions to foods: Food aversion, unpleasant reactions caused by emotional responses to certain foods rather than to the foods themselves, which are unlikely to occur in blind testing when the foods are disguised. Food allergy, physiological reactions to specific foods or ingredients is due to an immunological response. Antibodies to the allergen are formed as a result of previous exposure or sensitization, and cause a variety of symptoms when the food is eaten, including gastrointestinal disturbances, skin rashes, asthma and, in severe cases, anaphylactic shock, which may be fatal. Food intolerance, physiological reactions to specific foods or ingredients which are not due to immunological responses, but may result from the irritant action of spices, pharmacological actions of naturally occurring compounds or an inability to metabolize a component of the food as a result of an enzyme defect.

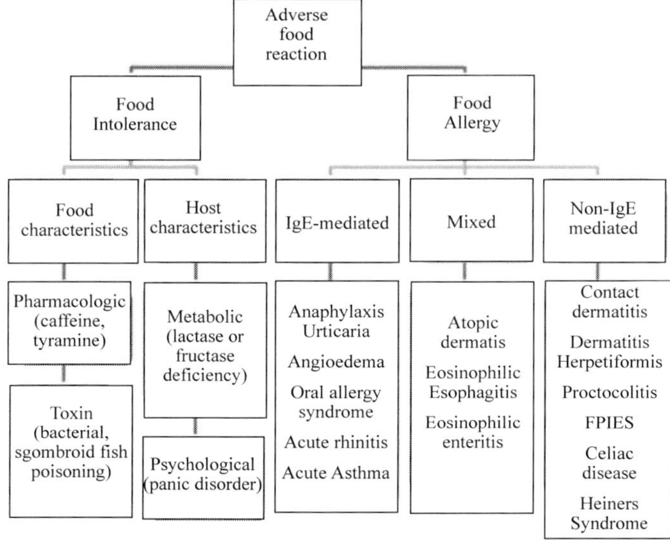

Aerated bread: Bread made from dough which has been mechanically mixed with carbon dioxide without using yeast.

Aerated confectionery: Confectionery produced with incorporation of air as an ingredient. Use of air adds bulk to the product without increasing its weight, improving product texture and flavour. Aeration of confectionery results in a range of products with densities ranging from 0.2 to 1.0 g/cm^3. Such products include chews, mallows, honeycomb, mousses and meringues.

Aeration: Introduction of air into a product to enhance texture, mouthfeel, rheology and visual appeal. The following methods are used to aerate foods: fermentation; whipping or shaking of low-medium viscosity liquids; mixing of doughs or high viscosity pastes, in which air bubbles are entrapped as surfaces come together; steam generation during slow to moderate cooking, baking or frying; entrapment of air between sheeted layers, as in pastries and croissants, or between pulled strands, as in pulled taffy and candy; frying in very hot oils, such that internal steam rapidly forms, causing the product to puff; use of chemical raising agents such as baking powders or sodium bicarbonate; rapid dry heating of small or thin products to induce blistering or slight puffing; gas injection (e.g. air, carbon dioxide, nitrogen and nitrous oxide); expansion extrusion; pres- sure beating (dissolution of air or gas under pressure in a syrup, fat mixture or chocolate); puffing, in which products such as breakfast cereals containing superheated moisture are subjected to a sudden release of pressure; and vacuum expansion, followed by rapid cooling to set the expanded products.

***Aeromonas hydrophilia*:** A food poisoning bacterium which will grow at temperatures below 10°C.

Aeromonas: Genus of Gram negative, facultatively anaerobic rod-shaped bacteria. Occur in salt and fresh water, sewage and soil. *Aeromonas hydrophila*, frequently found in fish and shellfish and occasionally in red meat and poultry meat, may cause septicaemia, meningitis and gastroenteritis in humans.

Aerosol packs: Containers for pressurized liquids, which are released in the form of a spray or foam when a valve is pressed. Aerosol propellants, usually liquefied gases, are used in the packs. Used as dispensers for a variety of foods.

Aerosols: The substances, including foods, stored under pressure in a container (for example in aerosol cans) containing a propellant and released as a fine spray or froth. Also, in a chemical sense, suspensions of submicroscopic particles dispersed in air or gas.

Affination: The first stage in processing of raw sugar, in which the layer of mother liquor surrounding the crystals is softened and removed. Raw sugar is mixed with warm, concentrated syrup of slightly higher purity than the syrup layer so that it will not dissolve the crystals. The resulting magma is centrifuged to separate the crystals from the syrup, thus removing the greater part of the impurities from the input sugar and leaving the crystals ready for dissolving before further treatment. The liquor which results from dissolving the washed crystals still contains some colour, fine particles, gums and resins and other non-sugars.

Affinity chromatography: Chromatography technique in which an immobilized ligand is used to retain an analyte that is later eluted under conditions where the binding affinity is reduced. The ligand, which may be a substance such as an enzyme, hormone or antigen, is bound to a matrix such as silica. It is used for isolation and purification of all biological macromolecule. It is used to purify nucleic acid, antibodies, and enzymes. It noticed which biological compounds bind to a particular substance. It reduces an amount of substance in a mixture. It is used in Genetic Engineering - nucleic acid urification. It helps in the Production of Vaccines - antibody purification from blood serum And Basic Metabolic Research - protein or enzyme purification from cell free extracts

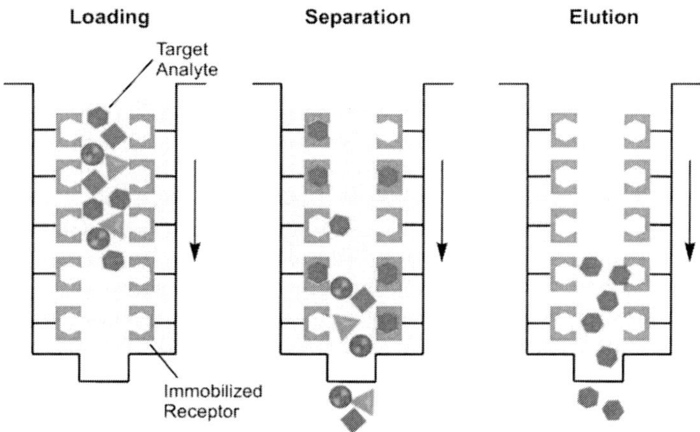

Aflatoxicosis: Mycotoxicosis caused by ingestion of aflatoxins in contaminated foods or feeds.

Aflatoxin B_1: A potent hepatocarcinogen. Toxic too many species, including humans, birds, fish and rodents.

Aflatoxin B_2: Moderate hepatocarcinogen compared with aflatoxin B_1.

Aflatoxin B_3: Hepatocarcinogen with a toxicity similar to that of aflatoxin B_1.

Aflatoxin D_1: Carboxylated product of aflatoxin B_1. Possesses less toxicity than aflatoxin B_1.

Aflatoxin G_1: Potent hepatocarcinogen with a toxicity similar to that of aflatoxin B_1.

Aflatoxin G_2: Dihydroxylated product of aflatoxin G_1. Possesses less toxicity than aflatoxin G_1.

Aflatoxin M_1: Metabolic product of aflatoxin B_1 in animals. Usually excreted in the milk of cattle and other mammalian species that have consumed aflatoxin B_1-contaminated foods or feeds. Possesses less toxicity than aflatoxin B_1.

Aflatoxin M_2: Metabolic product of aflatoxin B_2 in animals. Usually excreted in the milk of cattle and other mammalian species that have consumed aflatoxin B_2-contaminated foods or feeds. Possesses less toxicity than aflatoxin B_2.

Aflatoxin P_1: Demethylated and hydroxylated product of aflatoxin B_1 found in animals. Very weak toxin compared to aflatoxin B_1.

Aflatoxin Q_1: Main metabolite of aflatoxin B_1 found in humans and primates.

Aflatoxin: A toxin produced by certain moulds which grow on stored grains and nuts in tropical areas.

Aflatoxins: A toxin or mycotoxins produced by certain strains of the fungi/mould i.e. *Aspergillus flavus* and *A. parasiticus*. It formed during the growth of these fungi on stored commodities such as corn, peanuts, cottonseeds and soybeans. The toxin tends to cause low-grade illness and reduced food absorption and growth and is thus not easily noticed unless specially tested for. It is most serious in animal feeds.

Agar/Agar-agar: A natural gelling agent/extract obtained from various species of red seaweeds belonging to *Eucheuma, Gelidium* and *Graciliria* genera. It contains agarose and agaropectin polysaccharides. It sets following dissolution in warm water to form agar gels, which are widely used as thickeners and stabilizers in the food industry and sometimes used instead of gelatine. Additionally, used in gelling agents to prepare culture media for bacteriological plate counts. The gel has a very high melting point and it will only dissolve in boiling water. It is also called seaweed gelatine, seaweed jelly, Japanese gelatine, macassar gum.

Ageing: The process of leaving foods to age or artificially ageing them using chemicals, e.g. Changing meat and game to improve the flavour and tenderness, adding oxidizing agents to flour to produce a stronger dough. Ageing includes the intentional storage of foods and beverages to induce desirable changes in sensory properties, such as for wines and cheeses (also ripening). The term is also used to denote the artificial hastening of this process, such as treatment of flour with ammonium persulfate to produce a more resilient dough. Chemicals used to age (improve) flour include ammonium persulphate, ascorbic acid, chlorine, sulphur dioxide, potassium bromate and cysteine. In addition, nitrogen peroxide or benzoyl peroxide may be used to bleach flour, and chlorine dioxide both to bleach and age.

Agglomeration: The process by which particles or items are collected together and formed into a mass. Agglomeration by definition is a process during which primary particles are fixed together to form larger, porous secondary particles. Agglomeration has many applications in food processing and major applications include easy flow table salt, dispersible milk powder and soup mix, instant chocolate mix, beverage powder, compact cubes for nutritional-intervention program, health bars using expanded/puffed cereals, etc. The main purpose of agglomeration is to improve certain physical properties of food powders such as bulk density, flowability, dispersability, and stability. Agglomerated products are easy to use by the consumers and hence are preferred over the traditional non-agglomerated products that are usually non-flowable in nature.

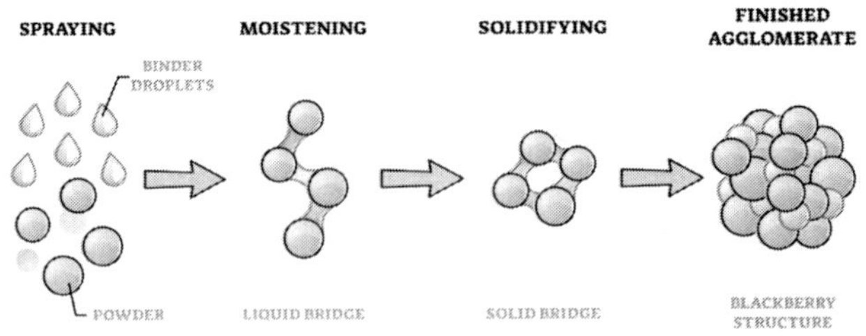

AGGLOMERATION

Aggregation: The process for forming a whole by combining several different elements or items.

Agitation: The process of stirring, shaking or disturbing briskly, particularly applied to a liquid. Agitation processing not only has the potential to improve quality retention, but also to reduce energy consumption during processing.

Agricultural produce/products: Collective name for crops and other commodities obtained as a result of agriculture and used for provision of food, fibre or livestock, other materials. Examples include fruits, cereals, cotton and livestock.

Air drying: Removal of moisture or liquid from a substance using air, or to preserve an item by evaporation. In air drying processes, the gaseous stream that leaves the dryer usually has an energy content that makes reuse possible, even when its moisture content is higher than the moisture content of the air entering the dryer. For this reason, the recirculation of air leaving the dryer is common practice and allows the global drying process to be cheaper.

Air quality: It is measure of the condition of the air, especially with respect to the requirements for specific environments. In food processing and packaging facilities, air quality is important for food safety and shelf life, and health of personnel. Special filtration systems are used to remove airborne hazards such as microorganisms, insects and dust from the atmosphere.

Airflow properties: It is the characteristics of the flow of air through, or across the surface of, a substance or piece of equipment. Airflow properties are utilized in designing ovens and driers and in determining the most appropriate ways of storing large quantities of foods such as fruits, vegetables, cereals and carcasses in order to minimize spoilage.

Albumen index: A measure of egg quality equal to the ratio of the height of a mass of egg white to its diameter as it sits on a flat surface. A higher index indicates higher quality. As the egg deteriorates, so the albumin spreads further, i.e. the albumin index decreases.

Albumen: The white of egg, the nutritive material surrounding the yolk of the eggs of higher animals, which also contains various anti-bacterial substances. Hen's egg albumen consists of the protein albumin (9 – 12%) and water (84 – 87%), in total about 58% of the weight of the egg. It forms a very stable emulsion with air and hence is used in a variety of air-raised dishes. It may be dried to a powder for reconstituting or deep-frozen. Like all proteins, these denature and coagulate when heated or treated with various chemicals e.g., Alcohol.

Albumins: Proteins which are soluble in water or dilute salt solutions and coagulable by heat. Albumins occurring in foods include conalbumin, lactalbumins and ovalbumins.

Alcohol free beverages: Beverages of types normally containing ethanol, which have been formulated or processed to be free from ethanol.

Alcohol The generic name for a class of chemical compounds which includes ethanol, the commonest mood-altering substance used in all cultures. Most are toxic, particularly methanol which is found in methylated and surgical spirits and many household liquids. Propanol and butanol, the other common alcohols, are used as solvents and in cleaning agents.

Alcoholic beverages: Beverages containing a significant concentration of ethanol. Major types include beer, wines, spirits, liqueurs and rice wines.

Alcoholic fermentation: Process by which certain microorganisms (mainly yeasts) metabolize sugars anaerobically to produce alcohols. In this process, glucose is converted to pyruvic acid, which is decarboxylated to acetaldehyde. The acetaldehyde is subsequently reduced to ethanol. A wide variety of substrates can be used to produce alcoholic beverages, e.g. grain for production of beer, and grapes and other fruits for production of wines. However, the constituent sugars must be released from these substrates prior to fermentation. Fermentation can be carried out by endogenous yeasts or by addition of starters. The most common yeasts used in the manufacture of alcoholic beverages are *Saccharomyces cerevisiae* and *S. carlsbergensis*. Synonymous with ethanolic fermentation.

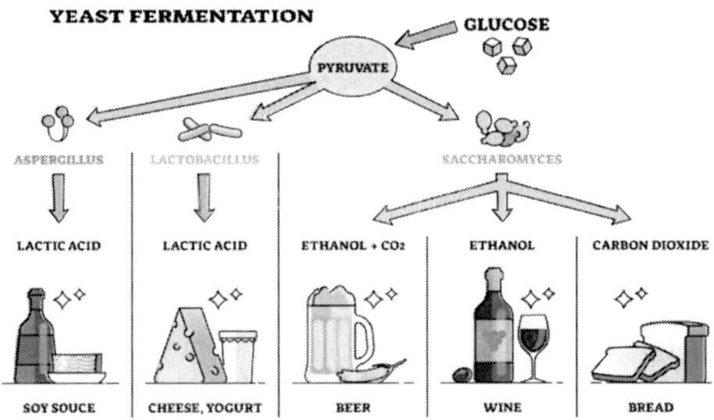

Alcoholic soft drinks: Beverages with flavour and other properties typical of soft drinks (e.g. fruits flavoured beverages), but with addition of a significant concentration of alcohol.

Aldoxime: An artificial zero-calorie sweetener, 450 times sweeter than sucrose.

Ale: An alcoholic drink (beer type) made from top fermented extract of malted grains. The name is generally nowadays applied only to crystal (lightly coloured) malts. Ale may be flavoured e.g. with hops.

Aleurone layer: Layer of cells found under the bran coat and outside the endosperm of cereal grains. Rich in cereal proteins and minerals as well as containing non-digestible carbohydrates and phytic acid. It forms about 3% of the weight of the grain, and is rich in protein, as well as containing about 20% of the vitamin B_1, 30% of the vitamin B_2 and 50% of the niacin of the grain. Botanically the aleurone layer is part of the endosperm, but in milling it remains attached to the inner layer of the bran.

Alfa-Acetolactate: Precursor of the flavour compound diacetyl.

Alfa-Tocopherol: The major contributor to vitamin E activity in foods. Rich sources of this fat-soluble vitamin include vegetable oils, margarines, wheat germ, nuts, seeds, sea foods, beef, eggs, fruits and vegetables. α-Tocopherol is a powerful antioxidant that protects polyunsaturated fats and vitamin A from oxidation in the gastrointestinal tract. α-Tocopherol also prolongs the life of red blood cells and protects lung tissue from the adverse effects of pollution. α-Tocopherol is included among GRAS substances and is one of the antioxidants used in the food industry to retard rancidity in foods containing polyunsaturated fats.

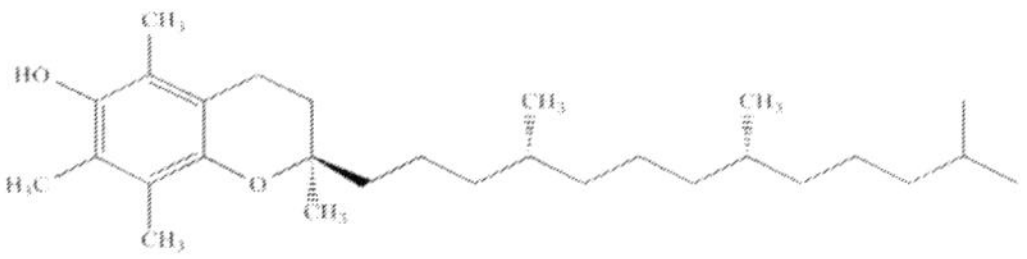

Algae: A heterogeneous group of unicellular and multicellular eukaryotic photosynthetic organisms which most occur in aquatic habitats. Certain algae are harvested for commercial production of thickeners (e.g. agar, alginates, carrageenans) or proteins (e.g. single cell proteins). A few algae produce toxins that may cause food poisoning in humans via con- sumption of fish and shellfish. Chlorella and Spirulina, which are used as health foods, and the larger seaweeds, used either as vegetables or as a source of various food-thickening agents. See also agar-agar, alginic acid, carragheen, laver bread, nori.

Algicides: Chemicals used to control growth of algae in water bodies or water containers. Examples include bethoxazin, dichlone, quinoclamine and simazine.

Alginates: Any of several derivatives of alginic acid (e.g. sodium, calcium or potassium salts or propylene glycol alginate). Used as stabilizers, thickeners and gelling agents in foods.

Alginic acid: A carbohydrate acid or a Polysaccharide (polymer of D- mannuronic acid), obtained from various sea weeds as well brown algae such as *Macrocystis pyrifera* or *Laminaria* and used as a thickening and gelling agent in commercial ice cream and convenience foods. It is also supplied as the sodium salt, sodium alginate, used mainly in ice cream, the potassium salt, and the calcium salt, which forms a much stiffer gel than the other salts. It is also available as an ester, propane 2-diol alginate. It possesses significant hydro-colloidal properties making it suitable for thickening, emulsifying and stabilizing applications. It is authorized for use in foods in various forms, including as sodium, calcium and potassium alginates.

Alitame: High intensity dipeptide sweetener formed from L-aspartic acid, D-alanine and a novel amine. It has good water solubility, and sweetness is approximately 2000 times that of sucrose at typical usage levels. Alitame has the potential for use in a broad range of foods and beverages, such as dairy products, frozen desserts and chewing gums, and in tabletop sweeteners.

Alkali: An alkali is a substance containing a deficiency of hydrogen ions (or an excess of hydroxyl ions) compared with pure water. An alkali (e.g. Sodium hydroxide) will combine with an acid (e.g. Alginic acid) to form a neutral salt (e.g. Sodium alginate). The most common culinary alkalis are sodium bicarbonate and ammonium carbonate both of which liberate carbon dioxide gas when heated or combined with an acid such as tartaric acid.

Alkalinity: A measure of the strength of an alkali on the scale of pH which goes from 7 (neutral) through 7.7 to 9.2 (egg white), 8.4 (sodium bicarbonate or baking soda), 11.9 (household ammonia) to 14 (highest value of alkalinity). Strong alkalis such as sodium hydroxide react with fat in the skin to form soaps which give a characteristic slimy feel.

Alkalization: Process by which the pH of a substance is increased to above 7 making it alkaline.

Alkaloids: The Organic nitrogenous base components of plants responsible in most cases for their bitter taste and therapeutic effects. Other well known alkaloids are cocaine, strychnine, atropine and morphine. Many have pharmacological activity. Some foods contain toxic alkaloids, e.g. solanine in potatoes. Some alkaloids are desirable food constituents, e.g. the purine alkaloids caffeine and theobromine in tea, coffee, chocolate and cocoa.

Allergens: Compounds, either proteins or smaller chemicals, which react with the proteins of the skin and mucous membranes (eyes, nose, lungs, gut, etc.). Of some persons to elicit the immune response of the body causing rashes, irritation, asthma, etc. Typical food allergies occur with strawberries, eggs, milk, cereal, fish, peas, and nuts. The response should not be confused with that produced by enzyme deficiencies as e.g. with the inability to metabolize lactose, found in most Chinese, and the inability to metabolize gluten, which is fairly common in the West.

Allergies: Hypersensitivity states induced by the body in reaction to foreign antigens that are harmless to other individuals in similar doses. Allergic reactions are of four basic types and can be immediate or delayed in their onset. Type I reactions, which involve release of histamine from mast cells by immunoglobulin E, can be induced by many food allergens often resulting in respiratory and dermatological symptoms. Severe type I reactions include anaphylaxis. Most foods have been demonstrated to produce allergic re- actions in certain individuals; however, common causes of food allergy in adults include shellfish, nuts and eggs. In children, the pattern of food allergy differs from that in adults, with allergies to eggs, milk, peanuts and fruits being common. In contrast to adults, children can outgrow allergies, especially to milk and soy infant formulas.

Allicin: One of the organic sulfur compounds occurring in onions and other *Allium* spp. vegetables. Important flavour compounds fraction with antibacterial properties.

All-purpose flour: All-purpose flour is white flour milled from hard wheat or a blend of hard and soft wheat. It gives the best results for many kinds of products, including some yeast breads, quick breads, cakes, cookies, pastries and noodles. All-purpose flour is usually enriched and may be bleached or unbleached. Bleaching will not affect nutrient value. Different brands will vary in performance. Protein varies from 8 to 11 percent.

Allyl isothiocyanate: Naturally occurring volatile organic sulfur compounds found in *Brassica* vegetables and some other plants, such as cassava. Largely responsible for the pungency of foods such as mustard and horse radish. Possess antimicrobial properties and are used in food preservatives and as anti-fermentative agents in winemaking. Like other isothiocyanates, display goitrogenic properties.

Almond biscuits: Ground almonds and caster sugar folded into stiffly beaten egg white, piped onto rice paper, decorated and baked at 180°C.

Almond butter: A compound butter made from freshly shelled sweet almonds pounded to a fine paste with twice their weight of butter and passed through a fine sieve. It is also called amandine butter.

Almond cream: A sweet made from a mixture of ground almonds with egg yolks and orange flower water, thickened in a bain-marie and garnished with almond praline.

Aloo bhurta: Boiled potatoes mashed with 3 teaspoon per kilogram of minced red peppers, chopped chives and molten butter, 2 teaspoon of chopped parsley or coriander leaves and a little powdered bay plus salt and pepper. It is served as hot or cold.

Aloo chat: A spiced potato dish made from large diced potatoes, fried in ghee until coloured, mixed with minced onions and garlic, chat masala, chilli powder, seasoning, crumbled bay leaf, chopped red chilli and chopped fresh coriander leaves, then cooked until tender in a covered pan with a little water if necessary.

Alpha-linolenic acid: A fatty acid found in ester form in high proportion in walnut oil, purslane and soya oil.

Alum: Double salts of aluminium sulfate and the sulfate of a monovalent metal. Alums are used as coagulating agents in purification of drinking water. Other food applications include use in colorants, manufacture of cured jellyfish products, modification of properties of starch, and use in processed fruit products and vegetable products.

Aluminium foil: Aluminium packaging materials which are used to decorate, protect and preserve foods, providing a barrier to external factors, such as light, oxygen and water vapour. Food applications include: foil containers and lids; metallized films; and wrappings. Also used in laminated packaging to enhance the barrier properties and rigidity of other packaging materials such as plastics and paper. There is very little migration of aluminium from aluminium foil containers into food. Environmental considerations include the importance of recycling and the use of aluminium foil laminates to fuel incineration processes. Aluminium metal rolled until very thin and flexible. Because it is relatively inert and does not easily corrode. It is also called cooking foil.

Aluminium phosphide: Synonym for phostoxin. Used in fumigants for stored grain, as it releases the toxic gas phosphine.

Aluminium potassium sulphate: A firming agent used in e.g. Chocolate-coated cherries. It is also used as a preservative and firming agent for pickles in Southeast Asian cooking. Also called alum.

Alveograms: Records of air pressure inside bubbles formed by inflating pieces of dough until rupture, a test performed on alveographs.

Alveograph: The instrument which measures the strength (stretching quality) of dough as an index of its protein quality prior to baking by injecting air into it at a fixed depth and flow rate through a nozzle and recording the pressure at which the resulting bubble bursts. A standard disc of dough is blown into a bubble and the pressure curve and bursting pressure are measured to give the stability, extensibility/distensibility and strength of the dough. In other words, it is used to analyze the physical properties of dough and the baking properties of wheat.

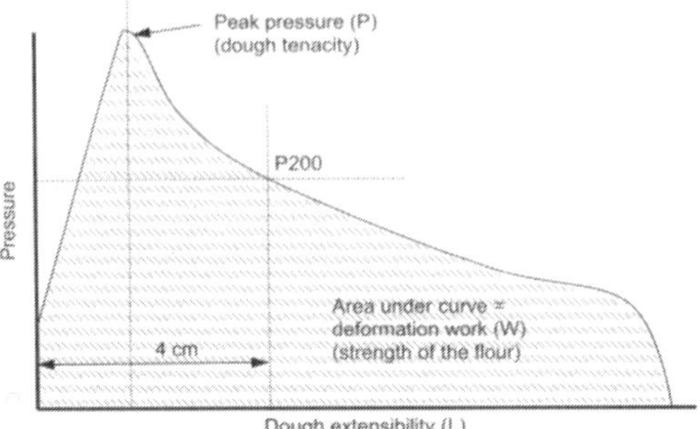

Amadori compounds: Intermediates of the Maillard reaction occurring between amino groups and reducing sugars. Amadori compounds are produced by rearrangement of nitrogen-containing carbohydrate ring structures and their fate is dependent on the conditions present in the reaction medium. Acid hydrolysis of these compounds can result in unsaturated ring systems that have a characteristic flavour and aroma, which under less acidic conditions may polymerize to form an insoluble dark coloured material.

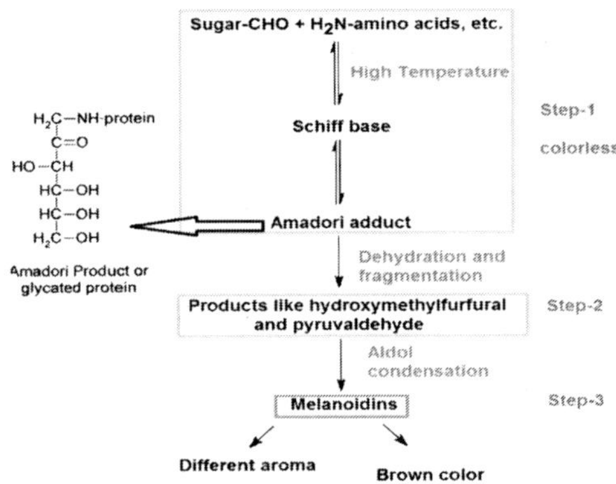

Amatoxins: Powerful toxins produced by several species of mushrooms of the genus *Amanita* (such as *Amanita phalloides* (Death Cap), *A. virosa* (Destroying Angel) and *A. verna* (Fool's Mushroom). Ingestion results in abdominal pain, persistent vomiting and watery diarrhoea, usually followed by death due to organ failure.

Amchoor: A flavouring similar to tamarind made by drying slices of unripe mango, *Magnifera indica*. It is also supplied in powdered form and is used for souring and tenderizing. It is known as aamchur, amchar, amchor, amchur, mango powder.

Amino acid: an organic acid which is a building block of proteins. The sequence of amino acids in a protein determines its properties. 20 amino acids are incorporated in human tissue. The basic units from which proteins are made. Chemically compounds with an amino group (—NH_2) and a carboxyl group (—COOH) attached to the same carbon atom. Thirteen of the amino acids involved in proteins can be synthesized in the body, and so are called non-essential or dispensable amino acids, since they do not have to be provided in the diet. They are alanine, arginine, aspartic acid, asparagine, cysteine, cystine, glutamic acid, glutamine, glycine, hydroxyproline, proline, serine and tyrosine. Nine amino acids cannot be synthesized in the body at all and so must be provided in the diet; they are called the essential or indispensable amino acids: histidine, isoleucine, leucine, lysine, methionine, phenylalanine, threonine, tryptophan and valine. Arginine may be essential for infants, since their requirement is greater than their ability to synthesize it. Two of the non-essential amino acids are made in the body from essential amino acids: cysteine (and cystine) from methionine, and tyrosine from phenylalanine.

Amino sugars: General term for sugars substituted with an amino group at the carbon-2 position. Examples of amino sugars include galactosamine, glucosamine and furosine, an important indicator of Maillard reaction in dairy products.

Ammonium compounds: Group of compounds containing the NH_4 radical. In the context of foods, important members include betaine, inorganic am- monium salts (e.g. ammonium bicarbonate used as a leavening agent, and ammonium salts used as nutrients for yeasts) and quaternary ammonium compounds used as disinfectants.

Amorphous sugar: Non-crystalline sugar made by melting sucrose and cooling it rapidly to form a transparent glass-like material.

Amydated pectin: A synthetic derivative of pectin which is more stable than the parent product.

Amyl acetate: An ester formed from amyl alcohol and acetic acid used to flavour the confectionery item, pear drops (NOTE: Some say it has a banana flavour.)

Amyl butyrate: An ester formed from amyl alcohol and butyric acid used as an artificial flavouring.

Amyl valeriate: An ester formed from amyl alcohol and valerianic acid with an apple-like flavor.

Amylase: An enzyme (a naturally occurring catalyst of chemical reactions) found in saliva and gastric juices which breaks down starch (a polysaccharide) into smaller subunits which are further broken down by other enzymes or which can be directly absorbed, usually as glucose, into the blood stream. Enzymes that hydrolyse the α-1,4 glycosidic linkages in both amyloses and amylopectins. Act on starch, glycogen, and related polysaccharides and oligosaccharides. Specific types are α-amylases and β-amylases. Strains obtained commercially by fermentation from *Aspergillus niger* and *Bacillus subtilis* are used in the manufacture of syrups, in the brewing industry and generally in food manufacturing. Also called diastase.

Amylograms: Records of results obtained using amylographs to investigate flour or starch viscosity as a function of temperature.

Amylographs: Instruments used to measure the viscosity of cereal flours/flour paste or other starch-based products during variations in temperature from 25°C to 90°C (as occurs in baking). Samples are mixed at a constant speed and viscosity is recorded on charts (amylograms). It is also a measure of the diastatic activity of the flour. The amylograph provides information on starch consistency and gelatinization behaviour of the flour. A low gelatinization maximum points to a stronger liquefaction of the starch and thus to a high alfa-amylase activity. Doughs from such flours have a sticky and damp upper surface and the baked goods made from these colour more strongly. Conversely, a very high gelatinization maximum is a sign of flours weak in enzymes and whick bake dry. Doughs from such flours are week in proof and the baked goods only have adequate colouring.

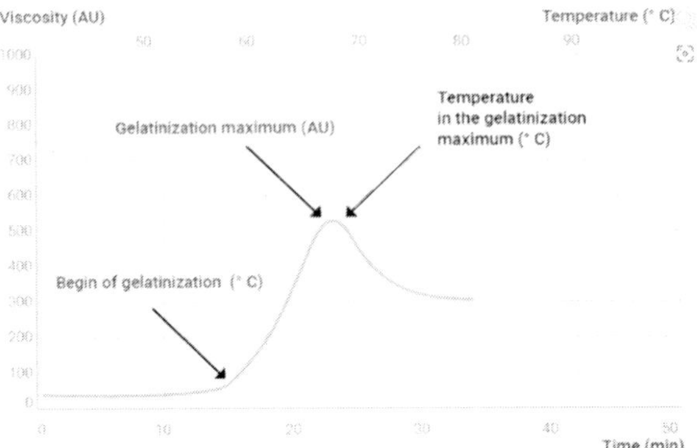

Amylolytic enzyme: Any of an important class of enzymes that degrade starch to mono-, di- and polysaccharides. It is important in bread-making.

Amylopectin: High molecular weight polymers that, together with amyloses, form starch. Composed ofα-1,4-linked glucopyranose chains connected by α-1,6- linkages. 3-6% of glucose residues are α-1,6-linked, giving rise to a highly branched polymer. Starch that is almost exclusively composed of amylopectin is termed waxy, e.g. waxy corn (>99% amylopectin and <1% amylose); in starch of this type, retrogradation is slow or absent, thus pastes of gelatinized waxy starch are non-gelling but gum-like. The main and more easily digested component of starch.

Amylose: Polysaccharides composed of chains of α-1, 4-linked glucopyranose residues that, together with amylopectins are constituents of starch. Amyloses have much lower molecular weights than amylopectins (at least 100-fold less) and are non-branched. In contrast to amylopectins, retrogradation of cooked amyloses is rapid, and thus gel formation occurs. The minor and less easily digested constituent of starch.

Anaerobe: A microorganism which will grow and reproduce in the absence of air. Facultative anaerobes will grow with or without air. Strict (i.e. Non-facultative) anaerobes are poisoned by the oxygen in the air.

Analogues: In relation to foods, products that are made to resemble and act as substitutes for specific commodities i.e. similar to simulated foods. Reasons for producing analogues include to provide alternatives to meat for vegetarians, for consumption by those with special dietary requirements or to reduce costs.

Anaphylaxis: A severe type I allergic reaction occurring rapidly in sensitized individuals following exposure to small amounts of allergens. Symptoms can range from itching and angioedema to widespread tissue oedema, airway constriction, respiratory distress and circulatory collapse. Foods that can induce anaphylaxis include peanuts, eggs and sea foods.

Anardana: Dried seeds of sour/wild pomegranates (*Punica granatum*). It is added as condiments or acidulants to a number of Indian foods including chutneys and curries and in fillings for bread and savoury pasties, and with braised vegetables and pulses.

Anatoxins: Neurotoxins produced in fresh water by some species of filamentous cyanobacteria of the genus *Anabaena*, especially *A. flosaquae*. Include the alkaloids anatoxina and anatoxina(s). Extremely poisonous, sometimes killing animals drinking contaminated water within a few minutes. It may represent a hazard for drinking water safety.

Anhydrous milk fats: Milk fats with a very high fat content and negligible moisture content. Sometimes it called water free milk fats.

Animal carcasses: Dead bodies of animals, particularly those used for meat production. The term is used by butchers to describe animal bodies after removal of the heads, limbs, hides and offal; these processed carcasses are also called dressed carcasses. Conditioning or ageing of carcasses results in breakdown of muscle glycogen into lactic acid, which tends to improve tenderness and shelf life of meat.

Animal fats: Lipid products derived from animal sources. Include butter, lard, tallow, suet and fish oils.

Animal foods: Foods derived from sources in the animal kingdom. Examples include aquatic foods (sea foods and aquaculture products), dairy products, eggs and egg products, animal fats, insect foods, meat and meat products and other animals such as worms.

Animal proteins: Proteins that are derived from animal sources, such as meat, fish, eggs and dairy products.

Animal rennet: The enzyme proteinase present in the abomasum of young ruminants, e.g. calves, and used for clotting of milk during cheese-making. Comprise a mixture of the main enzyme, chymosin, and pepsin, the ratio of the two enzymes affecting the final properties of the cheese. Due to shortages of animal rennets and the increasing popularity of vegetarian cheeses, microbial rennets, genetically-engineered enzyme preparations synthesized by various microorganisms and milk clotting enzymes of plant origin (vegetable rennets) have been developed.

Anisole: The phenolic compounds which occur naturally in a range of foods. Chlorinated anisoles derivatives may cause taints, e.g. in corks and wines.

Annatto: Yellowish red natural colorant obtained from seeds of the tropical tree *Bixa orellana*. It contains a fat soluble component (bixin) and a water-soluble component (norbixin). Bixin is the monomethyl ester of a dicarboxyl carotenoid. Norbixin is the saponified form, a dicarboxyl acid of the same carotenoid. The carboxylic acid portion of the molecule contributes to water solubility and the ester form contributes to oil solubility. Annatto is available in both water soluble and oil soluble liquids and powders. Annatto is somewhat unstable to light and oxygen but, technically, it is a good colorant. The principal use of annatto is as colorant for dairy products due to its water solubility but it is also used to impart a yellow to red color in a wide variety of products. It is also used to add colour to cheese, sausage casings and bakery products.

Anthocyanin/ anthocyan: A glucoside (of malvidin, pelargonidin, peonidin, cyanidin, delphinidin and petunidin) plant pigment extracted from grape skins which is red in acid and blue in alkaline conditions. It forms coloured complexes with metals e.g. Grey with iron, green with aluminium and blue with tin, hence the off colours in e.g. Canned pears where its reaction with tin gives the pears a pink colour. Extracted anthocyanins may be used as food colorants. Colour is pH sensitive, and stability differs from that of artificial colorants.

Anthoxanthan: A colouring compound in potatoes and onions which is colourless in acid and yellow in alkaline conditions.

Antibacterial activity: Ability to kill or inhibit the growth of bacteria.

Antibacterial compounds: The compounds that possess antibacterial activity, e.g. certain antibiotics, antiseptics and disinfectants.

Antibiotics resistance: Ability of microorganisms to be unaffected by treatment with specific antibiotics. Resistance can result from a range of mechanisms, including decreased permeability of the organism to the drug, modification of drug or receptor, and production of a modified protein that is unaffected by the antibiotic. Organisms can become resistant either by undergoing spontaneous mutations or by acquiring resistance genes from other resistant organisms through the processes of conjugation and transduction. Plasmids containing multiple resistance genes can be transferred not only amongst similar but also quite different bacteria.

Antibiotics: The chemicals/substances produced by living microorganisms that inhibit the growth of other organisms. The first antibiotic to be discovered was penicillin, which is produced by the mould *Penicillium notatum* and inhibits the growth of sensitive bacteria. Many antibiotics are used to treat bacterial infections in human beings and animals; different compounds affect different bacteria. Small amounts of antibiotics may be added to animal feed (a few grams per tonne), resulting in improved growth, possibly by controlling mild infections or changing the population of intestinal bacteria and so altering the digestion and absorption of food. To prevent the development of antibiotic-resistant strains of disease-causing bacteria, only those antibiotics that are not used clinically are permitted in animal feed (e.g. nisin, which is also used as a food preservative. The limits of these residues are strictly controlled as continuous ingestion could develop antibiotic-resistant microorganisms in the human gut.

Antibodies Proteins: It is also known as immunoglobulins that are produced by the body in response to foreign substances (antigens) and are capable of forming complexes with the antigens. Mechanisms by which antibodies protect the body include agglutination or precipitation of foreign antigens, lysis of foreign cells, and neutralization of toxins.

Anticaking agent: Anhydrous compounds that are added in small amounts to dry or powdered foods (e.g. salt, baking powders, pudding mixes) to prevent their particles sticking together and thus ensure the product remains dry and free flowing. Typical anticaking agents for the food industry include magnesium and calcium carbonates, magnesium stearate, calcium silicate and calcium stearate.

Anticarcinogenicity: Ability of a food or food component to slow, inhibit or reverse the process of carcinogenesis, in particular, the ability to attenuate carcinoma formation in response to application of known carcinogens. Anticarcinogenicity of a substance can be determined *in vitro* using cell culture or *in vivo* using animals treated with carcinogens or a carcinoma cell line.

Anticarcinogens: Substances that inhibit the formation of carcinomas induced by application of carcinogens. Potential dietary anticarcinogens include phytoestrogens (isoflavonoids, lignans), flavonoids, lycopene, glucosinolates, terpenes, allyl sulfides and simple phenols.

Antifoaming agents: It is used in a similar manner to defoaming agents to control foams formation during food processing. Examples include dimethylpolysiloxane.

Antifreeze proteins: Proteins occurring naturally in a range of organisms (especially cold water fish), which prevent or minimize freezing of tissues on expo- sure to low temperatures. Of potential use in the food industry for lowering the freezing point of foods and inhibiting recrystallization of ice. Possible applications include in ice cream, frozen foods or chilled meat products.

Antifungal compounds: Compounds that possess antifungal activity.

Antifungal activity: Ability to kill or inhibit the growth of fungi.

Antifungal agents: Substances that possess antifungal activity.

Antigenicity: Ability of substances to act as antigens by eliciting an antibody-mediated or cellular immune response.

Antigens: Substances that induce an immune response, either by stimulating formation of anti- bodies or eliciting a cellular response.

Antihypertensive activity: Ability of a substance to alleviate or reduce high blood pressure (hypertension). Food components that demonstrate antihypertensive activity often act as inhibitors of angiotensin I- converting enzyme (ACE; peptidyl-dipeptidase A). Potential dietary antihypertensive agents include bioactive peptides in dairy products, adenosine in plant foods and garlic constituents.

Anti-inflammatory activity: Ability to inhibit or counteract the inflammatory response, which is an innate immune response to tissue injury by stimuli such as chemicals, trauma, extremes of temperature or microbial attack. Many foods and food components possess anti-inflammatory activity. These include some fatty acids, tocotrienol, lactoferrin, colostrum, wines and honeys.

Antimicrobial activity: Ability to kill or inhibit the growth of microorganisms.

Antimicrobial packaging films: Antimicrobial packaging materials are food preservation systems with antimicrobial attributes that involve more than basic barrier properties. These attributes can be achieved by incorporating microbialcidal or microbialstatic agents into such a packaging system, or by using active antimicrobial polymeric materials. The introduction of antimicrobial agents into food packaging material helps to extend the shelf-life of food products by inhibiting the growth of microorganism. Antimicrobial packaging systems can extend the lag period and reduce the growth rate of microorganisms, prolonging a product's shelf stability and promote safety. Packaging films, e.g. polyethylene films that contain an antimicrobial substance such as chlorine dioxide.

Anti-nutritional factors: Substances that reduce the nutritional value of a food by reducing its nutrients content, bioavailability, digestibility or utilization. Anti-nutritional factors include enzyme inhibitors (proteinases inhibitors and amylases inhibitors present in a wide range of foods and microorganisms), inositol and its derivatives (including phytates and phytic acid present in legumes and cereals) and anti-vitamins such as thiaminase, dicoumarol, theophylline.

Antioxidants: the substances/chemicals used in preservation of foods by retarding deterioration, rancidity or discoloration due to oxidation. The most commonly used food antioxidants include BHA (butylated hydroxyanisole), BHT (butylated hydroxytoluene) and propyl gallate. Naturally occurring antioxidant compounds include tocopherols and ascorbic acid. Typical examples are vitamin E (from soya beans), vitamin C (from citrus fruit or made synthetically). Antioxidant extend the shelf-life of foods by preventing or retarding lipid oxidant through different means such as competitive binding of oxygen, retardation of initiation step, blockage of propagation by destroying or binding free radicals, stabilization of hydroperoxides, inhibition of catalysts etc., and also reduce food losses and help in cost control of the product. Fat oxidation leads to spoilage of foods, production of toxic chemicals, reduction in nutritive value, development of off-flavours and can be hazardous to human health. The use of antioxidant has been essentially in snacks, confectionery, and milk as well as oil industry.

Anti-oxidative activity: Ability of a substance to inhibit oxidation. Substances possessing antioxidative activity can be utilized in foods, such as oils, to inhibit oxidation, thus improving shelf life and quality. Foods possessing a high antioxidative activity have also been investigated as potentially health promoting foods, as lipid oxidation has been associated with a range of pathological processes, including atherosclerosis.

Antiviral activity: Ability to kill or inhibit the growth of viruses. Many food components possess antiviral activity. These include lactoferrin and other constituents of milk and dairy products, polyphenols, tannins and polysaccharides from some mushrooms.

Antivitamins: Anti-nutritional factors that destroy or inhibit the metabolic effects of vitamins. Examples of antivitamins in foods include thiaminase (antivitamin B_1, present in raw fish and other animal foods), caramel colorants (antivitamin B_6) and dicoumarol (antivitamin K).

Appa: Spongy breads made from a ground rice, rice flour and coconut milk batter or dough, raised using the fermenting sap of the coconut palm (although yeast may be substituted) and cooked in small, greased and covered wok-like pans. Eaten during breakfast.

Apparent density: Weight of a porous material per unit volume. Apparent density of a porous substance is always lower than the theoretical density of its constituents. The apparent density of a shape of regular geometry can be determined from the volume calculated from the characteristic dimensions and mass.

Appearance: Perception of the outward form of a substance. The appearance of a food contributes to its overall sensory properties. Some scientists maintain that product appearance is inclusive of product color and other appearance properties such as physical form (shape, size, and surface texture), temporal aspects (movement, etc.), and optical properties (reflectance, transmission, glossiness, etc.). For our purpose we will discuss color and appearance as separate entities, while keeping in mind that appearance attributes clearly affect perceived color. Usually, physical appearance characteristics can easily be measured through sensory techniques.

Appertization: The term for the heat-processing of foods at temperatures above 120°C, in particular, retorting and high temperature short time (HTST) processing. This process does not guarantee complete sterility but any spores present should be non-pathogenic and unable to grow in the processed food environment.

Appetite: A natural longing to satisfy bodily needs, particularly, but not exclusively, the recurring desire for food. Appetite is increased in the state of hunger and decreased during satiety. Appetite for foods, in general, and for particular foods, may become modified over time. A particularly intense appetite for certain foods occurs during cravings.

Appetizer: A general name given to small items of food served before or at the beginning of a meal or at cocktail parties.

Aquaculture products: Aquatic organisms (such as fish, shellfish and aquatic plants) produced by aquaculture for food or industrial purposes.

Aquaculture: Production of aquatic organisms under controlled or semi-controlled conditions; mainly for food purposes. A wide range of aquaculture products, including farmed fish, farmed shellfish, aquatic plants and algae are produced commercially across the world.

Aquatic foods: Foods derived from aquatic organ- isms, including fish, shellfish, aquatic plants and algae.

Arabinogalactans: Polysaccharides in which the main constituent sugars are arabinose and galactose. It occurs in the pectic substance fractions of a wide range of plant foods, including fruits, vegetables and cereals. It may be of importance for the processing properties of plant foods.

Arabinose: Monosaccharide of five carbon atoms (pentoses) found predominantly in plants as a component of complex polysaccharides, such as gums and pectins.

Arabinoxylans: Polysaccharides in which the main constituent sugars are arabinose and xylose. Form part of the pentosans fraction in cereals and cereal products, and may be of importance for tech- nological properties in processes such as baking and brewing.

Arabitol: Polyol synthesized by reduction of arabinose or produced by microbial fermentation of plant hydrolysates.

Arachidic acid: One of the saturated fatty acids with 20 carbon atoms. It occurs at low concentrations in a wide range of fats, oils and tissue lipids.

Arachidonic acid: One of the ω-6 polyunsaturated fatty acids with 20 carbon atoms. It is widely distributed in foods and essential in the human diet.

Arachin: One of the two major globulins present in peanuts, the other being conarachin. As well as having good nutritional quality, both globulins play an important role in flavour development during peanut processing.

Argemone oils: Oils derived from any species of the genus *Argemone* (prickly poppies) which are found in North America and the West Indies. The oil is largely used for burning purposes and in medicine for treatment of syphilis and certain skin diseases. Its properties are those of a drying oil and it is used to a certain extent as a substitute for linseed oil in South America. It has a nauseous and acrid taste and when administered causes purging and vomiting and these characteristics preclude its use for edible and culinary purposes. Consumption of adulterated mustard oil (*Brassica nigra*) with argemone oil (*Argemone mexicana*) even for a short duration leads to a clinical condition referred as epidemic dropsy. In humans, argemone oil contained in adulterated mustard oil causes oxidative stress and death of red blood cells via met-hemoglobin formation by altering pyridine nucleotide(s) and glutathione redox potential. Argemone oil contamination poses a serious threat to human health and should be checked by appropriate regulatory measures. Antioxidant therapy provides symptomatic relief and should be seriously considered for therapeutic interventions against argemone oil toxicity.

Arginine: One of the basic amino acids, present in most food proteins and essential in the human diet.

Aroma concentrates: Concentrates typically obtained by extracting and/or concentrating volatile compounds from a source material, e.g. fruit juices, coffee or butter. It can be used as flavourings in various foods or to restore aroma lost during processing. Other methods of producing aroma concentrates include fermentation and enzymatic modification (e.g. for cheese flavour concentrates).

Aroma: Physiological sensation, also known as smell, that results from stimulation of olfactory receptors in the nasal mucosae and the interpretation of this information by a specialized area of the cerebral cortex. Food aroma, which is generated by release of volatile aroma compounds from the food, makes a marked contribution to overall flavour.

Aromatic compounds: An organic compound is characterized by a cyclic, conjugated structure, such as occurs in benzene. Some aromatic compounds, such as polycyclic aromatic hydrocarbons (PAH), may occur as toxic or carcinogenic contaminants in foods. Also refers, more generally, to flavour compounds or aroma compounds present in foods and beverages.

Aromatization: Procedure for increasing the aroma of a food or beverage. Strategies include the addition of aroma compounds to the product or container, and the facilitation of aroma compound release through chemical or mechanical means. Also refers to the chemical conversion of non-aromatic compounds into aromatic compounds.

Arrowroot: A starch powder obtained from the root of a West Indian plant, *Maranta arundinacea,* used for thickening where a clear, glossy glaze is required. It is also called araruta. It loses its thickening power if overcooked. Neutral in flavour and easily digestible, it is used as a thickener in invalid diets, and also in fruit sauces, pie fillings and desserts, where it imparts a clear finish. It can also refer to starch obtained from roots or rhizomes of several other tropical plants.

Artificial colorants: Colorants which have been manufactured synthetically, as opposed to those extracted from natural sources (natural colorants). Tend to be less expensive and have better colour intensity, uniformity and stability than natural colorants. Examples include azo-dyes and FDC colours.

Artificial flavourings: Flavourings which contain one or more artificial components not yet identified in a natural material. Synthetic flavourings containing the same chemicals as those found in a natural product are known as nature-identical. Synthetic flavourings are usually less expensive than natural flavourings, and less likely to vary in quality, availability and processing stability.

Artificial neural networks: Systems of computer programs and data structures which are modelled on the human nervous system and brain. Incorporate large numbers of processors operating in parallel, each with an individual sphere of knowledge which has been fed into it along with rules about relationships. Networks can use this information to recognize patterns in large amounts of data. Used in the food industry in model- ling of processes and predicting the behaviour of foods under specific conditions. Also known as neural networks.

Artificial sweeteners: Synthetic non-nutritive sweeteners, usually many times sweeter than su- crose. Examples include aspartame, saccharin and alitame. Widespread applications include low calorie foods, soft drinks and sugar free foods.

Asafoetida: Bitter, strong smelling resin extracted from the stem or roots of the umbelliferous plant *Ferula foetida* which grow in Afghanistan and Iran. The pungent **garlic**-like **aroma** and **flavour** are due to the presence of sulfur compounds. It used in **spices** for Asian foods, **pickles** and **Worcestershire sauces**. The powdered form is used sparingly in Indian cookery, and the taste improves when it is briefly fried in hot oil. The related garden plant is highly poisonous. It is also called hing.

Ascorbates: Salts of ascorbic acid, including so dium ascorbate and calcium ascorbate, which can be used as food additives. Food uses include as antioxidants in products such as meat products, as browning inhibitors for fruits and vegetables, and as bakery additives.

Ascorbic acid: Synonym for **vitamin C**, an antioxidant nutrient present in a wide range of foods. Necessary for growth of bones and teeth, for maintenance of blood vessel walls and subcutaneous tissues, and for wound healing; dietary deficiency results in scurvy. It is used in **food additives**, with applications in **food antioxidants** and **bakery additives**. It used to prevent oxidation and thus browning reactions and colour changes in foods and also used as a flour improver. Strictly, this should be known as L-ascorbic acid, the biologically active laevo-form.

Ascorbyl palmitate: A chemical compound derived from ascorbic and palmitic acids used as an agent to prevent staling.

Ascorbyl stearate: A chemical compound derived from ascorbic and stearic acids used as an antioxidant.

Asepsis: The prevention of contamination of a sterile or appertized food with organisms that can cause spoilage.

Aseptic packaging: Packaging technique in which an aseptic product is placed into an aseptic container in an aseptic environment. The sealed container is designed to maintain aseptic conditions until the seal is broken. It is used to enhance shelf life of foods, e.g. fruit juices. Advantages over conventional sterilization techniques include high product quality, optimization of sterilization, minimum energy consumption and low production costs. Aseptic packaging is not suitable for use with products containing large particles, and shelf life stability is shorter than for sterilized foods.

Aseptic processing: High-temperature, short-time process which results in products with improved texture, colour, flavour and nutritional values compared with conventional canning. This technology involves filling of pre-sterilized containers with a commercially sterile cooled product, followed by aseptic hermetic sealing with a pre-sterilized closure in an atmosphere free of microorganisms.

Ash: Mineral content of foods, determined by combustion of the sample under defined conditions and weighing of the residue.

Aspartame: A dipeptide made from the two amino acids, aspartic acid and the methyl ester of phenylalanine, used as an artificial sweetener with roughly 160-200 times the sweetening power of sucrose. Low calorie artificial sweetener (chemical name aspartyl phenylalanine methyl ester, also known as NutraSweet). It is Non-cariogenic and without an aftertaste. It loses sweetness on prolonged storage and exposure to heat (unsuitable for baking). Safe for diabetics but not for individuals with phenylketonuria as phenylalanine is released during metabolism of aspartame. It is used in soft drinks, yoghurts drink mixes and sweetening tablets.

Aspergillus flavus: The mould which produces aflatoxin. It is a strictaerobe.

Aspergillus parasiticus: A mould which grows on cereals, potatoes and onions and produces aflatoxin.

Aspergillus niger: One of the principal moulds used in the production of food-grade citric acid, and for the production of enzymes such as glucose oxidase and pectinase.

Aspergillus oryzae: A fungus or mould used to ferment rice in the first stage of soya sauce and **miso** production.

Aspergillus sojae: One of the moulds used to ferment soybeans and rice to make miso.

Aspirators: Instruments or equipment for drawing fluids by suction from vessels or cavities.

Astaxanthin: Pigments fraction of the carotenoids group, occurring in crustacea.

Astringency: A sensation of dryness in the mouth combined with roughening of the oral epithelium and puckering of the muscles of the face and cheeks. It is induced by foods containing chemicals such as tannins and other polyphenols, acids and aluminium salts. Sensory perception of astringency has been attributed to binding of tannins to salivary proteins.

Astringent: A liquid such as lemon juice or a solution of tannin which tightens up the skin or mucous membranes of the mouth.

Atherosclerosis: A pathological process resulting in thickening and hardening of the walls of medium and large arteries due to formation of atherosclerotic plaques. Cardiovascular diseases produced by occlusion of the affected arteries can be of gradual onset (angina, peripheral vascular disease) or sudden onset (stroke, myocardial infarction). Rate of development of atherosclerosis is affected by many factors including lifestyle and diet.

Atmospheric steamer: A device used to cook foods in live steam at atmospheric pressure, Condensation of steam on the surface of the food or its container gives the highest rate of heat transfer available with the exception of high-temperature radiant heat. The food temperature is limited to 100°C.

Atomic absorption spectroscopy: Technique in which the mineral composition of a sample is determined from the absorption of light by atoms. A monochromatic source of light at a specific absorption wavelength is passed through the sample following atomization by various means. It is often abbreviated to AAS.

Atomic emission spectroscopy: Technique in which the mineral composition of a sample is determined from the emission of light from excited atoms at wavelengths characteristic of the atoms.

Atomic force microscopy: Imaging technique in which the surface of the sample is scanned using a small tip to construct a 3-dimensional image. The tip may be in contact with or just above the surface. Molecular forces exerted against the tip by the surface are used by image processing software to give information about the surface.

Atomizers: Devices that convert a substance into very fine particles or droplets.

Atta: A wholemeal flour used for making unleavened breads such as chapatis, puri and paratha.

Authenticity: The genuineness of foods and beverages; can be with respect to various factors, such as ingredient content, processing methods and geographical origin. For certain foods and beverages, labelling schemes have been implemented to indicate authenticity. A range of methods is used to test authenticity depending on the potential method of adulteration.

Autoclave A high-pressure-steamed and steam-heated chamber which cooks food very rapidly and which is also used to sterilize canned and bottled food. It employed in processes using high pressures and temperatures, e.g. steam sterilization.

Autolysed yeast: Yeast cells which have been subjected to limited and controlled autolysis to liberate cell contents and change their flavour. The result which looks like thick molasses in appearance is used in vegetarian cooking and as a sandwich spread.

Autolysins: Endogeneous enzymes found in cell walls which can hydrolyze certain structural cell components (e.g. peptidoglycans in bacteria) to bring about autolysis.

Autolysis: The process by which cells are broken down by their own enzymes or acid products when dead. In the early stages cell contents are released adding to flavour but later the material becomes putrid. The process occurs when meat is hung or when fish goes bad.

Autolysis: Process by which the structural components of cells are degraded by their autolysins. It is usually occurs after the cells have experienced a traumatic event such as injury or death. It may result in the release of intracellular enzymes from cells, which may play an important role in cheese ripening. It can be responsible for inactive cultures or for sensory defects (by autolytic products) in wines and beer.

Automatic boiler: A boiler with an automatic water feed designed to produce fresh boiling water on demand.

Autoxidation: An autocatalytic oxidation reaction that occurs spontaneously in the atmosphere. Initiators include heat and light. Unsaturated fatty acids present in foods are susceptible to autoxidation when exposed to the air, with the reaction proceeding by a free- radical mechanism. The reaction may result in production of stable non-propagating products that contribute to off flavour. In addition, radicals produced by autoxidation may cause bleaching of food colour and destruction of vitamin A, vitamin C and vitamin E. This type of deterioration is prevalent in fried snack foods, nuts, oils, and margarines.

Availability: The extent to which dietary nutrients are present in a form that can be absorbed and utilized. Similarly, bioavailability.

Available nutrients: Nutrients in food which can be digested and absorbed in the body. Some become unavailable when bound to another compound, e.g. Avidin in raw egg whites binds biotin to make it unavailable. Cooking frees the biotin.

Avenasterol : Sterols fraction which occurs in the unsaponifiable fraction of many vegetable oils. In combination with other sterols, avenasterol concentration may be used as an index for identification and for monitoring the authenticity of vegetable oils samples.

Avenins: Glutelins present in oats; the major storage proteins of this cereal.

Avidin: A component of egg white which impairs the uptake and utilization of the vitamin biotin in the body. One of a number of protective chemicals in the white whose function is to prevent bacterial attack of the yolk. It is inactivated by heat.

Aw (Water activity): The symbol for water activity, which is a measure of the water vapour generated by the moisture present in a hygroscopic product. It is defined as the ratio of the partial pressure of water vapour to the partial pressure of water vapour above pure water at the same temperature. In foods, it represents water not bound to food molecules. Level of unbound water has marked effects on the chemical, microbiological and enzymic stability of foods.

Azo dyes: Series of artificial colorants containing at least 1 chromophoric azo group. Examples include amaranth, tartrazine, Sunset Yellow and Carmoisine.

Azodicarbonamide: Oxidizing bakery additives used to age and bleach cereal flour, and to condition dough for bread-making.

B

Baby corn: Small ears of immature corn, generally harvested between 2 days before and 3 days after silking. Baby (dwarf) corn is sold fresh or canned and generally measures around 4-9 cm in length and 1-1.5 cm in diameter. It is popular in Oriental cuisine.

Bacilli: Generally refers to any rod-shaped bacterial cells. It may be used specifically to refer to a member of the genus *Bacillus*.

***Bacillus cereus*:** A bacterium which grows on rice, usually found in temperate countries in spore form. The spore will germinate after cooking and lead to growth of the vegetative form which can cause food poisoning. There are 2 strains, one has an incubation period of 1 – 5 hours, duration of 6 – 24 hours and the symptoms are nausea and vomiting and occasionally diarrhoea; the other has an incubation period of 8 – 16 hours, duration of 12 – 24 hours and the symptoms are abdominal pain, diarrhoea and occasionally nausea.

***Bacillus licheniformis*:** Bacteria which can produce a toxin causing diarrhoea and vomiting. The incubation period is 4 – 8 hours.

***Bacillus subtilis*:** Bacteria which can produce a toxin causing vomiting, abdominal pain and diarrhoea. The incubation period is 2 – 18 hours.

Bacillus: Genus of Gram positive, aerobic or facultatively anaerobic, rod-shaped, spore-forming bacteria of the family Bacillaceae. It occurs in soil and water. *Bacillus cereus* can cause spoilage of pasteurized milk and cream, while other species can cause spoilage of low acid canned foods. Some species are used commercially as sources of enzymes (e.g. glucose isomerases, subtilisins). *B. cereus* can cause a diarrhoeal type of food poisoning (associated with consumption of meat, milk, vegetables and fish) or an emetic type of food poisoning (associated with consumption of rice, potatoes, pasta, sauces, puddings and soups). *B. thuringiensis* is an important insect pathogen used as an agent for bio-control. Some species have now been transferred to the new genus *Geobacillus*, including *G. stearothermophilus* and *G. thermoleovorans*.

Bacon: The flesh of pork, divided in two halves along the backbone, then dried and preserved by treatment with salt, saltpetre and spices. These remove water from the meat and have some antibacterial action. Commercial curing methods leave more water in the flesh by including polyphosphates and do not give bacon which keeps as well as the traditionally cured or dry-cured variety. Bacon may be smoked after curing to enhance the flavour. It is usually sold after boning. In order to decrease the retail price per kilo gram, some bacon manufacturers increase the weight of their product using water, phosphates and other ingredients. Most bacon is sliced into rashers before retail; middle rashers have a round eye of lean meat, whilst streaky bacon is the tail end of the loin. A rasher of bacon can contain up to 40% fat.

Bacteria: Heterogeneous group of usually unicellular prokaryotic microorganisms, generally possessing a characteristic cell wall, and found in virtually all environments. Some cause diseases in humans and animals, while others are used in the manufacture of foods (e.g. dairy products). Important in the food industries are *Salmonella, Bacillus cereus, Campylobacter, Staphylococcus, Listeria*, all of which are food poisoning organisms and *Lactobacillus* spp. And *Streptococcus* spp., etc. Which are used in the production of various foodstuffs. See under each heading.

Bacterial count: The number of bacteria per unit weight or volume of a foodstuff, which is indicative of food quality in some cases

Bacterial spoilage: Spoilage caused by the action of bacteria.

Bacterial spores: Spores (either endospores or exospores) formed by bacteria (e.g. Bacillus and Clostridium species) under conditions of nutrient limitation. Endospores are resistant and may be disseminative, rather than reproductive, while bacterial exospores are characteristically reproductive and disseminative. They are generally more resistant than vegetative cells to heat, desiccation, antimicrobial com- pounds and radiation, and can remain dormant for long periods.

Bactericides: Biological, chemical or physical agents that kill bacteria, but not necessarily their endospores. Include formaldehyde, peracetic acid, hydrogen peroxide or activated carbon.

Bacteriocins: Peptides produced by specific bacteria that possess antibacterial activity. Both purified bacteriocins and bacteriocin producing bacteria are used in the food industry, applications including inhibition of the growth of pathogens and spoilage organisms.

Bacteriostatic: The quality of a substance, such as an antiseptic, which stops the growth of

Bacteriostats: Chemical agents that inhibit the growth and multiplication of bacteria. It includes several disinfectants, spices and antibiotics.

Bacterium: A single-celled micro-organism which replicates by growth and then division. Some are beneficial, some harmful. See examples under entry bacteria.

Bactofugation: High speed centrifugation process used to remove most bacterial endospores, yeasts and fungi from milk, thereby extending its shelf life. It is used to produce milk with a low spore count for cheese production to prevent late blowing of hard cheese.

Bag in box packaging: Packaging consisting of a flexible inner bag, which closely fits inside a box. The product is contained in the inner bag, which acts to keep out atmospheric oxygen. The rigid outer box protects the contents. It is used widely for breakfast cereals and also for storing and dispensing wines.

Bagasse: Cane sugar processing waste that is composed of unextracted sugar and the remains of the sugar cane after milling. Used as a fuel source, in feeds, as a substrate for microbial fermentation and for paper and board manufacture. It is called sugar cane bagasse and megass.

Bagging: Packing of substances, such as foods, into bags.

Baghar: A mixture of sliced onions and possibly other vegetables, flavoured with chopped garlic, chillies and whole spices, fried in ghee or clarified butter and used as a topping for dhal.

Bags Containers: With a single opening that are used for storing or carrying items. They are made from a variety of flexible materials. Bags for food use are usually made of paper or plastics. The term is also used for small perforated paper sacks in which tea leaves or coffee grounds are placed and which are used to make small quantities of tea or coffee beverages.

Baker's yeast: A particular strain of the yeast *Saccharomyces cerevisiae*. It is a leavening agents (e.g. *Saccharomyces cerevisiae*) used in making bread and other bakery products. It is sold as a crumbly, solid cake, usually containing about 75 to 80% free water and available in compressed, liquid and dried forms.

Bakeries: Facilities in which bakery products are manufactured. Also refers to retail outlets in which bakery products are sold.

Bakery additives: Ingredients used in making bakery products with the aim of prolonging shelf life or improving the quality of the finished products. It Include humectants, antifoaming agents, anti-staling agents, crumb softeners and texture improvers.

Bakery product mixes: Pre-mixed dry formulations which usually require the addition of liquid ingredients to make batters or dough.

Bakery products: Products in which flour based components are major ingredients, and which are cooked by baking. Include biscuits or cookies, bread, cakes, doughnuts, scones and tortillas.

Bakery: An establishment where cakes and bread are baked, often with a shop attached.

Baking powder: A chemical raising agent made from a finely ground mixture of two parts cream of tartar to one part bicarbonate of soda which liberates carbon dioxide gas on heating. Domestic baking powders replace some of the cream of tartar with an acid phosphate which is slower to act and gives more time for preparation and cooking.

Baking ovens: Enclosed chambers or compartments in which foods are cooked or heated by application of dry heat (baking).

Baking powders: Bakery additives comprising mixtures of sodium bicarbonate, starch and one or more acidic substance (e.g. cream of tartar). When moistened and heated, they act as raising agents by generating carbon dioxide, bubbles of which have a leavening effect.

Baking properties: Characteristics of cereals, bakery additives, flour or dough associated with their suitability for use in baking.

Baking quality: Extent to which flour is able to produce a well leavened bread which has optimal texture and an even distribution of air pockets formed during fermentation, or good quality bakery products.

Baking sheet: A heavy steel sheet, usually rectangular with raised edges on 3 sides, on which rolls, biscuits, etc. They are cooked in the oven. The unraised edge is to facilitate sliding the baked goods onto a cooling rack.

Baking: Cooking of foods in ovens by surrounding with dry heat. The temperature of the oven is varied depending on the type of food that is to be cooked.

Bamboo shoots: Emerging ivory-coloured shoots of several species of bamboo. These include *Bambusa oldhamii*, *Dendrocalamus latiflorus* and *Phyllostachys edulis*. An important component of Oriental cuisine, bamboo shoots are available fresh or canned and have a crispy texture. Bitter-tasting shoots require precooking due to the presence of cyanogenic glucosides.

Banana flour: A flour made from plantains which have been dried and ground. It is called also plantain flour.

Bar codes: Machine-readable codes which contain product specific information. Bar codes are formed by patterns of parallel lines of varying thickness with spaces of varying length between them. Information is usually read from bar codes using light pens or laser scanners. Standard international codes are used. Benefits of using bar codes include: rapid and efficient data capture; improved product traceability; the possibility of automated product storage; improved control of product storage and distribution; time and costs savings; and improved customer service. Consumer unit bar codes, which encode fixed information, are used on primary packaging of products intended for sale directly to consumers at retail outlets. Traded unit bar codes, which include fixed as well as supplementary product information (e.g. product weight, batch number and time of production), are often compulsory within product supply chains. Transport unit bar codes used to label pallets and encode shipping containers are used to track pallets through supply chains.

Barley flour: Ground hulled barley used to make unleavened bread and porridge.

Barley malt: Malt prepared from special malting barley cv; mainly used in brewing. Barley malt is the main malt type used in brewing worldwide.

Barley: Edible grain from *Hordeum vulgare* used as a cereal and livestock feed and in malt production. It contains little gluten, and so is unsuitable for bread-making. Most popular form is pearl barley in which the outer husk and part of the bran layer are removed by polishing. It provides a source of vitamins (e.g. niacin, folates) and minerals (e.g. zinc, copper, iron).

Barrel: A wooden cask with specially shaped wooden staves forming the sides which are held together with iron hoops which when knocked down from either end compress all the staves together to make the sides watertight. It is used for wine, beer, salted herrings, olives and the like.Cylindrical containers for liquids and dry materials. Traditionally made of wooden staves held together by metal hoops, but may also be made of cheaper and/or more durable materials, such as metal or plastics. Oak barrels are used for the ageing of wines and spirits; constituents of the wood (e.g. tannins, lignin and fragments, carbohydrates, acids and esters, volatile phenols, oak lactones, pyrazines, furfural and norisoprenoids) have major effects on flavour of wines and spirits. Barrels are also used as measures for liquids, e.g. beer and oils, based on the capacity of standard barrels.

Basic cake mixture: Flour, butter, caster sugar and eggs in the proportions by weight of 8:5:5:5 with 1 level teaspoon of baking powder (16 g per kg flour).

Basic foods: Seven classes of food are considered necessary for a balanced daily diet: green and yellow vegetables, starchy tubers, fruits, milk products, meat including fish or eggs, cereals and fats and oils. Vegetarians would substitute pulses for the meat products.

Basket: A steel mesh utensil (about 8 to 12 pitch) with a handle, which fits in a deep-fat fryer and is used to lower damp foods, especially potato chips, into the fat slowly so that the boiling off of the water does not cause the fat or oil to boil over. It is also used for blanching vegetables in boiling water.

Basmati rice: A dense thin long-grain rice from India with an aromatic and nutty flavour, more expensive than American long-grain rice, used in Indian cooking for biryani and pulao. Best boiled without salt for about 10 minutes, or in lots of boiling water for 8 minutes then drained, covered and finished on a very low heat for a further 10 minutes. (The name means fragrant, and the rice is grown in the foothills of the Himalayas).

Batters: Thin liquid mixtures of pouring consistency made from flour, milk and eggs. It may be used as coatings for foods such as fish prior to frying, or cooked on their own to make products such as pan cakes, waffles and Yorkshire puddings.

Beating: Vigorous stirring of cooking ingredients, usually in a circular motion with the intention of incorporating air.

Beef extracts: Water-soluble extracts prepared from beef, used widely as flavourings. Preparation involves immersion of beef mince in boiling water to leach out the water-soluble extractives, and concentration. Direct extract can be produced by exhaustive ex- traction of beef; it contains a high concentration of gelatin. Beef extracts are rich nutritional sources of the vitamin B group; they can be formulated for use as spreads for bread, as flavourings, and, when mixed with water, as beverages. Beef extracts can also be used in preparation of beef tea, an extract of stewing beef that may be used as a food for invalids.

Beef mince: Meat mince prepared from beef which is available in several grades; these primarily relate to the percentage of fat in the mince. For example, beef mince may be graded as: extra lean; lean, which has good flavour but does not shrink excessively on cooking; or regular, which is usually made from lower cost cuts of beef. Also known as ground beef or minced beef.

Beef: Meat from cattle, including bulls, calves, cows, steers and oxen. Quality is determined largely by breed, age and gender of the animal; it is also influenced by animal feeding, slaughtering technique and treatment of the meat post-slaughter. Tenderness and flavour are increased by hanging cattle carcasses (ageing/conditioning). Raw fresh beef is usually bright red in colour with creamy coloured marbling; however, meat from older cattle, particularly bulls, tends to be darker in colour. Composition varies with fat content and between different cuts, e.g. brisket, forerib, rump and silverside. Cuts which contain little connective tissues can be cooked by roasting, frying or grilling; however, tougher cuts should be cooked by stewing or braising, in order to soften the connective tissue. During the 1980s and 1990s, markets for beef have been affected negatively by consumer health concerns relating to high levels of saturated fats in red meat and to prion diseases, particularly bovine spongiform encephalopathy (BSE). Alternative term for beef muscles, bovine muscles, bull muscles, calf meat, calf muscles, cattle muscles and cattle tissues.

Beer: Alcoholic beverages manufactured by alcoholic fermentation of worts using either top or bottom fermenting brewer's yeasts. The malt is commonly barley malt, but other malt types, including wheat malt or sorghum malt may be used. Non-malted cereals or other brewing adjuncts may be used in combination with the malt. Beer is commonly, but not always, flavoured with hops.

Beeswax: Yellow-coloured substance secreted by bees to make honeycombs. Solid, but easily moulded when warm. It consists of esters, cerotic acid and hydrocarbons. It is used to make edible wax coatings for foods and edible films. Aqueous extracts may be used as flavourings.

Beet sugar juices: Aqueous solutions of beet sugar produced during processing of roots of sugar beets. Raw juices are solutions produced by direct hot water extraction of the roots and contain beet sugar and impurities. Thin juices are purified beet sugar solutions and thick juices are formed by concentration of the thin juices.

Beet sugar syrups: Highly concentrated aqueous solutions of beet sugar produced by evaporation of purified beet sugar juices (thin beet sugar juices).

Beet sugar: Sucrose purified from roots of sugar beets (*Beta vulgaris*). Stages of beet sugar manufacture include: cleaning and cutting of roots; hot water extraction of sugars; purification of beet sugar juices by precipitation of impurities with lime- phosphoric acid or lime-CO_2 treatments; filtration to remove solids; concentration of the purified beet sugar juices; and crystallization of the pure beet sugar. Commercially available beet sugar comprises 99.80% sucrose and <0.05% moisture content.

Bentonite: Type of absorbent clay (a colloidal hydrated aluminium silicate) formed by the breakdown of volcanic ash that has the ability to absorb water with an increase in volume. Bentonite uses in the food industry include fining agents for winemaking, clarifiers for fruit juices and vegetable oils, bakery additives to reduce staling, stabilizers and filter aids.

Benzyl alcohol: Aromatic alcohol which is a constituent of the flavour compounds and aroma compounds in various fruits and spices, and in plant- derived products such as alcoholic beverages.

Beta - Amylases: It acts on starch, glycogen, and related polysaccharides and oligosaccharides, producing β- maltose by an inversion reaction. It used for production of high maltose syrups.

Betacyanins: Red/violet pigments of the betalains group, which occur naturally in red beets and other plant foods. It used as natural colorants in foods.

Betaine: Soluble nitrogen compounds occurring in a range of foods, especially sugar beets, molasses and beet sugar factory wastes. It may be included in flavour compounds, and have antioxidative activity.

Beverages: Liquids intended for drinking. Types include alcoholic beverages, soft drinks, teas, coffee, cocoa beverages, dairy beverages, health beverages, fruit beverages, soy beverages and drinking water.

BHA: It means butylated hydroxyanisole, a controversial antioxidant allowed only for use in fats, oils and essential oils.

BHT: Butylated hydroxytoluene, a controversial antioxidant allowed only for use in fats, oils and essential oils.

Binding agents: Substances used to hold ingredient mixtures together, providing adhesion, solidification and correct consistency. For example, carrageenans and soy proteins can be used to bind free moisture and improve texture and juiciness in reduced fat or restructured meat products. Other applications for binding agents include chewing gums, extruded foods and confectionery.

Binding capacity: Ability of one substance to attach to another.

Bins: Large containers used for storing specified sub- stances or containers used for depositing rubbish. Also used to describe partitioned stands for storing bottles of wines.

Bio foods: Term used to describe biotechnologically derived foods or functional foods.

Bioactive compounds: Substances which display biological activity, e.g. immunomodulation, opioid activity, antihypertensive activity or hypolipaemic activity, upon ingestion. Found in a range of foods, and are of interest to the functional foods sector. Include bioactive peptides (occur widely in dairy products), many vitamins and fatty acids, flavonoids and phytosterols.

Bioactive peptides: Peptides produced from plant or animal proteins, which display biological activity (e.g. opioid, immunostimulatory or antihypertensive activity), and are of interest to the functional foods sector. Milk proteins are a particularly rich source of bioactive peptides, such as casein phosphopeptides and β-casomorphins. Peptides that inhibit activity of peptidyldipeptidase-A are found in a number of food sources and have potential use as antihypertensive functional foods ingredients.

Bioassay: Technique for measuring the biological activity of a substance by testing its effects in living material such as a cell culture.

Bioavailability: Extent to which a dietary component can be absorbed and utilized by the target tissue of the body. Nutrients with low bioavailability may be in a form that is poorly absorbed from the gastrointestinal tract (e.g. lysine combined with reducing sugars as a results of the Maillard reaction, minerals in the presence of anti-nutritional factors such as phytates) or may be biologically inactive once absorbed.

Biocatalysts: Substances that catalyze biochemical processes in living organisms. The most well known examples are enzymes, although RNA may also fulfill this function.

Biochemistry: Science of the chemistry of living organisms.

Biocide: A chemical agent, such as a pesticide, that is capable of destroying living organisms.

Bio-control: Deliberate exploitation by humans of one species of organism to eliminate or control another. Commonly involves introduction into the environment of parasites, insects or pathogens which can infect and kill or disable particular insect pests or weeds of crop plants. It is also known as biological control.

Bioconversions: Utilize the catalytic activity of living organisms to convert a defined substrate to a defined product in a process involving several reactions/steps. The term is often used interchangeably with bio-transformations. Advantages include the ability to operate under mild conditions, the ability to produce specific enantiomers and the ability to carry out reactions not possible using conventional chemical synthesis. Bio- conversions differ considerably from fermentation, since in the latter; the products often bear no structural resemblance to the pool of compounds given to the microorganisms.

Biodegradability: Ability of a substance to undergo biodegradation.

Biodegradability: This is the ability of certain substances to be broken down into relatively safe non-toxic or non-polluting products.

Biodegradation: Degradation of a substance as a result of biological (usually microbial) activity, rendering it less noxious to the environment.

Bio-deterioration: Deterioration (spoilage) of an object or material as a result of biological (usually microbial) activity. Bio-deterioration of foods causes them to become less palatable and sometimes toxic, and can involve alterations in flavour, aroma, appearance or texture. The organisms involved are typically bacteria and fungi, and their activity is dependent on factors such as nutrients present, water activity, pH, temperature and degree of aeration.

Biofilms: Films of microorganisms, usually embedded in extracellular polymers, which adhere to surfaces submerged in or subjected to aquatic environments. Frequently cause fouling of the surfaces of water pipes.

Bioflavonoids: Flavonoids present in a wide range of plant foods, which have beneficial effects on health.

Biogas: A mixture of gases produced by anaerobic digestion of organic wastes, comprising mainly methane and carbon dioxide with traces of hydrogen, nitrogen and water vapour. It is used as a fuel.

Biological activity: Activity of compounds, generally organic in origin, within living organisms. For food- derived chemicals, this is generally a non-nutritional property, such as antimicrobial activity, anti-oxidative activity, immune-modulation, or other physiological effects.

Biological contamination: This takes the form of microbiological contamination of food by pathogenic bacteria from a wide range of sources, such as food handlers, animals, birds, insects, food contact surfaces, water, waste materials, raw food, vegetables and fruit, meat and meat products, rice, cereals, dust and soil. This contamination produces the classic food poisoning symptoms, such as diarrhoea, vomiting, stomach pain and dizziness. In some cases, biological contamination can result in death of the consumer.

Biological membranes: Selectively permeable membranes containing mainly lipids and proteins that surround the cytoplasm in eukaryotic and prokaryotic cells. It can also contain carbohydrates and sterols. The precise composition depends on the species and, in some cases, on growth conditions and age of the cells. The lipids (phospholipids and glycolipids) usually form a bilayer within which proteins are partly or wholly embedded, some spanning the en- tire width of the bilayer. Artificial biological membranes (liposomes) are often used to transport biological molecules.

Biological oxygen demand (BOD): Amount of dissolved oxygen required for microbial oxidation of biodegradable matter in an aquatic environment containing organic matter, such as sewage, water or milk. It gives an indication of contamination by microorganisms which take up oxygen for their metabolism.

Biological values: Indication of the nutritional value of food proteins. Relative measure of the amount of absorbed proteins retained by the body, assuming no loss of protein nitrogen during digestion. Values are highest for egg proteins (0.9-1.0) and milk proteins (0.85), with meat proteins and fish proteins (0.7-0.8), cereal proteins (0.5-0.7) and gelatin (0) having lower values.

Biopolymers: Polymers which occur in living organisms. Included in this group of macromolecules are polysaccharides, proteins and nucleic acids.

Bioreactors: Reactors used for generating products using the synthetic or chemical conversion capacity of a biological system, such as enzyme reactions and fermentation. They can be fermenters, cell culture perfusion systems or enzyme bioreactors. For production of enzymes or proteins, recombinant microorganisms, such as bacteria, yeasts, fungi and plant cells, are usually chosen. Also used for bioremediation of industrial effluents, such as food industry waste water.

Biosensors: Biomolecular probes that can be used to measure a variety of parameters in biological systems by translating a biochemical interaction at the probe surface into a quantifiable physical signal. Immobilization of enzymes, antibodies, receptors, DNA, cells or organelles on the surface of a transducer forms the basis of various biosensors. Used widely in the food industry for measuring levels of various components in foods and beverages, detection of contamination and adulteration, and for monitoring and process control of fermentation processes, bioconversions and biotransformations.

Biosurfactants: Potent surface active agents produced by a variety of microorganisms, including *Pseudomonas*, *Rhodococcus*, *Candida*, *Corynebacterium*, *Mycobacterium*, *Acinetobacter* spp., *Bacillus subtilis*, *Serratia* and *Thiobacillus* spp. Low molecular weight biosurfactants are often glycolipids, and high molecular weight biosurfactants are generally either polyanionic heteropolysaccharides containing covalently-linked hydrophobic side chains or complexes containing both polysaccharides and proteins. Biosurfactants have a number of advantages over their chemical counterparts, such as biodegradability, effectiveness at extremes of temperature and pH, and lower toxicity. Biosurfactants are used in the food industry as emulsifiers and stabilizers.

Biotechnologically derived foods: Foods produced by means of biotechnology.

Biotechnology products: Products produced by microorganisms in biotechnological processes.

Biotechnology: In its broadest sense, any industrial process in which microorganisms are used. More commonly used for those industrial processes in which genetic engineering techniques have been used to construct novel strains to improve their properties and produce new products.

Biotin: A very stable vitamin necessary for yeast and bacterial reproduction. Its presence in bread dough is essential. Its absence, in an active form, in egg white is responsible for the preservation of the egg against bacterial attack. It is widely distributed in all animal and plant tissues, with high concentrations in yeast, liver and kidney. It is also synthesized by intestinal bacteria and easily absorbed in the gut. It is required for the metabolism of carbohydrates and has no known toxicity. Raw egg white taken in excessive quantities whilst under antibiotic treatment could lead to a deficiency state which would appear as skin disease. Currently this affects only a few babies and is easily cured. A water-soluble nutrient that is involved in the physiological biosynthesis of fats and the utilization of carbon dioxide. Rich dietary sources include egg yolks, cattle livers and yeasts. Avidin, a protein present in raw egg whites, can act as a biotin anti-vitamin by binding biotin and reducing its bioavailability. Also known as vitamin H and coenzyme R.

Bio-transformations: Specific modification of a defined compound to a defined product with structural similarity through the use of biological catalysts (enzymes, or whole dead or resting microorganisms). Advantages are the same as those for bioconversions.

Birefringence: The optical property of a substance, usually a crystal, in which a ray of light passing through the substance is separated into two plane polarized rays (double refraction). The effect can occur when the velocity of light in the material is not equivalent in all directions, resulting in different refractive indices for light polarized in different planes.

Biryani : A vividly coloured and aromatic pilau-type dish of rice and fish or meat. The meat or fish, precooked with spices, is mixed with uncooked, parboiled or cooked rice previously fried with spices as for risotto, caramelized onions, and possibly almonds, and then steamed in a tightly sealed container. The colour is derived from saffron or turmeric. Spices and herbs are selected from parsley, bay, mint, lovage, coriander leaves, garlic, fresh ginger root, chillies, cardamom, cloves, coriander seed, peppercorns, cumin, fennel, anise, cinnamon or cayenne pepper. Vegetables incorporated include carrots, green peppers and onions. Additional ingredients include butter and yoghurt.

Biscuit cutter: A light metal or plastic outline shape used to cut individual biscuits from rolled-out dough.

Biscuit dough: Dough used to make biscuits.

Biscuit factories: Factories in which biscuits are manufactured.

Biscuit press: A large syringe with variously shaped attachments at the end remote from the plunger. It is filled with biscuit dough which is pressed out to form the various shapes. It is also called **cookie press.**

Biscuit: Flour-based products that vary greatly in size, shape and texture, but are generally small, thin and short or crisp. Usually made with flour, butter or vegetable shortenings, sugar and sometimes a leavening agent; other ingredients, e.g. cocoa, chocolate chips, dried fruits, nuts, cheese or flavourings, are added according to the type of biscuits to be made. It is usually eaten as snack foods, often with beverages. It can be eaten as part of a meal along with cheese. Called cookies in the USA, where the term biscuits refers to soft, scone-like products.

Bisulfites: Hydrogen sulfite salts used in antimicrobial preservatives and antioxidants in foods and beverages.

Bitter almond oils: Oils rich in oleic acid derived from seeds of bitter almonds. Contain benzaldehyde and hydrocyanic acid; the latter compound, which is toxic, is removed during extraction. It used in flavourings.

Bitter almonds: Common name for nuts produced by *Prunus dulcis* (syn. *P. amygdalus*). Too bitter for fresh consumption and also contain highly toxic hydrocyanic acid, or hydrogen cyanide. Cultivated mainly for manufacture of bitter almond oils (principal component benzaldehyde), which are used as flavourings following removal of the hydrocyanic acid.

Bitter compounds: Compounds with a bitter taste; these may be used as flavourings in foods or beverages, e.g. hops bitter acids in beer, or quinine in soft drinks.

Bitter peptides: Peptides, formed during enzymic hydrolysis of proteins, which have a bitter taste and may impair the sensory quality of the food. Bitter pep- tides derived from casein may be a particular problem in cheese-making. Bitter peptides may also cause problems in soy products and protein hydrolysates. Treatment with peptidases may eliminate quality problems attributable to bitter peptides.

Bitterness: Flavour produced by bitter compounds such as caffeine and other alkaloids, often at low thresholds.

Bixin: A golden yellow food colouring and fat-soluble carotenoid obtained from the seeds of *Bixa orellena.* The main colouring component of annatto, and also used as a colorant in its own right, e.g. in margarines, salad dressings and baked goods.

Black tea: Tea prepared from *Camellia sinensis* leaves which have undergone a fermentation stage in which changes in phenols and flavour compounds occur.

Black vinegar: A dark, mild and sweetish vinegar similar to balsamic vinegar but made from fermented glutinous rice or sorghum. Presumably made in a similar way. Commonly added to many dishes as a flavouring in north and central China.

Blanching: Plunging of foods (particularly vegetables and fruits) firstly into boiling water (<100^{0}C) for a brief period, and then into cold water to stop the cooking process. Blanching is used to firm the flesh and to heighten and set colour and flavour (e.g. of vegetables before freezing), and to loosen peel prior to peeling (e.g. of peaches and tomatoes). It inactivates enzymes, thus prolonging the shelf life of frozen foods.

Blast chilling and freezing: Blast chilling is the process of using a fan-assisted refrigeration cabinet or room which lowers the temperature of cooked food very quickly to around 5^{0}C prior to going into refrigerated storage. In order to prevent bacterial multiplication, the process should be completed within 90 minutes. Attention must be paid to the specific size of the food commodities being chilled, particularly in the case of meat joints. Ideally these joints should be no more than 2.5 kg in weight and

Bleaching agents: Substances used to make a material whiter or lighter by a chemical process. Examples include fuller's earth, activated carbon and activated clays, which are used to bleach oils. Impurities are adsorbed onto such agents, which are then removed by filtration. Other bleaching agents used in the food industry include chlorine, chlorine dioxide, sodium metabisulfite, and hydrogen peroxide.

Bleaching: Making a material whiter or lighter by a chemical process or by exposure to sunlight. Bleaching agents may be employed. For example, oils can be bleached using fuller's earth, activated carbon or activated clays. Many impurities, including chlorophylls and carotenoid pigments, are adsorbed onto such agents and removed by filtration. When some oils are heated, heat bleaching takes place, in which pigments such as carotenoids are converted to colourless materials.

Blended flour: Two or more types of flour blended for making specific items, e.g. Pretzels, cookies.

Blender: 1. An electric machine consisting of a glass or plastic vessel with a fast rotating steel blade directly coupled to an electric motor below. This pulverizes, mixes, blends or emulsifies the contents of the vessel which can be wet or dry. 2. A hand-held electric machine with a fast rotating shrouded metal blade at the end of an extension on the motor, used for wet mixtures e.g. For making purées or emulsions such as mayonnaise. It also also called liquidizer.

Blending: The mixing and combining of two or more ingredients with a spoon, beater or electric blender to form a harmonious combination. This term also relates to a mixture of two or more flavours combined to obtain a particular character and quality, as in wines, teas and blended whisky.

Blond roux: A mixture of equal parts of fat (dripping, clarified butter, etc.) and flour cooked to a sandy texture for slightly longer than a white roux with no more than a hint of colour. Used for thickening liquids, soups, veloutés, sauces, etc. Also called roux blond.

Blood proteins: It includes both plasma proteins and proteins from the cellular components of blood, particularly albumins, globulins and fibrinogen. Commonly, they are recovered from blood by ultrafiltration. Blood proteins from animals have many food uses; for example, they are used as binders in sausages and as ingredients, particularly as emulsifiers, in other foods.

Blood: A red viscid liquid consisting of a straw coloured liquid portion, plasma, in which various types of blood cells are suspended, including: red blood cells (erythrocytes); white blood cells (leukocytes); lymphocytes; and platelets. Plasma also contains dissolved proteins, fats, minerals, salts and sugars. Blood collected hygienically from slaughtered food animal's forms an article of food. As a food, it must not be contaminated with stomach ingesta, urine or foreign material. Preliminary processing includes whipping or defibrination to prevent clotting; anticoagulants, such as sodium citrate or sodium phosphate, may be added to blood for human consumption. Both blood and blood components are used in foods; for example, they are used as ingredients in blood sausages, comminuted meat products and dry protein concentrates.

Blotting: Transfer of nucleic acids and/or proteins either directly or from a gel to a chemically reactive matrix (e.g. nitrocellulose), on which the nucleic acids/proteins bind covalently in a pattern identical to that on the original gel. After blotting, target molecules are detected through the use of complementary labelled nucleic acids or antibodies. It includes northern blotting, Southern blotting and western blotting.

Blowing: Defect of cheese, also known as late blowing, that leads to gas formation and abnormal flavour development during ripening. Caused by butyric acid fermentation by *Clostridium tyrobutyricum*, a pasteurization-resistant contaminant of milk that can occur when animals have been fed silage.

Blue green algae: Older term for cyanobacteria, a large group of prokaryotic, photosynthetic, unicellular or filamentous organisms, which differ from other bacteria in that they possess chlorophyll a and carry out photosynthesis. Occur in fresh, brackish, marine and hypersaline waters. Some freshwater bloom- forming strains (e.g. *Anabaena*, *Microcystis* and *Nodularia*) produce potent cyanotoxins (saxitoxin, microcystins and nodularin, respectively) which may contaminate reservoirs. Tainting of drinking water supplies by such cyanobacterial blooms may cause illness or death in humans and animals which drink the water.

Body mass index: A measure of obesity in human beings equal to a person's weight in kilograms divided by the square of their height in metres. If between 25 and 30, then the person is considered overweight, if over 30 they are obese. (Example weight 75 kg (168 lb), height 1.7 m (5 ft 7 in), body mass index = 75/(1.7 x 1.7) = 25.95, hence overweight). The normal range is usually quoted as 20-25 kg/m^2.

Body Texture: The term relating to the fullness of products; especially applied to wines.

Bodying agents: Additives used to impart desirable body, viscosity and consistency to foods. It is often used to improve texture of low calorie foods.

Boilers: Fuel-burning devices for heating water in which foods can be immersed and cooked.

Boiling point (BP): Temperature at which a liquid boils. This occurs when the vapour pressure of the liquid is equal to the atmospheric pressure of the surrounding environment. The boiling of water is 100^0C.

Boiling: Process of raising the temperature of a liquid, by application of heat, to the point where it bubbles and turns to vapour. The term also means cooking food in a boiling liquid.

Bologna: Smoked, cooked sausages prepared from finely minced, cured pork and/or beef. Their name originates from the city of Bologna in Italy, but true Italian bologna is known as mortadella. Bologna of various diameters can be purchased in rings, rolls or slices; they are retailed fully cooked and ready to serve. Types of bologna include chub bologna, beef bologna and ham-style bologna.

Bone marrow: Soft, gelatinous, highly vascular connective tissues that occur in certain long bones. It has responsible for producing red blood cells as well as many white blood cells. When bones of high marrow content are used during mechanical recovery of meat, the lipid and haem concentrations of the recovered product are increased, and the tendency of the meat to undergo oxidation is increased.

Bone mineral density: Level of mineralization of bone. Measurements can be taken by dual energy X- ray absorptiometry or ultrasound, and are used in the clinical assessment of osteoporosis risk. Many die- tary and lifestyle factors have been proposed to modulate bone mineral density, including positive effects being reported for dietary calcium, tea, soy isoflavones and vegetable proteins.

Bones: Components of the skeleton made from hard, rigid structural material. Bones are composed of an organic matrix of collagen, osseoalbumoid and osseomucoid; this is impregnated with mineral salts, particularly calcium phosphate, calcium carbonate and magnesium phosphate. Fluorides and sulfates are also present. The central cavities of most bones contain red bone marrow; however, the cavities of the long bones contain yellow marrow. Bones are processed to produce fats and gelatin. Chopped bones may be boiled to prepare bone broth. Degreased animal bones are used to prepare bone meal, which is used as a supplementary source of calcium and phosphates in foods and feeds, and as a source of phosphates in plant fertilizers. Bone charcoal is used in sugar refining and in bleaching. Specialized bone powders are used to re- move fluorine from drinking water.

Boning: Removal of the bones from meat or fish, usually before cooking.

Bordeaux mixture: Broad-spectrum fungicide originally developed in France to control disease in grapes. Made by mixing copper sulfate and hydrated lime in water. Used to control disease in a wide range of tree fruits, nuts and vine fruits.

Boric acid: An Acid used in preservatives for sea foods and wines.

Borneol: Member of the terpene alcohols class of flavour compounds present in many fruits, herbs and spices.

Bottled water: Any type of potable water which has been packaged in bottles for distribution and retail sale.

Bottles: Portable containers often made from glass or plastics, used to hold or store liquids. Typically with narrow necks, which can be closed with caps, corks or stoppers. The term also refers to metal containers which are used to transport and store liquefied gases.

Bottling: Process of putting substances into bottles for storage and preservation. Most commonly applied to beverages, such as wines, fruit juices and beer, but also used to describe a method of preserving fruits in syrups or in the form of jams. After the products have been placed in the bottles, the containers are sealed with corks or other closures to prevent air or microorganisms from entering and causing spoilage.

Botulinum cook: This is a process used in food canning operations whereby the presence of *Clostridium botulinum*, which is particularly resistant to heat, is used as an indication of the effectiveness of the operations. The presence or otherwise of *Clostridium botulinum* is a process indicator for foods which have a pH in excess of 4.5. Clear evidence that the organism is destroyed is an indication that other bacteria will also have been destroyed, thereby achieving true sterility of the product.

Botulism: Food poisoning/food borne disease caused by the botulotoxins excreted by *Clostridium botulinum*, which is strict spore-forming anaerobe. It can therefore only grow in sealed cans and jars from which air is excluded or rarely in the centre of cooked food. It is used to be responsible for deaths when home-bottling of vegetables was common. The toxin is destroyed and made harmless by boiling for few minutes. Symptoms include vomiting, abdominal pain, visual disturbances and difficulty in speaking and swallowing. Foods commonly implicated are low-acid, low-salt foods (e.g. improperly canned vegetables and soups, and fish and meat products).

Botulotoxins: Extremely potent neurotoxins produced by *Clostridium botulinum*, which cause botulism. Also known as botulinus toxins, botulinum tox- ins, botulins and botulismotoxins.

Braising: Light frying of foods (usually meat or vegetables) followed by stewing in a small amount of liquid at low heat for a lengthy period of time in a closed (tightly covered) container. The long, slow cooking develops flavour and causes tenderization of the food by gently breaking down fibres. Braising can be undertaken on stoves or in ovens.

Bran: Protective outer layer of the seeds of cereals that is separated from the kernels during milling. Often used in breakfast cereals and other products as a source of dietary fibre.

Branding: Process of applying a trademark or distinctive logo to a food or its packaging to identify its manufacturer or retailer.

Bratwurst: Fresh sausages, usually made from a mixture of highly seasoned pork, veal and onions. Seasonings may include ginger, nutmeg, coriander or caraway. Numerous different types of bratwurst are produced in Germany; many districts produce their own special varieties. Traditional bratwurst must not contain nitrites as curing salts. Al- though some pre-cooked bratwurst is sold, most requires cooking before it is eaten. Other product names may also include the term bratwurst (e.g. bauernbratwurst and smoked bratwurst), but these sausages are produced by hot smoking and fermentation and are classified as raw dry sausages, rohwurst.

Bread crumb: The soft inner part of bread, which is surrounded by the bread crust.

Bread crust: Crisp, outer part of bread, which is de- hydrated and browned during baking.

Bread dough: Unbaked thick, plastic mixture of flour and liquid (e.g. water) that is kneaded, shaped and rolled to make bread. The elasticity of bread dough is dependent upon the amount of gluten contained in the flour.

Bread rolls: Bread products formed from pieces of dough shaped as required before baking. May have a soft or crisp crust. Also commonly referred to as rolls.

Bread: Baked dough product made from cereal grains (mostly commonly wheat) ground into flour, moistened and kneaded into a dough and then baked. Often leavened by the action of bakers yeasts or by addition of sodium bicarbonate.

Breadcrumbs: Small fragments prepared by grinding bread. Used in coatings, usually for foods to be fried, stuffings and in some desserts.

Breading: Coating of foods with breadcrumbs or other crumbs usually before frying or baking. The food is dipped first into a liquid (e.g. beaten eggs, milk or beer), then into the crumbs, which may be seasoned with herbs or spices. The breaded product is then fried or baked. Breading serves to retain the moisture content of the food and forms a crisp crust after cooking.

Breadings: Breadcrumbs and other types of crumb used in breading foods, usually before frying or baking.

Breadmaking properties: Characteristics of cereals, flour or dough that determine their suitability for bread manufacture.

Breadmaking: Process by which bread is prepared from flour and other ingredients that vary according to the type of bread to be made. Steps involved include fermentation, kneading of the dough, proofing and baking.

Breakfast sausage: Equal parts of finely chopped beef and lean pork, mixed with breadcrumbs, 1 cm dice of pork back fat, ground mace, nutmeg and seasoning, packed into weasands, smoked, then boiled. May be eaten cold or heated using any cooking method.

Breakfast cereals: Breakfast food other than bread or cake made from cereal grains, principally, maize, rice, wheat, or oats. Some are cooked as needed, e.g. Porridge, boiled rice and congé, others are commercially processed by precooking, pressing, drying and coating, e.g. Cornflakes and rice crispies. Yet others use only a part of the cereal, e.g. Bran flakes, and others are used raw such as rolled oats in muesli. Generally eaten with milk and sugar, possibly fruit and other flavourings. Pre- cooked or ready to eat foods prepared from cereal grains by processes such as flaking, puffing, toasting or shredding. Often sweetened with sugar, syrup or fruits. May be enriched with bran as a source of extra dietary fibre.

Breakfast foods: General term applied to foods, especially cereal products and bakery products, which are commonly eaten as part of breakfast. Include breakfast cereals, bread, muffins, bagels, pancakes and waffles.

Breakfast: First meal of the day. Typical breakfast foods include breakfast cereals, toast, bread rolls and other morning goods.

Brewer's yeasts: Yeasts of the genus *Saccharomyces*, now generally classified as *S. cerevisiae* but some formerly classified as *S. carlsbergensis* or *S. uvarum*, used for brewing of beer. Brewers yeasts may be divided into top fermenting yeasts (used for ale and similar beer types) and bottom fermenting yeasts (used for lager and similar beer types) on the basis of their characteristics during the fermentation process.

Breweries: Industrial premises used for brewing of beer.

Brewers spent grains: By-products from the clarification stage of the brewing process, comprising the solid residue of substances such as malt grist, from which the soluble material has been extracted in the mashing process.

Brewing adjuncts: Fermentable material other than malt used in the brewing process. Brewing adjuncts may include unmalted cereals, syrups or sugars.

Brewing by-products: By-products of the brewing process, including brewers spent grains, surplus brewer's yeasts and carbon dioxide formed during fermentation.

Brewing: The process of manufacture of beer; important stages of the process include mashing to extract soluble material from malt and other ingredients, hopping, clarification, boiling the worts, alcoholic fermentation, ageing and filtration.

Brine: A solution of salt in water used to preserve food. Preservation occurs by diffusion of salt from the brine into the food and of water from the food into the brine. On a small scale this can make the brine too dilute. Brines should be between 80 and 100% saturation i.e. Between 290 and 360g of salt per litre of water. Eggs and potatoes will float in such a brine. However 200g of salt per litre of water is sufficient to inhibit bacterial growth.

Brined meat: Meat that is preserved by brining. Meat cuts may be immersed in brines, which contain salt or salt and other curing agents dissolved in water. However, it takes a long time for brines to diffuse the whole meat product; thus, in general, only specialty meat products are brined by immersion. A more rapid, uniform distribution of brines throughout meat products can be achieved by direct injection of brines into the meat cuts, methods for which include artery pumping, multiple injections and stitch pumping.

Brines: Water saturated or strongly impregnated with salt used for pickling or preservation of foods. Sweeteners such as sugar or molasses can be added to the brines to make sweet brines.

Brining: The immersion in brine (a solution of approximately 25% salt, 3% sodium nitrate or potassium nitrate, and other substances, such as 1% sugar) of slices, joints or sides of pork during the preparation by curing of bacon. Certain vegetables are also brined prior to pickling in order to reduce their water content.

Brining: The process of soaking vegetables in brine before pickling in vinegar, in order to remove some of the water, and retain a crisp texture. Dry brining is when the vegetables are covered with dry salt, rather than immersed in a salt solution. It may also a process of treating foods with brines for preservation or pickling.

Briquetting: Use of presses for compacting and processing organic waste or compacting plants to reduce volume for valuable material recycling and disposal. In the food industry, ice cream may be formed into bricks, blocks or slabs and sold as briquettes.

Brix scale: A scale of density based on sugar syrups and measured in degrees. The degree Brix (°Brix), is the percentage by weight of sugar in a sugar water solution at 20°C. Thus 100°Brix is pure sugar, 50°Brix is equal weights of sugar and water i.e. 1 kg of sugar per litre of water. Some hydrometers are calibrated in °Brix.

Brix values: Properties that give an indication of the density of sugar in a solution at a specific temperature. Brix values are frequently used to express the sugar levels in fruits, beverages and sugar juices. This is discovered by a German inventor A. F. W. Brix.

Broiler meat: Meat from broilers, specific types of chickens, also used as an alternative term for chicken meat.

Broilers: Fast-growing strains of chickens reared under intensive conditions before slaughter, for meat production at about nine to twelve weeks of age. Development of intensive systems to produce broilers has led to large-scale production at low prices.

Broiling: Cooking of foods (usually meat or fish) by exposure to direct heat. Food can be broiled in ovens, directly under a gas or electric heat source, or on a barbecue, directly over charcoal or other heat source.

Bromates: Salts of bromic acid, including potassium bromate, which is used in flour improvers. Bromates can also be formed as disinfection byproducts during ozonation of drinking water containing bromide ions.

Broth: (1) The liquid in which meat or fish had been boiled together with herbs, vegetables and spices (original.) 2. Stock, brunoise vegetables and either barley or rice simmered with a bouquet garni and seasoning and finished with chopped parsley.

Brown flour: Wheat flour containing between 80 and 90% of the dehusked grain. Part of the bran is removed but the wheat germ remains. It is also called wheatmeal flour.

Brown mustard seed: The hot and bitter brown seed of an annual, *Brassica juncea*, native to India. Used whole in Indian cooking or ground and mixed with white mustard in the various mustard-based condiments and flavourings available. It is also called Indian mustard.

Brown onion soup: Finely sliced onion and garlic well browned in butter, flour added and cooked to a brown roux; brown stock added, simmered for 15 minutes, strained and seasoned. The soup is served in bowls with the top covered with toasted French bread slices and the whole sprinkled with cheese and browned under a fierce heat.

Brown rice: Rice from which the husks have been removed, leaving the germ and outer layers containing the bran intact.

Brown roux: Four parts of fat (clarified butter, dripping or vegetable oil) cooked slowly but not overcooked with 5 parts of soft plain flour until the flour is light brown in colour. Used for espagnole sauce and soups and as a general thickening agent.

Brown sugar: Granulated white sugar that has been covered with a layer of cane sugar syrups to give it a brown appearance and a caramel-like flavour. Brown sugar has a higher moisture content and a higher ash content than white sugar.

Browning agents: Ingredients or additives that promote browning of foods during processing, thereby imparting a darker colour to the finished product. Examples include caramel, milk and certain sugars. Commercial browning agents are often applied to foods cooked in microwave ovens in order to produce surface browning.

Browning inhibitors: Substances used to prevent browning in foods, also known as anti-browning agents. Use of sulfur dioxide or sulfites is one of the most widespread chemical means of controlling both enzymic browning and nonenzymic browning. Other browning inhibitors include kojic acid, which inhibits tyrosinase activity, citric acid and cysteine.

Browning: The process by which foods become brown, resulting from either enzymic or nonenzymic reactions. Enzymic browning occurs in freshly cut fruits and vegetables due to the oxidation of phenols by catechol oxidases. However, nonenzymic browning is generally a consequence of the Maillard reaction. Browning can be either a favourable process and encouraged using browning agents, or unfavourable and reduced using browning inhibitors.

Bruising: Damage to the surface of foods, particularly fruits and vegetables, resulting from mechanical impacts.

Bubble gums: Sweetened products made from chicle (gum-like exudate consisting of coagulated milky juice from the bark of the evergreen sapodilla tree, *Achras zapota*) or similar resilient substances (e.g. plasticized rubber or polymers), and chewed for its flavour. Bubble gums differ from chewing gums in the user's ability to blow bubbles from them during chewing

Buckwheat: Grains of *Fagopyrum esculentum* used as a cereal. It is unsuitable for the manufacture of bread. Whole grains are cooked like rice and made into baked puddings. Also made into other products such as porridge, noodles or griddle cakes. Good source of protein, niacin and vitamin B_1.

Buffalo milk: Milk obtained from buffaloes. Compared with cow milk, buffalo milk has a higher fat content (approximately 8%), higher contents of protein, calcium and some other minerals, vitamin A and biotin, and lower contents of potassium, and vitamin B_2 and vitamin B_6. It contains no carotenes. It used in making a range of dairy products, including Mozzarella cheese and Domiati cheese.

Buffering capacity: Ability of a substance or solution to resist change in acidity or alkalinity.

Bulgar: Wheat which has been dehusked, parboiled, dried then milled into a coarse powder prior to sale. It is easily reconstituted with boiling water and may be used for made up dishes or to accompany a main dish. Also called bulgur, burghul bourgouri, pourgouri.

Bulk density: Weight per overall unit volume of a substance. Bulk density is used in particular for porous substances where density is affected by pore volume and can be increased by the presence of pore fluid.

Bulking agents: Originally used to describe inert products added to contribute bulk and act as inexpensive fillers/extenders for more expensive ingredients. Now also used in low calorie foods and low fat foods to produce a feeling of satiety and to replace functional properties, flavouring characteristics and other qualities of the sugar and/or fat which have been removed. Substances used as bulking agents include methylcellulose, fibre, polyols and polydextrose.

Bulk volume (V_{bulk}): It is the volume of a material when packed or stacked in bulk. It includes all the pores enclosed within the material (internal pores) and also the void volume outside the boundary of individual particles when stacked in bulk (external pores).

Buns: Small yeast-raised rounded pieces of bread which are sometimes sweetened or flavoured, and may contain dried fruits. Choux buns are made from choux pastry and usually filled with whipped cream.

Burgers: Round, flat cakes of meat mince, cooked by grilling or frying. Specific types of burger include baconburgers, beefburgers, cheeseburgers and hamburgers. They are commonly eaten in bread rolls, served with lettuce, slices of onion and tomato ketchups.

Business: This includes the undertaking of a canteen, club, school, hospital or institution, whether carried on for profit or not, and any undertaking or activity carried on by a public or local authority.

Butcher: A person who prepares slaughtered and dressed meat for sale. Sometimes also slaughters the animals and dresses the carcasses.

Butter oils: Milk fat products with a very high fat content, usually not less than 90%, and a very low water content (no greater than 0.5%). In anhydrous butter oils, moisture content is no greater than 0.1%.

Butter starters: Bacterial mixtures, usually *Streptococcus*, used in ripening of cream in manufacture of cultured cream butter. Responsible for formation of flavour/aroma compounds such as diacetyl and acetoin. Affect stability of milk fat globule membranes, facilitating their breakdown during churning, the next step in the buttermaking process.

Butter substitutes: Products intended to be similar to butter in appearance and properties, but which differ in composition.

Butter: The spreadable water-in-oil emulsion of milk in fat produced by churning (i.e. mechanically mixing) cream and containing about 80% fat and 17% water. There are roughly one billion separate liquid globules per gram. A variety of types of butter are produced depending on, the type of animal, the method of production, the season, added salt or acidification of the cream prior to churning. Salt is sometimes added as a flavouring and colour may be adjusted using annatto or β-carotene. Rich in vitamin A; also contains vitamin E and vitamin D. Main types of butter marketed are cultured cream butter (also known as lactic butter or sour cream butter) and sweet cream butter, the latter having a higher pH value.

Butterfat: The fat found in milk which forms the main part (82 percent) of butter. It contains about 40% saturated, and less than 4% unsaturated fats.

Butterine: Fat product that was developed originally a substitute for butter. Composed of approximately 80% vegetable, animal or marine fats and 20% water, together with additives such as emulsifiers, colorants and preservatives.

Buttermaking: Process by which butter is made from milk or cream. It consists of cream ripening, churning, washing and working. Cream ripening involves a series of temperature treatments, with or without incubation with butter starters, that affect the consistency and flavour of the final product. Churning breaks up the milk fat globule membranes, allowing formation of butter grains which eventually separate from the buttermilk. The butter grains are washed with water to remove proteins, sugar and microorganisms, and worked into a homogeneous mass by kneading.

Buttermilk: Tangy flavoured residue remaining after separation of butter grains during buttermaking. Low in fat but rich in phospholipids and proteins, resulting from breakdown of the milk fat globule membranes during churning of cream. Buttermilk remaining after manufacture of sweet cream butter differs slightly in composition from that resulting from cultured cream buttermaking. A commercial product called cultured buttermilk is made by adding lactic acid bacteria to skim milk, the lactic acid produced during fermentation giving the product a tart flavour similar to that of churned buttermilk. Butter- milk is used as a beverage and as an ingredient in baking.

Butterscotch: Hard confectionery made by boiling brown sugar and butter or corn syrups together in water. Generally distinguished from caramels by the absence of milk or milk substitutes among the ingredients.

Butylamine: Synonym for 2-aminobutane. Fungicide used to prevent spoilage of fresh fruits and vegetables.

Butylated hydroxyanisole (BHA): A synthetic fat-soluble phenolic antioxidant. It has good heat stability and is widely used in the food industry to add stability to fats and oils. Applications include cereals, confectionery products, bakery products and packaging materials. Often used synergistically with BHT and other antioxidants.

Butylated hydroxytoluene (BHT): A widely used synthetic antioxidant with similar properties to BHA, but less stable at high tem- peratures. Applications include bakery products, breakfast cereals, and food packaging materials. Displays good synergy when used in combination with BHA.

Butyric acid: A fatty acid which when in ester form is a constituent of butter fat. One of the short chain saturated fatty acids which occurs as one of the flavour compounds in a wide range of foods and beverages. The free fatty acid is liberated by oxidation of the butter and is responsible for the rancid flavour and smell. Especially characteristic of milk fats and dairy products. At high concentrations, it may be responsible for development of off flavour. Butyric acid is synonym for butanoic acid.

Butyric fermentation: Process by which certain bacteria (mainly *Clostridium* spp.) produce butyric acid. Although a useful industrial process, it can cause cheese spoilage (blowing). Production of butyric acid in the colon by fermentation of dietary fibre may reduce the risk of certain cancers.

Butyrometers: Apparatus used to measure the fats content of milk. Samples are mixed with sulfuric acid in special graduated tubes which are then centrifuged. Fat separates as an upper layer, the size of which is measured from the markings on the tube.

C

Cafeterias: Self service restaurants. Often located within larger establishments, such as department stores, schools or universities.

Caffeine: One of the xanthine alkaloids naturally present in several plant foods, including tea, coffee and cola nuts. Acts as a stimulant. Used as an ingredient in some soft drinks, including cola beverages and energy drinks.

Cake batters: Batters usually prepared from flour, eggs, butter or margarines, and sugar that are used to make cakes. Other ingredients are added ac- cording to the type of cakes to be made.

Cake flour: A very high-extraction low-protein white flour used for making very light-textured cakes such as angel cake.

Cake mixes: Powdered formulations containing all the ingredients required to make cakes.

Cake mixtures: The five important cake mixtures are basic, Victoria sponge, Genoese, pound and fatless. The first two use a chemical raising agent; the last three rely on air beaten into the egg for raising.

Cake: Soft bakery products produced by baking a batter containing flour, sugar, baking powders and beaten eggs, with or without shortenings. It may also incorporate root vegetables, principally carrot. According to the final product, other ingredients are also included, such as flavourings, nuts, chocolate and dried fruits.

Caking: Solidification of powders or granules into a mass. Caking can be a problem during the storage of dried foods and sugar.

Calciferol: It is synonym for ergocalciferol and vitamin D_2; one of the group of sterols which constitute vitamin D. It synthesized by irradiation of the plant provitamin ergosterol.

Calcium acetate: The calcium salt of acetic acid (vinegar) used as a preservative and firming agent.

Calcium carbonate: A natural mineral source of calcium produced from limestone. Used in a very finely ground form as an acidity regulator and as a source of calcium in processed foods, especially manufactured bread in the UK. It is also called chalk, ground chalk, and precipitated chalk.

Calcium chloride: Salt, chemical formula $CaCl_2$, widely used as a general purpose food additive. Applications include flavour preservation in pickles, as a firming agent in fruits and vegetables, and as a source of calcium for calcium alginate gels.

Calcium heptonate: A food additive used as a firming agent and sequestering agent in prepared food and vegetables.

Calcium tartrate: The calcium salt of tartaric acid. Calcium tartrate may precipitate in wines, forming an undesirable haze or sediment. Haze stabilization treatments may be required to prevent this problem.

Calcium: Mineral with the chemical symbol Ca. It is a constituent of most foods and an essential nutrient in the human diet, particularly important for strong bones and teeth of which it is a major component. Rich sources include milk and dairy products, oily fish and spinach; staple foods are sometimes enriched with calcium. Also important in the setting of pectins gels, and the firmness of processed fruit and vegetable products.

Calf rennets: Substance extracted from the abomasum of calves that is used in coagulation of milk for cheesemaking. The active enzyme is chymosin; pepsin is also present.

Calorie: The outdated measure of the energy content of foods, still used in popular parlance to estimate the fattening potentiality of food (since if not used for energy, food is usually stored as fat). The Calorie *(capital C)* or kilocalorie used in nutrition has 1000 times the value of the calorie *(small c)* used in science and is the energy required to heat 1 kg of water by 1°C.

Calorific value: The amount of heat produced when 1 g (small calorie) or 1 kg (large Calorie or kilocalorie) of food is completely burned or metabolized to carbon dioxide and water by oxygen. It is also amount of calories in foods or nutrients, indicating the levels of utilizable energy. It is also known as energy values.

Calorimetry: Technique for measuring the energy content of foods from the number of calories formed during combustion of a known amount of sample.

Camembert: A soft, small, wheel-shaped cheese (of France*)* about 250 g in weight made from full cream cows' milk, dry salted to 3% salt, surface-ripened for 10 to 14 days with *Penicillium candidum* and *P.camembertii*, which forms a white fungus on the rind, then wrapped in paper and boxed. It contains 57% water, 21% fat and 20% protein, half of which is hydrolysed by the fungus. Originating in Normandy but now widely produced.

Campylobacter jejuni **:** A food-poisoning bacteria found in chicken and milk which is a major cause of gastroenteritis. The incubation period is 2 to 10 days and the duration of the often severe illness, which is flu-like with abdominal pain and fever followed by diarrhoea, is 5 to 10 days.

***Campylobacter*:** A genus of Gram negative, microaero- philic rod-shaped, food-poisoning bacteria found in raw meat and poultry which cause diarrhoea, vomiting and fever. It occurs in the reproductive and intestinal tracts of animals and humans. Some species are pathogenic, e.g. *Campylobacter jejuni*, which frequently contaminates raw chicken meat. Raw milk is also a source of infection.

Campylobacteriosis: Any human or animal disease caused by infection with *Campylobacter* species. *C. jejuni* causes food poisoning in man characterized by diarrhoea, fever, abdominal pain, nausea, headache and muscle pain.

Cancer: A range of malignant diseases characterized by uncontrolled cell proliferation that result in tissue invasion and destruction. Dietary factors have been linked with increased risk for certain cancers (e.g. high intakes of dietary fats) and with reduced risk (e.g. increased intakes of fruits and vegetables).

***Candida*:** Genus of yeast fungi of the class Saccharo- mycetes. Occur in soil and on plants. It may be used in the production of fermented foods (e.g. *Candida kefir* in the production of kefir and koumiss, and *C. famata* in the production of fermented sausages). *C. lipolytica* and *C. zeylanoides* cause meat spoilage, while *C. valida* causes spoilage in wines. *C. utilis* and *C. lipolytica* may be used for production of single cell proteins.

Candied fruits: Fruits, usually whole, preserved by softening in water and then soaking in syrups of progressively increasing sucrose concentrations. After drying, the fruits are coated in sugar to make crystallized fruits or dipped in concentrated sugar syrups to make glace products, such as glace cherries. It is often regarded as luxury products, although glace cherries are frequently used as ingredients in bakery products. Used as a confectionery item and for decoration. Also called crystallized fruit, preserved fruit.

Candied peel: Citrons, halved, pulp removed, immersed in brine or sea water for 1 month to ferment then washed, dried and candied with sugar syrup. Usually candied in the country where used. Sometimes used as a confectionery item but more usually chopped and used in cakes.

Candling: Technique for determining the quality of eggs wherein the egg is held before a light which penetrates the egg and makes it possible to inspect the contents and shell.

Cane sugar factories: Factories containing process- ing lines equipped for extracting cane sugar from sugar cane (*Saccharum officinarium*). Sugar cane factories located close to where the sugar cane is cultivated (plantation factories) are involved with manufacture from sugar cane of pure white sugar or raw cane sugar. Sugar refineries are normally situated nearer to the markets for sugar and are involved in purification of raw or salvaged sugar to produce white sugar.

Cane sugar juices: Aqueous solutions of cane sugar produced during processing of sugar cane. Raw juices are produced by compression of the sugar cane stalks and contain cane sugar and impurities, thin juices are the purified raw juices and thick juices are concentrates of the thin juices.

Cane sugar syrups: Highly concentrated aqueous solutions of cane sugar produced by evaporation of purified cane sugar juices (thin juices).

Cane sugar: Sucrose extracted from stalks of sugar cane (*Saccharum officinarium*). Processing of sugar cane to produce cane sugar involves: washing and cutting the cane stalks; extraction of cane sugar juices by crushing the stalks using a series of heavy rollers; purification of the raw cane sugar juices by precipitation of impurities (liming and clarification); filtration to remove the precipitates; evaporation of the purified juices which results in concentration of the cane sugar juices and crystallization of sucrose. Dried purified cane sugar is composed of 99.80% sucrose and has $<0.05\%$ moisture content.

Cane syrup: The concentrated juice from sugar cane used in place of golden syrup or molasses.

Canned Foods: Canning was one of the first processes directed at preserving food for long periods. These goods should be stored at ambient temperatures around 10 –15°C in a dry area, such as a dry goods store. They can have extensive shelf lives from several months, in the case of canned tomatoes, to 20 or 30 years, as with corned beef. Defects in cans, such as leaking seams, can be a source of contamination, particularly if dirty water is used to cool the can after processing, which results in water being sucked in through the defective seam. This can result in spoilage of the contents due to bacterial growth and the potential for food poisoning. Past outbreaks of typhoid fever in the UK were caused by canned meats contaminated by Typhoid bacilli from polluted cooling water. Canned goods stored in moist atmospheres or for long periods may rust and develop minute pin holes through which organisms can pass. Blown cans, which bulge at each end, due to the production of gas inside may also be encountered, particularly with products such as canned meats, tomatoes and fruits. Where there is evidence of any the above faults in canned goods, the cans and contents should be discarded and not used.

Canning: A sterilization process in which spoilage and pathogenic microorganisms are eliminated in foods prior to hermetic sealing in containers (cans). Most commercial canning operations are based on the principle that bacterial destruction increases tenfold for each 10°C increase in temperature. Food exposed to high temperatures for short periods of time is known to retain more of its natural flavour. Canning is an important means of preserving food and canning processes have been in operation for over 200 years since the days of Napoleon. It is a method in which food is hermetically sealed in metal containers and heat treated until it is commercially sterile. The main objective of a canning process is to destroy pathogenic and spoilage micro-organisms through the application of heat in order to achieve commercial sterility. The quantity of heat applied to a food is directly related to its pH as micro-organisms do not survive or grow below a pH of 4.5. Heat treatment duration further takes into account can sizes and the nature of the contents, as solid contents require a longer period of treatment than liquids. The construction of a can is critical. Generally cans are manufactured from tinplated steel and incorporate a lacquer coating on the inside. Seams at the end of the can forming the lid must be correctly made and sealed as this is the principal point where leakage can occur. At the head of the can (the head space) a vacuum is created and this vacuum is critical. This vacuum can be measured and is an indicator of satisfactory processing and future shelf life.

Canning quality: Canning quality scores represent the sum of scores for colour (chroma, uniformity, and attractiveness), wholeness, smoothness, firmness, moistness, lack of fibre, mouthfeel and flavour of canned foods.

Cans: Rigid cylindrical metal containers made of steel sheet or plate, aluminium, copper or other metals. Used as packaging for foods and beverages; most are sealed hermetically for storage and retail over long periods of time.

Canteen meals: Meals served in canteens, i.e. restaurants catering for workers in establishments such as schools or factories. Food is usually prepared in large amounts and served from a central point

Canteens: Restaurants located in establishments such as schools and factories. Usually self service and designed to cater for large numbers of people. Also refers to vessels with caps or other closures used for carrying water or other beverages, especially while travelling.

Canthaxanthin: Red pigment of the carotenoids group. Occurs naturally in crustacea and salmonid fish and has antioxidative activity. Used as a feed additive to improve the colour of egg yolks, skin colour of broilers and flesh colour of aquacultured salmon or trout.

Capsaicin: One of the flavour compounds of chillies and other Capsicums, in part responsible for their pungent characteristics.

Capsaicinoids: Flavour compounds of chillies and other Capsicums related to capsaicin and partly responsible for the pungent characteristics.

Capsanthin: Pigment of the xanthophylls group which occur in peppers (Capsicums).

Caramel colorants: Colorants resulting from the carefully controlled heating of carbohydrates (e.g. sugars or malt syrups) in the presence of small amounts of food-grade acids, alkalis or salts. Widely used to impart a yellow or brown colour to numerous foods and beverages, including cola beverages and other soft drinks, beer, soy sauces, bakery products, browning agents and sausage casings. Both positively and negatively charged caramel colorants are available (particles of the caramel colorant must have the same charge as the colloidal particles of the product to be coloured, in order to avoid precipitation). Also reported to act as vitamin antagonists to vitamin B_6. Caramel is also used in flavourings.

Caramel: Complex mixture of golden brown flavouring/colouring substances produced when sugars are heated up to (182°C) during caramelization. Thermal degradation of the sugars results in a similar bitter-sweet flavour profile to that of molasses and maple syrups. Caramel is used in flavourings and flavour enhancers for a wide range of foods, including caramels, cakes and biscuits. Colouring properties are employed in caramel colorants.

Caramels Sugar confectionery: The products similar to toffees made from sweetened, condensed or evaporated milk, butter or vegetable oils, and sugar. Boiled at lower temperatures than toffees, and may be soft or hard.

Caraway: The plant from which caraway seeds are obtained. The young leaves may be used in salads and soups and the roots can be cooked as a vegetable.

Carbohydrates: One of the main classes of compounds present in foods, which includes monosaccharides, their derivatives such as glucosides, polyols, nucleotides and nucleosides, and their oligomers and polymers (oligosaccharides and polysaccharides). Important carbohydrates in foods include sugars, starch, pectins, fibre fractions, celluloses and their derivatives, and polysaccharides used as additives such as gelling agents and thickeners. It is the principal energy sources in human food which are simple sugars *(monosaccharides)* or chains of repeated sugar units *(disaccharides, trisaccharides and polysaccharides)*. The polysaccharides have many repeated sugar units and are starches or various types of cellulose. They are all with the exception of cellulose and a few starches broken down in the gut into absorbable monosaccharides either by acids and enzymes in the gastric juices or by the action of microorganisms. See also monosaccharide, cellulose, dietary fibre.

Carbon black: A very finely divided form of carbon used as a food colouring. It is also called **vegetable carbon**.

Carbon dioxide: A colourless, odourless gas (chemical formula CO_2) produced by the burning of carbon and organic compounds and by respiration, and absorbed by plants in respiration. The gas produced when the carbon in foods is oxidized either by the action of yeasts and other microorganisms, by combustion or by cellular processes in the body, or when chemical raising agents are heated or react in water. The gas is responsible for the raising of bread, cakes, etc. It is used in modified atmosphere packaging of foods.

Carbonatation: Process used in the manufacture of white sugar for purification (clarification) of sugar juices. Various carbonatation methods have been developed for specific purposes, but the basic principle is the same. The process involves addition of lime (CaO) to sugar juice followed by bubbling of carbon dioxide through this mixture. A precipitate of $CaCO_3$ forms that entraps suspended impurities within its crystalline structure and adsorbs soluble impurities. Soluble impurities may also react with the lime to form insoluble Ca salts.

Carbonated water: Water in which carbon dioxide has been dissolved. Sometimes stored under pressure. The gas is released when the pressure is reduced or when the water is warmed up on contact with the mouth. Also called soda water, mineral water

Carbonated beverages: Beverages, especially soft drinks, which have been impregnated with sufficient carbon dioxide to cause effervescence.

Carbonation: Conversion of a compound into a carbonate, or the impregnation of a liquid with carbon dioxide (CO_2) under pressure. CO_2 is added to beverages to make them effervescent. Examples of carbonated beverages include lemonade and sparkling mineral waters.

Carbonic maceration: A winemaking process in which whole grapes are macerated under a carbon dioxide atmosphere before alcoholic fermentation; it is used in manufacture of Beaujolais and similar wines. Carbonic maceration enhances the fruity character of the wine aroma.

Carboxymethyl cellulose: A non-nutritive cellulose derivative used for thickening and stabilizing ice cream and jellies. It is also used as wallpaper paste.

Carcass: The body of a slaughtered animal, prepared for use as meat. Dead bodies of animals and birds, especially those prepared for cutting up as meat. The term is used by butchers to describe animals' and birds' bodies after removal of the heads, limbs, hides (or feathers in birds) and offal; these types of carcasses are also called dressed carcasses. Bird carcasses are usually chilled whole, whilst animal carcasses are usually split longitudinally into sides before chilling. Many countries operate carcass classification schemes, which are designed to categorize carcasses with common characteristics such as carcass weight, fatness (fat class) and conformation. Usually, carcass classification schemes discriminate against very fat and very lean carcasses.

Carcinogenesis: Processes leading to the formation of cancer (tumours).

Carcinogenicity testing: Analyses, including the Ames test, to determine the carcinogenicity of suspected carcinogens. Also applied to other chemical compounds as part of routine safety evaluation studies. Tests can include the use of animal models, cell cultures or microorganisms.

Carcinogenicity: A measure of the relative activity of carcinogens.

Carcinogens: Substances that are able to induce carcinogenesis, encompassing direct-acting agents that possess genotoxicity and indirect-acting procarcinogens that require activation by cell metabolic path- ways. Food sources of carcinogens are widespread, and include heterocyclic amines formed in meat during cooking, nitrosamines in nitrite-treated meat products, urethane in fermented foods and alcoholic beverages, and agaritine in mushrooms.

Cardamom: The dried fruit of a perennial bush, *Elettaria cardamomum*, belonging to the ginger family, consisting of green or bleached pods (5 to 10 mm long) containing loose black seeds with an aromatic flavour and smell. Larger pods are in general brown to black and their seeds are of a lower quality and used in pickles and chutneys. Cardamom is used to flavour coffee in the Middle East as well as being common in Indian, Eastern and Scandinavian cuisines. Young cardamom leaves are used as a food wrapping and to flavour food in Southeast Asian cuisines. Cheap cardamom substitutes are Nepal cardamom, Chinese cardamom, Javanese winged cardamom and Ethiopian cardamom. Also called cardamon, cardamom.

Cardiovascular diseases (CVD): Congenital and acquired diseases of the heart or blood vessels including coronary heart diseases (CHD) and stroke. Many risk factors for cardiovascular diseases have been identified, including lifestyle (smoking, lack of physical exercise), diseases (obesity, hyperlipaemia) and diet. Cardiovascular risk may be modified by lowering intake of fats, modulating dietary fatty acids composition and in- creasing consumption of whole grains, dietary fibre and fruits and vegetables.

Cargo rice: A dark rice which has only been dehusked, rich in bran, protein, vitamins, etc.

Carotenes: Long chain unsaturated hydrocarbons with provitamin A activity found in green and yellow plant foods such as carrots, sweet potatoes, green leafy vegetables and yellow fruits. Carotenes are the simplest of the carotenoids and are cleaved *in vivo*, generating two molecules of vitamin A. It is converted into the antioxidant vitamin A in the body and used as a permitted orange food colouring. Lycopene in tomatoes and beta-carotene in carrots are two well-known examples. It is also called carotenoids, provitamin A.

Carotenoids: Pigments of the polyenoic terpenoids class, which are present in a wide range of plant foods and animal foods. Impart a yellow, orange, red or purple colour to foods, and may be used as food colorants. Many have antioxidative activity; some have vitamin A activity.

Carrageenans: Gums extracted from red seaweeds (mainly *Chondrus crispus* and *Gigartina stellata*). Used as stabilizers, thickeners and emulsifiers in a wide range of foods including milk beverages, processed cheese, ice cream, other dairy products, desserts and ready to feed infant formulas. It forms thermo-reversible carrageenan gels, which are also used widely in the food industry.

Cartoning: Process of packaging items such as foods or beverages in cartons.

Cartons: Lightweight containers made from carton- board. It is usually delivered to the user in the form of flattened, pre-cut and pre-creased carton blanks.

Casein whey: Liquid remaining after precipitation of casein by the action of acids or rennets. It is also called whey.

Casein: The main protein of milk, representing approximately 80% of the total milk proteins. A phosphorus-containing protein that is heat stable, but precipitated by alcohol, rennets and acids. Individual fractions are combined into larger units called casein micelles, structure and stability of which are related to calcium content.

Caseinates: Salts formed by acid precipitation of casein from milk followed by neutralization and drying. Some caseinates, including potassium, sodium and calcium caseinate are widely used as food ingredients due to their nutritional and functional proper- ties. Uses include binding agents, emulsifiers, whipping agents and protein supplements in foods.

Casings: Items used to give processed meat products a uniform or characteristic shape, to hold comminuted products together during further processing and to protect meat products. Casings are most commonly used as forms and containers for sausages; these types of casings are specifically known as sausage casings. There are two major types of casings: natural and manufactured. Natural casings are derived al- most exclusively from the gastrointestinal tract of cat- tle, sheep and swine. Natural casings are highly permeable to moisture and smoke; moreover, they shrink and thereby remain in close contact with the surface of a meat product as it loses water. Most natural casings are digestible and can be eaten. There are four major classes of manufactured casings, namely cellulose, in- edible collagen, edible collagen and plastic. Strength, shrinkage and permeability characteristics differ between the different types of casings, providing a range of products suitable for the preparation of many different types of meat products.

Cassava: Starchy tubers produced by the tropical plant *Cassava esculenta* (syn. *ultissima*), also known as manioc. An important staple food in many tropical regions, cassava tubers are a good source of carbohydrate and vitamin C, but low in protein, minerals and other vitamins. Tubers are the source of tapioca starch, while the leaves can be eaten as a vegetable in soups and stews. Fresh cassava roots and leaves (particularly those from bitter cultivars) contain the cyanogenic glycosides, linamarin and lotaustralin, and must therefore be detoxified prior to consumption in order to prevent cyanide poisoning. Detoxification is achieved by conventional grating, washing and cooking methods, or by fermentation into a variety of products including gari, fufu, attieke and tape ketela.

Catalases: This enzyme is responsible for breakdown of H_2O_2 to water and O_2. Used for removing the H_2O_2 added to cold- sterilized milk, improving the baking properties of dough and improving the flavour of fermented whey. Exhibit antioxidative activity and play an important role in preventing oxidation of lipids in meat. In conjunction with D-amino-acid oxidases, catalases can be used for production of α-ketoacids, which are gaining importance as nutraceuticals. The enzymes also protect microorganisms, including several foodborne pathogens, against various environmental stresses.

Catalysts: Substances that promote a chemical reaction by lowering the activation energy, but which are not consumed or altered during the reaction.

Catechin: Catechol which occurs in tea and many other foods and beverages. Catechins are thought to have beneficial effects on health, because of their ap- parent antimicrobial activity, antioxidative activity and anticancer properties.

Catechol oxidases: A group of copper proteins that act on catechol and a variety of substituted catechols. Also known as diphenol oxidases, phenolases, polyphenol oxidases and tyrosinases, these enzymes also catalyze the reaction of monophenol monooxygenases under certain conditions. It involved in enzymic browning in fruits, vegetables and cereal grains.

Catecholamines: Phenolic biogenic amines which occur in tissues of plants and animals. Some, e.g. adrenaline and noradrenaline, act as hormones and high pre-slaughter levels of these compounds (as a result of stress) may be associated with poor meat quality. Aerobic oxidation of catecholamines in the presence of catechol oxidases results in formation of melanins, and hence browning of plant foods.

Catechols: Flavan-3-ols which are present in a wide range of foods of plant origin. It may be polymerized to form tannins by the action of polyphenol oxidases (catechol oxidases). Catechols may contribute to the antioxidative activity and health benefits of plant-derived phenols.

Catering: Provision of foods and beverages in a commercial or institutional setting, or at a function. Occasionally accommodation is provided also. It includes services provided by hotels, restaurants, canteens and hospital kitchens. Also encompasses foods service.

Cathepsins: Enzymes important in meat tenderization during ageing and deterioration of fish proteins gels, with subsequent effects on sensory properties. Also exhibit proteolytic activity in dairy products.

Cellobiose: Reducing sugar composed of two molecules of glucose linked via a β-1,4-glycosidic bond. Although free cellobiose is not found in nature, it is the monomer unit for celluloses, one of the most abundant substances in nature. Cellobiose may be prepared from celluloses by hydrolysis with cellulases.

Cellophane: Thin, transparent material made from celluloses. Used as a wrapping for foods to protect against contamination and to preserve freshness.

Cellulases: It catalyse the endohydrolysis of 1,4-β-D-glucosidic linkages in celluloses, lichenin and cereal β-D-glucans. It produced commercially from a number of fungi and bacteria. These enzymes have many applications in the food industry, e.g. processing of fruits and vegetables and their juices, brewing, winemaking, improving the shelf life of bakery products, enhancing the quality of soy protein hydrolysates and hydrolysis of celluloses prior to ethanolic fermentation.

Cellulolytic enzymes: Enzymes that act synergistically to hydrolyse celluloses or chemically modified cellulose polymers. These enzymes are traditionally classified into three groups, cellulose 1,4-β- cellobiosidases, cellulases and β-glucosidases. True cellulase systems, produced by a number of fungi, are able to hydrolyse crystalline cellulose completely, while low-value cellulase systems can only hydrolyse amorphous cellulose. Cellulolytic enzymes can hydrolyse cellulose waste materials prior to ethanolic fermentation and, in conjunction with pectic enzymes, represent an alternative to chemical peeling of fruits and vegetables.

Centrifugal separators: Machines with rapidly rotating containers used to separate two liquids, solids from a liquid, or a liquid from a gas. In the food industry, these separators are used for clarification of beer and fermentation broths, during sugar processing to separate sugar crystals from syrups, and during food hygiene practices (e.g. cleaning in place).

Centrifugation: Process in which liquids are separated from solids, or heterogeneous liquids are separated, on the basis of differences in density using machines (centrifuges) with rapidly rotating drums.

Centrifuges: Machines with rapidly rotating drums used to separate liquids from solids or heterogeneous liquids on the basis of differences in density.

Cereal bran: Protective outer layer of the seeds of edible members of the grass family which is separated from the kernel during milling. It is often added to foods as a source of dietary fibre.

Cereal by-products: Secondary products of cereal processing, e.g. bran and germ removed during milling of cereals to produce refined flour.

Cereal products: Generic term for foods which have been formulated using cereals as their main ingredient.

Cereal proteins: Proteins found in cereal grains, which may be classed as biologically active enzymes or biologically inactive storage proteins. Storage proteins make up approximately 80% of total cereal proteins and are often used for varietal classification.

Cereal wines: Non-distilled alcoholic beverages made by fermentation of saccharified mashes made from cereals. Examples of cereal wines include sake and other rice wines.

Cereals: Plants and seeds from monocotyledonous plants of the grass family. The edible, starchy seeds are suitable for food use and are processed to make a wide range of products.

Champagnization: The specific winemaking process used for manufacture of champagne, involving in-bottle secondary fermentation under defined conditions.

Chaptalization: Addition of sugar to grape musts to increase alcohol content in the resulting wines. Legal in some winemaking countries, prohibited in others.

Char: Any of several trout-like fish species belonging to the genus *Salvelinus* within the family Salmonidae. Char species include *S. alpinus* (Arctic char) *S. fontinalis* (brook trout) and *S. namaycush* (lake trout). Flesh of most species is highly regarded. Usually marketed fresh or frozen.

Charcoal: Amorphous, usually impure, form of carbon produced by heating wood or other organic material in the absence of air. It can be used in absorbents (activated carbon), as a cooking fuel which produces a distinctive flavour, e.g. in barbecued foods, or in fermentation technology.

Cheddar cheese: Semi-hard cow milk cheese originally made in England but now made all over the world. Natural colour ranges from white to pale yellow, but some cheeses have colorants added to form a more orange colour. Generally matured for 9-24 months, the flavour getting sharper with time.

Cheddaring: Process used in manufacture of scalded cheese. Pressed curd is cut into pieces which are covered and left for 6-10 hours at 15-20°C during which the curd becomes elastic and develops a yellow colour and characteristic flavour.

Cheese making: Process by which cheese is made from milk. Depending on the type of cheese being made, steps include preparation of the cheese milk, coagulation of milk with addition of cheese starters and rennets, draining of whey, pressing, shaping of curd, salting and ripening.

Cheese starters: Microbial cultures inoculated into milk to produce acidity by fermentation during manufacture of cheese. Commercial starter preparations are available in liquid form, or as freeze-dried or deep-frozen powders or granules. Composition of the culture is varied according to the type of cheese being made.

Cheese whey: By-product of cheesemaking formed along with curd during coagulation of milk. Rich in milk proteins including α-lactalbumin and β- lactoglobulin. Whey is produced in large amounts, leading to disposal problems. As well as being utilized as a food ingredient, whey is used as a fermentation substrate and in animal feeds. Also known as lactose- rum or serum.

Cheese: Dairy products made by separating the solid component of milk (curd) from the liquid part (whey). It is made mainly from cow milk, but also from milk of many other mammals, commonly goats, ewes and buffaloes. An important part of the diet worldwide due to its nutritional properties and ease of preparation.

Chelating agents: Substances which form a stable chelate ring with free metal ions and can therefore be used in foods to help control the reaction of trace met als with other food components. They act as sequestrants to prevent meta-catalysed oxidation, unwanted crystal formation and loss of nutritional quality in a variety of foods, and can also be used for the con- trolled release of metal ions for nutritional purposes or for controlled gelation in thickeners. Examples of chelating agents include EDTA (ethylenedi- aminetetraaceticacid) and glucono-δ -lactone.

Chemesthesis: Complex sensation obtained from foods, regarded as a component of the sensory properties flavour and mouthfeel. Examples include the burn of capsaicin in chillies, the cooling sensation from menthol and the tingle associated with carbonated beverages.

Chemical oxygen demand (COD): Measure of the quantity of chemically oxidizable components present in water.

Chemical residues: These are chemical substances intentionally or accidentally brought into contact with, or added to, a food during its growth period. These include pesticides, herbicides and additives in feed and veterinary products.

Chemostats: Apparatus for maintaining a microbial population in the exponential phase of growth by regulating the input of a rate-limiting nutrient, and removal of medium and cells. The concentration of biomass in the culture vessel remains constant and the culture is normally grown at a sub-maximal growth rate. Under steady-state conditions, the relationship between growth rate and concentration of growth- limiting substrate can often be predicted using the Monod equation, while specific growth rate is numerically equal to the dilution rate.

Chevon: Alternative term for goat meat; the term is commonly used in India.

Chewiness: Texture term relating to the extent to which a product needs chewing, or a measure of the effort needed to chew, i.e. its toughness, rubberiness or leatheriness in the mouth.

Chewing gums: Sweetened products made from chicle (gum-like exudate consisting of coagulated milky juice from the bark of the evergreen sapodilla tree, *Achras zapota*) or similar resilient substances (e.g. plasticized rubber or polymers), sugar or similar sweeteners. It may also be made using a gum base, softeners and flavourings. Some chewing gums are specially formulated to promote dental health. Also known as chicle gums or gum balls.

Chhana: Indian style soft Cottage cheese analogue prepared by heating milk (usually cow milk) to nearly boiling, adding acid coagulants while the milk is hot and removing whey by filtration. Used as a base for various Indian sweets, such as rasogolla and sandesh. Also known as channa.

Chicken bones: Bones from chicken carcasses. During cooking, they darken in colour; this colour change is increased by freezing and thawing prior to cooking. Chicken bones are commonly used to prepare chicken soups or are processed into animal feeds. Hot-water extracts prepared from chicken bones are used in many types of products, especially in flavourings. Exposure of chicken meat containing bone to a dose of ionizing radiation results in the formation of long-lived free radicals which give rise to characteristic electron spin resonance (ESR) signals. The presence of these signals provides clear evidence that chicken meat has been irradiated. Mechanical boning of chicken meat remains a problem to the meat industry, as bone fragments often remain in chicken fillets, escaping manual or X-ray machine detection.

Chicken meat: Meat from chickens. Different pro- portions of red and white myofibrils produce light and dark meat in different parts of chicken carcasses. Chicken leg meat is darker than chicken breast meat. Composition of feeds influences flavour and colour of chicken meat. Compared with chicken meat produced in intensive systems, free-range chicken meat tends to have more flavour; however, it is tougher and, in developed countries, more expensive. Chicken meat can be roasted, grilled, poached or casserolled. Chickens are sold whole, or portioned into joints, including chicken breasts, wings, drumsticks and thighs.

Chicken sausages: Sausages prepared from chicken meat, often spent hen meat. Commonly they are made from mechanically recovered meat or chicken meat trimmings. They also tend to include chicken skin and the less preferred components of chicken offal, such as gizzards and hearts. Other ingredients may include water, salt, nitrites, pork fat, blood and phosphates.

Chicory: Common name for *Cichorium intybus*. Utilized in a number of ways, some cultivars being grown for the root, a powder or extract from which is used as an additive in coffee, making a more bitter beverage. Other cultivars are grown for the leaves, which are used in salads or cooked as a vegetable. Some cultivars, such as witloof, are used to produce blanched leafy growths called chicons, which are eaten raw or cooked. Similar nutritionally to lettuces and endives.

Chilled foods: Perishable foods that can be stored at chilled (refrigerator) temperature for a specified amount of time. Examples include chilled ready meals, pizzas, sandwiches and many dairy products.

Chillers: Cold cabinets or refrigerators that is capable of rapid cooling/chilling of foods to a few degrees above their freezing point in order to extend shelf life.

Chilling injury: Disorder of fruits and vegetables is induced by low temperatures. It may occur in the field, during transit or in retail or domestic refrigerators. Symptoms include surface lesions, water soaking of tissues, water loss, and internal discoloration, failure to ripen, and decay. Critical temperature for chilling in- jury varies with type of crop. Storage life of produce susceptible to chilling injury is short, as refrigeration cannot be used to preserve quality.

Chilling: Process of making foods colder to extend their shelf life, usually undertaken by application of refrigeration.

Chipping properties: Functional properties are relating to the ability of different cultivars or varieties of potatoes to be processed into good quality chips. The most important processing quality parameters for chips are colour, flavour and texture.

Chips: Small pieces of food prepared by chopping or cutting, which are then usually fried. Include potato chips (French fries), corn chips and tortilla chips. The term is frequently used to refer specifically to potato chips in the UK and to potato crisps in the USA and continental Europe.

Chitin: Homopolysaccharide, consisting of β(1→4)- linked D-*N*-acetylglucosamine. It occurs in shells of crustacea and cell walls of fungi, and may be recovered from crustacea shell wastes. It has potential use in functional foods.

Chitosan: Polysaccharide derived from chitin by partial deacetylation with a strong base. Its potential use is in functional foods.

***Chlorella*:** Genus of unicellular green algae of the family Oocystaceae. Occur in fresh water and soils. Species (e.g. *Chlorella pyrenoidosa*) may be used in the production of single cell proteins, or as food additives owing to their nutritional composition (high protein, vitamin B_{12} and iron contents) and beneficial physiological effects. Some species are added to foods (e.g. cakes, cheese, mayonnaise, ice cream and rice) to improve their flavour. Due to their high contents of carotenoids, they are used as feed additives for the enhancement of the colour of rainbow trout flesh. *C. protothecoides* produces lutein, which is used in food colorants for foods such as pasta.

Chlorides: Salts of hydrochloric acid. Occur widely in foods and beverages, the most important being common salt, NaCl, which is used in food additives such as flavourings, preservatives and bulking agents.

Chlorination: Insertion of a chlorine atom into a compound, or treatment of an item with chlorine gas (Cl_2). For example, chlorine gas can be used in sterilization of water.

Chlorine dioxide: Gaseous chlorine compound which is used in oxidizing agents - type disinfectants, used for sterilization of foods and water.

Chlorine: Member of the halogens group, chemical symbol Cl. Chlorine and it compounds have strong microbicidal activity and are used in the food industry as disinfectants and sterilizing agents. Chlorine gas is toxic.

Chlorites: Salts of chlorous acid, used as disinfectants in the food industry.

Chlorofluorocarbons (CFC): Any class of synthetic compound of carbon, hydrogen, chlorine and fluorine used as refrigerants and aerosol propellants. Commercial CFC is nonflammable, non- corrosive, nontoxic and odourless, but is known to be harmful to the ozone layer. The most common commercial CFC, marketed as Freons, is trichlorofluoromethane (CFC-11) and dichlorodifluoromethane (CFC-12).

Chloroform: Colourless, heavy, volatile, toxic liquid. Used as a solvent, fumigant and insecticide. Also known as trichloromethane.

Chlorogenic acid: It is synonym for caffeoylquinic acid. Phenol present in many foods of plant origin. It plays an important role in enzymic browning of fruits and vegetables. It has antioxidative activity, and may contribute to possible health-promoting or protective actions of dietary phenolic compounds.

Chocolate beverages: Hot or cold beverages in which chocolate is a main ingredient.

Chocolate chips: Small pieces of chocolate used as ingredients in confectionery and bakery products.

Chocolate coatings: Chocolate preparations used to coat various products such as sugar confectionery, bakery products, fruit or ice cream. It formed by pre-crystallization of chocolate, coating of the food and cooling. Pre-crystallization and cooling affect the gloss, degree of solidification and coat thickness of the coatings produced.

Chocolate liquor: Fermented and roasted cocoa beans, which are ground finely to form a paste used in the manufacture of chocolate and cocoa powders. Grinding releases fats (cocoa butter) from the cells of the cocoa beans which helps the chocolate to flow. Also called chocolate mass, cocoa mass and cocoa liquor.

Chocolate mass: It is alternative term for chocolate liquor, produced by grinding dehusked cocoa beans, or nibs, to a paste from which chocolate and chocolate products are made. Also called cocoa mass and cocoa liquor.

Chocolate powders: Manufactured from cocoa powders which are agglomerated to form larger particle sizes. It is used in the manufacture of chocolate based beverages.

Chocolate: A confectionery product made from hulled, fermented and roasted cocoa beans (nibs), blended with sugar, fats (cocoa butter or cocoa butter substitutes) and lecithins. Milk solids may be added to produce milk chocolate. Fat is an important component since its particular melting profile contributes to the mouthfeel of the product. Chocolate contains theobromine, an alkaloid with effects similar to those of caffeine.

Cholecalciferol: It is synonym for vitamin D_3; one of the group of sterols which constitute vitamin D. Fat soluble vitamin necessary for formation of the skeleton and for mineral homeostasis. It produced on exposure to UV light from the sun from the provitamin 7- dehydrocholesterol, which is found in human skin. Alternative recommended name is calciol.

Cholera toxin: Toxin produced by ***Vibrio*** *cholerae*

Cholera: Acute infectious human disease characterized by profuse diarrhoea leading to extreme dehydration that can result in shock, renal failure and death. It caused by cholera toxin produced by *Vibrio cholerae*. Spread by the faecal-oral route, usually via faeces- contaminated water and food.

Cholesterol: One of the sterols, and the major sterol found in vertebrate mammals. Present in all plasma membranes, but found especially in blood, liver, nerve tissue, brain tissue and animal fats. A precursor of many steroids, including the bile acids and steroid hormones. Not an essential dietary requirement; consumption of high levels have been associated with atherosclerosis and coronary heart diseases. Several health foods are claimed to reduce serum cholesterol levels; production of cholesterol-reduced products, especially dairy products and eggs, is increasing.

Choline: An amino alcohol and biogenic amine precursor with activity similar to that of vitamin B group members. It occurs widely in living organisms as a constituent of certain types of phospholipids (lecithins and sphingomyelin) and in the neurotransmitter acetylcholine. Choline is synthesized in the body, is a ubiquitous component of cell membranes and therefore occurs in all foods. Rich sources include egg yolks, meat, livers and cereals.

Chopping boards: Boards made of wood or plastics on which food is placed while being cut with a knives or cleavers (chopping). For safety reasons, it is best to use a separate board for vegetables and another (preferably wood) for raw meat. Hot water and detergents should always be used in conjunction with scrubbing to wash a chopping board after each use. Plastics boards may be cleaned in dishwashers.

Chopping: Cutting of foods into bite-sized (or smaller) pieces with repeated, sharp blows with knives or cleavers, usually on chopping boards. A food processor may also be used to chop foods.

Chromatography: Techniques in which components of a gaseous or liquid mixture are separated on the basis of differences in the rate at which they migrate through a liquid or solid stationary phase under the influence of a gas or liquid mobile phase. Once separated, individual components can be measured or identified by various methods. Types of chromatographic techniques include gas chromatography, thin layer chromatography, affinity chromatography and ion exchange chromatography, classified according to characteristics of the method.

Churning: The Process is used in buttermaking. Agitation or churning of cream breaks down the milk fat globule membranes, allowing individual milk fat globules to coalesce into grains which eventually separate from the buttermilk.

Chutneys: Fruit or vegetable pickles, containing ingredients including spices and sugar. Originally an Indian delicacy.

Cider: In the UK, alcoholic beverages made by fermentation of apple musts. In the USA, this alcoholic beverage is termed hard cider, and the term cider refers to unfermented apple juices.

Ciguatera: Food poisoning caused by consumption of tropical marine fish containing a neurotoxin (ciguatoxin) produced by certain dinoflagellates. Symptoms include abdominal pain, nausea and vomiting and multiple, varied neurological disorders. Ciguatera poisoning is the most common nonbacterial, fish- borne poisoning in the USA (mainly Hawaii and Florida) and is a significant health concern in tropical areas worldwide. Species of fish most frequently implicated in ciguatera outbreaks include grouper, amberjack, red snappers, eels, sea bass, barracuda and Spanish mackerel.

Ciguatoxin: Neurotoxin produced by dinoflagellates associated with coral reefs, which can accumulate in fish and cause ciguatera poisoning in consumers. *Gambierdiscus toxicus* is the dinoflagellate most nota- bly responsible for production of ciguatoxin, although other species have been identified recently. At least five types of ciguatoxin have been identified and are noted to accumulate in larger and older fish higher up the food chain.

Cinnamon oils: Essential oils obtained from either cinnamon bark or cinnamon leaves. Cinnamon leaf oil has high eugenol content and is used as an alternative to clove oils in seasoning blends. Cinnamon bark oil is characterized by high cinnamaldehyde content and is used as the source of cinnamon essences for cooking.

Cinnamon: It is widely-used aromatic spice obtained from the dried inner bark of trees belonging to several species of *Cinnamomum*. True cinnamon (also known as Ceylon cinnamon) is *C. zeylanicum,* while much of the cinnamon sold in North America is actually cassia (*C. cassia*). Cinnamon is used in stick (quill) or ground form for flavouring both sweet and savoury foods, including confectionery, meat dishes and cola beverages.

Citral: Member of the terpene aldehydes class of flavour compounds. It occurs in a wide range of plant foods, especially coriander, pepper, lemon peel and ginger.

Citrates: Salts of citric acid which occur naturally in many foods, and may be used as acidulants in foods and beverages.

Citric acid: Commercially important, versatile organic acid, widely used, along with its salts (citrates), in the food and beverage industries. Highly soluble in water and used in acidulants, antioxidants, flavourings, antimicrobial compounds and chelating agents. Usually obtained commercially by extraction from citrus fruits (it is the predominant acid in lemons, oranges and limes), or by fermentation of sugar or fruit processing wastes by *Aspergillus niger*. Isomer of isocitric acid.

Citric fermentation: The process by which certain organisms produce citric acid. *Aspergillus niger* is the organism mostly used in industrial processes. Substrates include molasses and starch hydrolysates.

Citrinin: Yellow pigmented mycotoxin produced by *Penicillium citrinum* and a few species of *Aspergillus*. It is used as an antibacterial agent against Gram positive bacteria.

Citronella essential oils: Yellow aromatic oils obtained from lemon-scented tropical grasses of the *Cymbopogon* genus (particularly *C. nardus*). Used in the food industry and as an aromatic/deodorizer in per- fumes, cosmetics, soaps and insect repellents. Contains geraniol, citronellol and citronellal.

Citronella: Tropical Asian grass (*Cymbopogon nar- dus*). Lemon-scented leaves are used in flavourings in cooking and as a tea. Source of essential oils that are used in commercial flavourings as well as in perfumery and insect repellents.

Citronellol: Member of the terpene alcohols group of flavour compounds. It occurs in a wide range of plant foods, including fruits, essential oils, ginger and wines.

Citrons: Long fruits produced by *Citrus medica*, with thick peel and acid flesh. Used in production of can- died peel, preparation of which involves fermenting immature fruit in brines and then soaking in a strong sugar solution. The candied peel is used in confectionery products.

Citrulline: One of the non-essential amino acids, which does not occur in proteins.

Citrus beverages: Beverages based on citrus juices and/or whole homogenates of citrus fruits.

Citrus essential oils: Essential oils obtained from citrus fruits, e.g. bergamot oils. Typically produced by pressing the oil from citrus peel, although leaves, fruit or juice may also be used as the source. Applications include as flavourings for soft drinks, ice cream, chewing gums and puddings. Limonene and other terpenes are major components.

Citrus oils: Essential oils obtained from citrus fruits, e.g. bergamot oils. Typically produced by pressing the oil from citrus peel, although leaves, fruit or juice may also be used as the source. Applications include as flavourings for soft drinks, ice cream, chewing gums and puddings. Limonene and other terpenes are major components. However, these are frequently removed prior to use of the oils, due to their susceptibility to off flavour production as a result of oxidation. Also called citrus essential oils.

Citrus peel: Outer skin of citrus fruits, consisting of the outer coloured flavedo (also called the epicarp or zest) and the white inner pith (also called the albedo or mesocarp). The flavedo is the source of citrus essential oils, while the albedo is used as a source of pectins. Peel is also rich in fibre and phytochemicals. Often candied and used in baking, or used in making flavourings.

Clarification: Process in which sediment and impurities are separated out of a liquid to make it clearer. Rendered fats can be clarified by adding hot water and boiling. The mixture is then strained and chilled. The resulting top layer of fat should be almost entirely clear of residue. Other products to which clarification is ap- plied include fruit juices, wines and beer.

Clarified butter: Pure butter fat without any solids, liquid or foam. It should be transparent when molten.

Clarifiers: Equipment used for the process of clarification, in which sediment and impurities are separated out of a liquid to make it clearer.

Clarifying agent: A substance that removes suspended impurities in liquids. Egg white is the most common in food preparation, e.g. For consommés.

Clarity: Optical properties relating to the extent to which an item is clear and transparent.

Clays: Sticky impermeable earth that can be moulded when mixed with water and baked to make containers. Clay is plastic when moist and becomes permanently hard and retains its shape when baked or fired. Of widespread importance in industry, clays consist of a group of hydrous alumino-silicate minerals. Individual mineral grains are microscopic in size and shaped like flakes. This makes their aggregate surface area much greater than their thickness and allows them to take up large amounts of water by adhesion, giving them plasticity and causing some varieties to swell. Clays are effective filter aids and are used during ad- sorption bleaching of oils.

Clean room technology: Technology that incorporates use of a sterile, dust-free environment. Objectives of a clean room are to isolate a controlled area from the outside, and to control movement of materials and personnel. Parameters requiring control in a clean room are temperature, relative humidity, water activity, pressure, noise and lighting. Sources and parameters of potential contamination include air quality, type and geometry of air intake systems, personnel, machinery and equipment, waste produced and packaging materials.

Cleaning : To make a surface free from dirt, pollutants or harmful substances.

Cleaning agents: Agents, such as disinfectants, used in the cleaning process.

Cleaning in place: A process in which processing equipment is cleaned using an in-place cleaning system that is usually computer controlled. Cleaning in place (CIP) systems are useful for equipment that is not easily accessible to the operator, and when opening the equipment would be harmful to the operators or the environment, and detrimental to product quality. Clean in place Alternative term for cleaning in place.

Cleaning-in-Place (CIP) systems: CIP systems are commonly installed in those premises manufacturing and/or bottling liquid products, such as beer, milk and mineral waters. The system

***Clostridium botulinum*:** The bacterium found in home canned and bottled vegetables which causes botulism. The spores of this organism are not killed by cooking at normal temperatures. The incubation period is 18 to 36 hours and death occurs in 1 to 8 days or there is a slow recovery lasting 6 to 8 months. The symptoms are disturbance of vision, dry mucous membranes of the mouth, tongue and pharynx, which cause difficulty in speaking and swallowing, and progressive weakness and respiratory failure. Immediate medical attention should be sought if there is any suspicion of the condition.

***Clostridium perfringens*:** A food-poisoning bacterium found in cooked and reheated meats and meat products. The incubation period is 8 to 12 hours, the duration 12 to 24 hours and the symptoms are diarrhoea, abdominal pain and nausea. There is rarely vomiting and no fever.

Closures: It is a devices or packaging components used for closing or sealing of containers. It includes caps, corks, crown corks, lids, stoppers and tamper evident closures.

Clotting: The process of coagulation to produce a thick mass of cohesive material, e.g. formation of curd upon coagulation of milk.

Cloud: Turbidity or haze within a product, usually applied to beverages.

Cloudiness: Extent to which an item is turbid, i.e. hazy in appearance. It is usually applied to liquids such as beverages.

Clouding agents: Substances used to impart the appearance of turbidity to foods and beverages. Soy proteins and citrus fruit processing wastes are frequently used as clouding agents in citrus beverages.

Cloves: Pungent, aromatic spices obtained from the dried, unopened flower buds of the tropical evergreen tree *Syzygium aromaticum* (syn. *Eugenia caryophyllata, E. caryophyllus*). It is used whole or ground in a range of foods and beverages, including cakes and biscuits, bread sauces, curries and mulled wines.

Coagulants: Substances or agents that causes separation or precipitation of solids from a solution, a process known as coagulation or clotting.

Coagulation: Precipitation of solids from a solution, usually upon addition of specific agents, producing material of a solid or semi solid state. Coagulation is a process particularly applicable to cheesemaking. It is also known as clotting.

Coagulum: It formed by precipitation of casein by the action of acids or rennets, as in cheese curd.

Coal tar dyes: Artificial colorants originally obtained from coal tar hydrocarbons. The term is now used to refer to any artificial organic dyes or pigments, regardless of source. It is known as aniline dyes.

Coating batter: A thick, viscous mixture of flour, water or milk, with possibly egg, cream or sugar, and seasoning used to coat items of food, generally prior to frying, deep-frying, or baking.

Coating consistency: (A liquid) with sufficient body or viscosity such that when it coats a solid it will not drain off. It is tested by inverting a spoonful of the mixture which should not leave the spoon.

Coating sauce: A thick sauce of coating consistency used to cover foods to improve their appearance and flavor.

Coating: Covering food with a layer of coating material. For example, chicken pieces may be dipped or rolled in seasoned breadcrumbs or flour prior to cooking. The food can be dipped into beaten eggs, milk or beer before being coated with the dry mixture, to aid adhesion of the coatings to the food. Coating food in this manner usually precedes frying or baking. Products such as mayonnaise or sauces can also be used to coat food.

Coatings: Materials which form thin continuous layers or coverings over the surface of foods. Used to enclose and/or protect the food, and may be eaten along with the food or removed before consumption. Include batters, breadcrumbs, breadings, carnauba wax, chocolate coatings, shellac and wax coatings.

Cobalamins: Term that covers several chemically related compounds, members of the vitamin B group, that are essential for cell division in tissues where this process is rapid, e.g. in formation of red blood cells. Deficiency leads to pernicious anaemia when immature red blood cells are released into the- bloodstream, and there is degeneration of the spinal cord. This type of anaemia is the same as seen in folates deficiency.

Cochineal: Water-soluble natural red colorant obtained from the dried bodies of South American insects (*Coccus cacti*). The red colour is due to carminic acid, whose aluminium lake is known as carmine.

Cocoa butter extenders: Vegetable fats that may be mixed with cocoa butter to a limited degree without significantly affecting its physicochemical properties.

Cocoa butter substitutes: Fractionated **fats** based on various oils (palm, palm kernel, coconut or hydrogenated soybean) designed to replace **cocoa butter** in **confectionery** applications. It is known as cocoa butter replacers.

Cocoa butter: Edible vegetable fat obtained by pressing or solvent extraction of ground, roasted dehulled cocoa beans (*Theobroma cacao*). It composed of symmetrical disaturated oleic glycerol esters resulting in brittleness at room temperature and a sharp melting point at 31-35°C. It is used primarily in the food industry for manufacturing chocolate.

Cocoa butter: The highly saturated white or yellow fat pressed out of cocoa beans, mainly used for chocolate manufacture

Cocoa mass: This is produced by grinding of cocoa beans nibs (beans from which the shell or husk has been removed) to release the cocoa butter from the cells. Used in the manufacture of chocolate and chocolate products. It is called cocoa liquor, chocolate liquor and chocolate mass.

Cocoa powder: Defatted cocoa beans treated with an alkali, further processed, dried and ground to a fine powder for use as a flavouring or to make drinks. It is a product obtained by extracting a predetermined amount of cocoa butter from chocolate liquor using hydraulic presses, and grinding the resulting press cake. Cocoa powders produced are classified according to fat contents.

Coconut cream: Product similar to coconut milk, but richer. Relatively high fat content, with level varying among commercial brands. For the canned product, coconut milk is filtered, mixed with emulsifiers and stabilizers, and emulsified to give a creamy consistency, before pasteurization and canning. Used in the same way as cream in many recipes and also in beverages.

Coconut milk: (i) The liquid obtained by soaking grated or desiccated coconut in water and straining off the solid particles. It used in Indian and Eastern dishes and available as a dried powder for reconstituting. (ii) The white translucent liquid inside a coconut used as a refreshing drink or as flavouring. It is also called coconut water.

Coconut oil: The semi-solid white fats or pale yellow highly saturated oil extracted from dried coconut (copra), used in the food-manufacturing industry and in Asian cooking as well as for soap and cosmetics. It contains about 75% saturated fat and is solid at ambient temperatures in temperate climates. It is rich in lauric acid and myristic acid and used extensively in the food industry. It is known as copra oils.

Coconut toddy: Alcoholic beverages made by fermentation of the sap of coconut palms (*Cocos nucifera*).

Coconut water: The liquid enclosed within the kernels of coconuts (*Cocos nucifera*), which may be used in beverages.

Cod liver oils: Pale yellow oils derived from the livers of Atlantic cod (*Gadus morhua*) and other species of the family Gadidae. It has a typical fish-like flavour which is intensified on exposure to light. Rich in vitamin A and vitamin D. and the oil contain saturated, monoe- noic and polyunsaturated fatty acids, such as eicosapentaenoic acid and docosahexaenoic acid.

Codex Alimentarius: A food code which provides an opportunity for all countries to join the international community in formulating and harmonizing food stan- dards and ensuring their global implementation. The Codex Alimentarius Commission, established in the early 1960s by the Food and Agriculture Organization (FAO) and the World Health Organization (WHO), is the body responsible for compiling the standards, codes of practice, guidelines and recommendations that constitute the Codex Alimentarius. Membership of the Commission is open to all Member Nations and Associate Members of FAO and WHO; the Commission meets every two years. National delegations are led by senior officials appointed by their governments. Countries that are not yet members of the Commission attend in an observer capacity.

Coeliac disease: Life-long intolerance to wheat gluten, characterized by inflammation of the proximal small intestine. The disease is often manifested as persistent diarrhoea, malabsorption and malnutrition. Aetiological mechanisms include genetic predisposition, dietary exposure to wheat and immunological factors; prevalence of the disease is high in geographical areas where wheat is a dietary staple. Management of the condition involves consumption of a gluten free diet, which has been facilitated by the development of gluten free foods, especially gluten free bread.

Coenzymes: Low molecular weight non-protein organic molecules, whether freely dissociable or firmly bound, necessary for the activity of certain enzymes.

Co-extrusion: The process of producing continuous multilayer products in sheet, film, tubing, filament, or other forms, and for production of filled foods. Separate polymer or ingredient streams are fed from different extruders to a die feed block, where they are combined in the die, emerging in combined form as a continuous multilayer extrudate.

Co-fermentation: Fermentation of two or more substrates by a single microorganism or fermentation of a single substrate by two or more microorganisms.

Coffee substitutes: Materials for preparation of beverages with sensory properties resembling those of coffee. It is commonly based on roasted plant materials, e.g. grains or chicory roots.

Coffee whiteners: Whiteners used in coffee and tea beverages as an inexpensive alternative to milk. Typically it is made from vegetable fats, casein, carbohydrates, emulsifiers and stabilizers. Available in liquid or powdered (shelf stable) forms.

Cold boning: Cutting of meat (muscle) from animal carcasses that have been refrigerated at 1-2°C for 48 h *post mortem.*

Cold chain: A term used to describe the continuing and uninterrupted storage requirements of certain high risk chilled foods from the final production stage to actual consumption. The cold chain incorporates a number of stages of storage, including that after production, during transport and at the premises of wholesalers, retailers and ultimate consumers.

Cold shock proteins: Protein fractions which are synthesized in various bacteria in response to cold shock, and which contribute to cold tolerance and psychrophilic properties of these bacteria.

Cold shortening: Contraction of muscle fibres in raw meat at low temperatures. It is related to toughness in the meat once cooked.

Cold shortening: The process by which carcass meat becomes tough if chilled too rapidly after slaughter, especially with beef and sheep. The chilling rate should not lead to temperatures in any part of the carcass below 10°C in 10 hours or less after slaughter.

Cold storage: Storage of foods at refrigeration temperature in order to extend shelf life.

Cold store: Any premises, not forming part of a cutting plant, game-handling establishment or slaughterhouse, used for the storage, under temperature controlled conditions, of fresh meat intended for sale for human consumption.

Cold stores: Refrigerated rooms or cabinets used for storage of foods at low temperatures, to extend shelf life.

Coliforms: Coliforms or, more correctly, faecal coliforms are a group of bacteria commonly found in the colon (large intestine) and include *Eischerichia coli* (*E. coli*), *Vibrio cholerae* and *Shigella dysenteriae*. Evidence of their presence in food may be an indication of sewage contamination. Coliforms are also responsible for much food spoilage, causing the production of 'off' flavours.

Coliforms: Gram negative, anaerobic, lactose- fermenting, rod-shaped bacteria, typically found in the gastrointestinal tracts of humans and animals (e.g. species of the genera Citrobacter, Enterobacter, Escherichia and Klebsiella). May loosely refer to any Gram negative, rod-shaped enteric bacteria.

Collagen: Insoluble animal proteins, with high contents of the amino acids glycine, hydroxyproline and proline. Collagen is the main fibrous component of skin, tendons, connective tissues and bones. Networks of collagen are also present in tissues and organs including the muscles. Thermal denaturation of collagen occurs between 60 and 90°C. When collagen is boiled it is converted into soluble gelatin. Collagen is important in relation to meat texture. Collagen cross-links link together molecules and fibrils of collagen, increasing its tensile strength; thus, the greater the number of cross-links the tougher the meat. In cooked meat, the presence of collagen cross-links contributes to shrinkage and tension development, with a subsequent increase in meat toughness. Collagen is used to form edible, biodegradable films and coatings for the packaging of foods.

Colloidal stability: A measure of the longevity of colloids; the ability to maintain the suspension of one material in another.

Colloids: Mixtures containing small particles of one material suspended in another, often of a different phase. Colloidal particles are generally 1 to 100 nm in size and so are larger than the individual solution molecules, but smaller than particles found in precipitates, which can be removed by filtration. Examples of colloids include aerosols, foams, emulsions and gels. A colloid containing solid particles suspended in a liquid is more accurately called a sol. The colloids is very finely divided particles of any phase, gas, liquid or solid, which are dispersed in another phase without settling by surface forces alone. Food examples are milk solids in milk and oil droplets in mayonnaise.

Colony count: A microbiological technique which, fundamentally, involves counting the number of colonies of micro-organisms present in or on a particular food sample after a specified period of time.

Colony counting: Enumeration of cell colonies in a given sample cultured on a solid medium.

Colony counts: Numbers of cell colonies in a given sample cultured on a solid medium.

Colorants: Substances that impart colour, such as dyes or pigments. Added to foods to improve visual appearance, replace colour lost during processing and ensure colour consistency. Broadly classified into natural colorants and artificial colorants, depending on whether they are substances extracted from natural sources or manufactured for use as a food additive.

Colorimeters: Instruments that measure the contents of components in a sample solution by comparison of colour with that of standard solutions.

Colorimetry: Analytical technique based on comparison of the colour of a solution with that of a standard solution.

Colostrum: Mammary secretion produced during the first 4-5 days *post partum*. It differs from mature milk mainly in the high content of immunoglobulins, which provide passive immunization of the suckling infant or animal. Other differences in composition include increased contents of milk fats, short-chain fatty acids, lactoferrin, minerals, most vitamins, some hormones and some organic acids in colostrum, and reduced contents of medium-chain fatty ac- ids, lactose, orotic acid, and some vitamins and hormones. Sale of colostrum is prohibited.

Colour: Optical properties relating to the subjective appearance of the wavelength or wavelengths present in a beam of light perceived by the eye. Although it is actually continuous, the visible spectrum is usually split into seven major colours - red, orange, yellow, green, blue, indigo and violet - in order of decreasing wavelength. Colour of foods not only helps to deter- mine quality, but is also an index of ripeness or spoilage. Various types of spectrophotometers or colorimeters can be used for colour measurement.

Colours: Artificial or natural colours are used to replace the natural colour lost during food processing or storage, or to make products a consistent colour. Commonly used colours are caramel (E150a), in products such as gravy and soft drinks, and curcumin (E100), a yellow colour extracted from turmeric roots.

Column chromatography: Chromatography technique in which a column or tube is used to hold the stationary phase.

Combination oven: A standard fan-assisted electric oven combined with a microwave energy source to give very fast cooking together with the surface browning and hardening of a conventional oven

Commensal: A biological relationship whereby one organism gains benefits without necessarily

Commercial freezing: The quick freezing and storage of food products at a temperature below -18°C. In some cases, products may be stored at -22°C or below that temperature.

Commercial sterility: A term used in canning of goods, in particular, to describe the virtual total destruction of all microorganisms and spores. Whilst some non-pathogens may survive the sterilisation process, the product is generally considered safe.

Composite flours: Products made by blending wheat flour with flour of other origins. It is often used to make bakery products that are conventionally made with wheat flour alone.

Compound butter: Butter flavoured by pounding it with herbs, shells, spices, etc. Then sieving out any inedible or unwanted bits. Usually formed into a roll, refrigerated and cut in slices to finish hot dishes for presentation. Usually written as beurre followed by the name of the flavouring e.g. Beurre d'amande, beurre de paprika. All are listed under beurre and/or the English name of the flavouring.

Compressed yeast: Small greyish cakes of compressed live yeast weighing about 18 g, used in baking.

Compressibility: One of the rheological properties, and a measure of the degree to which matter can be squashed or crushed by an externally-applied force. It indicates the hardness, firmness or sponginess of a material.

Compressimeters: Apparatus used for determining compressibility.

Computerized data processing: Analysis and organization of data by the repeated use of one or more computer programs.

Conalbumin: Iron-binding protein found in egg whites. It is also known as ovotransferrin.

Conarachin: One of the main proteins in peanuts. Present in two forms (I and II) that differ in size. Along with arachin, these make up 75% of the total protein in peanuts.

Concentrated milk: Product resulting from removal of a considerable proportion of the water from milk. It includes evaporated milk and condensed milk. A vacuum is applied to reduce the boiling point of milk and thus maintain its quality during evaporation. Evaporated and unsweetened condensed milk are sterilized by heat. Usually it sold in cans in a range of fat contents. It can be reconstituted by addition of water in amounts stated on the packaging.

Concentration: The process by which the strength of a solution or substance is increased. It achieved by a variety of means, including evaporation, filtration and dialysis.

Conching: The final step in chocolate manufacture, in which machines with rotating blades slowly blend heated chocolate liquor, ridding it of residual moisture and volatile acids. Conching continues for 12 to 72 hours (depending on the type and quality of chocolate), while small amounts of cocoa butter and sometimes lecithins are added to give chocolate its smooth texture.

Condensation: The conversion of a vapour or gas to a liquid. In physics, condensation is the process of reduction of matter into a denser form, as in the liquefaction of vapour or steam. Condensation is the result of the reduction of temperature by removal of latent heat of evaporation, the liquid product being known as condensate. Condensation is an important part of the process of distillation. In chemistry, condensation is a reaction involving the union of atoms in the same or different molecules. The process often leads to elimination of a simple molecule such as water or alcohol to form a new and more complex compound, often of greater molecular weight.

Condensed milk: Milk is thickened by evaporation of a considerable amount of its water content. In this about 85 to 90% of the water has been removed to leave an evaporated milk containing 45 to 50% water, to this is added about 70 g of sugar per 100 g of evaporated milk to give a thick sticky liquid containing about 55% sugar including lactose, and 28% water which is thus resistant to bacterial contamination. It is available in both full cream and skimmed versions. The skimmed version containing about 60% sugar in the resulting mixture. Unsweetened milk is similar to evaporated milk, i.e. sterilized by heat, sold in a range of fat contents and may be reconstituted by addition of water in amounts stated on the packaging. Sweetened condensed milk contains sugar in amounts high enough to act as a preservative, and is used in baking and to make sweet products such as confectionery, puddings and pies.

Condiment: A seasoning or flavored products, usually salt, pepper, nutmeg, various pasty or dry mixtures of herbs and/or spices, sometimes pickles, individually added to food by the eater after it is served. It refers in particular to items that are added to foods at the table immediately prior to consumption (e.g. sauces, relishes and mustard), rather than items added during cooking.

Confectionery cream: Water-in-oil or oil-in-water emulsions used mainly as fillings for bakery products.

Confectionery fillings: Products such as fondants or crèmes which may contain nuts, flavourings or other ingredients and are used to fill sugar confectionery or bakery products.

Confectionery pastes: Products containing ingredients such as glucose syrups, sugar, fats, colorants and flavourings that are used in the production of extruded sugar confectionery.

Confectionery: Generic term for sweetened food products. Sugar confectionery refers to products such as sweets, candy and chocolates, while baker's confectionery refers to bakery products such as cakes and pastries.

Confections: Sweet food products, particularly sugar confectionery.

Connectin: Elastic protein found in muscle foods (meat and fish), in which it is important for texture and tenderness. It is similar to titin.

Connective tissue: The structural material of the animal body normally seen as cartilage, sinew or gristle in meat but also present in bone and as the inner layer of skin, etc. It consists mainly of proteins principally collagen fibres interlaced with elastin fibres embedded in a gel. The more elastin the less soluble is the tissue. Tissues that is connect, bind, support or separate other organs or tissues. Connective tissues include cartilage, ligaments, tendons, adipose tissues, the non-muscular structures of blood vessels and the matrix of bones. In fish, collagen connective tissues separate the myotomes (muscle segments). In animals, connective tissues consisting of both collagen and elastin bind muscle fibres into bundles and support the blood vessels. Toughness of meat is correlated with connective tissue content. On cooking, the collagen component of connective tissues is converted into gelatin, thereby making meat more tender; however, the elastin component is unchanged on heating. Frying or roasting has little effect on meat tenderness, but tough meat is made more tender by stewing.

Consistency: Texture term relating to the degree to which a product, usually a thick liquid, is viscous or dense. The simplest method to determine consistency is to measure the time it takes for the food to run through a small hole of a known diameter. Alternatively, measurements can be made of the time it takes for more viscous foods to flow down an inclined plane using Bostwick consistometers. These devices might be used with tomato ketchups, honeys or sugar syrups. The word is usually qualified as e.g. Coating consistency, dropping consistency, thick consistency, jam-like consistency, buttery consistency, etc.

Consistometers: Instruments used to measure the uniformity and consistency of a manufactured material, such as a food product. The Bostwick consistometer is widely used to evaluate consistency of food suspensions. The Bostwick measurement is the length of flow recorded in a specified time.

Consumer acceptability: Extent to which a commercial product is considered satisfactory by consumers. In the case of foods and beverages, overall acceptability is judged on the basis of a number of factors, including sensory properties, physical properties such as colour, appearance and texture. Evaluation of consumer acceptability is important in development and marketing of new products.

Consumer complaints: Expression of dissatisfaction made by consumers regarding a commercial product or service. With respect to foods and beverages, the term covers complaints made at a local level, e.g. regarding the acceptability of a meal served in a restaurant, through to those reported to official agencies,

Consumer education: Provision of a variety of forms of training to consumers so as to increase their knowledge of a product or service.

Consumer information: Information, such as guidelines and details of use, given to consumers so as to increase their awareness of products and services.

Consumer panels: Groups of consumers employed during sensory analysis tests who are not specifically trained but can provide a good insight into consumer preference.

Consumer preference: Extent to which consumers like one commercial product more than others. In the case of foods and beverages, preference is governed by a range of factors, including sensory properties, appearance, physical properties, texture, health concerns, price and type of packaging. Evaluation of consumer preference variables is important in development and marketing of new products.

Consumer research: Any form of marketing research undertaken using the final consumers of a product or service from which to gather data. For example, consumer preference data are obtained and information is gathered on the way in which consumers in a free market choose to divide their total expenditure in purchasing goods and services.

Consumer response: Behaviour that consumers exhibit when provided with information in areas such as product purchase, new product development and product labelling. It covers concepts such as consumer attitudes, consumer awareness, consumer choice, consumer complaints, consumer expectations and consumer preference.

Consumer surveys: Marketing research tools used to gather data on consumer response to a particular product or service.

Consumption: As well as being the action or process of eating foods, this term also means the using up of goods created by production in an economic sense.

Containers: Receptacles for holding, storing or transporting substances such as foods. Of many different types, and made from a variety of materials. The term is also used to describe large, portable, standard-sized metal boxes, which are used in the transportation of cargo on lorries or ships.

Contaminant: Any extraneous or unwanted physical, chemical or biological substance, or object, encountered in food.

Contaminants: Agents that contaminate. It may be undesirable substances (e.g. residues of pesticides, fungicides, herbicides or fertilizers) or undesirable or harmful microorganisms.

Contamination: The ingress of impurities, either microorganisms or compounds such as detergents, bleach, dust etc. into a foodstuff, usually by contact with surfaces or other foods.

Continental sausage: Sausages usually made from 100% meat which are traditionally preserved by the addition of small amounts of glucose on which species of *Lactobacillus* grow and reduce the ph by producing lactic acid. They are also ripened by the surface growth of *aspergillus* and *penicillium* moulds.

Continuous phase: That phase in a two or more phase mixture which is continuously interconnected, e.g. Vinegar is the continuous phase in mayonnaise and hollandaise sauces, butterfat in butter and milk in cream. See also dispersed phase, emulsion

Continuous processing: Automated processing systems which operate in a continuous fashion. Such systems allow improved product consistency and reduced manufacturing costs, and are designed to meet the demand for high output.

Control systems: Systems in which inputs and out- puts are progressively altered in a well-planned way to cause a process or mechanism to conform to some specified behaviour under a set of given constraints. Computer-based control systems for large industrial plants can involve control of hundreds or thousands of individual variables. Recent developments in control engineering include self-tuning and adaptive control systems, in which controller settings are modified automatically in response to changing process and/or disturbance conditions, and the application of neural networks and artificial intelligence techniques, which mimic the actions of skilled human operators.

Controlled atmosphere packaging: Packaging technique in which specified concentrations of gases, including water vapour, are maintained throughout storage to achieve the desired atmosphere. It is used to extend the shelf life of foods, particularly fresh fruits and vegetables.

Controlled atmosphere storage: Storage of fruits and vegetables in sealed warehouses where temperature and humidity are closely controlled, and the composition of gases in the atmosphere is altered to minimize spoilage. Usually, the concentration of oxygen is reduced, the concentration of carbon dioxide (CO_2) is increased, and ethylene, a gas naturally produced by plants that accelerates ripening, is removed from the atmosphere. This controlled environment helps slow the enzymic reactions that eventually lead to decomposition and decay, and may increase the time that produce can be stored by several months. Ripening rooms, in which ethylene gas is added to the atmosphere, also help produce higher quality fruits and vegetables. This technology enables produce to be picked before it is ripe, for easier handling, and then ripened quickly and uniformly under controlled conditions.

Controlled atmosphere: This process is used in the storage of fruit, vegetables and meat whereby carbon dioxide is introduced to delay ripening and slow down the process of the growth of microbes.

Convenience food: Food which allegedly needs little preparation prior to serving. It may be a chilled, non-sterile cooked meal or vacuum-packed fully cooked meal. Other forms require the addition of eggs, milk, water, etc. Or elaborate mixing, heating and stirring, often taking as long to prepare as the same dish cooked with fresh ingredients purchased in semi-processed form. Examples include ready to eat meals, cooked sliced meat, sauces for pasta and pizzas and microwaveable foods.

Converted rice: Rice which has been soaked and steamed before being hulled. This preserves more of the nutrients from the outer coat in the polished grain which is yellow but whitens on cooking. The process does not shorten the cooking time. It is also called parboiled rice.

Conveying: It is a process by which items are transported or carried to a particular place.

Conveyors: A continuous moving band used for transporting objects from one place to another. Conveyors include simple chutes, unpowered roller conveyors, and a range of powered systems in which materials are carried along by belt, bucket, screw, trolley, or other arrangement. Pneumatic conveyors are tubes in which goods - usually in a finely divided form are moved along by blowers.

Cook chill foods: Foods, particularly ready meals, produced by cook chill processing and kept at low temperatures (<5°C) from manufacture to point of sale. The minimal processing involved results in high- quality convenience foods with a short shelf life. Generally it is packaged in plastics trays, as either ready to eat foods or easily reheatable products.

Cook chill processing: A method of catering that involves cooking of foods in batches to a just done status followed by immediate, fast chilling (using blast chilling or water bath chilling techniques) to just above freezing point. Products are then stored for reheating at a later time. The cook chill process offers a cost effective means of providing quality food while reducing overhead costs.

Cook-chill: This is a method of mass food preparation used by hospitals, airlines and supermarkets, in particular, where there is a high demand for ready-prepared meals on a continuing basis. In this process food is prepared, cooked to a core temperature of 70ºC and then chilled rapidly to 3ºC. It is then stored at that temperature and, on demand, regenerated (reheated) to 70ºC. The success of this process is dependent on effective temperature control at all stages. This entails regular temperature checks using probe thermometers and should the food, during its storage period, reach 5ºC, it must then be consumed within 12 hours. Where the food commodity temperature reaches 10ºC, it must be rejected. This is, fundamentally, a five day process although some operators of the system may reduce this time. The storage period lasts three days with a further day for regeneration and eventual consumption on the fifth day.

Cookers: Appliances for cooking foods, domestic cookers typically consisting of an oven, hob and grill.

Cook-freeze: This process is similar in theory to that for cook-chill products. In this case, the food commodity is stored at -18ºC giving it an extended shelf life.

Cookies: Biscuits prepared and baked as needed by slicing them from an uncooked roll of thick biscuit dough usually wrapped or cased in plastic and kept in the refrigerator.

Cooking fats: Fatty substances such as butter, margarines and vegetable shortenings which are solid at room temperature and are used to moisten, enrich, tenderize and flavour foods during cooking.

Cooking of food: One of the principal objectives of cooking fresh food is the prevention of bacterial growth which could cause food poisoning. Food cooked at a high temperature for a short period of time is generally safer than food which has been prepared over a long period of time at lower temperatures. This is due to the fact that the temperature at which bacterial growth can take place is quickly passed and the food reaches a temperature where bacteria actually die. Thus the time/temperature relationship is crucial to ensure safe cooking. Similarly, in the subsequent cooling of food, bacterial growth may take place if the food is allowed to cool too slowly.

Cooking oils: Fatty substances which are liquid at room temperature and have usually been refined, bleached and deodorized. It may be used for **deep frying**, or in **baking**, **frying** and **grilling** of foods.

Cooking properties: Ability of a food product to have acceptable properties upon cooking, particularly relating to texture, flavour and colour.

Cooking: It is a Process of preparation of foods by mixing, combining and heating the ingredients. Heat activated cooking methods take five basic forms. Food may be immersed in liquids such as water, stocks, or wines (boiling, poaching, stewing); immersed in fats or oils (frying); exposed to vapour (steaming and, to some extent, braising); exposed to dry heat (roasting, baking, broiling); and subjected to contact with hot fats (sauteing).

Coolers: Devices or containers for making or keeping items cool.

Cooling of food: The risk of bacterial growth can be significant where cooked food is to be prepared for eventual cold service. To reduce this risk, cooked food should be cooled to between 10ºC and 15ºC within 90 minutes before being refrigerated. This prevents bacterial multiplication. In the case of meat joints and other large items, such as turkeys, they should be cooked in pieces weighing no more than 2.5 kg. Similarly, in the case of liquids, such as soups and gravies, these should be split into smaller quantities of no more than 25 litres and cooled in wide shallow containers which assist rapid cooling.

Cooling: A Process by which the temperature of items is lowered, usually after some form of cooking.

Copra: The white inner meat of the coconut. It is used fresh or grated and dried in cooking. The dried copra is traded internationally as a source of oil and animal feed.

Coring: Process by which the tough central part of various fruits and vegetables is removed, often using corers.

Corks: Closures for bottles, particularly wine bottles, or jars. Made from cork, as it is a material which can be compressed to a smaller size and resists absorption of liquids. A sensory defect, known as corking, may occur in wines due to growth of microorganisms on corks. There is also the risk of release of substances such as trichloroanisole, tannins and peroxides from the corks into the food or beverage contained in the bottle or jar. Synthetic closures made from plastics have been developed as an alternative to natural cork closures for wine bottles. Crown corks are closures made from metal.

Corn bran: Outer protective coating of corn kernels; removed during milling.

Corn fibre oils: By-products of the corn processing industry which are rich in cholesterol-lowering phytosterols and of potential use in the manufacture of nutraceutical foods. It is produced during wet milling of corn.

Corn flour: Flour often ground from a variety of corn with large, soft grains and friable endosperm, from which the germ and outer hull are first removed. Also known as maize meal. Distinct from cornflour, which is used as an alternative term for corn starch.

Corn oil: A light pale yellow delicately flavoured vegetable oil extracted from the germ of maize kernels. It contains 15% saturated, 35% monounsaturated and 50% polyunsaturated fat. Typically it is bland in flavour and widely used as a cooking oil and salad oil and in the manufacture of margarines. Palmitic acid, oleic acid and linoleic acid are the major fatty acids. Oxidative stability is high, despite the highly unsaturated nature. It is also known as maize oils.

Corn syrups: Nutritive sweeteners manufactured by partial hydrolysis of corn starch. Corn syrups are a mixture of glucose, maltose and maltodextrins produced by acid hydrolysis of corn starch at >100°C, followed by enzymic hydrolysis if required. Degree of hydrolysis required depends on the final application of the syrup; commonly, 40-60% of glycosidic bonds are hydrolysed, i.e. the corn syrups have a dextrose equivalent of between 40 and 60, although corn syrups of 24-80 dextrose equivalents are also produced. Corn syrups are approximately half as sweet as sucrose, although sweetness increases with increased hydrolysis. In addition to their use as sweeteners, corn syrups are used as thickeners, humectants, carbon sources for microbial fermentation and to provide body to soft drinks and beer.

Corn: Grains, also known as maize, from any of nu- merous varieties of a tall, annual cereal plant (*Zea mays*), which are borne on large ears. It is low in tryptophan. Niacin is present in bound form, making development of the deficiency disease pellagra a possibility in those eating corn as a staple. Corn is processed into a great many products, including corn oils, corn starch, corn syrups, flour and corn masa. It is also used in making some kinds of beer, whisky and gin. Some types of corn have a hard endosperm and kernels that burst on heating; these are used to make popcorn.

Cornflakes: Breakfast cereals made from corn, often enriched with vitamins.

Coronary heart diseases (CHD): Diseases of the heart resulting from narrowing of the coronary arteries resulting in myocardial ischaemia and/or infarction. Narrowing or occlusion of the arteries can be a consequence of atherosclerosis or thrombosis. Risk factors for coronary heart diseases have been identified, and include dietary, genetic and lifestyle factors.

Coronation chicken: A cold dish made from diced, cooked chicken meat mixed with mayonnaise, chopped tomatoes, onions and apricots, whipped cream and flavoured with curry powder, lemon and bay.

Corrective action: Corrective action implies any form of action directed at preventing contamination of food and is an important feature of the HACCP process. Other food safety monitoring techniques, such as food safety risk assessments, food hygiene inspections and audits, may identify deficiencies in cleaning procedures, inadequate storage temperatures and other breaches of food hygiene legislation. Inspection reports should specify corrective action required and the time scale for such action.

Cottage cheese: A low-fat, very loose-textured, soft mild cheese made of small white curds which have been repeatedly washed and drained and not pressed or matured. It contains 75 to 80% water, 3 to 5% fat and 15 to 16% protein. It is soft white cheese made from cow milk. An acid curd cheese, made without rennets.

Cottonseed oils: Pale yellow oils derived from cottonseeds (*Gossypium* spp.). Used as a salad oil and cooking oil and, on hydrogenation, in the manufacture of margarines. Rich in palmitic acid, oleic acid and linoleic acid; gossypol is a minor constituent.

Countercurrent chromatography: Form of liquid partition chromatography in which no solid support is required and two immiscible solvent phases are used. Partition takes place in an open column in which one phase (the stationary phase) is retained and the other (the mobile phase) passes through continuously. The stationary phase is retained in the column as a result of column configuration and gravitational or centrifugal force fields. The technique is used in the food industry for preparative separation of food constituents, such as polyphenols from tea, and anthocyanins from fruits and vegetables, and for analysis of food components and contaminants.

Coxiella: Genus of Gram negative, anaerobic, rod-shaped bacteria of the Coxiella group family. *Coxiella burnetti*, the causal agent of Q fever in humans, may be transmitted from infected animals to humans via their milk. Sheep, cattle and goats may act as reservoirs for the disease.

Cracked wheat: Coarsely crushed grains of wheat, dry cooked for 25 minutes. It served hot as a breakfast cereal, served as an accompaniment to other dishes or sprinkled on rolls or bread prior to baking. It is also called kibbled wheat.

Cracking: Breaking of an item with little or no separation of the component parts. It can be used to refer to damage to a commodity, e.g. freeze cracking of foods, or a processing step, e.g. cracking of eggs to remove them from their shells.

Crackling: The skin of a pork, bacon or ham joint which has been scored to 3 or 4 mm with a sharp knife in strips or a diamond pattern prior to roasting and which becomes crisp and golden brown if basted with water; it is served as an accompaniment to the roast

Crates: Re-usable, slatted, wooden or plastics containers used for transportation of goods, including various foods. Crates subdivided into units are used for holding individual items, such as bottles. The term is also used to describe containers for the transportation of live animals, particularly poultry.

Cream cheese: A soft, acid curdled spreading cheese made from a mixture of cows' milk and cream. The curds are spun off and milled with stabilizers and preservatives prior to packaging for immediate sale and consumption. It contains between 45% and 65% butter fat.

Cream liqueurs: Liqueurs in which cream is combined with alcohol to produce a thick and shelf stable blend. Cream is homogenized to break down the milk fat globules to a size suitable to encase the alcohol molecules, preventing the cream from being curdled and the product from separating. The milk fat globules are made as small as possible to give a smooth taste and long shelf life. Cream liqueurs are generally packed in dark glass bottles to protect from UV radiation and are best kept refrigerated after the bottle has been opened.

Cream of tartar: Acid potassium tartrate (potassium hydrogen tartrate), a crystalline substance which is precipitated from wine as it ages. Now made synthetically and combined with sodium bicarbonate to make the raising agent baking powder.

Cream: Fatty product prepared from whole milk by centrifugation. It is marketed in a range of types differing in fat content. In the UK, half cream contains approximately 12% fat, single cream and extra thick single cream 18%, whipping cream 34%, double cream and extra thick double cream 48%, and clotted cream 55%. In the USA, light cream contains 20-25% fat and heavy cream 40% fat.

Creaminess: Consistency term relating to the extent to which a product is creamy, i.e. smooth, glossy and uniform. As an attribute, creaminess has viscous, flavour and taste aspects.

Creaming: Natural formation of a layer of milk fat on the top of milk left to stand for some time. This happens because of the lower specific gravity of milk fat and is dependent on the size of the milk fat globules. Clustering of the milk fat globules is also affected by globulins in the milk fat globule membranes. Creaming can be controlled by homogenization of milk and heating to cause denaturation of the globulins.

Creep: One of the rheological properties, describing a deformation with time of materials under continual stress. An important parameter in a wide range of foods, including fruits and vegetables, bakery products, dairy products and extruded starch based foods, as well as in gels and films, and packaging materials.

Cresols: Methylphenol flavour compounds found in a range of foods, especially smoked foods. Cresols residues from lacquers applied to cans may occur as contaminants in canned foods. Cresols and cresol derivatives may cause taints in foods and beverages.

Crispbreads: Thin cracker-like products having considerably lower water content than bread. It is made from rye flour or wheat flour and water. Commonly it is eaten as an alternative to bread by those on a weight reducing diet.

Crispness: Term relating to perception of product texture in the mouth; the extent to which a product is brittle when bitten.

Crisps: Popular savoury bagged snack foods comprising very thinly sliced vegetables or extruded cereals that have been fried and flavoured, e.g. potato crisps. Also known as chips in some countries.

Critical Control Point (CCP): A factor or stage in processing when a loss of control would result in an unacceptable safety or quality risk.

Critical Control Points (CCPs): These are stages in HACCP at which operators of food preparation processes should exercise controls directed at preventing or minimising the risk of food hazards.

Critical control points: Points, steps or procedures at which quality control can be applied and a food safety hazard prevented, eliminated, or reduced to acceptable levels. The selection of critical control points (CCP) is aided by the use of a CCP Decision Tree, which is designed to help determine what should be used in a Hazard Analysis Critical Control Point (HACCP) plan to control hazards.

Critical limit: In HACCP, a dividing line between what is safe and unsafe with respect to food

Critical moisture content: The moisture remaining in a food at the end of a period of drying at a constant rate.

Crocetin: Dicarboxylic carotenoid pigment derived from saffron that is used as a natural food colorant. Forms red rhomboid crystals. Slightly soluble in water and organic solvents, and soluble in pyridine and dilute sodium hydroxide.

Crocin: Yellow water-soluble carotenoid pigment which is found in the fruits of gardenia (*Gardenia jasminoides* Ellis) and in the stigmas of saffron. Purified crocin (purity >99.6%) has antioxidative activity comparable to that of BHA at some concentrations. Used in colorants for foods, e.g. smoked haddock and cod, and has potential for use in antioxidants.

Cross contamination: Cross contamination implies the transfer of, particularly, bacterial contamination from one contaminated food product, such as uncooked poultry, to another uncontaminated food product, such as cooked meats.

Crown corks: Metal closures for bottles. Comprise preformed caps and a sealing pad. These are placed over the mouth of the container to be sealed, and the edges are crimped to secure them to the containers.

Crude fibre: The indigestible matter left in foods after successive digestion with ether, acids and alkalies, and subtraction of ash. It is commonly determined as part of the proximate composition of foods. Values are related to, but not equivalent to, the dietary fibre component of foods.

Crumbing: Process by which foods are coated in crumbs, usually fresh or dried breadcrumbs. Alternatively, breaking a food down into crumbs.

Crumbliness: Texture term relating to the extent to which a product is brittle, fragile and liable to break up into fragments (crumbs).

Crumbling: The process by which goods fall apart into small fragments, or the process by which foods are broken up (usually with the fingers) into small pieces.

Crumpets: Small round bakery products made with flour, water and milk with added sodium bicarbonate. The resulting batters are leavened with yeasts and baked on a griddle. Usually served toasted and spread with butter.

Crunchiness: Texture term related to the extent to which a product is crunchy, i.e. hard and crisp.

Crushing: To deform, squash or pulverize an item by compressing forcefully. When applied to foods, this can result in products such as crumbs, pastes or powders. Crushing is often accomplished with a pestle and mortar, or with a rolling pin.

Crust: Crisp, outside portion of bakery products, e.g. bread, that has been caramelized or dehydrated during baking.

Cryogenic freezing: This is a form of quick freezing whereby food is immersed in liquid nitrogen. It is mainly used for the preservation of commodities such as prawns and strawberries.

Cryogenics: Branch of physics concerned with the production and effects of very low temperatures. Cryogenic temperatures are achieved either by the rapid evaporation of volatile liquids or by the expansion of gases.

Cryopreservation: Preservation of foods using very low temperatures.

Cryoprotectants: Compounds used to protect frozen foods from quality deterioration (cryopreservation). Include sucrose, which prevents muscle pro- tein from denaturation during frozen storage of surimi, and sorbitol, starch or starch hydrolysates, which can be used to restrict undesirable changes in the **functional properties** of meat proteins. **Glycerol** is also commonly used as a cryoprotectant.

Cryoscopes: Instruments used for measuring the boiling point and freezing point of liquids.

Cryoscopy: Technique for determining the molecular weight of a substance by measuring the amount by which the freezing point of a solvent drops upon addition of a known quantity of that substance.

Cryovac: Trade name for a process for packaging of meat, in which it is sealed in plastics films and the air is removed by vacuum pump.

Cryptoxanthin: Garnet-red carotenoid pigment with vitamin A activity which occurs naturally in egg yolks, butter, blood serum and in many plants. Slightly soluble in ethanol and methanol, and soluble in chloroform and benzene. It has many nutritional and medical uses. Also known as provitamin A.

Crystallization: Formation of crystals, particularly used to purify a material or extract it from solution. Extensively used in sugar processes, and also in the processing of butter and margarines, chocolate and ice cream. Also used to prepare proteins, including enzymes, for structural analysis by X-ray diffraction spectroscopy.

Crystallography: Measurement of the shape and structure of crystals.

Crystals: Solid materials formed by crystallization, in which the atoms are arranged in a single regular arrangement called a lattice. Sugars and fats readily form crystals under favourable conditions such as temperature and concentration. Starch can also crystallize, and in bread this retrogradation process is associated with staling. The presence of crystals in foods strongly influences their texture.

Culture media: Liquid or solid preparation of nutrients specifically designed to support the growth or maintenance of microorganisms or other cell types.

Cultured buttermilk: Commercial product made as a substitute for buttermilk produced during churning as a by-product of buttermaking. Made by adding lactic acid bacteria to skim milk. The lactic acid produced during fermentation gives the product a tangy flavour similar to that of churned buttermilk, but composition of the two products differs. Used as a beverage and as an ingredient in baking.

Curcumin: Phenolic pigment which exists as a yellow- orange powder or needles and is derived from rhizomes of plants of the genus *Curcuma*, e.g. turmeric (*Curcuma longa*

L). Insoluble in water but soluble in ethanol. Curcumin is a powerful antioxidant, and shows antitumour activity in animal studies and anticarcinogenicity *in vitro*. Used as a food dye, a biological stain and an analytical reagent. Also known as turmeric yellow.

Curd cheese: A semi-soft white cheese with a creamy texture and mild flavour made from the curds which separate from a lactic acid fermentation and coagulation of milk. The cheese used especially in cooking. It is not pressed and is slightly sour tasting. Often used for making cheesecake.

Curd: Protein (casein) gel formed by coagulation of milk, e.g. during cheesemaking. Other milk proteins are retained in the liquid portion (whey).

Cured meat: Meat preserved with the aid of salt and colour fixing ingredients, e.g. sodium nitrate and/or some sodium nitrite. Other curing agents may be added to accelerate curing (reducing agents), to modify flavour (e.g. sweeteners), to modify texture, to retard development of oxidative rancidity, and to in- crease water binding capacity and decrease shrink- age during subsequent processing. The curing process may involve: rubbing dry curing ingredients into the meat; immersion of meat in curing brines; or injection of the meat with solutions of curing ingredients. In the past, curing was used primarily to preserve meat, but with increases in the use of refrigeration and freezing, the major purpose of curing has changed. Meat curing ingredients are now mainly used to impart unique colour, flavour, palatability and texture properties to cured meat products. During the curing process, nitrates are converted into nitrites. Nitrosomyoglobin, formed from myoglobin and nitric oxide during curing, is responsible for the red colour of cured meat. Health concerns relating to use of nitrites and NaCl in cured meat have led to reductions in use of both ingredients.

Curing agents: Ingredients is used in the curing of foods. Examples include salt (sodium chloride; NaCl), nitrates and nitrites, sugar and spices.

Curing brines: Brines is used for curing of foods, such as meat products. Curing brines are often injected into meat to produce the final cured meat products.

Curing of food: This is a method of preserving food, in particular, meat and fish. In this case, the food is immersed in a solution of salt and sodium or potassium nitrate or alternatively, the brine solution is injected into the carcase, as with the preparation of bacon. The subsequent increase in osmotic pressure reduces the available water and produces conditions which are difficult for bacterial growth to take place.

Curing: Preservation of foods such as meat, cheese and fish by salting, drying, pickling or smoking. Smoking can be carried out by the cold smoking method (in which the food is smoked at 22 to 32°C) or by the hot smoking method (which partially or totally cooks the food at 37 to 88°C). Pickled foods are soaked in flavoured, acid-based brines. Cheese curing can be undertaken by methods such as injecting or spraying the cheese with specific bacteria or by wrapping the cheese in flavoured materials.

Curry powder: A mixture of ground spices used for making curry especially in the West. Made from a selection of coriander seeds, cumin seeds, mustard seeds, black peppercorns, fenugreek seeds and chillies to produce varying degrees of hotness, all of which are dry-roasted, together with unroasted dried ginger, turmeric, cinnamon and cloves. There are many variations, e.g. Madras and West Indian curry powder. Rarely used in Indian cooking, but useful for those not wishing to stock individual spices.

Cutting: Process by which an opening or incision is made in an item, or by which a slice is taken from an item, using sharp objects such as a knives or cleavers.

Cyanocobalamin: It is synonym for vitamin B_{12}. Member of the vitamin B group, found in foods of animal origin such as livers, fish and eggs. Vitamin B_{12} is the coenzyme for methionine synthase (EC 2.1.1.13), an enzyme important for the metabolism of folic acid, and methylmalonyl coenzyme A mutase (EC 5.4.99.2). Absorption of this vitamin requires the presence of an intrinsic factor. Failure of absorption, rather than dietary deficiency, is the major cause of pernicious anaemia.

Cyanogenic glycosides: Cyanogens which are capable of liberating large amounts of toxic cya- nides, which can be metabolized to goitrogens (thiocyanates). Include linamarin, linustatin and neolinustatin. Occur naturally in many plants, including cereals, pulses, fruits, root crops, nuts and oil- seeds, usually in parts that are not eaten or at such low concentrations that they do not present a health risk to consumers. However, in cassava, they occur in high levels both in the edible roots and leaves. Readily detoxified by appropriate processing of plant materials.

Cyclamates: Salts of cyclamic acid, also known as sulfamates. Used as non-nutritive sweeteners in foods, usually in the form of calcium cyclamate or sodium cyclamate. Cyclamates are approximately 30 times as sweet as sucrose and display good solubility and heat stability characteristics for a variety of food applications, such as low calorie foods. Use of cyclamates was banned in the USA, UK and Canada due to concerns about possible carcinogenicity. However, later studies have failed to confirm this and use is still permitted in many countries.

Cyclodextrins: Dextrins containing at least six glucose units in the form of a ring. Can associate with a range of substances and are therefore used as complexing agents, particularly in the β-cyclodextrin form. Used in the food industry as emulsifiers, stabilizers and masking agents for off odour and off flavour.

Cyclones: Processing equipment used for separation of solids from air. It consists of a conical chamber into which the air and solid, such as a food powder, is added tangentially at high speed producing a whirl or cyclone. Particulate matter is forced to the sides of the chamber, decelerates and drops down to the conical end of the chamber from which it is removed. The air stream remains in the central region of the cyclone. Used for separation of powders, e.g. milk powders, from air after spray drying. Deposits. Most CIP systems tend to be fully automatic and controlled from a central location. Harming another organism. It is a form of symbiosis. Involves extensive pipework connected to all items of equipment which come into contact with the product at various stages of the process. At the completion of the food manufacturing process, or following the completion of a particular batch, the whole of the plant is flushed with a series of detergents and/or detergent/sanitisers followed by hot and cold water rinses. Detergents may be held within the system at a particular temperature for a specified length of time with a view to clearing soiling, food residues and other preparation.

D

Dahi: Fermented milk product popular in India. Dahi made from buffalo milk is generally preferred to that made from cow milk. A sweet variety of dahi, misti dahi, is prepared by adding cane sugar to milk during heating, giving a caramelized flavour and brown colour.

Daidzein: One of the two isoflavones of particular importance in soybeans, the other being genistein. Both compounds are structurally similar to oestrogenic steroids and possess both oestrogenic and antioestrogenic activities, the principal functions responsible for the health benefits associated with consumption of soybeans and soy products.

Dairies/Dairy Factories: Premises in which dairy products are manufactured. It is called creameries or dairy factories.

Dairy beverages: Drinks based on milk or other dairy products, e.g. whey.

Dairy desserts: Ready to eat desserts based on dairy products, such as cream, milk or yoghurt. It is available as chilled, frozen and shelf-stable products. It includes mousses, custards, milk puddings and ice cream products.

Dairy products: Products manufactured from milk. It includes as major product groups, cheese, yoghurt, butter, cream, fermented milk, ice cream and whey products. It is also called milk products.

Dairy science: Division of food science dealing with the characteristics, manufacture and quality of dairy products as well as the production, management and distribution of dairy animals such as cows, goats and sheep.

Dairy spreads: Spreads based on milk fats and containing other, sometimes non-dairy, ingredients to give a lower fat content than butter.

Dairy starters: Microbial cultures used in manufacture of fermented dairy products, including fermented cream, fermented milk and cheese.

Dalia: Type of porridge made from wheat grits.

Dark cutting defect: A defect of beef often associated with bull beef. Dark cutting meat, also known as black beef or dark cutter beef, has a darker colour, and poorer flavour and texture than normal beef; moreover, the high pH value of dark cutting meat encourages the growth of spoilage bacteria and reduces shelf life. Physiological stress and exhaustion pre- slaughter deplete muscle glycogen stores, ultimately increasing the pH of meat and leading to the development of dark cutting defect. In young bulls, incidence of dark cutting defect can be decreased by low stress handling and prevention of bull behaviour (mounting, mock fighting and butting) in abattoir pens prior to slaughter.

Dark firm dry (DFD) defect: A condition associated with pork in which meat has a high pH value and darker than normal lean colour. The defect results from a decreased glycogen content in swine muscles prior to slaughter; it is often associated with pre-slaughter stress. In beef, the term dark cutting defect or dark cutter is used to refer to the same condition.

Darkening: Discoloration of a substance by becoming dark or darker. Red colour is often used by consumers as an indicator of the freshness of meat. Darkening of the product, which occurs during storage due to pigment shifts, is perceived as being a negative event, even though this is not a true indicator of wholesomeness or nutritional value. Because of consumer concerns, packaging films are designed to protect meat colour, largely by controlling diffusion of oxygen. Darkening is also a problem during repeated use of frying oils.

Date marking: Marking of food or beverage containers with a date that may be the date of manufacture, the sell-by date and/or the use-by date (expiry date). The sell-by date is the date by which the manufacturer recommends that a perishable product should be sold. Use-by dates are chiefly used in the UK instead of sell- by dates, and indicate the recommended date by which a perishable product should be eaten or used, after which it is no longer deemed to be safe, desirable or effective. Date marking is often required by law, particularly on packs of foods which should be maintained at low temperature, e.g. cheese, pates and ready meals, and on foods in which spoilage organisms are likely to multiply or cross contaminate other foods, e.g. fresh meat and fish. Other foods, such as bread and cakes, which tend to deteriorate in quality rather than safety do not require date marking by law, but are often labelled voluntarily by the manufacturer or retailer.

DATEM: Anionic oil in water emulsifiers used as improvers in breadmaking. Acronym for diacetyl tartaric acid esters of mono and diglycerides.

Dating: Process of marking a product or its outer packaging with date information, such as date of manufacture or date by which the product should be consumed to ensure quality.

Deacidification: Neutralization process whereby the acidity of a substance is reduced. Deacidification is often used in conjunction with the processing of apple juices, cider, vegetable oils, wines and grape musts. Deacidification of grape musts is crucial for the production of well-balanced wines, especially in colder regions of the world. Malolactic fermentation is widely used to reduce the acidity of grape juices. Young wines can also be deacidified with calcium carbonate and potassium hydrogen carbonate. Deacidification of vegetable oils (such as rice bran oils and corn oils) can be carried out using solvent extraction and membrane processing. Nanofiltration has been used for deacidifying and demineralizing cottage cheese whey, ready for use in ice cream and other frozen dairy desserts.

Deaeration: Removal of air or oxygen from a solution, for example by bubbling with an inert gas. It is also known as degassing.

Debittering: Removal of bitter compounds from foods such as citrus fruits, chocolate, soybeans and cruciferous vegetables, and beverages such as wines, fruit juices, cider and beer, to make them more palatable. Debittering can be achieved biologically, using enzymes or immobilized bacteria. Lactone hydrolases are used commercially for debittering triterpenes present in citrus juices. Correction of excessive naringin bitterness in citrus fruits can be achieved through use of adsorbents or cyclodextrins to form less bitter inclusion complexes. Deliberate aeration of the pulp during apple juice extraction for cidermaking promotes the removal of bitter and astringent flavonoids through their binding to the pomace. Fining with gelatin decreases contents further still by co-precipitation. Proline-specific amino-peptidases can be used for debittering of food protein hydrolysates. Enzymic hydrolysis of oleuropein by β-glucosidase from *Lactobacillus plantarum* offers a potential alternative to chemical debittering treatments for table olives.

Deboning: A process for cutting of meat from the bones, which can be done either manually or mechanically.

Debranning: Process of removal of bran from cereals. It may be achieved by milling or by soaking in a solution of an alkali such as sodium hydroxide. It used to enhance milling performance of cereals as well as to provide by-products with potential as food ingredients. However, debranning may also affect the nutritional quality and functional properties of the cereal and subsequent products.

Decaffeinated coffee: Coffee from which caffeine has been removed by a solvent extraction process using aqueous, organic or supercritical solvents.

Decaffeinated tea: Tea from which caffeine has been removed by a solvent extraction process using aqueous, organic or supercritical solvents.

Decaffeination: Removal of caffeine from a substance such as coffee or tea. Caffeine is removed from coffee by soaking coffee beans in chemical solvents or water. The resulting decaffeinated product contains approximately 3 mg caffeine per 150 ml cup, compared with 75,150 mg for normal coffee.

Decanter: A glass container with a stopper, often decorated, in which decanted wine or other drinks are kept for serving

Decanting: To gently pour off the clear liquid from the top of a mixture of liquid and heavier particles, where the latter have settled to the bottom. It is usually applied to wine to separate it from the crystals of tartaric acid which are precipitated on long standing.

Decarboxylation: Chemical modification involving the removal of carboxyl groups from organic compounds, generating CO_2. It can be due to the influence of enzymes (decarboxylases) or other catalysts, or can occur spontaneously. Several aroma compounds, including diacetyl, are formed by decarboxylation reactions.

Dechlorination: Process of removing residual chlorine from a substance. In the food and beverages industries, chlorination usually cannot be considered without the added expense of dechlorination, as residual chlorine must be removed to prevent chemical changes affecting flavour, aroma and colour of the final product. Activated carbon is usually used in the beverages industry to dechlorinate and remove trace levels of outside flavour compounds from water to be used in producing beer and soft drinks. A non-chemical means of dechlorination involves use of a high energy ultraviolet system. This cost effective process reduces free chlorine levels by up to 99%.

Decoction: A liquor containing the concentrated essence of a substance, produced as a result of heating or boiling.

Decoloration: Removal of the **colour** from an item. It is also known as decolorization.

Deep fat: Hot fat or oil of sufficient depth to completely cover the item being fried. It is usually held in a special cooking utensil designed to prevent accidents.

Deep freezing: A method for preservation of foods by rapid freezing and storage at -18°C. Freezing preserves foods by preventing microorganisms from multiplying. Enzymes in the frozen state remain active, although at a reduced rate. Commercial freezing is usually undertaken by one of the following methods: blast freezing, where air is circulated at -40°C; contact freezing, in which refrigerants are circulated through hollow shelves; immersion freezing, where, for example, fruit is frozen in a solution of sugar and glycerol; and cryogenic freezing, using, for example, liquid nitrogen spray. Rapid freezing avoids structural change that would affect flavour or appearance of foods, as in the shrinkage and distortion of cells by formation of enlarged ice crystals in the extracellular spaces. Some quick frozen foods require thawing before use, and cooking must then be prompt. This method of preservation is widely used for a great variety of foods, including bakery products (both ready to eat, and to be cooked when desired), soups, and precooked complete meals.

Deep frying: Cooking of foods at a temperature between 175 and 195°C in an amount of hot fats or oils sufficient to cover them completely during frying. The food is quickly sealed, crisped and cooked if not too thick.

Deep-freeze: An insulated and refrigerated cabinet or chest for freezing and storing frozen food at temperatures below –20°C to prevent deterioration

Deep-fryer: The cooking utensil which contains sufficient depth of fat and a means of heating it, used for deep-frying. It should have a cool zone below the heating element and hot fat, to collect burnt fragments of food, and a thermometer and/or thermostat to control the temperature.

Deep-poaching: Large cuts of fish or whole fish cooked in simmering court bouillon or acidulated and salted water until cooked. Cooking is finished when the flesh leaves the bone. Usually garnished with picked parsley, plain boiled potatoes, slices of blanched aromatic vegetables and a little cooking liquor and served separately with a suitable sauce.

Defeathering: Removal of feathers from the carcasses of meat-producing birds, such as poultry, during processing. If defeathering is not performed properly, carcasses can be mechanically damaged or microbially contaminated, both of which are of economic importance to the poultry industry.

Defecation: Removal of impurities, usually applied to the stage of purification of sugar juices during sugar manufacture. Defecation involves clarification of sugar juices by heat and lime. The lime is added to neutralize the organic acids present, after which the temperature is raised to approximately 95°C. This lime and heat treatment forms a heavy precipitate of complex composition, which contains insoluble lime salts, coagulated albumin, and varying proportions of fats, waxes and gums. The flocculent precipitate carries with it most of the finely suspended material of the juice that has escaped mechanical screening. Separation of this precipitate from the juice is undertaken using a juice clarifier. Degree of clarification has a great bearing on the boiling house operations, and on yield and refining quality of raw sugar.

Deficiency diseases: Conditions arising due to the absence of a dietary nutrient, such as one of the essential vitamins or minerals. Include various types of anaemia, rickets, scurvy, pellagra, beriberi and goitre. Strategies to counteract these disorders and improve nutrition often combine direct dietary intervention (provision of food supplements, food fortification, dietary diversification) with agricultural measures (development of foods of improved nutritional values and bioavailability, development of improved agricultural practices) and economic measures for improving food security.

Defoaming agents: Substances, often silicon-based, is used to minimize formation of foams during food processing. These foams would otherwise cause problems for both the processing operation and final product quality. Typical applications where foaming problems occur include freeze drying, sugar processes and manufacture of fruit and dietetic soft drinks. Similar to antifoaming agents.

Deformation: Persistent change in shape or size of a substance in response to an externally applied force. Routinely determined for foods during analysis of rheological properties, and can include puncture deformation, torsional deformation, breaking deformation and maximal (peak) deformation.

Defrosting: Thawing of frozen foods, or alternatively the freeing of an item, e.g. freezers, of accumulated ice.

Degreening: Process of ripening or improvement of skin or peel colour, usually by application of ethylene to citrus fruits (such as satsuma mandarins and lemons), bananas, rapeseeds and mustard seeds. Decay tends to be more severe in degreened fruit because the degreening process itself promotes decay, and because packaging line fungicide treatments have to be delayed until after degreening. Uneven degreening of bananas is a ripening disorder characterized by either partial or delayed yellowing or by permanent greenness after treatment with exogenous ethylene. Green seed is a significant economic problem in rapeseeds because the rapeseed oils extracted from such seed contains chlorophyll-type pigments. Seed crushers can remove the green colour from rapeseed oil with bleaching clays, but this involves an added expense and poses an environmental problem.

Degumming agents: Processing aids used to remove phospholipids, trace metals and mucilaginous gums during the initial (degumming) stage of oils and fats refining. Examples include water, phosphoric acid and citric acid.

Degumming: The first stage in the purification of crude oils, which involves removal of phospholipids and colouring materials. Degumming is necessary to prevent separation and settling of gums (sticky, viscous oil-water emulsions stabilized by phospholipids) during transportation and storage of crude oils, to reduce oil losses in the subsequent phases of refining, and to avoid excessive darkening of the oils in the course of high-temperature deodorization. Degumming agents, such as phosphoric acid, may be used together with a flocculation agent such as alumina. During water degumming, phosphatides in seed oils are removed by centrifugal separation, after precipitation with water. Acid degumming involves removal of gums and impurities via centrifugal separation after precipitation with acid and water. By-products of the degumming process are known as lecithins.

Dehulling: Removal of the hulls from fruits or seeds prior to consumption. It is also called hulling or husking. This term also relates to removal of the cluster of leaves from the tops of strawberries prior to consumption.

Demineralization: Removal of minerals from substances. It includes processing steps in food manufacture, such as for sugar syrups, drinking water, musts, whey and food factories effluents. Processes used to achieve demineralization include electrodialysis, reverse osmosis and nanofiltration. It also covers the undesirable removal of selected minerals from previously healthy tissues such as bone and tooth enamel, which may be caused by a variety of factors including nutritional imbalance and excess acidity, respectively.

Denaturation: Structural change, especially in proteins or nucleic acids, in response to extreme conditions of temperature, pH, pressure, salt concentration or chemical action, which renders the molecule incapable of performing its original biological function. It used in food processing to inactivate detrimental enzymes, or to alter the gelation properties of proteins such as gelatin or whey proteins. However, it can also be deleterious, leading to impairment of functional properties such as water holding capacity in proteinaceous foods, and to reduced product yields in enzyme catalysis. Generally denaturing reduces the solubility and changes the colour and appearance of protein, e.g. as when heating egg white or blood.

Densitometry: Technique for measuring the optical density of a material by recording transmission of light.

Density: One of the physical properties of a sub stance, defined as the mass contained in a given volume. For example, water has a density of 1 kg per litre, saturated brine about 1.22 kg per litre and vegetable oil about 0.8 kg per litre. Routinely determined for a wide range of foods, including fruits and vegetables (sometimes related to ripeness and composition), fats and oils, foods produced by extrusion, and cereals. Density determinations can also be used as process control steps in food processing.

Deodorization: Removal or concealment of an unpleasant smell in an item. Deodorization is usually the last step in edible oil refining, involving vacuum-steam distillation at elevated temperature, during which free fatty acids and odoriferous volatile compounds are removed in order to obtain a bland and colourless product. Deodorization can be conducted under continuous, semi-continuous or batch conditions.

Designer foods: Functional foods targeted towards a certain purpose such as the prevention of certain diseases, or provision of tailored health benefits.

Desorption: Physical or chemical sorption process by which a substance (gas, liquid or solid) that has been adsorbed or absorbed by a liquid or solid material is removed from the material. Desorption isotherms of foods during drying are commonly studied to quantify reductions in moisture content. An O_2 adsorption- desorption process has been observed in dough during breadmaking. A thermal desorption step is used in analyte separation during GC analyses.

Dessert cream: Mixtures of cream with custard and/or fruit purée, often set with gelatine and with a smooth creamy texture. Used as the basis of many desserts.

Dessert mixes: Dried instant foods used to prepare desserts, typically by adding water or milk. Also called pudding mixes.

Dessert rice: Rice prepared for compounded sweets, puddings and desserts, etc. Long-grain rice is washed and blanched twice, cooked with boiling milk, sugar, butter, salt, and vanilla or orange zest flavouring in the oven without stirring, when cooked removed from the oven and enriched with egg yolks. Nowadays dessert rice more often made with short-grain rice cooked in sweetened milk, drained and sweetening adjusted.

Desserts: Sweet foods usually served as the last course of a meal. The term encompasses many different types of food, including dairy- and fruit-based products, cooked or raw. Available frozen, chilled or shelf- stable, as well as in the form of dessert mixes.

Desulfitation: Removal of salts of sulfurous acid, usually sulfites, and SO_2. Microbes can be used for desulfitation of waste water (effluent) from food factories. Wines for distillation can be desulfited using $CaCO_3$. Musts that are preserved by heavy sulfitation, and used for adjustment of sweetness of wines, require desulfitation before use. In the Brim-stone winemaking system, clarified grape juices are preserved with high levels of SO_2 (1200-2000 mg/l) and then desulfited just before fermentation.

Detoxicants: Substances which inactivate, neutralize, or render harmless toxins or poisons.

Detoxification: Process of removing poisons or toxins (e.g. from foods), or process of inactivating, neutralizing or rendering harmless toxins or poisons. It can be affected by the use of solvents, chemical reactions, enzyme systems or microbial action.

Dewatering: Process of removing excess water from a substance, e.g. after washing of a food. It is used in processing of foods and in treatment of wastes. In the case of foods, water can be removed by various procedures including passing over vibrating screens, using specially designed rotary screens or centrifugation.

Dewaxing: Process in which solvents are used to dissolve waxes from oil solutions. During the procedure, the wax solution is chilled and removed by filtration.

Dextran: Polysaccharide obtained from bacterial cell walls. It formed by fermentation of sugars. It used as a substitute for blood plasma and intravenously as a plasma expander under emergency conditions.

Dextrin: A water soluble polysaccharide made by reducing the number of repeating glucose units in starch or by partial hydrolysis of starch, including maltodextrins and cyclodextrins. It is used in commercial food preparations and when dissolved in water as a glaze for bread and rolls, and prevention of crystallization or as thickeners. Their sticky consistency also makes them suitable for use as edible adhesives. Cold-water soluble dextrins are used as carriers for flavourings in products such as dry mixes, soups and gravy.

Dextrose : Name given to the dextrorotary stereoisomer of glucose (D-glucose).

Dhal: Term used for dehusked and split pigeon peas, lentils or similar small pulses. Or A thin gruel-like dish of cooked dehusked split pulses with aromatic flavourings and spice, onions and other vegetables, very commonly served at Indian meals. Alternative spellings include dal, dahl and dhall.

Dhokla: Popular fermented foods of India. Typically prepared by soaking meal from chick peas or other legumes in water with buttermilk or curds for several hours, seasoning with ginger and chillies, and steaming the batter. The steamed cake is cut into squares, garnished with grated coconut and coriander and served hot.

Diabetes: Group of two diseases (diabetes mellitus and diabetes insipidus) of disparate pathology, both characterized by excessive urine production. Diabetes mellitus, the key feature of which is raised blood sugar levels or impaired glucose tolerance, is classified into two types: type 1, juvenile-onset or insulin-dependent diabetes; and type 2, maturity-onset or non-insulin dependent diabetes. Type 1 disease is a result of insulin deficiency and type 2 disease is due to insulin resistance. Control of blood sugar levels can be achieved by dietary manipulation in some cases, particularly in mild forms of type 2 disease, by reducing consumption of foods with high glycaemic index values. Diabetes insipidus is due, in general, to reduced ability of the kidney to concentrate urine, possibly caused by impairment in the hypothalamus/antidiuretic hormone system.

Diabetic foods: Dietetic foods manufactured specifically for individuals suffering from diabetes. Generally formulated to be low in absorbable carbohydrates, e.g. by replacing sucrose with fructose, sorbitol or other sweeteners that do not induce a large increase in blood glucose level.

Diacetyl: Yellow, flammable liquid with a strong aroma and buttery flavour derived from fermentation of glucose. It is soluble in water and alcohol. Used as an aroma carrier in foods and beverages.

Diafiltration: Extension of the ultrafiltration process in which water is added back to the extract during the concentration process. During diafiltration, both diffusive and convective mass transfer take place simultaneously as a result of two driving forces: a concentration gradient and a transmembrane pressure gradient. This is useful in selectively removing lower molecular weight materials from a mixture, and offers a useful alternative process to ion exchange or electrodialysis for removal of anions, cations, sugars, alcohol or anti-nutritional factors. Diafiltration is an accepted method for production of alcohol free, low calorie and low alcohol beer.

Diallyl disulfide: Organic sulfur compound which is a major component of garlic and garlic oils and a major contributor to their aroma. In addition to its sensory properties, the compound also possesses health benefits including antitumour activity and protection against the risk of cardiovascular diseases.

Dialysis: Separation of particles in a liquid on the basis of differences in their size and thus ability to pass through a membrane. Membranes are chosen that will allow small particles to pass through, but retain larger particles. The process can be used to remove unwanted particles and enrich or concentrate a solution.

Diarrhoea: Disorder characterized by loose watery stools which are often evacuated at increased frequency. Diarrhoea may be an indicator of many diseases of the gastrointestinal tract, including food- borne diseases, food poisoning, gastroenteritis, food intolerance, colitis and colorectal cancer.

Diastatic activity: Total activity of starch degrading enzymes in grain malts. An important quality characteristic for malting and brewing.

Dicing: Cutting of materials, such as foods, into small cubes.

Dielectric constant: One of the electrical properties, describing the ability of a material to store electrostatic energy when a unit voltage is applied. It is also known as relative permittivity. Dielectric constants have been used to determine changes in foods, such as moisture content changes in sugar confectionery, or degradation of frying oils, and also to monitor processing steps such as the use of microwaves in thawing and cooking.

Dielectric heating: Heating of electrically non- conducting materials, such as foods, by subjecting them to high frequency electromagnetic fields. The material to be heated is placed between two electrodes, to which a source of high-frequency energy is connected. In homogeneous materials, the resultant heating occurs throughout.

Dielectric properties: Electrical properties of dielectric materials, i.e. non-conducting materials which can sustain electric fields and act as insulators. These properties include the dielectric constant, dielectric relaxation and dielectric loss. Examples of their use in food analysis include assessment of the stability of dough during frozen storage, and comparison of the quality of musts from different cultivars of winemaking grapes.

Diet: Selection by individuals or population groups of foods and beverages for consumption. Dietary composition is the major factor affecting nutrition status and can have profound effects on health and risks for a range of diseases.

Dietary fibre: Complex mixture of long-chain carbohydrates, also known as non-starch polysaccharides, derived from plant cell walls and resistant to digestion in the intestinal tract. It is not digested in the gut but which add to the bulk of bowel contents, reducing constipation and decreasing the transit time of waste matter thus thought to reduce the incidence of bowel cancer. Little nutritional value in its own right, but considered to have beneficial effects on health. High-fibre diets can help control obesity and constipation; reduce the risk of cancer development and lower blood cholesterol. Fibre-rich foods include wholegrain products, wholemeal cereal products, fruits and vegetables. It is also called **roughage**

Dietary supplements: Alternative term for specific types of food supplements usually taken in tablet or capsule form as a supplement to the normal diet, with the aim of increasing an individual's intake of a specific nutrients, e.g. vitamins or minerals.

Dietetic foods: Products intended for consumption by individuals with metabolic disorders or allergies, such as diabetic foods or gluten low foods. It is also used to refer to foods providing specific nutritional benefits to healthy individuals with particular dietary requirements, such as infants or athletes.

Differential scanning calorimetry (DSC): Technique in which a sample and thermally inert reference material at the same temperature are heated using a temperature programme and the rate of heat flow is measured independently for each. The differential heat flow is monitored as a function of temperature. It can be used to measure heat capacity.

Diffusion: Spontaneous and random movement of molecules or particles in a fluid (gas or liquid) from a region of high concentration to a region of low concentration. Once a uniform concentration (or dynamic equilibrium) is achieved, net diffusion ceases and motion is random throughout the fluid. Diffusion rates are widely used in food analyses, and two common examples include moisture diffusion, which is routinely determined in foods during drying, and salt diffusion, which will affect the curing rate of foods.

Diffusivity: Measure of the ability of a substance to diffuse. It includes thermal diffusivity, which describes the diffusion of heat through a material.

Digestibility: A measure of the degree to which food can be converted to nutrients which can be used in metabolism. Generally it is measured subjectively in humans though animal measurement techniques could be used. True digestibility is measured as the difference between intake and faecal output, with allowance being made for that part of the faeces that is not derived from undigested food residues. Apparent digestibility is an approximate measure, which is simply the difference between intake and output. Nutrition term relating to the proportion of a food absorbed from the digestive tract into the bloodstream.

Digestion: Human physiology term relating to the breakdown of large polymeric molecules into their monomeric constituents, achieved chemically or enzymically. In particular, the term is applied to the breakdown by digestive enzymes of complex food molecules, e.g. proteins to amino acids, starch to glucose, fats to glycerol and fatty acids, so that they may be absorbed through the gut lining. Digestion can include the mechanical processes, such as chewing, churning and grinding of food, as well as the chemical action of digestive enzymes and other substances such as bile. Chemical digestion begins in the mouth with the action of saliva on food, but most takes place in the stomach and small intestine, where the food is subjected to gastric juices and pancreatic juices. The process makes use of acid, enzymes and emulsifying agents secreted into the mouth, stomach and bowel and of microorganisms in the gut which convert food into other compounds necessary for health.

Dipeptide sweeteners: Sweeteners based on dipeptides or their derivatives. It is usually more sweet, more stable and lower in calories than conventional sweeteners.

Dipeptides: Peptides consisting of two amino acid residues.

Disaccharide: A sugar consisting of two simple sugars such as glucose, fructose or galactose chemically bonded together. The most common is sucrose, ordinary beet or cane sugar, which consists of glucose and fructose.

Disaccharides Sugars: The sugar i.e. maltose, sucrose or lactose, consist of two linked monosaccharide molecules. It is dietary source of carbohydrate. Some individuals show disaccharide intolerance, e.g. lactose intolerance, and are unable to absorb disaccharides due to an enzyme deficiency.

Diseases: Abnormalities of the structure or physiological function of an organism which are regarded as being detrimental to its health.

Disinfectants: Chemical agents used for kills growing bacteria, yeasts etc but not necessarily spores including quaternary ammonium compounds, alcohols, phenols, halogens (chlorine and iodine), halogen compounds, and mercury compounds.

Disinfection by-products: It is by-products of drinking water disinfection. Trihalomethanes are associated with chlorination, while chlorites and chlorates are associated with chlorine dioxide disinfection. Ozonation may cause formation of bromates. It may be responsible for an increased risk of kidney and bladder cancer in humans and other long term health effects.

Disinfection: Destruction, inactivation or removal of pathogens or spoilage microorganisms. Commonly refers to the use of disinfectants for the treatment of inanimate objects and surfaces (e.g. surfaces in food processing plants and kitchens).

Disinfestation: Destruction of insect pests and other parasites of animals or plants. Generally involves the use of insecticides, applied either topically or as a spray.

Dispensers: Devices that supply or release a product, such as foods and beverages, by dispensing.

Dispensing: The process of supply or release of a product, such as foods and beverages, sometimes from special devices (dispensers).

Dispersibility: Measure of the ability of materials to form dispersions, in which one substance is suspended in a second material. Often determined for dried foods or ingredients such as powders to illustrate how well they can be rehydrated.

Dispersions: Two-phase systems consisting of particles (the disperse phase) suspended in a second sub- stance (the continuous or bulk phase) which generally present in relative excess. It includes colloids, emulsions and aerosols.

Display cabinets: Units in which items, including foods, are displayed in an appealing manner. Food should be displayed such that its quality is maintained (e.g. lighting and temperature are optimum), and so that it is protected from contamination and is attractive to potential customers.

Distillates: Spirits or their intermediate products manufactured from ethanol-containing mashes or other materials by distillation.

Distillation: Process for manufacture of spirits by condensation of vapour (containing ethanol and flavour compounds) liberated from ethanol-containing mashes during heating in a still.

Distilleries: Factories used for manufacture of spirits by distillation.

Distillers spent grains: Waste product from distilleries where cereals are used as the raw materials, comprising grain solids remaining after extraction of soluble material in the mashing process.

Distillers yeasts: Yeasts (*Saccharomyces* spp.) used for fermentation of mashes to be distilled in manufacture of spirits.

Distribution: The physical movement of commodities, including foods, into the channels of trade and industry. It can involve distributors, wholesalers, retailers, dealers and agents.

DNA fingerprinting: Technique that allows discrimination at the genomic level, utilizing the fact that the genetic material of an individual is unique. DNA sequences from regions of the genome of related organ- isms are isolated and analyzed either by Southern blotting or PCR in order to determine the organism's unique DNA fingerprint. Can be used to differentiate between different species, strains or cultivars of bacteria, yeasts, fungi and plants.

DNA hybridization: Formation of double-stranded DNA or DNA/RNA sequences by base-pairing between complementary single-stranded sequences. Can be carried out in solution or with one component immobilized on a matrix (e.g. nitrocellulose). The latter is known as Southern blotting. Hybridization can also be performed *in situ* using fluorescently-labelled DNA molecules to localize genes to specific chromosomes.

DNA probes: DNA or RNA sequences that have been labelled with radioactive isotopes, dyes or enzymes and are used to detect complementary sequences in DNA or RNA molecules by hybridization.

DNA: Abbreviation for deoxyribonucleic acid, a nucleic acid consisting of linked deoxyribonucleotides, each of which contains one of four nitrogenous bases (adenine, thymine, cytosine and guanine), a phosphate group and the pentose sugar deoxyribose. DNA is the genetic material of most organisms and usually exists as a double-stranded molecule in which two anti-parallel strands are held together by hydrogen bonds between adenine and thymine and between guanine and cytosine.

Dopamine: Important neurotransmitter in both the central and peripheral nervous systems, and metabolic precursor in adrenaline and noradrenaline biosynthesis. It also occurs in bananas where it is largely responsible for enzymic browning.

Dosimeters: Instruments used to measure doses of ionizing radiation such as X-rays.

Double cream: Cream with a high fat content (approximately 48%).

Dough conditioners: Ingredients added to yeast dough to improve its processing characteristics and/or the quality of the finished bakery products.

Dough: A thick, plastic mixture of flour and liquid (e.g. water or milk) that may contain yeasts or baking powders as leavening agents. It may be shaped, kneaded, rolled and baked to make bakery products.

Doughnut: A ball of slightly sweetened yeasted or chemically raised dough, deep-fried, drained and coated with caster sugar often with jam in the middle. The ring doughnut was the original form made by wrapping the dough around a stick which was suspended in hot oil, the ball form followed but was in turn generally superseded by a ring formed by extrusion because of the difficulty of cooking the centre of the ball without overcooking the surface. In simple way, it is rounded bakery products made from rich, sweetened dough leavened with either yeasts or baking powders and deep fried. It may be either ring-shaped with a hole in the centre or filled with jams, whipped cream or sweet pastes. It often coated with sugar or topped with icings.

Downstream processing: The processing steps involved in separation and purification of the products of fermentation processes and bioconversions. It can be performed either simultaneously with the process or after its completion.

Dressing: Post slaughter process, including the step of skinning, evisceration, trimming and washing, that follows the stage in which animals are bled. The head, feet, hides (in the case of sheep carcasses and cattle carcasses), excess fat, viscera and offal (edible and inedible) are separated from the bones and meat. With automated dressing lines, a series of mechanical devices stun the animals, remove the pelt (first from the brisket, then completely), eviscerate the carcass and process the head. Other devices debone the loin and thoracic regions. With respect to dressed meat, video image analysis can be successfully applied to grading for speedy online determination of the fat/lean ratio, and fibre optic probes permit objective prediction of such textural defects in the meat as excessive paleness or darkness. The term can also be applied to the

Dressings: Condiments used to coat and add flavour to foods prior to consumption. Most common types are salad dressings.

Dried fruit: Fruit that has had the water content reduced by solar or other drying methods to give a hard almost leathery texture and to reduce the water activity so that it is not degraded by microorganisms. Apricots, peaches, apples, bananas, grapes, tomatoes, plums and similar can all be dried.

Dried dairy products: Dairy products dried to a low moisture content, giving powders with a long shelf life. Packaged in materials that are impermeable to water vapour, oxygen and light to protect them during storage.

Dried egg products: Powders made by drying eggs or egg components. It includes dried egg whites, dried egg yolks and dried whole eggs. Utilized in the manufacture of foods where fresh eggs would be used, such as bakery products, bakery product mixes, mayonnaise, salad dressings and egg noodles. Their long shelf life and *Salmonella*-free status make them ideal for use by food manufacturers and caterers.

Dried eggs: Eggs which have been dehydrated, usually by spray drying, to form powders. Also called egg powders. It may be used in a range of foods, including bakery products, bakery product mixes, mayonnaise, salad dressings, confectionery, ice cream, pasta and convenience foods. Their long shelf life and *Salmonella*-free status make them ideal for use by food manufacturers and caterers.

Dried foods: Foods in which the majority of water present has been removed by drying, resulting in lighter weight products of extended shelf life, e.g. dried eggs, dried fruits, dried milk, mixes and powders. Sometimes rehydrated before consumption, although some, such as dried fruits, are consumed in their dried state. Rehydration properties are affected by the type of drying process used. Also known as dehydrated foods.

Dried meat products: Dried foods produced from meat. They include: pemmican, produced by sun drying strips of lean meat; and biltong and charqui, both produced by a combination of brining and air drying. Dried meat products differ considerably from fresh meat and are generally of lower eating quality.

Dried meat: Meat preserved by drying, a process that reduces water activity and so limits bacterial activity and enzymic changes. Traditional drying methods for meat include sun drying, air drying, oven drying and dry curing. A high surface to volume ratio is needed to allow effective drying of meat. Drying produces changes in nutritional value, particularly the vitamin content and sensory properties of the meat. Today, only a small proportion of meat is preserved by drying. It is mainly prepared when a light weight, high protein product with a good shelf life is required. Some specialty products are, however, prepared. For example, dried beef prepared by dry curing or sweet pickling followed by air drying is an expensive product, usually prepared from very lean beef.

Dried milk: Whole milk dried to a low moisture content, giving a powder with a long shelf life. It is also called milk powders.

Dried peas: Peas preserved by drying. Reconstituted in water before cooking as a vegetable, stir-fried or added to dishes including soups, stews and sauces.

Dried skim milk: Skim milk dried to a low moisture content, giving a powder with a long shelf life. Also called skim milk powders and non-fat dried milk.

Dried vegetables: Vegetables is preserved by drying. Commonly it is used types include peas, carrots, peppers and onions. Often reconstituted in water before use or added to dishes such as soups and stews.

Dried whey: Whey dried to a low moisture content, giving a powder with a long shelf life. Also called whey powders.

Dried yeasts: Active yeasts is preserved by drying for ease of handling, transport and storage. It is used in baking, brewing and winemaking, and as ingredients of soups, health foods, sauces and gravy.

Dripping: The fat which is extracted from fatty animal tissue and bones by heating or boiling. Originally it is referred only to the fat from roast meat or bird which had a fine flavour.

Drum drying: A process in which drying is under- taken continuously on the external surface of an internally steam heated rotating cylinder. A thin film of the product to be dried is applied at one location and re- moved at another, usually after less than one complete revolution of the cylinder. These driers may be atmospheric or vacuum types, and are classified as single drum, double drum, or twin drum (in which the two drums function almost as single drums).

Dry matter (DM): Measure of the proportion of a material remaining after the removal of water, also referred to as dry weight.

Dry sausages: Sausages which are dried during preparation. Often the sausages are hung in a drip room for 2-10 days at 21-27°C before they are transferred to a dry room, which has lower temperature and relative humidity levels, for a further 10-120 days. Natural casings are often used as they shrink and thereby remain in close contact with the surface of the sausage as it loses moisture.

Drying: Removal of moisture or liquid from an item to a level of <5%, a process also known as dehydration. Three basic methods of drying are applied to foods: sun drying, a traditional method in which foods dry naturally in the sun; hot air drying, in which foods are exposed to a blast of hot air; and freeze drying, in which frozen food is placed in a vacuum chamber to draw out the water.

Dryness: Sensory properties relating to the extent to which a product is perceived as being dry.

Durum wheat: Species of hard wheat (*Triticum durum*), the flour of which is glutinous and yellow and used to produce semolina from which pasta is made.

Dust explosions: Explosions caused by clouds of flammable particles at an appropriate concentration coming into contact with an ignition source. Dust and powders present a potential explosion hazard in processing plants, including those in the food industry. Common processes generating explosible dust include flour and provender milling, sugar grinding, spray drying of milk and instant coffee, and conveyance/storage of whole grains and finely divided materials. Parameters influencing explosions include nature of the combustible material, reactivity, particle dimensions, powder concentration, humidity, ignition energy and presence of inflammable gas. Powders can be classified on the basis of their explosion hazard, with explosible dusts including flour, custard powder, instant coffee, sugar, dried milk, potato powder and soup powder. Methods to control dust explosions include containment, suppression, inerting and venting.

Dyes: Natural or synthetic colorants used in foods. In contrast to pigments, dyes can usually be solubilized using an appropriate solvent or binder.

E

Eating quality: The extent to which a food is assessed as being edible, i.e. possessing acceptable sensory properties.

Eclairs: Finger-shaped bakery products made with choux pastry which is baked and filled with whipped cream or custards and topped with fondant icing, usually flavoured with chocolate or coffee. Also a name given to confectionery products comprising toffees filled with chocolate.

Ecology: Biological science, involving the study of interactions of organisms with their environment, including interrelationships between organisms.

e-commerce: Buying and selling of products and services transacted electronically via the Internet. It includes dealings among businesses and between companies and consumers. It is also called electronic commerce.

Edestin: One of the vegetable proteins present in certain plant seeds, including barley and hemp seeds.

Edible containers: Holders for foods which are intended to be consumed along with the food they contain. It is mainly made from dough. Examples include ice cream cones and taco shells.

Edible oils: Lipid-rich substances which are liquid at room temperature and are used in preparing foods. Usually have a high content of triacylglycerols and those of plant origin can be a source of bioactive phytochemicals. Should be of high quality, pale in colour, free from off odour and off flavour, and of high nutritional values. It includes vegetable oils and marine oils.

Edible packs: Packages for foods made from films and coatings that are suitable for consumption along with the products they enclose. The films and coatings are made from natural ingredients such as proteins, carbohydrates or lipids, or their combinations.

EDTA: It is abbreviation for ethylenediaminetetraacetic acid. Commercially available in the form of sodium and calcium salts, EDTA is one of the best known sequestrants and chelating agents, controlling the reaction of trace metals present in foods, and thus providing a variety of functions in foods. Applications include prevention of discoloration in canned corn, avoidance of crystals formation in canned sea foods and prevention of rancidity and microbial spoilage in mayonnaise and fatty spreads.

Egg coagulation: Separated white of egg coagulates between 60 and 65°C and the yolk between 62 and 70°C. These temperatures are raised when eggs are mixed into liquids with other additions. Coagulation and thickening of an egg, milk and sugar mixture, as in custard, will take place between 80 and 85°C and it will start to curdle at 88 to 90°C. It is also called egg setting.

Egg shells: Exterior hard coverings of eggs, which are composed mainly of calcium carbonate. It is vary in colour according to breed and species of bird. The shells are responsible for permitting gaseous exchange, conserving water, inhibiting microbial penetration and providing mechanical protection.

Egg white: The transparent gel surrounding the yolk of an egg which amounts to about 60% of the weight of a hen's egg. When fresh the egg white is viscous and stands high on a plate, as it ages it becomes thinner and is better for whisking.

Egg whites lysozymes: Lysozymes found in egg whites with good foaming properties and emulsification properties, particularly after modification or thermal treatment. Their antibacterial properties make them useful for preventing spoilage in foods and beverages (e.g. in meat, dairy products and beer). It is also potentially useful as sweeteners.

Egg whites: Portions of eggs which surround the egg yolks. It composed mainly of water and albumins. It form foams upon incorporation of air during whipping. Used in this form to make light products such as meringues and sponge cakes. It is also known as albumen.

Egg yolk: The central part of the egg usually light yellow to orange, semi-liquid and held separate within a fragile membrane. It contains about 16% protein mainly lecithin, 33% fat including cholesterol plus vitamins, trace elements and water. It represents about 30% of the total weight of a hen's egg. Colour of yolk may be enhanced by incorporation of pigmented feeds (e.g. yellow corn, alfalfa meal, corn gluten meal, dried algae meal and marigold petal meal) which contain carotenoid xanthophylls (e.g. lutein, zeaxanthin, carotenes and cryptoxanthin) into the poultry diet. Separated egg yolks may be used as emulsifiers in mayonnaise and salad dressings.

Eggs: External reproductive structures produced by the females of certain animals, such as birds, reptiles and fish. The term is used without qualification usually to refer to eggs laid by hens, although eggs produced by other birds, some reptiles (e.g. turtles) and fish (roes) are also eaten. Generally composed of egg yolks and egg whites surrounded by hard egg shells. It eaten raw or cooked in a variety of ways, e.g. scrambled, fried, poached or boiled. Also incorporated into a range of foods and beverages, and can be used as thickeners, emulsifiers, binding agents and foaming agents.

Elasticity: Rheological properties relating to the ability of a substance to return to its original size and shape after being deformed. The deforming force is known as a stress, and the resulting deformation is the strain. A body is elastic only below a certain stress; above this point, known as the elastic limit, the body is permanently deformed. The point at which the material begins to give is called the yield point.

Elastin: One of the animal proteins present in mammalian connective tissues, and thus a compo- nent in meat and meat products. Particularly rich in glycine residues and also contains high levels of proline, alanine and valine.

Electric fields: A region of space characterized by the existence of a force generated by electric charge. The magnitude of the electric field around an electric charge depends on how the charge is distributed in space. Each point in space has an electric property associated with it, the magnitude and direction of which are expressed by the value of the electric field strength. The value of the electric field has dimensions of force/unit charge. In the SI system, units are New- tons/Coulomb, equivalent to Volts/Metre.

Electrical conductivity: Ability of a substance to transmit an electric current. One of the electrical properties commonly determined in food analyses. It can be used, for example, as an indicator of *post mortem* changes in meat quality and to monitor the composition of food factories effluents.

Electrical properties: Generalized term for the physical properties of a food relating to its ability to conduct electricity. It includes capacitance, dielectric properties, conductivity/resistance and electrostatic interactions.

Electrical resistance: One of the electrical properties commonly determined in food analyses, electrical resistance is a measure of the extent to which a material withstands passage of an electric current. It is inversely related to electrical conductivity. Heat is generated as a consequence of resistance and this characteristic is exploited in some cooking or heating methods, an example being ohmic heating.

Electrical stimulation: Controlled application of an electrical current to animal carcasses immediately after slaughter. It is used to increase meat tenderness, and also to give meat a lighter, brighter colour. In particular, it is used to achieve accelerated conditioning (ageing) of animal carcasses, and to decrease cold shortening and subsequent toughness, which accompany very rapid chilling of meat. Electrical stimulation of carcasses breaks cross-linkages between actin and myosin filaments in the muscles, increases enzyme activity and causes some tissue damage; all of these effects increase meat tenderness. It may considerably improve the quality of beef, veal, lamb and goat meat, but has negative or negligible effects on the quality of pork. Electrical stimulation is well established in lamb slaughtering practice and has also been widely used in deer slaughtering.

Electrical stunning: A form of stunning, which is used during slaughter to immobilize animals and birds before bleeding. It is widely used during the slaughter of swine, sheep and poultry, but can also be used effectively during cattle slaughter. Before consciousness returns, bleeding can be carried out humanely and effectively. As well as improving animal welfare during slaughter, the method has beneficial effects on meat quality; for example, it reduces the incidence of the PSE defect in pork. There are two basic types, namely high voltage and low voltage. Electrical stunners include: pillar types; electrically charged knives; stunning tongs; and electrified water baths.

Electrodialysis: Technique in which dialysis is accelerated by application of a potential across the compartments of the apparatus.

Electrolysed water: Salted water which has been passed through an oxidizing unit, causing it to undergo ionic changes. Depending on which electrode the water is passed over, either acidic or alkaline electrolysed water is formed. Acidic water is lethal to food borne microorganisms and is considered more efficient for washing food, especially fruits and vegetables, during preparation than using chlorine-containing solutions or, in some cases, heat treatment. Its use has little effect on food sensory properties. Alkaline water is useful as a sanitizer, as it functions like a soap to remove substances from food preparation surfaces.

Electrolytes: Liquid or solid compounds which, when dissolved in or in contact with water, will dissociate into ions and conduct electricity. In physiological use, the term refers to certain inorganic compounds, e.g. those containing sodium, potassium or calcium, which dissociate into ions that conduct electrical currents and play an important role in controlling body fluid balance.

Electromagnetic fields: Fields of force associated with electric charge in motion, having both electric and magnetic components and containing a definite amount of electromagnetic energy. The mutual interaction of electric and magnetic fields produces an electromagnetic field, which is considered as having its own existence in space apart from the charges or currents with which it may be related. Under certain circumstances, this electromagnetic field can be described as a wave transporting electromagnetic energy. In the food industry, electromagnetic fields are utilized in dielectric heating.

Electron microscopy (EM): Microscopy technique which utilizes extremely short wave radiation from electrons in a vacuum tube to give high resolution.

Electron paramagnetic resonance: Spectroscopy technique for studying the structure and bonding of a paramagnetic substance based on microwave- induced transitions between the energy levels of un- paired electrons. Synonym for electron spin resonance.

Electronic noses: Apparatus, consisting of arrays of semiconductor metal sensors coated with polymers, used for characterization of aroma compounds. The polymers in the sensors adsorb volatile compounds from aromas, vapours and gases. Each polymer adsorbs a different combination of ingredients, so that conductivity changes and variations may be processed electronically to produce visual fingerprints.

Electronic tongues: Apparatus, consisting of arrays of lipid/polymer membrane based sensors, which can quantify the taste of substances such as amino acid mixtures, foods and beverages. The lipid/polymer membranes are fitted onto a multichannel electrode, and electric signals from the sensors are fed into a computer; voltage differences between the multichannel electrode and a reference electrode are measured. Output from the sensors varies for chemical substances with different taste qualities but is similar for sub- stances with similar tastes. The sensor array detects the five types of taste quality, i.e. sourness, saltiness, bitterness, sweetness and umami.

Electrophoresis: Technique in which charged electrical species are separated by migration in an electrolyte through which a current is passed, with cations moving towards the cathode and anions to the anode. Separated species are identified by staining or radioactive labelling. Usually conducted on paper or in a gel (gel electrophoresis), although faster methods using capillary columns (capillary electrophoresis) have been developed that have other advantages, such as the possibility of on-line detection of separated species.

ELISA: It is abbreviation for enzyme linked immunosorbent assay, a very sensitive immunological technique which can be used to detect and measure the presence of antigens or antibodies in a wide variety of biological samples. In the assay, protein antigens or antibodies are labelled with enzymes, after which one of the reactants is immobilized onto a support material. As soon as the immunochemical reaction has taken place, unbound substances are washed out and the bound material is quantified by measuring the activity of the enzyme by spectroscopy. The immobilization is preferentially performed in the wells of polyvinylchloride or polystyrene microtitre plates, and the colour forming enzymes used are normally peroxidases, alkaline phosphatases or glucose oxidases.

Emulsification properties: Functional properties relating to the ability of food components to form emulsions, suspensions of small globules of one liquid in a second liquid with which it will not mix.

Emulsification: Process for forming fine dispersions (emulsions) of minute droplets of one liquid in another in which it does not dissolve or form a homogeneous mixture.

Emulsifier: A compound which aids the dispersion of one liquid or semi-solid in another with which it is immiscible, e.g. Oil and vinegar to make mayonnaise, butter with milk and sugar in cake mixes. The emulsifier is usually a long-chain molecule, one end of which dissolves in water and the other end in fat. Egg yolk and French mustard are culinary emulsifiers but many others, synthetic as well as natural, are used in commercial made-up dishes and mixes. It is widely used in the food industry, where applications include manufacture of **bakery products**, **confectionery**, **ice cream**, **mayonnaise** and **margarines**. Types of emulsifiers used in foods include **carrageenans**, **lecithins** and **glycerides**.

Emulsifying capacity: Functional properties relating to the extent to which food components can be form emulsions.

Emulsion: A mixture of two or more immiscible liquids or semi-solids in which one is divided into very small droplets (less than 0.01 mm diameter) dispersed in the other(s). Mayonnaise, hollandaise sauce and cake batters are typical examples. Processed foods based upon emulsions include sauces, salad dressings, soups, spreads, coatings, mayonnaise, sausages and some dairy products. Emulsions display variable stability, and most require the addition of emulsifiers to maintain emulsion structure.

Enamels: Semi-transparent or opaque ceramics substances applied as protective or decorative coatings to the surface of metals, pottery or glass. Often applied to the surfaces of food containers, e.g. cans and cooking pots. Enamelled objects that come into contact with food or beverages may release lead or cadmium, posing a health risk. Also used to describe paints or varnishes which become smooth and hard when dried. The enamel can be chipped, crazed by temperature shock or scratched with harsh cleansers or metal spoons, etc.

Enantiomers: Stereoisomers of a compound which are mirror images of each other. The left and right handed forms of these chiral isomers are optically active and generate a racemate when mixed in equal proportions. Chirality may affect the biological activity and functional properties of the compound; for example, D-amino acids but not L-amino acids are useful as sweeteners.

Enantioselectivity: Preferential formation of one enantiomer over another in a chemical reaction, expressed quantitatively as enantiomer excess. Enantiomers formed may affect the biological activity and functional properties of the product, e.g. D- amino acids but not L-amino acids are useful as sweeteners.

Encapsulation: A technology that allows sensitive ingredients to be physically enveloped in a protective matrix or wall material in order to protect these ingredients or core materials from adverse reactions, loss of volatile compounds, or nutritional deterioration. Spray drying is a microencapsulation technique readily used in the food industry. Carbohydrates, such as maltodextrins, starch and corn syrup sol- ids, and acacia gums are widely used examples of encapsulating agents.

Endosperm: The white 80% to 90% inner portion of the wheat grain consisting of spherical starch granules surrounded by a protein matrix and some cell wall material which remain after the bran and wheat germ have been removed.

Endotoxins: Lipopolysaccharide toxins of Gram negative bacteria, or any microbial toxins which are released only upon cell lysis.

Endpoint temperature indicators: Indicators showing the adequacy of heating of foods, particularly meat and meat products, in relation to destruction of pathogens. The bovine catalase test and tests based on protein solubility, enzymes activity, colour, electrophoresis patterns of proteins, differential scanning calorimetry (DSC) of muscle proteins, near infrared spectroscopy (NIR spectroscopy) and enzyme linked immunosorbent assays (ELISA) can be used for this purpose.

Endpoint temperature: Temperature to which a food product, particularly meat, needs to be heated to ensure destruction of pathogens.

Energy conservation: Planned management of energy supplies by various means. One type of energy conservation is curtailment (doing without). A second type is overhaul (for example, using less energy intensive materials in production processes, and decreasing the amount of energy consumed by certain products). Another type involves the more efficient use of energy and adjusting to higher energy costs (for example, capturing waste heat in factories and reusing it).

Energy drinks: Soft drinks containing ingredients intended to enhance or maintain physical energy of the consumer.

Energy foods: Health foods designed for people, such as sportsmen and sportswomen, requiring a source of high energy. Energy foods are frequently available in the form of carbohydrate-rich energy food bars. Energy drinks and isotonic drinks are popular for the same purpose.

Enrichment techniques: Procedures which specifically promote the growth of a particular microorganism, thereby increasing its proportion in a mixed population.

Enrichment: Improvement of the quality or nutritional value of a food, usually by addition of nutrients. Like to add some or all of butter, milk, sugar and eggs to a bread dough to produce buns, etc. or to add butter or cream to a batter or sauce, or similar additions. To add back vitamins, trace elements or minerals to processed foods such as white flour, cereals, margarine, fruit juice, etc., sometimes to replace losses in processing, sometimes as a matter of public health policy.

Enrobing: Coating of a centre material, for example nougat, biscuit, fondants or caramel, in chocolate. It is necessary to use tempered chocolate for enrobing processes. The centres for coating are placed on a continuous moving wire chain belt, which transports them underneath a flow of chocolate. Below the belt is a bottoming trough that retains the chocolate that falls through the chain belt and recirculates it, forming a layer of chocolate on the undersides of the centres. Sometimes two chocolate streams are used in enrobers; this is particularly useful when the product to be enrobed has an uneven surface. The first coating flows into all the crevices and provides a good moisture barrier to the product. The second coating gives the chocolate a more rugged appearance. Products finally pass through a cooling tunnel to set the chocolate.

Enterotoxins: Bacterial toxins (e.g. cholera toxin) which, upon ingestion or production by microorganisms within the gastrointestinal tract, cause disturbances of the gastrointestinal tract. Diarrhoea is a common symptom.

Enterovirus: This is a virus which can be transmitted by faeces. Transmission may be through

Enthalpy: Measure of energy (heat) commonly used to study the thermodynamics of chemical reactions. Changes in the structure of food macromolecules, such as denaturation, gelatinization and crystallization, are often associated with changes in enthalpy.

Entoleters: Machines used in disinfestation of cereals and other foods. Food is fed to the centre of a high-speed rotating disc which bears studs. The impact of the food being thrown against the studs kills insects and destroys their eggs.

E-numbers: The term 'E-number' applies to the incorporation in food of a range of additives,

Environment friendly packaging materials: Materials developed for packaging of products including foods and beverages, with special consideration given to biodegradability and recycling.

Environment friendly processes: Processing procedures that are not harmful to the environment.

Enzymatic browning: Browning of fruit and vegetables that occurs when cut surfaces are exposed to air. This is inhibited by acids, e.g. Lemon juice or vinegar. Formation of brown coloration of cut fruits and vegetables due to the action of catechol oxidases (polyphenol oxidases). In the presence of oxygen, the enzymes break phenols down into quinones, which polymerize to form brown coloured melanins. It is also called Enzymic browning.

Enzyme electrodes: Type of ion selective electrodes in which the electrodes are coated with a layer containing an enzyme that reacts with the analyte to form a product to which the electrodes respond. Commonly used examples include glucose sensitive electrodes, which are coated with glucose oxidases.

Enzyme inhibitors: Substances which reduce the activity of enzymes and, when present in foods, may act as antinutritional factors. Certain proteinases inhibitors such as calpastatins and cystatins play a role in development of meat tenderness and also may be useful for maintaining the quality of fish and surimi by inhibiting proteolysis. However, trypsin inhibitors and chymotrypsin inhibitors present in plant foods, particularly legumes, can reduce the digestibility and nutritional values of these foods.

Enzyme: A natural protein which speeds up the rate of chemical reaction under moderate conditions of temperature, such as oxidation, browning of apples, decolorizing of vegetables, etc. They are easily inactivated by extremes of pH, high concentrations of salt, addition of alcohol or most importantly by temperatures of 80°C or above. It acts as highly efficient and specific biological catalysts. Increase the rate of reactions by decreasing the activation energy but do not alter the equilibrium constant. Divided into six main groups: oxidoreductases, transferases, hydro lases, lyases, isomerases and ligases.

Enzymic techniques: Analytical techniques in which enzyme reactions form a major part.

Ergocalciferol: Synonym for calciferol and vitamin D_2; one of the group of sterols which constitute vitamin D. Synthesized by irradiation of the plant provitamin ergosterol. Alternative recommended name is ercalciol.

Ergosterol: Sterol which occurs naturally in algae, bacteria, fungi, yeasts, higher plants and animals. When exposed to UV radiation it is converted into vitamin D_2 (ergocalciferol), a potent antirachitic substance. Used in synthesis of oestradiol.

Ergotamine: One of the alkaloids produced by the ergot fungus *Claviceps purpurea*, which attacks cereals, predominantly rye. Also a secondary metabolite of some strains of *Penicillium*, *Aspergillus* and *Rhizopus*. Can cause poisoning (ergotism) if contaminated grain is used for food, but modern grain cleaning and milling procedures remove most of the ergot, leaving low levels of ergotamine in flour. Baking and cooking usually cause destruction of remaining alkaloid. Ergotamine is commonly used, in combination with caffeine, for treatment of migraine.

Erucic acid: Monounsaturated fatty acid, which exists as a combustible solid with low toxicity. It is insoluble in water, but soluble in alcohol and ether. Occurs naturally as a minor component of many plant seeds and is obtained from plant seed oils, particularly hydrogenated mustard seed oils and rapeseed oils. It uses include manufacture of waxes, plasticizers, water resistant nylon and stabilizers and as an additive in polyethylene films. Alternative term for docosenoic acid.

Erythrosine: Artificial red colorant used for colouring **cherries**, **meat** products, **candy** and **confectionery**. It is known as **FDC red** 3.

Essence: An extract of flavouring compounds usually from fruit or vegetable matter but occasionally from meat. The extractant is usually alcohol. Extracts may be of natural origin (e.g. essential oils) or may be synthetic.

Essential fatty acids: Fatty acids which cannot be synthesized in the body and must be obtained from food sources. The important ones are: linolenic, linoleic, arachidonic (AA), gamma linolenic (GLA) and eicosopentanoic acid (EPA). All are involved in the synthesis of prostaglandins which have a variety of important bodily functions and in the construction of the myelin sheaths around nerve fibres and the synthesis of cell walls.

Essential amino acids: Amino acids which are necessary for health and cannot be synthesized in the body and must therefore be obtained from food. They are: isoleucine, leucine, lysine, methionine, phenylalanine, threonine, tryptophan and valine, plus histidine which are required by growing infants and may be required by adults.

Essential oils: Volatile aromatic oils of complex composition extracted from plant material, usually by distillation, although supercritical CO_2 extraction and cold pressing may also be used. Widely used as flavourings, either by adding their characteristic flavour to an end product or in the creation of natural flavouring blends. Some of the most widely used essential oils are citrus essential oils, peppermint essential oils and cinnamon oils.

Ester: Esters is organic compounds which are formed by combination of an acid with an alcohol. Most esters have pleasant fruity flavours and aromas. Some are extracted from fruits but many are now made synthetically. Some esters have a pleasant, generally fruity, aroma and occur in plant essential oils. Uses vary widely according to type of ester, but include synthesis of flavourings and perfumes.

Esterase: An enzyme which breaks down esters into their component acids and alcohols, thus destroying or modifying the flavour or aroma.

Esterification: The process by which acids and alcohols react to form alkyl or aryl derivatives. It can be catalysed enzymatically by esterases or chemically by mineral acids.

Ethanal: Aldehyde (systematic name for acetaldehyde) which in pure form exists as a volatile, colourless liquid with a pungent, fruity aroma. It is produced by oxidation of ethanol and soluble in water and alcohol. Fruits and vegetables produce ethanal during ripening. It is also produced during fermentation, and is present in foods such as fermented dairy products and alcoholic beverages. It used in food flavourings and in the manufacture of acetic acid. It is also known as acetic aldehyde.

Ethanol: The alcohol produced by the fermentation of sugars which is responsible for the intoxicating effects of alcoholic drinks/beverages. It formed by fermentation of sugars by yeasts. It is used as a solvent for food colourings and flavourings, and also called alcohol or ethyl alcohol.

Ethnic foods: Foods belonging to the traditional cuisine of other ethnic groups. For example, Chinese, Indian and Mexican foods are all popular ethnic foods in the UK and USA. There is an increasing tendency for consumers to try foods from other countries as cultural diversity increases. This is reflected in the continuing increase in international sales of ethnic foods, including ethnic ready meals, flavourings and take away foods.

Ethyl vanillin: Artificial flavouring, approximately 2 to 4 times stronger than vanillin. It is synthesized from eugenol, isoeugenol or safrole. It used to enhance fruit and chocolate flavour notes in ice cream, beverages and bakery products.

Ethylene dibromide: Colourless, non-flammable liquid with a sweetish aroma. Toxic and carcinogenic. Slightly soluble in water and miscible with most organic solvents and thinners. Used in fumigants for grain and tree crops, as a general solvent and as a water-proofing preparation.

Ethylene glycol: One of the glycols or polyols. A colourless, viscous, hygroscopic liquid commonly used as a solvent, osmotic solute, antifreeze or plasticizer. It has been used as an additive in edible films.

Ethylene oxide: Highly flammable, colourless gas which liquefies at temperatures below 12°C. It soluble in organic solvents and miscible with water and alcohol. It has sporicidal and viricidal activities, and is probably carcinogenic. Sometimes used in fumigation of spices. Also known as epoxyethane or oxirane.

Ethylene: Highly flammable, colourless hydrocarbon gas with a sweetish aroma and flavour. It is slightly soluble in water and alcohol. It occurs in natural gas and coal gas, and is produced by fruits and vegetables during ripening. Removal of ethylene from food packages is used to delay ripening of fruits. As a plant growth regulator, ethylene has many horticultural uses, e.g. as a fruit ripening accelerator.

Eucalyptol : Monocyclic terpene distributed widely in plants. It occurs as a colourless liquid with a characteristic aroma and pungent flavour. Major food sources include eucalyptus oils, spices including sage, rosemary and basil, and essential oils extracted from herbs and spices. It is used in flavourings for foods and beverages. Cough candy contains particularly high levels of eucalyptol due to a high content of eucalyptus oil.

Eugenol: Combustible, colourless or pale yellow phenol with a spicy aroma and flavour which is derived from oil of cloves and cinnamon oils. It is only very slightly soluble in water, but soluble in alcohol, ether and volatile oils. Used in flavourings, perfumes, essential oil preparations, as a dental analgesic and local anaesthetic, and in the manufacture of isoeugenol for production of vanillin.

Eukaryotes: Organisms in which the cells have a distinct nucleus containing the genetic material (DNA). It includes all organisms except bacteria. Alternative spelling is eucaryotes.

Evaporated milk: Milk is concentrated by partial removal of water with the aid of a vacuum to reduce the boiling point and thus maintain the quality of the milk during the process. In this, homogenized cows' milk from which 60 to 65% of the water has been removed under a vacuum at not more than 66°C. It may have a range of fat contents depending on the concentration ratio used. After evaporation, the product is homogenized, mixed with stabilizers and sterilized in cans, or is UHT (ultra high temperature) treated combined with aseptic packaging in cartons. It may be reconstituted by addition of water in amounts stated on the packaging.

Evaporation: Gradual change of state from liquid to gas that occurs at a liquid's surface. The average speed of particles within a liquid depends on the liquid's temperature. Fast-moving particles striking other particles near the liquid's surface may impart enough speed, and therefore enough kinetic energy (energy of motion), to cause the surface particle to leave the liquid and be- come gas atoms or molecules. As particles with the most kinetic energy evaporate, the average kinetic energy of the remaining liquid decreases. Because a liquid's temperature is directly related to the average kinetic energy of its molecules, the liquid cools as it evaporates.

Evaporative cooling: Evaporative cooling is the process by which the temperature of a substance is reduced due to the cooling effect from the evaporation of water. The conversion of sensible heat to latent heat causes a decrease in the ambient temperature as water evaporated provide useful cooling. This is the most economical way of reducing the temperature by humidifying the air. It has many advantages over refrigeration system, as it does not use refrigerant so it is environmental friendly (reduces CO2). It does not make noise as there is no moving part. It does not use electricity so its saves energy. It does not require high initial investment as well as operational cost in design and maintenance. It can be constructed with locally available materials in remote areas and most importantly, it is eco-friendly as it does not need chlorofluorocarbons. Evaporation of moisture from vegetables causes wilting and shriveling resulting in weight loss. The process of evaporative cooling is an adiabatic exchange of heat when ambient air is passed through a saturated surface to obtain low temperature and high humidity, which are desirable for extending the storage life of vegetables.

Evaporators: Equipment used in turning a liquid into a vapour by evaporation.

Evisceration: The process of disembowelment, the cutting open and removal of the inner organs or entrails of animal carcasses. Similar to gutting of fish.

Exotoxins: Potent extracellular toxins secreted by certain species of bacteria (e.g. *Clostridium botulinum* and *Staphylococcus aureus*).

Extensibility: Extent to which a material can be distorted or stretched without breaking. It is often ex- pressed as a proportion of the material's original size. A decrease in extensibility resulting from shortening of muscles has traditionally been used to define *rigor mortis*. It is also commonly measured during assessment of the rheological properties of dough.

Extensographs: Instruments used to investigate the physical properties of dough. It is similar to alveographs.

Extraction: The process of extracting one component from a mixture. Or The percentage of flour which is obtained from the original dehusked wheat grain. For white flours this is normally around 80%.

Extractive fermentation: Simultaneous extraction of fermentation products during fermentation processes. Organic solvents are frequently used for extraction, and the process often results in higher yields since it eliminates the problem of product inhibition.

Extractives: The flavouring components in meat and browned vegetables which dissolve in the cooking liquor to give it flavour extracts Flavourings, aromas and colours in concentrated form produced by extraction

Extracts: Term usually applied to concentrated flavourings obtained by solvent extraction or supercritical extraction of substances such as herbs, meat, yeasts or fruits. It may also more generally apply to any product obtained by extraction.

Extrinsic contamination: This is the contamination arising from sources other than the raw food itself, namely external sources of contamination, such as foreign bodies. During manufacturing and catering processes, food can be exposed to extrinsic contamination from items such as glass particles, wood splinters, hairs, small personal items, such as pens, rings and hair grips, rodent droppings, dead insects, cigarette ends, plastic items and metal objects falling from badly-maintained plant and equipment. Contamination may also be associated with poor standards of building maintenance resulting in flaking paintwork, rust from overhead pipes and contaminated water leaking from pipes, all of which could enter the product at some stage of preparation. Food safety management systems should explore the potential for extrinsic contamination at the various stages of processing.

Extrudates: Items which have been shaped by forcing them through a die (extrusion).

Extruders: Die equipment that is used to shape items during extrusion. The basic concept behind an extruder is to force the product through a small diameter hole of constant area. Extruders come in a wide variety of shapes, sizes and methods of operation, but can be categorised into one of three main types: piston, roller and screw extruders. The simplest of these is the piston extruder, which consists of a single piston or a battery of pistons that force the material through a nozzle onto a wide conveyor. The pistons can deliver very precise quantities of materials and are often used in the confectionery industry to deposit the centre fillings of chocolates. Roller extruders consist of two counter-rotating drums placed close together. The material is fed into the gap between the rollers that rotate at similar or different speeds and have smooth or profiled surfaces. A variety of product characteristics can be obtained by altering the rotation speeds of the rollers and the gap between them. This process is used primarily with sticky materials that do not require high-pressure forming. Screw extruders are the most complex of the three categories of extruder and employ single-, twin-, or multiple-screws rotating in a stationary barrel to convey material forward through a specially designed die. Amongst the screw type extruders, it is common to classify machines on the basis of the amount of mechanical energy they can generate. A low-shear extruder is designed to minimise the mechanical energy produced and is used primarily to mix and form products. Conversely, high-shear extruders aim to maximise mechanical energy input and are used in applications where heating is required.

Extrusion cooking: The process of cooking of foods by forcing them through a die. Extrusion cooking is an important process for the gelatinization of starches. The degree of starch gelatinization of extrudate mainly depends on extrusion conditions, such as the initial moisture content of materials, screw speed, and heating temperature. However, excessive initial moisture content may reduce the degree of starch gelatinization of the extrudate. The heating level and retention time in an extruder are also important factors affecting the gelatinization of starches.

Extrusion: Extrusion can be defined as the process of forcing a pumpable material through a restricted opening. It involves compressing and working a material to form a semisolid mass under a variety of controlled conditions and then forcing it, at a predetermined rate, to pass through a hole. Extrusion is predominantly a thermo-mechanical processing operation that combines several unit operations, including mixing, kneading, shearing, conveying, heating, cooling, forming, partial drying or puffing, depending on the material and equipment used. During extrusion processing, food materials are generally subjected to a combination of high temperature, high pressure and high shear. This can lead to a variety of reactions with corresponding changes in the functional properties of the extruded material. The ingredients used in extrusion are predominantly dry powdered materials, with the most commonly used being wheat, maize and rice flours. The conditions in the extruder transform the dry powdered materials into fluids and therefore, characteristics such as surface friction, hardness and cohesiveness of particles become important. In the high solids concentration of the extruder melt, the presence of other ingredients, such as lipids and sugars, can cause significant changes in the final product characteristics. In addition to starch-based products, a range of protein-rich products can be manufactured by extrusion, using raw materials such as soya or sunflower, fava beans, field bean and isolated cereal proteins.

F

F2 toxin: Mycotoxin produced by *Fusarium graminearum*, *F. culmorum* and other *Fusarium* species. It may be formed when the fungus grows on damp cereal grain (e.g. wheat, barley and corn) used as animal feeds. It has oestrogenic activity and can cause hyperoestrogenism in swine, cattle and poultry. Also known as zearalenone and zearalenol.

Falling number: Indicator used to measure the activity of α-amylases in cereal flours. In wheat, a low falling number may signal reduced grain quality and poor breadmaking properties.

Farina: A fine flour or meal which is prepared from cereals, particularly wheat, or other plant foods with a high starch content. It can be used in the manufacture of foods such as pasta.

Farinographs: Instruments used to investigate the physical properties of dough.

Fast foods: Prepared foods obtained from restaurants and other catering establishments, where the aim is to provide a fast service and rapid customer turnover at reasonable prices. Examples of fast foods include burgers, pizzas, sandwiches and French fries.

Fast-freezing: The food is freeze very rapidly so as to prevent the formation of large ice crystals within the cell which would disrupt them and cause the contents to leak out on thawing, thus destroying the structure of the food.

Fat extenders: Food additives that allow the fat content of food to be reduced without affecting the texture.

Fat mimetics: Alternative term for fat substitutes. Fat replacers Alternative term for fat substitutes.

Fat substitutes: Substances of various types and origins that show similar properties to triacylglycerols in that they have a creamy and fat-like texture but have low calorific values. Used in the complete or partial replacement of fats in foods, e.g. low fat foods. Also known as fat mimetics or fat replacers.

Fat: An ester of fatty acids and glycerol formed in the bodies of animals or plants as a long term energy store, and in the case of animals as a source of energy for their young prior to weaning and as an essential component of cell walls. It may be heated to high temperatures without decomposition and generally adds flavour, texture and succulence to most foods and dishes. Oils are fats with low melting points. It is non-volatile, water insoluble substances that are usually solid at room temperature and are greasy to the touch. It composed of esters synthesized by reaction of fatty acids with glycerol in a ratio of 3 to 1 to form triacylglycerols or triglycerides. Arrangement and type of the fatty acids in the glycerol molecule affect the physical properties of the fat. Also called lipid.

Fatty acid: A long-chain hydrocarbon with an acid group at one end or organic acids consisting of a chain of alkyl groups containing between 4 and 22 or more carbon atoms with a terminal carboxyl group. This combines with an alcohol such as glycerine by the esterification reaction to form natural fats and oils, or with a metal such as sodium to form the soap. They are also esterified with a variety of higher alcohols for use in convenience foods. Generally produced from natural fats and oils, and they are the initial breakdown product of fats in the body. In saturated fatty acids, e.g. butyric acid, palmitic acid and stearic acid, the carbon atoms of the alkyl chains are connected by single bonds. Unsaturated fatty acids, e.g. oleic acid, linoleic acid and linolenic acid, contain at least one double bond. Fatty acids occur naturally and are derived from animal fats, fish oils and vegetable fats. Some, e.g. linoleic, linolenic and arachidonic acid, are essential nutrients (essential fatty acids) that are not synthesized in the human body and must be obtained from the diet.

Fermentation products: Products of microbial fermentation processes, e.g. alcohols, flavour com- pounds, food additives, surfactants and organic acids.

Fermentation technology: Technologies and methods used for production of specific products by means of microbial fermentation.

Fermentation: The process whereby a microorganism such as a yeast, fungi, lactobacillus, etc. breaks down an energy source (starch, sugar, protein, etc.) in the absence of air, into smaller molecules such as alcohol, acetic acid, lactic acid, etc. Energy-yielding process in which organic compounds are metabolized, usually under anaerobic or micro-aerobic conditions, to simpler compounds without the involvement of an exogenous electron acceptor. Commonly refers to processes carried out by microorganisms, regardless of whether fermentative or respiratory metabolism is involved. It is used frequently in the food industry, e.g. for production of alcohols, bread, vinegar, yoghurt, cheese, soya sauce, tempeh, sauerkraut, flavour compounds, and a wide variety of fermented foods and fermented beverages.

Fermented beverages: Beverages whose manufacture involves a fermentation process, generally alcoholic fermentation and/or lactic fermentation.

Fermented cream: Cream acidified naturally or artificially by the action of lactic acid bacteria. Lactose in the cream is converted to lactic acid by fermentation. Types of fermented cream include sour cream, ripened cream and smetana. Used in cooking and baking, and in dips.

Fermented dairy products: Produced by fermentation of liquid dairy products by lactic acid bacteria (starters). During fermentation, lactose is converted into lactic acid and sometimes flavour compounds such as diacetyl, depending on the organisms used and fermentation conditions. Fermentation is allowed to proceed until the required acidity is achieved. In some cases, where yeasts are also present, alcohol is formed in the final product, e.g. kefir, koumiss. Fermented dairy products include fermented cream, some types of butter, cheese, cultured buttermilk, and fermented milk, a popular type of which is yoghurt. Many traditional fermented dairy products exist throughout the world. Consumption of fermented dairy products, especially those containing specific organisms or probiotic bacteria, can enhance intestinal health.

Fermented foods: Foods subjected to fermentation by beneficial microorganisms in order to bring about desirable changes. These changes are mainly concerned with preservation (e.g. manufacture of cheese and yoghurt from milk), enhancement of nutritional value (e.g. removal of anti-nutritional factors from legumes), or alteration of flavour and texture (e.g. manufacture of soy sauces from soybeans). Many different types of fermented foods and fermented beverages are available, and play a major part in the human diet. Fermentation is favoured as an inexpensive method of preservation in developing countries and many items, such as fermented dairy products, are attracting increasing attention as functional foods due to the beneficial actions of the microorganisms and/or enzymes involved in fermentation. Also known as fermented products or cultured foods.

Fermented milk: Produced by fermentation of milk (of various species) by lactic acid bacteria (starters). During fermentation, lactose is converted into lactic acid, aroma compounds are formed and milk proteins are partly decomposed to peptides and free amino acids, improving digestibility of the milk. If yeasts are included in the starter mixture, alcohol is also present in the final product. Consumption of fermented milk may have many health benefits including alleviation of the symptoms of gastrointestinal disorders. Many types of fermented milk are produced throughout the world, including yoghurt, kefir, dahi, shubat and shrikhand.

Fermented red rice: The yeasty rice grains (brewer's grains) left after making rice wine, coloured red. Alternatively cooked rice fermented with *Monascus purpureus*.

Fermented sausages: Traditionally produced by chance contamination of sausages with local microorganisms. In modern practice, however, starters are usually added to sausage emulsions in order to produce a more uniform product. Lactic starters are often included, but other microorganisms, particularly those with good nitrate reducing abilities, are also used. As well as affecting sausage colour and consistency, fermentation has major effects on flavour. Raw fermented sausages are prepared from un- heated raw meat, which is fermented and then held at a controlled temperature and relative humidity until the desired degree of dryness is obtained. In contrast, heat-treated fermented sausages are pasteurized after fermentation and then dried, usually for a brief period.

Fermenters: Vessels in which aerobic or anaerobic fermentation processes can be carried out, in either batch culture or continuous culture. Typically vertical, closed, cylindrical steel vessels that can range in volume from less than one litre to several thousand litres. Usually have means for ensuring adequate heat transfer, mixing and aeration.

Ferulic acid: An acid which concentrates in the bran fraction of wheat and can be used to measure the bran content of flour.

Fibre Non-digestible, non-starch, complex mixture of carbohydrates of plant origin and resistant to digestion in the intestinal tract. Fibre mean crude fibre is the indigestible part of foods, determined from the residue remaining after extraction under specified conditions which add bulk to the bowel contents causing a more rapid transit through the body, curing constipation and reducing the incidence of bowel cancer. It is available from most whole unprocessed natural vegetables, seeds and fruit and in processed form from the bran by-products of grain milling. Soluble fibre is the part of dietary fibre that is soluble and forms a gel in water. Also refers to long thin thread-like structures, such as muscle, glass, nylon and nerve fibres.

Fibre optics: Application of superfine glass fibres as light conduits in sensors where measurement is based on light transmission. Used in analytical procedures and also in monitoring food processing operations.

Fibreboard: Strong material made from wood or other plant fibres which are compressed, with or without binders, into boards. Used to make containers, such as crates, and panelling. May be corrugated to improve cushioning characteristics. Compared with plastics, fibreboard has both cost advantages and environmental benefits, particularly as it may be made from recycled fibres.

Fibrin: Insoluble animal protein which is produced on hydrolysis of fibrinogen by thrombin. It forms a network of fibres during blood clotting.

Fibrinogen: Soluble animal protein which is secreted into blood plasma by the liver. It converted into fibrin by the action of thrombin during blood clotting.

Fillers: Devices used to transfer products into containers or casings for storage or retail. These are alternatively referred to as filling equipment. Also refers to substances, e.g. buttermilk powders, whey protein concentrates or corn starch, used to ex- tend meat batters or emulsions to add bulk or functional properties.

Filleting: Removal of the bones from a piece of meat or fish, so creating a fillet.

Filling equipment: Devices used to transfer products into containers or casings for storage or retail. Also sometimes called fillers.

Filling: It is process of transferring products, e.g. foods and beverages, into containers or casings for storage or retail. Also refers to the developmental stage in cereals in which kernel dry matter increases.

Fillings: Sweet or savoury ingredients used to fill a cavity within or between layers of a food, e.g. confectionery fillings, pie fillings and stuffings.

Filter aids: Materials such as bentonite, clays, diatomaceous earths and kieselguhr employed to facilitate the course of filtration.

Filter coffee: Coffee made by letting near boiling water flow through ground coffee held in a filter paper supported on a metal or plastic mesh or perforated container

Filters: Porous devices for removing solid particles from a liquid or gas passed through them.

Filth tests: Microscopic examinations of food for the presence of contaminants (e.g. hairs, rodent and bird faeces, insect and feather fragments). Used as an index of hygienic food processing and handling.

Filth: Food contaminants such as hairs, rodent and bird faeces, and insect and feather fragments.

Filtration: Process of removing suspended solids from a liquid by straining it through a porous medium that can be penetrated easily by liquids.

Fining agents: Substances used for clarification of wines or beer, including gelatin, isinglass and diatomaceous earths.

Fining: Clarification of beer or wines by removal of the minute floating particles that prevent these products from being clear.

Finometers: Instruments used to measure the tenderness of foods, especially peas and beans, by determining mechanical resistance.

Fish cakes: Cooked fish products made from fresh or salted fish, mixed with potatoes and seasonings; sometimes eggs, butter and onions are added. Fish content may range from 35 to 50% by weight. A variety of fish are used, including cod, haddock, coal fish and salmon.

Fish crackers: Fish products popular as snack foods in some Asian countries (known as keropok in Malaysia). Commonly made by mixing minced fish flesh with sago flour, tapioca flour, salt and monosodium glutamate. The mixture is then moulded into cylinders, steamed, cooled, sliced and sun-dried.

Fish fingers: Fish products consisting of rectilinear portions cut from a block of frozen fish flesh; typically the length is about three times the breadth and product weight is approximately 18 g. It is fften coated with batters or breadcrumbs and fried in oil.

Fish hydrolysates: Products formed from minced or comminuted fish (often processing wastes) after treatment with hydrolytic enzymes, filtration and drying. Average product consists of 85% hydrolysed protein (mainly small peptides and free amino acids), 10% inorganic material and 5% water. It is mainly used as flavourings in soups and in animal feeds.

Fish meal: Dried, powdered or granular product obtained from cooked whole fish or fish processing wastes; fish species frequently used for its production include anchoveta, capelin, sand eel, herring and mackerel. It constitutes a valuable ingredient of animal feeds. Sold on the basis of its protein content and rated according to the percentage of protein contained in the product.

Fish noodles: Fish products consisting of minced fish flesh mixed with wheat flour (or other cereal flour), cassava starch and various additives; the mixture is extruded through tubular holes to form long strands which are dried in hot air. Products are boiled prior to consumption. A popular meal accompaniment is some parts of South East Asia.

Fish nuggets Fish products comprising pieces of **fish** flesh (not minced) formed into small irregular shapes. May be formed from fillets, fillet pieces or fish blocks; normally occurs in breaded form. Marketed frozen.

Fish oil: Oil/lipids obtained from muscle, livers or other organs of fish, principally from the livers of herring, menhaden, anchovy, sardine and cod. and halibut, which contains essential fatty acids (EFAs) as well as vitamins A and D which are all necessary for health. Often sold in gelatine capsules as a vitamin supplement, but the required amounts can be obtained by eating oily fish once a week. Oily fish can contain up to 30% fat of which about 20% is saturated. Fish oil contains *n*-3 polyunsaturated fatty acids, the principal ones being eicosapentaenoic acid and docosahexaenoic acid, which are reported to protect against heart disease. Used in manufacture of margarines and cooking oils.

Fish paste: Fish preserved either by salting or fermenting or both, sometimes with added grain, and processed to make a paste of variable consistency with a fishy salty flavour. The liquid which drains off is fish sauce.

Fish preserves: Fish products consisting of fish (whole, headed or filleted) preserved in oils, brines, sauces or pickling solutions and stored in cans or sealed glass containers. Fish with high fat content in the flesh (such as mackerel, sardine and tuna) are often used to make preserves.

Fish protein concentrates (FPC): Dried fish products pre- pared from ground whole fish, which contain enhanced protein content (around 80%) and are marketed in powdered and granular forms; used as food ingredients.. Occur in two forms, A and B; type A is odourless and colourless, while type B has an odour and flavour associated with its higher fat content.

Fish proteins: Proteins extracted from fish bodies, often from processing wastes such as fish heads and offal. Some fish proteins form useful food ingredients due to their functional properties. The major components of fish protein concentrates and fish hydrolysates.

Flaking: Process of breaking an item up into small, flat, very thin pieces.

Flame photometry: Spectroscopy in which a solution of the substance to be analyzed is vaporized by introduction into a flame. Spectral lines resulting from a light source going through the vapours are analyzed for characteristic bands, the intensity of which are related to the quantity of analyte present in the sample.

Flaming: A method of food presentation. Warmed spirits, such as brandy or rum, are sprinkled over foods and ignited just before the product is served.

Flavanols: Flavonoids which contain a hydroxyl group. In pure form they exist as yellow, needle-like crystals. They include the hydroxyflavones, kaempferol, chrysin, fisetin and quercitrin. It occurs in many plant foods and wines, and associated with bitterness and astringency. They may be used to reduce the perceived sweetness of foods and beverages. It is responsible for the ivory and yellow colours of many flowers. Dietary sources include fruits, vegetables, red wines, green tea and black tea. It believed to protect against cardiovascular diseases and cancer.

Flavanones: One of the major groups of flavonoids derived from flavone. In pure form they are colourless, crystalline solids. Found in the tissues of higher plants, including fruits, particularly citrus fruits and apples, and vegetables, either in free form or as glucosides. Flavonones occurring in foods include hesperidin, naringenin and eriodictyol.

Flavins: Naturally occurring yellow pigments which have a tricyclic aromatic molecular structure. It is soluble in water. Include riboflavin and its products, e.g. flavin mononucleotide (FMN) and flavin adenine dinucleotide (FAD).

Flavones: Flavonoid pigments which in pure form exist as colourless, crystalline solids and insoluble in water. It occurs in higher plants, including fruits and vegetables, and is responsible for ivory and yellow colours in plants and flowers. Some plants contain high levels of flavones, e.g. parsley (*Petroselinum crispum*) contains high levels of apigenin. Dietary flavones are believed to have various health benefits, e.g. flavones in tea and red wines may protect against cancer and cardiovascular diseases. Flavones are used to derive various yellow dyes.

Flavonoids: Large group of aromatic, oxygen- containing, heterocyclic pigments. Include various subgroups of compounds, such as catechins, flavanols, flavanones, flavones, flavonols, anthocyanins and leucoanthocyanidins. Occur widely in higher plants and are responsible for the majority of yellow, red and blue colours in fruits, vegetables and flowers (with the exception of colours produced by carotenoids). Major dietary sources are apples, onions, red wines and tea. Believed to protect against cancer and cardiovascular diseases. Mechanisms of these inhibitory effects are not fully understood, but they are thought to involve inhibition of low density lipoprotein oxidation.

Flavour compounds: Compounds present in sub- stances that give foods their characteristic flavour; components capable of stimulating the sense of taste.

Flavour enhancer: A substance which having little or no flavour of its own intensifies other flavours with which it is combined. Monosodium glutamate is the most common and is used to enhance the flavour of vegetables and meat, especially in Chinese cooking. Flavour enhancers are common in processed and convenience foods. It is used to enhance the original flavour and/or aroma of a food, without imparting a characteristic taste or aroma of their own. It is similar to flavour modifiers.

Flavour modifiers: Flavourings used to modify the original flavour and/or aroma of a food. Similar to flavour enhancers.

Flavour thresholds: Term used in sensory analysis relating to the levels at which perception of increasing concentrations of flavour compounds begins.

Flavour: It is a sensory propertiy of foods. The tongue can distinguish five separate tastes (sweet, salt, sour, bitter and savoury/umami) due to the stimulation of the taste buds. The blend of taste and smell sensation experienced when food and drink are placed in the mouth. Taste is a direct nervous transmission from the tongue and is the same for all humans comprising salt, sweet, sour and bitter. Smell is far more complex, involves the nose and a large area of the cerebral cortex and is to some extent culturally determined. The overall flavour of foods is a combination of these components, together with astringency in the mouth, texture, and aroma.

Flavourings: Substances whose primary purpose as food additives is to impart flavour and/or aroma when added to foods. They can be of natural origin and add little bulk, such as herbs, spices, seasoning, extracts, essences, etc., or those which also add bulk, such as aromatic vegetables and meats. It includes natural flavourings, such as essential oils and spices, artificial flavourings, seasonings, condiments and extracts. Others, particularly fruit flavours, are purely synthetic in origin. Processed and convenience foods generally contain mixtures of natural and synthetic flavours. It also known as aromatizing agents or flavours.

Flavoxanthin: A natural carotenoid yellow food colouring.

Flavr Savr: A genetically engineered tomato designed to maintain its flavour over an extended shelf life. It contains anti-ripening genes.

Flesh: The name for edible muscular parts of animals, for some other tissues of shellfish and crustaceans and for the edible parts of fruits

Flexible packs: Packs which are capable of bending easily and repeatedly without breaking.

Floc: A small clump or mass of colloidal particles formed in a fluid by the process of flocculation or aggregation of fine suspended particles.

Flocculants: Substances that induce the formation of a mass of colloidal particles (floc) in a dispersion of solids in a liquid.

Flocculation: Formation of a mass of colloidal particles (floc) in a dispersion of solids in a liquid, or alternatively, removal of suspended solids by coalescence.

Florentines: Chewy, thin biscuits, often chocolate- coated on one side, containing nuts and dried fruits or candied fruits.

Flotation: Technique in which different types of solid particles in a liquid are separated out, the principle being that some particles will absorb water while others will not.

Flour improvers: Substances added to milled flour to improve its colour and/or baking properties. These include oxidizing agents to accelerate flour ageing (e.g. potassium bromate, ascorbic acid) and bleaching agents such as chlorine dioxide.

Flour mills: Machines or devices for grinding grain into flour and other cereal products.

Flour: Finely ground cereal grains, dried pulses and dried starchy tubers, consisting mainly of starch, often treated to remove fibre and oils or modified to change its cooking or baking properties. The term is most commonly used of wheat flour. It is used as a basic ingredient for bakery products and many other products. Unless specified otherwise, the term usually refers to the product of wheat grains.

Flow cytometry: Technique for sorting, selecting or counting individual cells in a suspension as they pass individually through a small hole or tube in a flow cytometer. May refer specifically to such a technique which involves the detection of a cell-bound fluorescent or fluorochrome label.

Flow meters: Devices for measuring the flow of a gas or liquid through pipes or other types of equipment.

Flow: Rheological property concerned with the characteristics of movement of a substance. Flow behaviour affects processing properties of a substance and texture of the final product. It is affected by properties such as cohesion and internal friction.

Fluidization: Process in which finely divided solids (e.g. catalysts) are made to behave in the same way as fluids by suspending them in moving gases or liquids. The principle is used in fluidized beds.

Fluidized beds: Novel class of heat transfer media. A fluidized bed is produced by passing a stream of gas or liquid upwards through a bed of particles at sufficient velocity to suspend the particles (fluidization). In this state, the mixture of particles and fluid behave like a liquid having density equal to the bulk density of the particles. Circulation of particles in the bed, particularly by the vigorous mixing action of bubbles rising through the bed, results in large heat transfer rates between the bed and immersed surfaces. In some in- stances, the heat transfer rate may be orders of magnitude greater than achieved using the same fluid flow conditions in the absence of particles. Fluidized beds are used in many chemical engineering processes where small solid particles must be brought into intimate contact with a gas stream. Examples are drying of finely divided solids, adsorption of solvent vapours from air, and heterogeneous catalytic reactions.

Fluorescence microscopy: Microscopy in which samples are illuminated with UV or blue light causing them to emit light of longer wavelengths.

Fluorescence: Absorption of radiation to produce radiation of a longer wavelength, a phenomenon exploited in fluorescence microscopy.

Fluorescent light: Visible light produced by fluorescence, especially that from a discharge tube in which a phosphor on the inside of the tube is made to fluoresce by ultraviolet light from mercury vapour. Fluorescent light can accelerate oxidative deterioration of foods such as oils, nuts and milk during storage.

Fluoridation: Addition of traces of fluorides, particularly to water supplies and toothpastes as a means of preventing tooth decay. Fluoridation of water supplies is a controversial issue due to possible health hazards associated with long-term ingestion of high levels of fluorides.

Fluorides: Salts which contain fluorine. Often added to toothpastes or drinking water in order to reduce the incidence of dental caries. Fluoridation of water supplies is a controversial issue due to possible health hazards associated with long-term ingestion of high levels of fluorides.

Fluorometry: Technique used to identify a substance from the wavelength of the light that it emits during fluorescence. Also called fluorimetry.

Flushing: Cleansing by passing large quantities of water through an object.

Foamed plastics: Lightweight plastics made by solidifying plastic foams. Plastic foams are produced from liquid plastics, and contain many small bubbles.

Foaming agents: Substances that promote foaming. Foaming capacity Functional properties relating to the extent to which an item is able to form foams.

Foaming properties: Functional properties relating to the ability of food components to be formed into foams.

Foaming: Formation of a mass of small bubbles on or in a liquid (foams).

Foams: Light textured colloidal dispersions of a gas, such as air, in a liquid or solid, typically achieved by whipping or frothing. Foams are unstable, requiring the presence of stabilizers to form the gas bubble membranes. Egg whites have good foaming capacity and are used to produce foamed foods such as meringues and souffles. Gelatin and modified milk proteins are also widely used to produce foams, e.g. in manufacture of foamed confectionery.

Foils: Thin, flexible metallic sheets or strips. Commonly, thickness is specified as being less than a given amount. Used widely as packaging materials, e.g. aluminium foils.

Folic acid: Water soluble member of the vitamin B group. In its active form, tetrahydrofolate, it is a coenzyme in various reactions involved in the metabolism of amino acids, purines and pyrimidines. It is synthesized by intestinal bacteria and widespread in food, especially green leafy vegetables. Deficiency causes poor growth and nutritional anaemia. Daily intake should be increased prior to conception and during the first three months of pregnancy to prevent neural tube defects (e.g. spina bifida) and other congenital malformations (e.g. cleft lip and cleft pal- ate) in the foetus.

Fondant: Low moisture sugar syrups which are made by boiling concentrated sugar solution, adding glucose syrups or inverting agents, and cooling rapidly while mixing, to produce fine sugar crystals in a saturated sugar solution. It is boiled to 116 to 118°C, i.e. between the soft and hard ball stage, used when cooled as a basis for cake icing. It must be diluted with water and used at a precise temperature. If mixed with glycerine and gums it remains temporarily malleable at room temperature and may be rolled out for covering cakes and used to mould decorations. It is used to make fondant sweets, as fillings in chocolates and biscuits, and as toppings for cakes.

Food and Agriculture Organization: The Food and Agriculture Organization (FAO) was founded in October 1945 with a mandate to raise levels of nutrition and standards of living, to improve agricultural productivity, and to better the conditions of rural populations. Today, FAO is the largest autonomous agency within the United Nations (UN) system, with 180 Member Nations plus the EC (Member Organization) and more than 4300 staff members around the world. FAO works to alleviate poverty and hunger by promoting agricultural development, improved nutrition and the pursuit of food security. The Organization is active in land and water development, plant and animal production, forestry, fisheries, economic and social policy, investment, nutrition, food standards, and commodities and trade; it also plays a major role in dealing with food and agricultural emergencies. FAO aims to meet the needs of both present and future generations through programmes that do not degrade the environment and are technically appropriate, economically viable and socially acceptable. FAO offers direct development assistance, collects, analyses and disseminates information, provides policy and planning advice to governments, and acts as an international forum for debate on food and agriculture issues.

Food and Drug Administration: US agency within the Department of Health & Human Services which was formed in 1927 by division of the Bureau of Chemistry (established 1862) into the Food, Drug and Insecticide Administration (name shortened in 1930) and the Bureau of Chemistry and Soils. The name is commonly abbreviated to the FDA. It is a scientific, regulatory and public health agency including under its jurisdiction most foods, animal and human drugs, therapeutic agents of biological origin, medical de- vices, radiation emitting devices, cosmetics and animal feeds. The FDA evaluates applications for new drugs, foods, food additives, infant formulas and medical devices, as well as monitoring manufacture, import, transport, storage and sale of these products. Its mission is to promote public health by helping safe and effective products to reach the market in a timely manner, and monitoring products for continued safety once in use. With respect to foods, the agency aims to ensure that they are safe, wholesome, sanitary and properly labelled.

Food bars: Hand-held snack foods, usually in the shape of a rectangular block, e.g. cereal bars, chocolate bars, ice cream bars and meal replacement bars.

Food chopper: A food processor in which the bowl revolves in the opposite direction to the rotating cutting blades, used for high volume work in commercial kitchens

Food colourings: Permitted dyes either of natural or synthetic origin used to colour food especially by manufacturers of processed food.

Food emulsifiers: Emulsifiers used specifically in foods.

Food emulsions: Colloidal suspensions in which a substance is dispersed in another, e.g. oil in water emulsions. Emulsions can be formed from immiscible water and oil phases with the aid of emulsifiers; stabilizers are used to maintain structure. Examples of food emulsions include milk, cream, margarines and mayonnaise.

Food enrichment: Addition of nutrients, e.g. vitamins and minerals, to processed foods, such as cereal products, to correct for losses occurring during processes such as milling. The term fortification is used to indicate addition of nutrients to foods to overcome dietary deficiencies in specific population groups.

Food factories effluents: Liquid wastes (waste water) often discharged into a river or the sea from food factories.

Food factories wastes: Solid wastes generated in food factories during food processing operations.

Food flavourings: Flavourings used specifically in foods.

Food handlers: Personnel involved in the preparation, processing or handling of foods.

Food hygiene certificate: A certificate of competence in food hygiene issued after appropriate training.

Food hygiene: That science and/or craft which determines how food should be grown, picked or harvested, slaughtered if appropriate, transported, stored, handled, cooked and served to avoid contamination or deterioration which could be harmful to health, and also the design and cleaning of premises, vehicles, implements and equipment with which food is associated

Food intolerance: Group of diseases in which there is inability to digest a particular food or food constituent properly, often resulting in malabsorption syndromes. Examples include lactose intolerance, resulting from lack of a gastrointestinal tract brush border enzyme, and coeliac disease, in which an immunological response to wheat gluten results in histopathological changes to the intestinal mucosa. Exclusion of the appropriate food from the diet can result in elimination of the symptoms of the disease, and also, in cases such as coeliac disease, reversal of intestinal pathology.

Food labeling: Under EU rules, packaged food must be labelled with the name, the list of ingredients in descending order of weight, the net quantity, a best-before or best-before-end date, special conditions of storage or use, the name and address of the packager or seller and, under certain conditions, the particulars of the place of origin. The location of the factory or packaging centre is required in some countries.

Food laws: The legislation which lays down requirements for food hygiene, for the purity of food, for the description of food and for the weights and measures associated with it.

Food mixer: An electrically operated machine that processes solids and liquids by means of various attachments to a powered head, which operate within or over a bowl. It can mix, beat, knead, whip, sieve, chop, grind, mince, liquidize etc.

Food poisoning bacteria: The main bacteria responsible for food poisoning are, in alphabetical order: *Bacillus cereus, Campylobacter, Clostridium botulinum, Clostridium perfringens, Escherichia coli, Salmonella, Staphylococcus aureus, Vibrio parahaemolyticus* and *Yersinia enterocolitica.* See under each heading.

Food poisoning causative agents: Agents that can cause food poisoning are are: substances added during preparation, e.g. Lead; substances synthesized by the commodity during growth and storage e.g. Poisons in potatoes and red beans; substances formed by processing or storage under poorly controlled conditions e.g. Some kinds of fish poisoning; substances stored by a commodity e.g. Paralytic shellfish poisoning caused by a poison in some of the plankton or algae that they consume; allergens, e.g. Strawberries; bacteria; and unidentified agents, probably viruses. It is also known as **food poisoning bacteria**

Food poisoning: Any disease of an infectious or toxic nature caused by or thought to be caused by consumption of food or water regardless of presenting symptoms and signs (i.e. Not necessarily gastrointestinal) and including illness caused by toxic chemicals associated with food but excluding illness due to known allergies or food intolerances. It may range in severity from mild to life-threatening.

Food policy: A broad term used to encompass those programmes, usually governmental, that most directly affect the food chain. Issues encompassed by this term include the role of food in international trade, agricultural pricing policies, food security, food aid, nutrition planning and food control.

Food powders: Alternative term for dried foods or powders made for use as foods.

Food preservation science: The study of the means used to destroy or inhibit the growth of those microorganisms with the potential to cause deterioration in food. The means involve intrinsic properties of the food, extrinsic properties of the environment and physiological properties of the microorganisms which enable them to flourish because of the interaction of the intrinsic and extrinsic properties.

Food preservatives: Preservatives used specifically in foods.

Food processor: A kitchen appliance consisting of a cylindrical bowl with cover through which projects a vertical shaft onto which various circular cutting, slicing and beating implements may be attached. The shaft is rotated by an electric motor at varying speeds but generally lower than a blender Used for chopping, slicing, mixing, puréeing and pulverizing tasks for all kinds of ingredients and mixes.

Food reference materials: Reference samples comprising food materials of certified composition (e.g. bovine liver and skim milk powder) that are used as standards in analytical procedures.

Food safety: Encompasses activities and policies which are essential for ensuring that food will not cause injury or illness upon consumption.

Food science: Study of the characteristics of foods, including chemical properties, biochemical properties, physical properties, physicochemical properties and biological properties, and effects of these on the quality of products. Also covers application of this information to development of new products and efficient food processing techniques.

Food security: Access (both physical and economic) by all people at all times to sufficient food for an active, healthy life.

Food store: A room or cabinet set aside for storing foods, generally with good ventilation, vermin proof and with easy clean slotted or mesh shelving.

Food supplements: Substances consumed as extra sources of specific nutrients in the diet with a view to improving nutritional status or health. It may be added to foods during processing, e.g. calcium or iron fortification of cereal products, as a form of food enrichment. Alternatively, taken separately from foods in the form of tablets, capsules, liquids or oils, depending upon the nature of the dietary constituent. Such preparations may be single or multi-component. Dietary components commonly consumed in this form include vitamins, minerals, proteins, oils (e.g. fish oils and evening primrose oils), phytochemicals and plant extracts, such as those from garlic.

Food technology: Application of a diversity of scientific and practical disciplines, including chemistry, biology, physics and engineering, to the development of food products and to their worldwide distribution.

Food thickeners: Thickeners used specifically in foods.

Foodborne diseases: Diseases whose causative agents are transmitted through food.

Foods: Substances which can be ingested by living organisms. Usually solid and contain a range of nutrients which can be metabolized to produce energy, and sustain life and growth.

Fortification: Increasing the nutritional quality of a food by addition of nutrients such as vitamins and minerals.

Fortified wines: Wines to which ethanol has been added, as spirits or neutral alcohol. Important types include sherry and port.

Fouling: Accumulation of unwanted materials on the surfaces of processing equipment. The fouling layer has a low thermal conductivity, so increasing resistance to heat transfer and reducing the effectiveness of heat exchangers. Chemical reaction fouling involves deposits that are formed as the result of chemical reactions at the heat transfer surface. This kind of fouling is a common problem in chemical process industries, oil refineries and dairy plants. Biological fouling is the development and deposition of organic films consisting of microorganisms and their products. Effective fouling control methods should involve: prevention of foulant formation; prevention of foulants from adhering to themselves and to heat transfer surfaces; and removal of deposits from the surfaces.

Fourier transform IR spectroscopy: Type of IR spectroscopy that utilizes the Fourier transform mathematical technique, in which samples are irradiated with polychromatic radiation and the entire range of frequencies is recorded at the same time, giving an interferogram. Fourier transformation then sorts the interferogram into its components, which can be represented as a traditional spectrum. Advantages over conventional IR spectroscopy include increased speed and sensitivity. Usually abbreviated to FTIR spectroscopy.

Fractionation: Separation of the components of a mixture into fractions, using techniques such as **gel filtration** and **electrophoresis**. This term also relates to precipitation and phase-separation methods used to determine the molecular weight distribution of polymers; these techniques are based on the tendency of polymers of high molecular weight to be less soluble than those of low molecular weight.

Fracture properties: Mechanical properties governing the way in which and conditions under which a structure will break down when an external force is applied.

Franchising: Authorization granted by a government or company to an individual or group enabling them to carry out specified commercial activities, for example to market a company's goods or services, in a designated territory. In return for a specified fee and usually a share of the profits, the franchiser provides the product, the name, and sometimes the plant and advertising.

Frankfurters: Mild flavoured, smoked, cooked sausages originally produced in Frankfurt, Germany. Varieties include hot dogs and wieners. They can be made from beef, chicken meat, pork, turkey meat or veal; typically, they are prepared from a blend of 40% pork and 60% beef. Frankfurter seasonings include coriander, garlic, mustard, nutmeg, salt, sugar and white pepper. They tend to have high contents of fat and salt. Some are retailed in natural sausage casings, but most are prepared in cellulose casings, which are later removed. Most commonly, frankfurters are about 15 cm long, but they are produced in a wide range of sizes. When tradition- ally made, frankfurters are smoked over hardwood, in order to improve colour and flavour; however, now smoke flavourings are mostly applied as a paint. Despite being precooked, frankfurters taste better after reheating; usually, they are boiled, fried, grilled or steamed immediately before serving.

Free radicals: Highly reactive molecular entities, containing one or more unpaired electrons that are usually short-lived and capable of initiating or mediating a wide variety of chemical reactions. It is often formed by the splitting of a molecular bond.

Freeze concentration: Concentration of a liquid by freezing out pure ice, leaving a more concentrated solution. This process requires less input of energy and causes less loss of flavour than concentration by evaporation; it is used primarily in the concentration of fruit juices, vinegar and beer. Limitations of freeze concentration are its high cost, the difficulty in separation of ice from solid, and the degree of concentration that can be achieved.

Freeze dried foods: Foods dehydrated by freeze drying and used to make various types of products, including instant soups, dried herbs, instant coffee granules and meat products. Dried foods obtained in this manner are light, porous, easy to rehydrate and tend to have better shape and colour retention than foods obtained by other drying processes.

Freeze driers: Apparatus for preservation of foods by applying rapid freezing followed by a high vacuum which removes ice by sublimation (freeze drying).

Freeze drying: Preservation of foods by rapid freezing followed by subjection to a high vacuum, which removes ice by sublimation. Adequate control of the processing conditions contributes to satisfactory subsequent rehydration, with substantial retention of nutrients, and colour, flavour and texture characteristics. Freeze-dried vegetables are used in instant soups and the technique is most widely used for instant coffee.

Freezer burn: The discoloration which occurs on the surface of food where it comes into direct contact with a cold surface because of inadequate packaging.

Freezers: Refrigerated cabinets or rooms for preserving frozen foods at very low temperatures. Foods are usually frozen to an internal temperature of -18°C in freezers; the food must be maintained at this temperature or slightly lower during transport and storage. Commercial freezers include the following types: blast freezers, where air is circulated at -40°C; contact freezers, in which refrigerants are circulated through hollow shelves; immersion freezers, where, for example, fruit is frozen in a solution of sugar and glycerol; and cryogenic freezers, which use, for example, liquid nitrogen spray.

Freezing point: The freezing point is the temperature at which the vapor pressure of the liquid is equal to that of the solid crystal. While freezing uses temperatures well below the freezing point, conventionally below - 18°C. The stronger preserving action of freezing is due not only to the lower temperature but also and mainly to the depression of water activity as a result of conversion of part of the water to ice. Freezing point measurement can be used to detect the **adulteration** of **milk** with water, since the value increases when water has been added.

Freezing: Method of preservation in which micro-organisms are prevented from multiplying by application of freezing temperatures. Foods are usually frozen to an internal temperature of -18°C in freezers; the food must be maintained at this temperature or slightly lower during transport and storage. During freezing, a proportion of the water in the food changes from liquid to solid to form ice crystals, so lowering its water activity. Because the process does not kill all types of bacteria, those that survive reanimate in thawing food and often grow more rapidly than before freezing. Enzymes in the frozen state remain active, although at a reduced rate. Freezing does, however, cause the water in foods to expand, and tends to disrupt the cell structure by forming ice crystals. With quick- freezing, however, the ice crystals are smaller, producing less cell damage than with slowly frozen products. Freezing has been a key technology in bringing convenience foods to homes and restaurants; it causes minimal changes in quality of food in terms of size, shape, texture, colour, flavour and microbial load.

French fries: Potato products made by cutting potatoes into thick or thin strips, soaking in cold water, drying and deep frying in oil. Also called chips (UK), pommes frites (France), fries or French-fried potatoes.

Freons: Series of nonflammable, non-explosive fluorocarbons (FC) or chlorofluorocarbons (CFC) widely used as refrigerants.

Freshness: Extent to which a product is fresh and of good eating quality.

Friabilimeters: Devices used in assessment of malt quality on the basis of friability, a measure of the breakdown of malt endosperm cell wall components. Malt samples are subjected to an abrasive action for separation of hard and ripe constituents which are then weighed. The presence of hard and glassy components, which will cause problems with brewing, is detected in this way. It may be used in quality control of the malting process or to determine malt quality in breweries.

Friabilins: Water-soluble proteins which control wheat kernel hardness and are located on the surface of wheat starch granules.

Fried egg: A shelled egg shallow-fried in hot fat. The white on the top of the egg is coloured either by basting with hot fat or by turning the egg over when almost finished. Tastes differ as to the amount of cooking required.

Fried foods Foods fried in fats or oils, e.g. French fries, fritters and doughnuts. It is often coated in batters or breadings prior to frying.

Fried rice: Cooked rice stir-fried in oil with chopped vegetables usually containing spring onions, and beaten egg which is dribbled into the mix as it is being stirred and fried.

Fritters: Pieces of food (e.g. fruits, meat, fish) that have been dipped in batters and deep fried. bubbles in a liquid by agitation or fermentation.

Frozen beverages: Beverages, generally soft drinks or fruit beverages, which have been frozen. May be served and consumed in a soft-frozen (slush) state.

Frozen desserts: Desserts is preserved by freezing and requiring frozen storage. These are often premium quality products, such as ice cream products, gateaux and cheesecakes. Some require cooking before consumption, but others can be eaten immediately after thawing or while still frozen. Unlike many other frozen foods, texture is not usually compromised by freezing.

Frozen dough: Dough prepared at a lower temperature than conventional dough (in order to minimize fermentation activity of yeasts), followed by immediate freezing to extend its shelf life. Used for production of bakery products in in-store bakeries.

Frozen food: Particular foods, made-up or partially made-up dishes which have been frozen to –20°C and which can be stored in frozen condition for a least 6 weeks and generally much longer. It is generally stated that once defrosted, frozen foods cannot be refrozen but there is no justification for this rule providing the period at greater than 0°C is less than an hour. Frozen foods have generally been prepared for use prior to freezing. It is usually of higher quality than canned foods or dried foods, with any losses in quality being due to texture deterioration. A wide variety of foods can be frozen, either cooked (ready meals and some desserts) or uncooked (vegetables, fish fillets and poultry). Some products are thawed before use, while others can be cooked/ reheated directly from the freezer.

Frozen meals: Frozen foods in the form of complete dishes. Usually it reheated directly from frozen form prior to consumption. Common types include pizzas, ready meals, entrees, vegetarian foods and savoury pies.

Frozen storage: Storage of foods at freezing temperatures (below 0°C).

Frozen yoghurt: Fermented low-fat dairy desserts served in a similar manner to ice cream.

Fructans: Group of oligosaccharides and polysaccharides which consist of fructose residues attached to a single glucose molecule. Depending on the source, chain lengths can range from 3 to 50 residues. In cereals, shorter fructans predominate, while Jerusalem artichokes contain high levels of inulin, a fructan of about 35 residues. Onions, garlic and asparagus are other dietary sources of fructans. In the stomach and small intestine, hydrolysis of fructans is negligible; any trisaccharides which are absorbed directly are usually excreted in urine. The majority of dietary fructans reach the large intestine where fermentation occurs.

Fructose high corn syrups: Syrups containing 40- 90% fructose produced from glucose syrups which have been manufactured by hydrolysis of corn starch. Glucose can be converted to fructose by the action of glucose isomerases to produce syrups composed predominantly of glucose and fructose. Higher purity fructose syrups are produced using gel filtration chromatography to separate fructose from glucose and other sugars present. Applications for these syrups include soft drinks, marmalades, jams, canned fruits, fruit juices, dairy products and bakery products.

Fructose syrups: Aqueous solutions, containing predominantly fructose, which are used as sweeteners.

Fructose: Monosaccharide ketose sugar comprising six carbon atoms. It constituent of **sucrose** which occurs naturally in **fruits** and **honeys**. Commercially produced from **glucose** by **isomerization**, a reaction catalysed by glucose isomerases. It may be crystallized from **fructose syrups** by addition of an organic solvent, such as **ethanol**. Fructose is the sweetest natural saccharide and is approximately 1-1.5 times as sweet as sucrose. It is also known as laevulose and fruit sugar. Also called fruit sugar, laevulose, levulose. One of the sugars found in corn syrup.

Fruit beverages: Beverages derived from fruit juices, fruit extracts or fruit homogenates.

Fruit bread: Bread made by adding up to 50% (flour basis) raisins to the dough mixture. It may also contain other dried fruits such as currants, dates or bananas.

Fruit cheese: A solid preserve made by boiling a purée of fruit with sugar to form a mixture which when cooled and set can be sliced like cheese

Fruit cocktail: A mixture of various diced, possibly parboiled fruits, with cherries and other whole small fruit, in a sugar syrup. Used in desserts, pies, tarts, etc.

Fruit compotes: Desserts made from fruits stewed in sugar or cooked in syrups. It is eaten hot or cold.

Fruit cordials: Term referring to fruit juice beverages, often presented as concentrates for dilution, or to sweet fruit-based liqueurs.

Fruit curd: Mixtures of sugar and butter thickened with egg yolks over a double boiler and flavoured with various fruits. The most common is lemon curd.

Fruit desserts: Desserts based on fruits. Include fruit salads, fruit cocktails, fruit compotes, mousses, flans and sorbets.

Fruit extracts: Preparations obtained from fruits by a variety of means that can be used as flavourings in foods and beverages.

Fruit fritters: Largish slices or pieces of raw fruit, coated in fritter batter, deep-fried until golden brown, drained, dusted with icing sugar and eaten hot

Fruit gums: Sugar confectionery products made with sucrose, glucose, fruit flavourings and gum arabic either alone (to produce hard gums) or mixed with gelatin (to produce soft gums).

Fruit jellies: Semi-solid foods with an elastic consistency, made either by setting of fruit juices containing pectins or gelatin, or by addition of gelatin to fruit juices.

Fruit juice beverages: Beverages containing fruit juices, together with other ingredients such as water, sugar or flavourings.

Fruit juice concentrates: Fruit juices which have been concentrated by evaporation, membrane processes or freezing. May be diluted to make reconstituted juices, or used as ingredients in a wide range of foods and beverages.

Fruit juices: Juices extracted from fruits consumed as drinks or used as ingredients in a wide range of foods and beverages.

Fruit leathers: Products made from fruit purees, sometimes sweetened with sugar or honey that are spread in a thin layer on a baking sheet and dried in a slow oven. The dried sheets may be cut into strips or rolled into cylinders.

Fruit nectars: Beverages manufactured from fruit juices by addition of water and/or sugar, optionally with addition of other ingredients.

Fruit peel: Rind or skin of fruits i.e. rich in fibre. It may be removed before consumption of the fruits or eaten at the same time. Some types are removed and used in garnishes or as ingredients of various dishes. Peel most commonly used in cooking is that from citrus fruits.

Fruit pies: Dishes, usually served as desserts, having one or more crusts and fruit-based fillings. Crusts, generally made from pastry, can be on the bottom or top of the dish only, or on both the bottom and top.

Fruit preserves: Prepared by cooking pieces of fruits with sugar and sometimes pectins. Similar to jams, except that the fruit pieces tend to be larger in preserves.

Fruit pulps: The soft, succulent part of fruits or a preparation made from them by mashing and concentration. Used in the manufacture of a range of foods and beverages, including syrups, milkshakes, fruit juice beverages and ice cream.

Fruit purees: Fruit flesh that is mashed to a smooth, thick consistency by various means, such as forcing through sieves or blending in food processors. Used as garnishes and side dishes or as the base of many types of product, including beverages, parfaits, ice cream, mousses and souffles.

Fruit syrup: Fruit flavoured sugar syrup made from fruit juices and water, used for flavouring and as a drink base.

Fruit syrups: Syrups produced by concentration of fruit juices. Used as flavourings and sweeteners.

Fruit yoghurt: Yoghurt containing pieces of fruit, fruit pulps or fruit purees, either as a separate layer or stirred in to give a homogeneous product.

Fruit-flavoured yoghurt: Yoghurt mixed with fruit juice, fruit flavouring or essences, usually sweetened, often coloured and thickened with starch

Fruitiness: Extent to which a product has the aroma or flavour of fruit.

Frying fats: Lipids which are usually solid at room temperature and used as a medium in which to cook foods by frying. Heating of the fat results in it acting as a thermal transfer agent, with some of it remaining in the fried foods. Repeated use of the fat for frying may result in its degradation by means of autoxidation, cyclization or polymerization.

Frying oils: Lipids which are usually liquid at room temperature and used as a medium in which to cook foods by frying. Heating of the oil results in it acting as a thermal transfer agent, with some of it remaining in the fried foods. During frying, the heated oil may undergo several degradative changes.

Frying properties: Ability of foods to maintain or develop acceptable properties upon application of frying procedures.

Frying: Cooking of foods in hot fats or oils over a moderate to high heat. In deep frying, the foods to be cooked are immersed in the fats or oils.

Fudge: A confection of butter, sugar and milk which forms a soft sweet pasty mixture, used as a filling for tarts or as a sweet or candy on its own or enrobed in chocolate

Fufu: Unfermented or fermented product usually made from cassava, but also from other tubers and corms, such as taro, yams or cocoyams. The unfermented form is prepared by boiling or steaming and pounding the vegetables, either individually or in combination. Fermented fufu is prepared from roots which have been soaked for 3-4 days before being formed into pastes. In some areas, fufu is sold as a convenience food. Usually served as an accompaniment to dishes with sauce, such as stews.

Fumigants: Gaseous pesticides used for fumigation.

Fumigation: Use of gaseous pesticides to rid an area of insect pests.

Functional foods: Term originally introduced in Japan to mean foods with a physiological function or activity. Foods containing additional components which provide specific medical or physiological benefits over and above those provided by the naturally occurring nutrients, including the more traditional supplemented foods such as bread, breakfast cereals etc. As well as those with a more ostensibly medicinal purpose, e.g. Benecol, Yakult, etc.. fucntional foods used for products containing biologically active components (such as nutrients, bioactive peptides or phytochemicals) at levels that may confer specific health benefits. Examples include bifidus yoghurt, eggs incorporating ω-3 fatty acids and fibre-enriched breakfast cereals. Also known as nutraceutical foods, designer foods, medical foods and probiotic foods.

Functional properties: Characteristics of a substance that affect its behaviour and that of products to which it is added. Influence potential applications of a sub- stance in the food industry, as a particular functional property may be especially useful for the manufacture and stability of specific types of foods. Include a wide range of characteristics, such as buffering capacity, emulsification properties, foaming properties, gelling capacity, water binding capacity and whipping properties.

Fungal proteins: Fungal mycelia which are used as foods or food ingredients, e.g. Quorn (produced by the continuous fermentation of *Fusarium graminearum*). Also known as mycoprotein.

Fungal spores: Spores produced by fungi, e.g. ascospores, basidiospores, chlamydospores, sporangio- spores and zygospores.

Fungi: Eukaryotic microorganisms of the kingdom Fungi, that possess cell walls and lack chlorophyll. Some species are pathogens of humans, animals and plants. Certain fungi are used commercially (e.g. in the production of enzymes and fermented foods). Species such as *Penicillium* and *Aspergillus* are important agents of food spoilage, while other spe- cies (e.g. *Penicillium camemberti* and *P. roqueforti*) are desirable and essential in the ripening of certain types of cheese.

Fungicides: Chemical substances used to kill or inhibit the growth of fungi that cause diseases of plants and animals. Most are applied as sprays or dusts and either have a systemic or protectant effect. Examples of fungicides commonly applied to crops include benomyl, captan, dichlofluanid, maneb and zineb. Residues in foods and the environment can represent a health hazard.

Furcellaran: Gums produced from the red alga *Furcelleria fastigiata*, also known as Danish agar. It is used as thickeners and gelling agents in flans, jellies and dairy desserts. Some similarities to carrageenans.

Furfural: Viscous, colourless volatile liquid aldehyde, which has a distinct aroma and is unstable in air, exposure to which results in polymerization to a reddish brown colour. Composed of a furan ring and aldehyde side chain, it is derived from the thermal breakdown of pentoses from cornstalks, corn cobs and bran distillation. Often used as a solvent. Alternative terms include fural and furaldehyde.

Furosine: Amino sugar generated during acid hydrolysis of fructosyl-lysine. It may be a useful indicator of the extent of damage that occurs during the early stages of the Maillard reaction.

Fusarin C: Mycotoxin produced by *Fusarium* species during growth on foods. It is strongly mutagenic and possibly carcinogenic in humans.

Fusarium: Genus of fungi which occur in soil and decaying organic matter. Some species may cause plant diseases, and the spoilage of stored fruits and vegetables. May also cause diseases in humans and animals through the production of mycotoxins on foods and feeds.

Fusel oils: Colourless viscous liquids with an unpleasant aroma and flavour. It composed of a mixture of amyl alcohol with higher alcohols and traces of other components. Present in distilled spirits as by-products of alcoholic fermentation. It is more toxic than ethanol.

Fusion proteins: Proteins containing amino acids sequences from two distinct proteins, formed by expression of a recombinant gene in which two coding sequences have been joined together in-frame. Fusion of proteins with affinity tags can be used to facilitate purification, while fusion with signal peptides can be used to facilitate secretion of proteins from cells.

Fuzzy control: Control processes based on the theory of fuzzy logic, an artificial intelligence concept used in expert systems for estimating the degree of certainty of conclusions.

Fuzzy logic process control: A form of logic used in process control in which statements can be given fractional values rather than simply true or false.

G

Galactose: Monosaccharide with six carbon atoms which occurs naturally as a component of many complex plant-derived polysaccharides, such as pectins and gums. It constituent of lactose from which it may be produced by hydrolysis. It has approximately 40% the sweetness of sucrose and is used in sweeteners. It is this monosaccharide liberated in the body which is responsible for lactose intolerance. One of the common monosaccharides which in combination with glucose form lactose, the sugar in milk.

Galactosides: Glycosides formed from mixing galactose with an alcohol; on hydrolysis, galactose is produced.

Galgals: Type of lemons produced by *Citrus pseu- dolimon*, which are indigenous to and cultivated on a commercial scale in India. Used in manufacture of pickles and as a source of fruit juices, peel, pectins and essential oils.

Gallates: Esters of gallic acid used as antioxidants in oils, fats and essential oils only. Those available are propyl gallate, E310, octyl gallate, E311 and dodecyl gallate, E312.

Gallic acid: Organic acid with antioxidativeactivity which occurs naturally as a component of tannins., e.g. in tea. Gallic acid esters, such as octyl gallate and propyl gallate, are used as antioxidants in the food industry.

Gamma irradiation: Exposure of foods to gamma rays, generated by radioactive decay of cobalt-60 (^{60}Co) or caesium-137 (^{137}Cs). It is used for sterilization or preservation purposes. Irradiation delays ripening of fruits and vegetables, inhibits sprouting in bulbs and tubers, causes disinfestation of grain, cereal products, fresh and dried fruits and vegetables, and destroys bacteria in fresh meat. Continued consumer concern over the safety of irradiation and irradiated foods has limited its full-scale use.

Gamma rays: Penetrating electromagnetic radiation of shorter wavelength than X-rays. For food irradiation, sources used for generation of gamma rays include cobalt-60 (^{60}Co) and caesium-137 (^{137}Cs).

***Ganoderma*:** Edible fungi used in health foods and medicines, especially in China and Japan. Most common example is *Ganoderma lucidum*.

Garam masala: A blend of ground spices with many variants meant either to be sprinkled on food just before it is served, to be added towards the end of cooking, or to be used to form an aromatic crust on foods which are bland and simple. Typically it will contain black pepper, black cumin, cinnamon, cloves, mace, cardamom, coriander seed and bay leaf, all dry-fried or roasted then dried and ground. Blends include: Moghul masala, Gujarati masala, Kashmiri masala, parsi dhansak masala, chat masala, char masala and green masala.

Gari: Meal produced by roasting and drying fermented cassava mash. It is major food source in West Africa. Protein content is low. It may contain potentially toxic levels of residual cyanogens, depending on the processing techniques used.

Garlic bread: A long thin French loaf cut almost through transversely at 2 to 4 cm intervals, a pat of garlic butter put in each cut and the whole, wrapped loosely in aluminium foil and baked in the oven until the butter has melted, and the outside become crisp (180°C for 15 minutes.)

Garlic butter: A compound butter made by processing butter, garlic, salt and chopped parsley to a smooth paste.

Garlic oils: Highly pungent essential oils obtained from garlic. Used in spice mixes and other flavourings. Major constituent is allyl sulfide.

Garlic: Pungent, edible bulbs of *Allium sativum*. One of the world's most widely used spices, used to flavour many different dishes. Each bulb comprises a number of cloves, which release a characteristic aroma when peeled and crushed. This aroma is due to the presence of allicin, which is believed to play a key role in the beneficial health effects reported for garlic. As well as being used fresh, much of the crop is further processed to yield garlic powder, garlic salt or garlic oils.

Garnishes: Decorative and edible accompaniments to sweet or savoury dishes, usually added just before serving. May be placed on the plate beside the dish or applied to the surface of the food. Vary greatly in size and content, including sprigs of parsley or other herbs, salad vegetables, croutons, slices of fruit, whole fruits and chocolate shapes. Garnishes often indicate the main ingredient or flavour of a dish.

Gas chromatography (GC): Chromatography technique, usually abbreviated to GC, in which the sample is vaporized and injected into a carrier gas (mobile phase) that moves through a column, the inner surface of which is coated with a stationary phase. Sample components are separated on the basis of their affinity for the stationary phase, and identified by the time they are retained by the stationary phase. A range of detection techniques can be used in combination with gas chromatography, including mass spectroscopy (GC MS).

Gas liquid chromatography (GLC): Chromatography technique in which the mobile phase is a gas and the stationary phase is a liquid adsorbed on a porous solid in a tube or on the inner surface of a capillary column. Components of the sam- ple are partitioned between the gas and liquid phases, the rate at which they are eluted from the column de- pending on their partition coefficients. They are identi- fied by the time taken to reach the detector for the system.

Gastrointestinal tract: The organ commencing at the mouth and finishing at the anus, including the stomach and intestines, into which foods are taken and digested, and from which nutrients and non-nutrients are absorbed into the body, and waste is excreted.

Gel /Jelly: Technically any liquid which is in a solid state by virtue of the network of strands of protein or carbohydrate which gives it a solid structure.

Gel electrophoresis: Electrophoresis technique in which separation is performed in a gel, usually com- prising agarose or polyacrylamide.

Gel permeation chromatography: Chromatography technique in which separation is based on the size and shape of molecules. Samples are applied to a column of gel, e.g. polyacrylamides, cross-linked dex- trans, large polysaccharides, and components are sepa- rated on the basis of their ability to penetrate the pores of the gel beads while being washed through with a liquid mobile phase. The technique may be used to estimate relative molecular mass (M_r). It is also known as gel filtration chromatography.

Gelatin: Soluble protein extracted from animal collagen, bones or connective tissues using hot water and acid or alkaline treatment. Widely used in the food industry in gelling agents, e.g. in aspic, jellies, ice cream, yoghurt and canned meat, and can also act as an emulsifier or stabilizer, e.g. in marshmal- lows and confectionery fillings. Lacks the essen- tial amino acid tryptophan, but is a source of several amino acids lacking in plant foods such as wheat, oats and barley. It is alternatively spelt gelatine.

Gelatinization: Process involving disruption of molecular order within starch granules as a result of heating in water. Occurs over a temperature range and is also affected by granule size. Alterations caused include irreversible swelling, loss of birefringence, leaching of amylose and reduced crystallinity. Prolonged heating of the starch granules will eventually lead to total disruption.

Gelation: Process of gels formation by coagulation of sols or aggregation of particles. It formed in a variety of ways according to the type of material con- cerned. In the case of polymer molecules, gelation is caused by formation of intermolecular crosslinks dur- ing heating or cooling. Aggregation of particles may be induced by a variety of stimuli including changes in pH or ionic strength. It also called gelling.

Gelling agents: Additives used to promote gelation. It used in manufacture of jellies and other food gels. Commonly it is used gelling agents include pectins, agar, guar gums and gellan gums.

Gelling capacity: One of the functional properties of a substance concerned with its ability to form a gel.

Gels: Solid or semi-solid jelly-like colloids, such as those formed when gelatin is mixed with hot water and allowed to cool. Products such as pectins and agar are well known for their gel forming ability. Gels, including agar gels, are widely used as food stabilizers and thickeners.

Genetically engineered foods: Foods that have been modified or that have been prepared with agents, e.g. enzymes, or contain ingredients that have been modified using genetic techniques. It is used to confer new properties such as enhanced nutritional values and prolonged shelf life. More commonly it referred to as genetically modified foods or GM foods.

Genetically modified microorganisms: Microorganisms that have been modified by genetic techniques to enhance their properties or confer upon them new properties. It is abbreviated to GM microorganisms.

Genistein: Yellow isoflavone which occurs in free or glucosidic form and has a weak oestrogenic effect. It is found in soybeans, chick peas, lucerne and clover.

Germ: Germinating portion or embryo of a cereal grain which is extracted and discarded when the grain is milled to make white flour. It is high in fats and several vitamins.

Germicides: Antimicrobial chemical agents is used for disinfection, antisepsis or sterilization.

Germination capacity: Ability of a seed to germinate.

Germination: Sprouting of a seed, spore or other re-productive body. Influenced by a number of factors, including temperature, light and oxygen supply. It is used commercially in preparation of cereals for manufacture of alcoholic beverages, and in production of mushrooms.

Ghee: Product made from butter; originally produced in India but now more widespread. Butter is melted at a high temperature, during which moisture is evaporated. Proteins are then removed from the melted butter by centrifugation.

Ginseng: Root of the plant *Panax ginseng*, used for preparation of beverages. It is widely considered to have health-promoting properties, possibly related to the presence of saponins (ginsenosides).

Glass bottles: Bottles made from glass which are commonly used as containers for beverages and other liquids. Available in a range of shapes, capacities and colour.

Glass containers: Containers made from glass which may be used to store or package a range of foods. Include glass bottles, beakers, jars and pots.

Glass transition temperature: Temperature range at which the glass transition (change from a glassy to a flexible condition) of polymers takes place. Value varies according to the polymer and the range is relatively small.

Glass transition: Reversible sudden transition of an amorphous polymer from a glassy condition to a flexible condition when it is heated to a specific temperature range (glass transition temp.). Due to a change in the arrangement of the polymer molecules from a coiled and motionless state to one where they are free to move.

Glassiness: Optical properties relating to the extent to which a product appears to have the surface properties of glass, i.e. smoothness, uniformity, shini- ness and glossiness.

Glazes: Substances, such as milk, beaten eggs or thin jams, which are used to create a shiny appearance or provide protective coatings on foods. Also, smooth, glossy, glass-like materials fused onto the surface of pottery, where they form hard, impervious decorative coatings.

Glazing: Application of a liquid, such as milk or beaten eggs, to hot or cold foods to produce a smooth, shiny coating after setting. For example, milk or beaten eggs can be brushed onto pastry before baking to add colour and shine. To give a glossy coating to sweet and savoury dishes so as to produce a decorative finish. Beaten egg, milk, and lard are used with bread or buns. Sugar and fruit syrups are used with sweetened doughs and cakes. Arrowroot and sugar thickened fruit juices are used with fruit tarts, butter with vegetables, and aspic jelly is used with savoury items, canapés, etc. To brown food under a very hot grill e.g. Crème brûlée and sauces containing a liaison of eggs and cream used on fish.

Gliadins: Cereal proteins from the endosperm of wheat or rye. The elastic constituent of gluten.

Globulins: Any of a class of spherical or globular shaped high molecular weight proteins which are rela- tively insoluble in water and soluble in dilute salt solu- tions.

Found widely throughout nature; they include lactoglobulins, serum globulins and immunoglobulins. Subdivided into α-, β and Υ-globulins.

Gloss: Optical properties relating to the surface lustre or sheen on a product. Gloss is important to the attractiveness of specific products such as gelatin desserts and buttered vegetables.

Glucans: Soluble, indigestible polysaccharides composed predominantly of D-glucose residues and found in cereals such as oats, barley and rye.

Glucose oxidase: An enzyme obtained from *Aspergillus niger*, used to remove traces of glucose from egg white that is to be dried glucose syrup.

Glucose syrups: Syrups consisting predominantly of glucose. Produced commercially by hydrolysis of starch; corn starch is the most commonly used substrate.

Glucose: Monosaccharide with six carbon atoms. A monosaccharide which forms the building block of many starch and cellulose carbohydrates and is one of the constituents of the disaccharides sucrose (with fructose), maltose (with itself) and lactose (with galactose). It inhibits the oxidation of foods and facilitates browning in fried sausages. Often added to continental sausages to act as the substrate for a lactic acid fermentation which improves their keeping qualities. Free glucose is present naturally in fruits and honeys and it is the monomer unit from which starch and celluloses are synthesized; commercial manufacture of glucose is by hydrolysis of starch. It is the main energy source for living cells. Glucose is a constituent of sucrose and is used in sweeteners. Free glucose has 0.7-0.8× the sweetness of sucrose. The D- stereoisomer of glucose is known as dextrose.

Glucosides: A range of glycosides found mainly in plants, the sugar component of which is glucose. These compounds may be useful as aroma precursors, pigments and surfactants, and may exhibit antioxidative activity. However, cyanogenic glucosides found in several plants are a potential source of cyanides and are therefore potentially toxic.

Glucosinolates: Class of toxic glucosides which are found in *Brassica* (e.g. broccoli, cabbages, radishes). Degraded by thioglucosidases to produce mustard oils, accounting for the pungent flavour of these compounds. May have anticancer properties, since they increase the rate at which potential carcinogens are excreted.

Glutamates: Salts of glutamic acid used as flavourings, e.g. the flavour enhancer monosodium glutamate.

Glutamic acid: Amino acid which is believed to play a part in thehigh-quality flavour of young fresh vegetables and in the enhancement of other flavours in general. Salts of glutamic acid (glutamates) are widely used as flavourings.

Glutelins: Group of globulins present in the seeds of wheat, rice and barley. Soluble in dilute acids or alkalies and insoluble in water, they are a constituent of gluten.

Gluten flour: Dry powdered gluten made by washing the starch from high protein flour, drying and grinding

Gluten free bread: Bread formulated to contain no gluten by excluding wheat and rye proteins to make it suitable for consumption by people suffering from coeliac disease. Alternative term is Gluten low bread.

Gluten free foods: Foods formulated to contain no gluten by excluding wheat and rye proteins to make them suitable for consumption by people suffering from coeliac disease.

Gluten low foods: Alternative term for gluten free foods, foods made with the exclusion of wheat and rye proteins to make them suitable for consumption by people suffering from coeliac disease.

Gluten: The general name given to the proteins in common wheat consisting mainly of glutenin and gliadin. These are responsible for the viscoelastic behaviour of and the baking properties of dough made from wheat flours and for the ability of the dough to entrap bubbles of carbon dioxide. It is extracted from wheat, especially in East Asia and when cooked forms a spongy soft mass which absorbs flavours and is used rather like bean curd and for making cakes. It is water insoluble protein complex found in the endosperm of wheat and rye. When mixed with water, forms cohesive, elastic, cross-linked molecules. These confer elasticity to bread dough, allowing the dough to trap carbon dioxide during breadmaking and causing the bread to rise.

Gluten-free food: Food which contains no gluten. Necessary for the small fraction of people who are allergic to gluten.

Glutenin: Glutelin found in the endosperm of wheat and one of the major components of gluten. A bread made with the exclusion of wheat and rye proteins to make it suitable for consumption by people suffering from coeliac disease. It is insoluble in water or salt solution. It contains many cross linkages which are broken or rearranged during the kneading of dough and is responsible for its elasticity.

Glutinous rice flour: A fine white flour made from glutinous rice and used to make soft cakes, buns and dumplings especially in Asian and Southeast Asian cooking.

Glutinous rice: Black or white, short- or long-grain rice which is particularly sticky when boiled and easy to handle with chopsticks. Used in Chinese and Japanese dishes especially boiled and sweetened in sushi. It should be soaked in cold water for 8 hours then cooked with an equal weight of water for about 12 minutes and left to stand covered for a further 10 minutes. See also black rice, mochigome, pinipig. Also called sticky rice

Glycaemic index values: Nutritional values relating to the ability of carbohydrate to increase the level of glucose in the blood compared with the same amount of glucose.

Glycation: Modification involving nonenzymic reaction of sugars with proteins (or sometimes lipids), as in the Maillard reaction. Results in alterations in physicochemical, biological and functional properties, such as foaming properties, emulsification properties or antioxidative activity, of proteins.

Glycerides: It is synonym for acylglycerols. Fatty acid esters of glycerol, such as monoglycerides, diglycerides and triglycerides. Major components of natural fats and oils (particularly as triglycerides); also used as emulsifiers.

Glycerol: Clear, odourless, sweet-tasting, viscous, hy- groscopic liquid produced by fat saponification. Functions as a humectant to prevent candy and other foods from drying out, as a solvent for flavourings and colorants, as an emulsifier, and as a texture improver for cakes. Also used to control crystallization and in the formulation of fat substitutes. It is used as a humectant to prevent foodstuffs drying out, especially royal icing. Synonym for glycerin and glycerine. Glycerin Synonym for glycerol; alternative spelling glycerine.

Glyceryl monostearate: Ester formed by reaction of stearic acid with glycerol. Used in food emulsifiers, in the manufacture of products such as coffee whiteners and ice cream. It included as bakery additives in manufacture of bread and other bakery products due to the antistaling properties of the glyceride component. Also called glycerol monostearate.

Glycine betaine: One of the soluble nitrogen compounds and a derivative of betaine occurring in a range of foods, especially sugar beets, spinach and molasses; also found in some shellfish, where it is important for flavour. An effective osmoprotectant, glycine betaine is also synthesized by microorganisms living at very high osmotic pressures. Accumulation of glycine betaine in some pathogens, e.g. *Listeria monocytogenes*, allows them to survive under conditions of extreme temperature, leading to food safety problems. The compound may also be added to increase thermal tolerance and osmotolerance in bac teria used in food manufacture.

Glycine: Non-essential achiral amino acid, structurally the simplest of the amino acids. Sweet-tasting and used to retard the onset of rancidity in fats, as well as being a nutrient. Gelatin is a particularly rich source of glycine.

Glycinin: One of the main soy proteins. An 11S storage protein that, along with β-conglycinin (7S globulin), makes up approximately 70% of storage proteins in soybeans.

Glycogen: High molecular weight branched polysaccharide comprising D-glucopyranose residues (**glucose** in the ring conformation) and produced in the animal body from glucose and stored in the liver and muscles. Very quickly broken down into glucose and thus a very important quickly available energy store. Animals that are killed after stress e.g. of the hunt or from poor handling have little or no glycogen in their muscles which are thus less flavoursome than otherwise. It formed predominantly in muscle and liver tissues and is the main store of energy in animals and humans.

Glycolic acid: Colourless, hygroscopic chemical in- termediate of the conversion of glycine to ethanolamine. Constituent of cane sugar juices and unripe grapes.

Glycolipids: Compounds consisting of lipid moieties which are glycosidically linked to one or more mono- saccharide residues. Also known as glycosphingolipids.

Glycols: General term for diols, organic compounds with two alcohol groups. Include ethylene glycol and 1, 2-propanediol (propylene glycol).

Glycolysis: Series of reactions which take place in most living cells by which glucose is converted into pyruvic acid and then to lactic acid.

Glycoproteins: Conjugated proteins composed of polypeptide backbones to which carbohydrates are covalently attached. Present in ovalbumins, mucins and fish antifreeze proteins.

Glycosides: Compounds occurring abundantly in plants in which a sugar is combined with a non-sugar entity (aglycone); this may be an alcohol, phenol or sterol, and replaces the hydroxyl group on the carbon-1 atom. Often found in fruit pigments, e.g. anthocyanins.

Glyoxal: Dicarbonyl compound found as an aroma precursor/compound in wines. Also one of the Maillard reaction products in nonenzymic browning.

Goat meat: Meat from goats; also known as chevon, particularly in India. It resembles mutton, but includes very little intermuscular fat. During the dressing process, goat carcasses nearly always become tainted with the typical aroma of goat, which transfers from the goat skin. The most tender meat comes from young goats, also known as kids, capretto or cabrito; meat from older goats is tougher. Goat meat is widely consumed in North Africa and the Middle East. It is often produced from goats managed traditionally, as free-foraging herds; consequently, goat meat tends to be fairly lean.

Goat milk: Milk produced by dairy goats. Similar in composition to cow milk, but with slightly higher contents of calcium, niacin and vitamin A, and significantly lower concentrations of folic acid and vitamin B_{12}. Goat milk contains almost no carotenes. It is often used in cheesemaking.

Goitrogens: Compounds found in foods (especially *Brassica* species, peanuts, cassava and soybeans) that can cause goitre, especially when dietary intake of iodine is low, by inhibiting synthesis of thyroid hormones (glucosinolates) or uptake of iodide into the thyroid gland (thiocyanates).

Golden syrup: An invert sugar mixture (i.e. One which will not crystallize) made either by clarifying and decolorizing the remaining uncrystallizable portion of the sugar after production of sugar crystals or by treatment of sugar syrup with acids or the enzyme invertase. It contains sucrose, glucose and fructose and is used as a sweetening agent.

Good Manufacturing Practice: (GMP): Part of quality assurance which ensures that products, including foods, are consistently produced to the quality standards appropriate to their intended use and as required by the marketing authorization or product specification. It concerned with both production and quality control. It contains the following ten principles: writing procedures; following written procedures; documenting for traceability; designing facilities and equipment; maintaining facilities and equipment; validating work; job competence; cleanliness; component control; and auditing for compliance. Basically it is a method used in the food industry to control contamination with food poisoning organisms, their toxins or food spoilage enzymes.

Gossypol: Yellow, potentially toxic phenolic substance composed of four benzene rings attaching to isopropyl, hydroxyl, aldehyde or ketone side chains. Occurs in some varieties of cottonseeds from which it is removed during the refining process for cottonseed oils.

Grading: Establishing the degree or rank of an item within a scale. In the food industry, grading is the classification of a food by variables such as quality, size and colour.

Grain: The edible seeds of various cultivated plants usually, but not always, monocotyledons (grasses). The most common are wheat, rice, maize, barley, oats, rye and millet. Sold as whole, kibbled, cracked, ground into flour, flattened into flakes or partially cooked.

Graininess: Consistency term relating to the extent to which a product is grainy, i.e. granular, sandy and gritty.

Gram negative bacteria (GNB): Bacteria that, following staining with crystal violet, are decolorized by organic solvents (e.g. ethanol or acetone) but stain red with the counterstain (safranin) in the Gram stain procedure. Their cell walls are composed of a thin layer of pepti- doglycans covered by an outer membrane of lipoproteins and lipopolysaccharides.

Gram positive bacteria (GPB): Bacteria that resist decol- orization by organic solvents (e.g. ethanol or acetone) to retain their original purple crystal violet stain in the Gram stain procedure. Their cell walls are composed of a thick layer of peptidoglycans with attached teichoic acids.

Granulated sugar: Crystalline solid comprising at least 99.8% sucrose. Granulated sugar is produced by crystallization or graining of concentrated sugar syrups and is the most pure form of sugar manu- factured from sugar beets and sugar cane.

Granulation: Processing of a food into small compact particles (granules). Granulators are often used in cane and beet sugar manufacture to remove unbound moisture from the sugar by driers and coolers. Moisture is removed in driers by blowing hot air through a stream of cascading sugar or through a bed of wet sugar.

Granules: Small particles or grains. Starch exists as granules which are insoluble in cold water but form a viscous solution when heated. Some food ingredients and instant foods are provided in the form of dried granules, which are reconstituted with water before use or consumption.

Granulometry: Technique for measuring particle or granule size distribution.

Grating: Reduction of a piece of firm food, e.g. cheese or vegetables, to small shreds by rubbing on a coarse, serrated surface, usually a kitchen utensil called a grater.

Gravimetry: Technique based on weighing of the sample. Examples of its use include weighing of a sample before and after heating to indicate the content of volatile compounds and quantitative analysis of a substance following precipitation.

Gravity: In a broad sense, the gravitational force which acts on any object within the earth's gravitational field. Also, the attraction between two massive bodies. The force of gravity acting on a body determines its mass. The centre of gravity of an object is the point at which the total weight acts. Specific gravity is a number equal to the ratio of a substance's weight to that of an equal volume of water.

Gravy granules: Instant foods in the form of free flowing granules that produce a ready to serve gravy when reconstituted with boiling water.

Gravy powders: Instant foods in the form of a powder from which a ready to serve gravy is pro- duced on addition of boiling water.

Gravy Sauces: It produced using fats and juices that exude from meat during cooking. A roux is pro- duced from the meat fats and flour, then liquid (e.g. wines, cider, stocks) is added and the mixture is thickened by heating. Browning agents may also be added to colour the sauce if required.

Greasiness: A sensory property relating to the extent to which a product is greasy, i.e. smeared or covered with grease, slippery or fatty, or is perceived to have a greasy quality in the mouth.

Green tea: Tea made from tea leaves which have not undergone fermentation before drying.

Green vegetables: Plants with edible green leaves. Good sources of a range of vitamins and minerals. Include lettuces, cabbages, spinach and kale.

Grilling: Cooking of food on a grill, using radiated heat. Considered by some consumers a healthier way of cooking than frying, as no fats or oils are needed.

Grinding: Reduction of a food to small particles or powders by crushing in grinders. Grinding can be undertaken to varying degrees, producing food that is fine, medium or coarse in texture, as desired.

Grits: Hulled, degerminated and coarsely ground grain, especially corn. Often boiled and served at breakfast or as a side dish. Also called hominy grits.

Grittiness: Mouthfeel term relating to the extent to which a product is perceived to be grainy or sandy.

Groundnut oils: Pale yellow oils extracted from peanuts (*Arachis hypogaea*). Rich in palmitic acid, oleic acid and linoleic acid with good oxidative stability. Due to the desirable flavour, often used in cooking and as a substitute for olive oils and other edible oils. Also known as arachis oils or peanut oils.

Guaiacol: Member of the phenols group. Guaiacol has an antiseptic or medicinal-type aroma and is present as one of the aroma compounds in beer, wines and whisky. Also occurs as a taint in fruit juices caused by bacterial spoilage, and is used as a substrate for analysis of peroxidases.

Guar bean: A variety of green bean, *Cyamopsis psoraloides*, with long thin pods that grow in clusters. It is used in India as a vegetable and in the USA as cattle fodder. The dried beans are not eaten. The beans are a source of guar gum (E412) used as a thickener and stabilizer in processed foods. Also called gaur bean, cluster bean

Guar gums: High viscosity gums isolated from ground endosperms of the legume *Cyamopsis tetragonoloba*, also known as guar beans or cluster beans. It composed of repeating (1→4)-β-D- mannopyranosyl units with branches of α-D-galactopyranosyl units linked via (1→6) linkages. Mannose:galactose ratio is 2:1. These gums are used as thickeners (thickening capacity is approximately 8- fold that of starch), stabilizers and emulsifiers in foods, e.g. in low fat foods and ice cream.

Gum acacia: Dried exudates from African species of the genus *Acacia*, particularly varieties of *A. senegal* and *A. verek*. Forms low viscosity aqueous solutions that are used as thickeners and coatings for products such as confectionery, jellies and chewing gums, stabilizers of beer foams and flavourings. It is synonym for gum arabic.

Gum Arabic: A polysaccharide vegetable gum obtained from the dried exudate of a Middle Eastern and Indian tree, *Acacia senegal*, used as a thickener, emulsifier and stabilizer, in the manufacture of chewing gum, marshmallows and fruit gums and as a glaze on cake decorations. It forms low viscosity aqueous solutions that are used as **thickeners** and **coatings** for products such as **confectionery**, **jellies** and **chewing gums**, **stabilizers** of beer **foams** and **flavourings**.

Gums: High molecular weight polysaccharides that form viscous solutions or gels when dissolved or dispersed in a solvent, usually water. Obtained from plant exudates and seaweeds or produced as exopolysaccharides by bacteria. Gums have many applications in the food industry: low viscosity gums, i.e. gum arabic and gum ghatti, are used as water binding agents for prevention of syneresis and as encapsulating agents for flavourings; medium viscosity gums, i.e. gum tragacanth and alginates, provide body and are useful emulsifying agents, e.g. in salad dressings; and high viscosity gums, i.e. guar gums and locust bean gums, are good thickeners and stabilizers and improve mouthfeel in re- duced fat or low fat foods. Gel forming gums, i.e. carrageenan and gellan gums, are employed as gelling agents to produce semi-solid structures, e.g. in jellies or fruit fillings. They also improve freeze-thaw stability and thus are included as ingredients of ice cream and frozen desserts.

Gur: Unrefined brown coloured sugar produced mainly in India by evaporation of sugar cane juices. Also known as jaggery.

Gushing: Phenomenon occurring in beer in which there is violent foaming when the bottles or barrels are opened. Associated with formation of calcium oxalate crystals in the beer and contamination of malting barley with *Fusarium graminearum*.

Gutting: Removal of the internal organs of fish before cooking.

H

Haem: Iron-containing compounds in which the iron is complexed in a porphyrin ring. Component of pigments such as haemoglobin, myoglobin and cytochromes. The iron atoms can bind oxygen in a reversible fashion or conduct electrons. Alternative spelling heme.

Haemoglobin: Oxygen-carrying protein which is found in the blood of animals. Haem groups within the protein bind oxygen to form oxyhaemoglobin, which is carried to oxygen-depleted cells where the oxygen is released. Other inorganic compounds, including carbon dioxide, can also be bound by the haem groups. Alternative spelling hemoglobin.

Haemoproteins: General term for haem-containing proteins, including haemoglobin, myoglobin, cytochromes, catalases and peroxidases.

Halal foods: Foods permitted under Islamic dietary law, particularly meat from animals that have been slaughtered according to accepted Islamic procedures. For foods to be certificated as halal (lawful), they must be free from haram (unlawful, prohibited) substances, such as pork and swine by-products, carrion and intoxicants such as alcohol.

Half butter: Butter product with a low fat content of approximately 39-41%.

Half cream: Cream product with a fat content of approximately 12%.

Ham: Meat from the upper part (between the hip and hock) of swine hind-legs; usually it is cured. It may be cooked, raw, smoked or unsmoked, dried by me- chanical means or by air drying, or stored in vacuum packaging. Common types include: whole leg ham on the bone; single-muscle ham; boiled ham; and baked ham. Some highly valued speciality hams are dry cured, including prosciutto crudo, jambon de Bayonne and serrano. However, more commonly, ham is cured by brining and then hung to dry before it is smoked, if applicable. There are many different styles of ham, often particular countries and regions within countries are well known for a particular style. Traditionally, the names of hams, for example Parma or Bayonne, refer to geographic localities and techniques developed there. Lean ham has a fairly low fat content, but even low salt ham has a high content of sodium.

Hamburgers: Round, flat patties of meat mince, cooked by grilling or frying. Hamburgers are typi- cally prepared from meat mince with a 15-20% fat content. They are commonly eaten in bread rolls, served with lettuce, slices of tomatoes and onions, and tomato ketchups.

Handling: Broad term referring to manipulation of goods during manufacture, distribution and storage, as well as control of live animals. Proper handling of sensitive foods, such as fruits and vegetables, frozen foods and refrigerated foods, is important from economic and hygienic perspectives. Robotic systems may be used for bulk handling of foods. Correct pre-slaughter handling of animals is important, as stress prior to slaughter can decrease meat quality.

Hard cheese: Cheese with a high dry matter content. Usually aged for a number of years and pressed with weights during ripening to extract whey. Examples include mature Cheddar cheese and Parmigiano Reggiano cheese.

Hard to cook defect: Irreversible condition that de- velops in legumes during storage at high temperature and under high humidity. Affected legumes absorb water during cooking but do not soften within a reasonable time.

Hard wheat: A tough wheat which resists cracking and with the starch grains tightly bound. It contains no triabolin and it fractures across the starch grains. Durum wheat is one of the hardest.

Hardening: Making or becoming solid, firm and rigid. May be problematic or necessary e.g. common beans are susceptible to hardening during storage, giving problems for cooking, while the hardening stage is important in the manufacture of good quality ice cream. Hardening is also a stage in fats and oils processing, e.g. manufacture of margarines, usually referred to as hydrogenation and involving treat- ment with hydrogen.

Hardness: Extent to which water is perceived as being hard, i.e. containing high levels of minerals. Also relates to mechanical properties of foods related to flexibility, such as strength, firmness, solidity, im penetrability, resistance, density, toughness, stiffness and rigidity.

Harvesting: Gathering of agricultural crops, aquaculture products or cultured cells. Agricultural produce may be harvested manually or using special machinery (mechanical harvesting).

Hazard analysis critical control point (HACCP): Comprehensive systematic approach to identifying and minimizing the occurrence of microbiological, physical and chemical hazards, which can affect food safety and quality during all stages of the food chain, including processing operations and during subsequent storage, distribution and retailing.

Hazard Analysis Critical Control Point (HACCP): Hazard Analysis Critical Control Point (HACCP) is recognised as a system for the identification, assessment and control of food hazards in food production and has been defined as 'a systematic approach to the identification and assessment of the microbiological hazards and risks associated with food and the definition of means for their control' (ICMSF). The central feature of HACCP analysis is the determination of the Critical Control Points (CCPs), those stages in the process which must be controlled to ensure the safety of the product. Once the CCPs have been identified, a monitoring system is established for each CCP to ensure that correct procedures are maintained and actions taken if CCP criteria are not achieved. HACCP is a systematic approach to the control of potential hazards in a food operation. A hazard is anything that could harm the consumer. It may be of a physical, chemical or biological nature. HACCP aims to identify problems before they occur, and establish mechanisms for their control at the stages in production critical to ensuring the safety of food.

Hazards analysis: Identification of areas within an HACCP flow diagram, for the production of a food, where unacceptable microbial, chemical or physical health risks may occur.

Haze: Decreased visibility in the air or clarity of solutions caused by suspended particles. In beer, haze can develop as a result of chilling, when proteins are pre- cipitated. This can be prevented by chill proofing, in which the proteins are absorbed or broken down by enzymes.

HCH: Insecticide used for control of a wide range of plant-eating and soil-dwelling insects on crops. Also used for control of insect pests in food storage facilities and as an ectoparasiticide in farm animals. Classified by WHO as moderately toxic (WHO II). Also known as hexachloran and lindane.

Headspace analysis: Technique for analysis of volatile compounds in samples not suitable for direct injection into a gas chromatograph. Samples are heated in a closed chamber and the surrounding atmosphere is swept with a stream of inert gas, compo- nents of the sample being collected for analysis by GC MS.

Health beverages: Beverages formulated with ingredients claimed to enhance the health of the consumer and/or protect against diseases.

Health claims: Claims made by manufacturers about the health benefits of their products. They form a part of the consumer information, which is provided on food labelling. Due to consumer concerns about health, addition of health claims to labelling provides manufacturers with a powerful tool for marketing foods. Increasingly, regulations and legislation are being introduced to ensure that health claims are the result of appropriate scientific trials and are clear, measurable and distinct from nutrition claims.

Health foods: Loosely defined term usually taken to encompass foods perceived as healthy by the consumer, such as organic foods, natural foods, whole grain cereal products, royal jelly and energy foods.

Health hazards: Microbial, chemical or physical ele- ments which may cause injury to health.

Hearts: Hollow muscular organs composed of cardiac muscle; animal hearts are a part of edible offal. They are often inexpensive because they lack popularity, although in some cultures they are considered to be delicacies. Lamb and calf hearts are tender and have a very delicate flavour. They are generally cooked by sauteing or grilling until they are medium rare, or are cooked slowly using moist heat. Cattle and swine hearts are generally too tough to be cooked by sauteing or grilling, but become very tender if cooked slowly using moist heat.

Heat distribution: The extent to which heat energy is transmitted throughout an item during thermal processing. Non-uniform distribution of heat during processing can lead to non-uniform destruction of target microorganisms, which could compromise product safety. Heat distribution studies are therefore crucial to ensuring effective heat treatment of the product.

Heat exchangers: Devices that transfer heat between fluids on either side of a barrier without bringing them into direct contact. In many engineering applications, heat exchangers are used to increase the temperature of one fluid while cooling the other.

Boilers, evaporators, superheaters, condensers and coolers may all be considered heat exchangers. Heat exchangers are manufactured with various flow arrangements and in different designs. The simplest is the concentric tube or double-pipe heat exchanger, in which one pipe is placed inside another; the fluids run in parallel flow. Heat is transferred from the warm fluid through the wall of the inner tube to the cold fluid. A heat exchanger can also be operated in counterflow, in which the two fluids flow in parallel but opposite directions. Concentric tube heat exchangers are built in several ways, such as a coil or in straight sections placed side by side and connected in series. The most common type of heat exchanger is the shell-and-tube type, which utilizes a bundle of tubes through which one of the fluids flows; the tubes are enclosed in a shell in which the other fluid flows. Here, the free fluid flows approximately perpendicular to the tubes containing the other fluid, in a cross-flow exchange.

Heat resistance: Thermophysical properties relating to the ability of materials, especially microorganisms, to withstand various temperatures of applied heat. Acquired heat resistance of bacteria such as *Listeria* can cause food safety problems.

Heat shock proteins: Proteins that are synthesized by an organism in response to the stress of a sudden rise in temperature. May be necessary for survival of the organism at high temperatures. May be produced in response to other stresses, e.g. exposure to UV radia- tion. Also called stress proteins and heat stress proteins.

Heat stability: Thermophysical properties relat- ing to the ability of materials to maintain stability when subjected to various temperatures of applied heat. If food ingredients or additives are heat stable, it is possible for them to be used successfully in prod- ucts which have to be thermally processed. It is synonymous with thermal stability.

Heat transfer: Exchange of heat energy between a system and its surrounding environment, resulting from a temperature difference between the two. The energy exchange occurs by thermal conduction, mechanical convection, or electromagnetic radiation.

Heat treatment of curds: Curds for cheese may be uncooked, i.e. Not subjected to a temperature greater than 39°C, scalded, i.e. Subjected to temperatures between 40 and 48°C or cooked i.e. Subjected to temperatures in excess of 48°C

Heating: Treatment of an item to make it hot or warm, most commonly by conduction, convection or radiation. Used to modify the properties of a material.

Heavy metals: Collective term for metals of high atomic mass. It includes the minerals mercury, cadmium, chromium, lead, nickel and arsenic. Common pollut- ants of land and water, generally as a result of industrial activity, and are consequently present as contaminants in plant and animal foods, where, if present in excess, they may cause toxicity problems. Maximum permitted levels have been defined for heavy metals in specified food groups to ensure food safety.

Heifers: Young, usually sexually mature female cattle, especially those that have not borne a calf, or have borne only one calf. The term is generally used until the end of an animal's first lactation.

Hemicellulases: Enzymes that hydrolyse the hemi- celluloses of plants (which include polymers of hex- oses (glucose, rhamnose or mannose) and pentoses (xylose and arabinose), as well as plant mucins). These enzymes have numerous applications in the food industry, including processing of fruit juices, fruits and vegetables, winemaking, brewing, breadmaking and extraction of vegetable oils.

Hemicelluloses: Polysaccharides tightly associated with lignin in cell walls of all plants and some seaweeds. Composition of hemicelluloses differs between plants and is influenced by environmental factors, and plant growth and maturation. Predominant sugars present are: D-xylose, D-glucose, D- galactose, D-mannose, L-arabinose, D- glucuronic acid, D-galacturonic acid, L- rhamnose, L-fucose and 4-*O*-methyl-D-glucuronic acid. Hemicelluloses are produced as waste from proc- essing of cereals and other crops. Hemicelluloses or hemicellulose hydrolysates (mixtures of oligosaccharides and saccharides produced by enzymic, acid or alkali hydrolysis) are used as substrates for mi- crobial fermentations. They are also a source of dietary fibre. A polysaccharide carbohydrate found in fruit and vegetables. Some forms known as pectin are responsible for the setting of jams.

Hen meat: Meat from female chickens. Often, hen meat is derived from spent hens, which have completed a period of egg laying. Spent hen meat is commonly used as an ingredient in chicken sausages and in restructured meat products, such as chicken nuggets.

Hepatitis viruses: Viruses labelled A to E that cause inflammation of the liver (hepatitis). Hepatitis A and E viruses can be transmitted through faecal contamination of food or water.

Hepatitis: Inflammation of the liver which can be a result of infections or non-infectious pathology. Certain causes of infection, such as hepatitis A virus, can be borne in foods and water supplies.

Hepatotoxicity: Quality or property of having a poisonous or destructive effect on liver cells.

Heptachlor: Non-systemic insecticide used for control of termites, ants and soil-dwelling insects in a wide range of crops. Classified by WHO as moderately toxic (WHO II).

Herb tea: Tea-type infusion beverages prepared from dry plant material other than tea leaves (*Camellia sinensis*).

Herb: A generic name for low growing plants and shrubs but more generally reserved for those plants with flavouring and medicinal properties.

Herbal beverages: Beverages in which herbal material is a significant source of flavour and/or active ingredients.

Herbicides: Chemical substances used to kill or inhibit growth of unwanted plants around crops. Most are applied as sprays and have either a systemic or contact effect. Examples of herbicides commonly applied to crops include atrazine, diuron, glyphosate and propham. Residues remaining in foods and the envi- ronment can represent a health hazard.

Heterocyclic amines: Amines with a cyclic molecular structure containing atoms of at least two different elements in the ring or rings. It formed particularly in meat and fish during grilling or frying. Some are of concern because of their mutagenicity or carcinogenicity.

Heterocyclic aromatic amines: Heterocyclic amines containing ring structures with conjugated double bonds and delocalized electrons. It formed particularly in meat and fish during grilling or frying. Some are of concern because of their mutagenicity or carcinogenicity.

Hexoses: General term for sugars comprising six carbon atoms, e.g. glucose, mannose, galactose, fructose, sorbose and tagatose.

High amylose corn starch: Starch manufactured from hybrid corn plants that have been selected for the high amylases and amylopectins ratio of their starch. Amylose content in high amylose corn starch is usually 55%. Due to the high amylose content, the starch produces firm gels on heating.

High calorie foods: Any foods that have a high calo- rie content in relation to bulk, such as peanut butter or chocolate syrup. Also includes dietetic foods and energy foods which have been specifically manufactured to give an increased calorie content. These are designed for weight gain and may be targeted at individuals with specific nutritional requirements, e.g. athletes, invalids, low birth-weight infants. Light weight, calorie-dense foods are also used as space flight foods and military rations.

High density polyethylene (HDPE): Polyethylene of high-density grade. It is used as a packaging material in many food and beverage applications.

High gravity brewing: Brewing process in which worts of higher than normal concentration are fermented, and the resulting high concentration beer is diluted to normal beer strength.

High performance liquid chromatography (HPLC): Column chromatography technique with a liquid mobile phase in which high column inlet pressure, narrow bore columns and small particle size stationary phases are used to achieve rapid separation. It can be applied to separation of a wider range of compounds than is possible with gas chromatography. It is also called high pressure liquid chromatography.

High pressure processing (HPP): Nonthermal preservation technique used to inactivate vegetative microorganisms in foods by isostatic pressure pasteurization (1000-9000 atmospheres). High pressure processing affects only noncovalent bonds, enabling phase transitions, permeabilization of biological membranes, denaturation of proteins, gelatinization of proteins and starch, increasing reaction rates, and compacting of materials. Bacterial spores are con- siderably more resistant to high pressure processing than vegetative or germinating cells.

Homogenization: Creation of emulsions by reduc- ing all the particles to the same size. For example, in homogenized milk, the milk fat globules are emulsified, preventing the cream from separating out. Commercial salad dressings are also often homogenized.

Homogenized milk: Milk treated in a homogenizer to break up the milk fat globules and reduce creaming, thus increasing shelf life. Modifications to casein structure improve digestibility of the milk; smaller milk fat globules and increased surface area increase contact with the taste buds, giving a fuller flavour. Homogenized milk has a greater whitening power in coffee. It is more sensitive to light-induced off flavour but less sensitive to development of flavour defects caused by oxidation.

Homogenizers: Apparatus used in homogenization of foods, such as milk.

Honey beverages: Beverages in which honeys are major constituents, as sweeteners, flavourings or sources of fermentable material.

Honeys: Natural syrups produced by honeybees pre- dominantly from nectar but also from honeydew and fruit juices. Honey consists of approximately 20% (w/w) water and 80% sugars, mostly fructose and glucose. Honeys also contain the flavour com- pounds and aroma compounds present in the nectar or fruit juices collected, composition of which is dependent on its botanical origin, and it is these minor components that give honeys their individual flavour. Honeys are collected from honeycombs, where they are stored, and may be used directly as both foods and sweeteners.

Hop substitutes: Substances used in place of hops to impart flavour and bitterness in beer. Required particularly in situations where climatic and economic considerations prohibit the use of conventional brewing materials, e.g. in Nigeria where malted or unmalted sorghum has been used instead of malted barley to produce lager. Materials which have been used suc- cessfully as hop substitutes include seeds from *Gar- cinia kola* and extracts from bitter leaf (*Vernonia amygdalina*).

Hoppers: Large containers for grain, typically those that taper downwards and discharge their contents through valve-like openings at the base. In general, it used as temporary receptacles for grain.

Hopping: It is a process used in brewing. It is the addition of hops to fermenting worts to impart flavour and bitterness. Hops may also be added to the finished beer (dry hopping) to enhance hop flavour.

Hordein: Prolamin found in barley.

Hordenine: One of the biogenic amines. It is found in germinated barley, sorghum and millet, and in malt and beer.

Hordeumin: High molecular weight anthocyanin/polyphenol complex formed during ethanolic fermentation of uncooked barley bran. It exists as a purple pigment at low pH values and has potential for use in colorants for foods.

Hordothionins: Antifungal proteins which occur in barley kernels.

Horticultural products: These are the products of horticulture, such as fruits, vegetables and flowers.

Horticulture: Cultivation of fruits and vegetables for human consumption, and of flowers and other plants for ornamental purposes. It practiced on a small scale as a pastime (gardening) or on a larger, commercial scale (also market gardening).

Hot boning: It is process of cutting of meat (muscle) from animal carcasses that have first been conditioned at 16°C for varying time periods *post mortem*.

Hot dogs: Hot frankfurters served in long, soft bread rolls, with added mustard, tomato ketch- ups or other condiments. Hot dogs are particularly popular in the USA.

Hot peppers: Fruits produced by various members of the *Capsicum* genus. It is vary in size, shape and colour, but always with numerous seeds. It is very pungent, due to the presence of capsaicin in the seeds and veins. It includes chillies which is rich in vitamin A and vitamin C; good source of vitamin E, potassium and folic acid. It is also used as a dried powder in many dishes, such as stews, and to make hot sauces.

Hot water dips: Treatment used to protect fruits and vegetables from conditions such as chilling injury, pests infestation and decay during cold storage.

Hot-smoking: It is smoking of food at a temperature between 40 and 105°C thus cooking it at the same time. Food is usually cold-smoked first, and fish is restricted to 80°C.

HTST pasteurization: High temperature, short time (HTST) pasteurization treatment used widely in the food industry, but particularly applied to liquid foods such as raw milk and fruit juices to reduce substantially the total bacterial count for improved shelf life and to eliminate any pathogens. For milk, heat treatment is accomplished using plate heat exchangers. Cold raw milk held in a cool storage tank is pumped into pasteurizers, where it is heated to a temperature of at least 72°C. The milk, at pasteuriza- tion temperature and under pressure, flows through the holding tube where it is held for at least 16 seconds. At the end of the tube is an accurate temperature-sensing device that checks if any of the heated milk has not reached the pasteurization temp. If any milk has not, a diversion device is activated, and the product is made to flow back through the heat exchanger. Properly heated milk continues to flow through the system and is cooled to 4°C or less. Cold, pasteurized milk passes through a vacuum breaker then on to storage tank filler for packaging.

Hulling: It is removal of the hulls from fruits or seeds prior to consumption. It is also called dehulling or husking. It is also, removal of leaves from the tops of strawberries prior to consumption.

Hulls: The outer (usually fibrous) coverings of some fruits or seeds, that are removed by dehulling or hulling prior to consumption. It is also known as husks or shells.

Human milk substitutes: Preparations for feeding to infants and young children as a replacement for human milk, designed to meet their specific nutritional requirements. It is also called infant milk formulas. It may be based on cow milk or soymilk.

Human milk: Milk produced by women during human lactation. Composition differs considerably from that of cow milk. Although fat contents of human and cow milks are similar, fatty acids composition varies. Human milk contains less protein than cow milk; pro- portions of individual proteins and amino acids also differ. Contents of lactose, oligosaccharides and some vitamins, and activities of some enzymes are higher in human than in cow milk, while human milk contains a lower amount of minerals in total. It is also called breast milk or mothers' milk.

Humanized milk: Milk, in which the nutrients composition is adjusted to that of human milk as far as possible, making it suitable for feeding to infants.

Humectants: Ingredients added to increase or maintain the water activity of foods. Examples of humectants include, gums, which possess water binding activity, and NaCl, glycerol and sucrose, which increase water activity by altering the osmotic pressure of foods.

Humidification: Process whereby the level of moisture in the air is increased. By circulating air of higher humidity, the moisture content of hygroscopic products can be increased. This process, known as conditioning, is applied to some grain prior to milling or other processing.

Humidity: Moisture content of the atmosphere. Relative humidity (RH) is the moisture content of the air at a given temperature as a percentage of the level required to cause saturation at that temperature.

Hurdle technology: Food processing technique employing a combination of preservation procedures or hurdles to inhibit growth of microorganisms in the product. These include manipulation of factors such as temperature, water activity and acidity, as well as processes such as gas packaging and high pressure processing. The aim is to interfere with several dif- ferent mechanisms within microorganisms simultaneously. This multitargeted approach allows effective use of mild techniques.

Husbandry: The breeding, care and cultivation of crops and animals. It may also include the management and conservation of plant or animal resources.

Hybridization: Formation of double-stranded nucleic acid molecules by base-pairing between complementary single-stranded molecules. Used to detect specific sequences and for determining the degree of sequence identity, and can be carried out in solution or with one component immobilized on a suitable matrix (e.g. nitrocellulose). Hybrids can be detected by EM or by labelling one of the components, e.g. fluorescently or radioactively. Hybridization can also be performed *in situ* using fluorescently-labelled DNA molecules to localize genes to specific chromosomes.

Hydration: Process by which water is added to a dried substance to hydrate it.

Hydrocarbons: Any organic compounds that contain only carbon and hydrogen.

Hydrochloric acid: Solution of hydrogen chloride gas in water, chemical formula HCl. Strong mineral acid widely used in the food industry as a processing aid.

Hydrochlorofluorocarbons (HCFC)**:** Organic compounds consisting of carbon, hydrogen, chlorine and fluorine. Used as refrigerants. HCFC are less destructive to the ozone layer than chlorofluorocarbons (CFC). HCFC are currently used as replacements for CFC, but their use is to be phased out, as specified by the amended Montreal Protocol, when they are expected to be replaced by hydrofluorocarbons (HFC).

Hydrocolloids: High molecular weight polymers of animal, plant or microbial origin that form viscous solutions or gels on addition of water, e.g. gums and gelatin.

Hydrocooling: Precooling method for heat sensitive products, such as certain fruits and vegetables. During hydrocooling, fruits and vegetables are cooled by direct contact with flowing cold water, which absorbs heat directly from the produce. Hydrocooling allows the grower to harvest produce at optimum maturity with greater assurance that it will reach the consumer at maximum quality. Hydrocooling benefits the produce by slowing the natural deterioration that starts shortly after harvest, slowing the growth of decay organisms and reducing wilt by retarding water loss.

Hydrocyanic acid: Toxic, colourless gas with a boiling point of 26°C. It is synonym hydrogen cyanide and chemical formula HCN. It occurs as a hydrolysis product of cyanogenic glycosides in a range of foods, especially cassava, but also including edible fungi flax seeds and wines. It is used as a fumigant to control pests in stored foods.

Hydrocyclones: Cyclones used for clarification of liquids, such as for removal of dust and soil particles from thin sugar juices and extraction of casein particles from whey. Liquid is added tangentially at high speed to a conical chamber to produce a spinning motion (the cyclone). Particulate matter is forced to the sides, decelerates and falls to the bottom of the chamber from which it is collected. A liquid column is formed in the centre of the cyclone and rises to an outlet at the top of the chamber.

Hydrofluorocarbons: Hydrofluorocarbons (HFC) are organic compounds that contain hydrogen, carbon and fluorine. HFC, which do not contain chlorine, are not harmful to the ozone layer, and so are suitable replacements for chlorofluorocarbons (CFC) in refrigeration.

Hydrogen peroxide: Strong oxidizing agent and antimicrobial compound with chemical formula H_2O_2. Used in foods at low concentrations (e.g. maximum limit is 0.05% in milk) as preservatives, dough conditioners, bleaching agents, and for artificial ageing of wines and spirits, and refining of fats and oils. At concentrations greater than those used in foods and beverages, it is used in disinfectants.

Hydrogen sulfide: Toxic, colourless gas, chemical formula H_2S, with a distinctive odour of rotten eggs. Formed by reduction of organic sulfur compounds, including proteins, during microbial fermentation and can occur in musts and worts as an undesirable by-product of alcoholic fermentation by yeasts giving rise to sulfide taints in the resulting wines and beer. It is also produced by spoilage bacteria during decomposition of high-protein foods such as meat or fish. Dietary protein from meat is an important substrate for hydrogen sulfide generation by bacteria in the human large intestine and this process has been implicated in the development of ulcerative colitis.

Hydrogenated fats: Oils from an animal or vegetable source that have been subjected to hydrogenation, which hardens and stabilizes the oil by reducing unsaturated double bonds in the fatty acids.

Hydrogenation: Chemical reaction in which molecular hydrogen reacts with unsaturated fatty acids, usually in the presence of catalysts to make them more saturated and harder. It is improves their oxidative stability. In this hardening process, hydrogen reduces carbon atoms linked by a double bond, decreasing the level of saturation of the fatty acids. Often it is used in the manufacture of margarines and lard substitutes.

Hydrolysed lactose syrups: It is syrups manufactured by acid or enzymic hydrolysis (treatment with β- galactosidases) of lactose syrups or whey. It consists of an aqueous solution of glucose and galactose; whey-derived hydrolysed lactose syrups also contain salts and oligosaccharides.

Hydrolysed protein: Protein which has been broken down into smaller subunits (peptides) by treatment with water at high temperature and pressure or other hydrolysing agents. This produces highly flavoured compounds as in yeast and meat extracts and also occurs during fermentation of proteins (soya sauce, miso, tempeh, etc.) and to a limited extent when cooking proteins.

Hydrolysed starch syrups: It is a syrups manufactured by acid and/or enzymic hydrolysis of starch slurries. The starch may be derived from any source, although commonly corn starch is used due to advantages of cost and availability. Examples of hydrolysed starch syrups include corn syrups, glucose syrups and maltose syrups.

Hydrolysis: Reaction in which a substance is split into two or more component parts by the action of water in the presence of **catalysts** such as **enzymes**, **acids** or **alkalies**, acting at specific points within the molecules. Types of hydrolysis include **proteolysis**, in which **proteins** are broken down to component **peptides** or **amino acids**, **lipolysis**, in which **lipids** are broken down into constituent **fatty acids**, and **saponification**, in which lipids are hydrolysed in the presence of alkalies to form soaps. It is used e.g. to convert starch into simpler sugars and is the process which occurs when starch solutions and sauces are thinned by boiling them too long.

Hydrometer: A short stubby cylinder surmounted with a long graduated tube. The whole is sealed and weighted so that it floats upright in a liquid always displacing its own mass of liquid thus indicating by the depth to which it sinks the density or specific gravity of the liquid. Used for measuring alcohol,.

Hydrometry It is a Measurement of specific gravity of a liquid or salt and sugar concentrations in solution or strength of alcoholic beverages. Usually it is performed using a sealed graduated tube weighted at one end, which sinks in the liquid to a depth that indicates the specific gravity.

Hydroperoxides: Organic compounds in which one hydrogen atom of a hydrocarbon is replaced by an -O-OH group. Lipid hydroperoxides are formed by lipoxygenases during oxidation of lipids and these are further degraded enzymically or thermally to pro- duce acids and aldehydes which can be associated either with flavour and aroma development or with decreases in lipid quality in fats and oils.

Hydrophobicity: State in which a substance has low affinity for water. It is extent to which molecules are insoluble in water.

Hydroponics: Cultivation of plants in a nutrient solution rather than soil.

Hydroquinone: It is a member of the phenols group of aromatic compounds with antioxidative activity and synonyms include 1,4-benzenediol, *p*- dihydroxybenzene and quinol. It occurs naturally in several foods and beverages, including fruits, vegetables, grain, coffee, tea and beer. It can also include any member of the aromatic *p*-diols derivable from *p*- quinones or any compound with a quinol nucleus.

Hydrothermal processing: Application of heat and moisture treatments, such as steam infusion processes used for cooking, puffing or flaking of foods.

Hydroxybutyric acid: One of the short chain fatty acids, with four carbon atoms. It is synonym of hydroxy- butanoic acid. Not widely identified as a lipid component of foods, but does occur in an esterified form as an aroma compound in sake and cheese. 3- Hydroxybutyric acid has been used as a marker for fertile incubated eggs in which the embryo has died, and which are not permitted to be used in foods.

Hydroxycinnamic acid: One of the aromatic phenols widely distributed in plant foods including fruits and cereals, and plant-derived beverages including fruit juices, wines, whisky and sake. Three isomers exist, including 4-hydroxycinammic acid (synonym coumaric acid). Also more widely used as a general term to describe hydroxy-substituted forms of cinnamic acid, including ferulic acid (4-hydroxy-3-methoxycinnamic acid) and caffeic acid (3,4-dihydroxycinnamic acid).

Hydroxymethyl furfural: It is a member of the heterocyclic organic compounds composed of a furan ring with aldehyde and hydroxymethyl substituents. It includes 5-(hydroxymethyl)-2-furaldehyde and 5- (hydroxymethyl) furfural. It is found as a natural component in honeys and as a thermal breakdown product of sugars in heat-treated products such as UHT milk and pasteurized fruit juices. It is determined chemically as a marker of nonenzymic browning.

Hydroxyproline: Member of the amino acids with eight possible structural isomers, of which only the L- isomers are known to occur naturally. It found in several animal proteins including collagen and gelatin, and in extensin, a plant protein.

Hydroxypropylcellulose: Non-ionic ether of cellulose that forms a viscous liquid when solubilized in water. It uses in foods include as emulsifiers, stabilizers, encapsulating agents and thickeners.

Hydroxystearic acid: A C18 member of the fatty acids family of aliphatic compounds, synonym hydroxyoctadecanoic acid. It produced in microbial bio-conversions of oleic acid as an intermediate in the formation of lactones.

Hygiene The general term for those procedures and practices which prevent the transmission of disease between and to humans, the multiplication of harmful microorganisms in the body or environment and the accumulation of dirt, poisons and toxins where they could be harmful to living creatures. It a salso cience of health and its preservation, or a practice or condition that is conducive to the preservation of health.

Hygienic quality: Extent to which something is conducive to health.

Hygrometers: Instruments used to measure the humidity of the atmosphere.

Hygromycins: Antibiotics which exhibit relatively poor antibacterial activity, but are effective ant helmintics. It is used to control parasitic worm infections in swine and poultry.

Hygroscopic properties: Extent to which a substance absorbs moisture from the atmosphere without dissolving in the moisture. Highly hygroscopic substances, e.g. silica gels, can be used as desiccants.

Hygroscopic: It describes a substance that absorbs water from the atmosphere without necessarily showing any signs of damp, e.g. Silica gel which is used to maintain a dry atmosphere over some foods

Hyperactivity: Psychiatric condition, also known as ADHD (attention deficit hyperactivity disorder) characterized by inattention, restlessness and impulsiveness. It has been linked anecdotally to consumption of refined sugar and food additives, particularly colorants such as tartrazine. However, scientific evidence indicates that most cases are not a result of dietary factors.

Hypercholesterolaemia: Condition in which abnormally high levels of cholesterol are present in the blood. A very high cholesterol level is a known risk factor for coronary heart diseases (CHD) and stroke. Blood cholesterol levels may be controlled by diet or functional foods containing cholesterol lowering constituents, such as stanol esters.

Hypermarkets: Very large self-service shops selling foods and household goods, and sometimes clothing.

Hypertension: Prevalent disease in which blood pressure is elevated. In the majority of cases the cause is unknown; a rare cause is excessive consumption of liquorice rich in glycyrrhizic acid. High blood pressure is a risk factor for other diseases such as cardiovascular diseases (CVD) and has been shown to be improved by reduction in body mass index. The association between hypertension and consumption of salt is controversial.

Hyphae: The filaments and fungal strands which make up the main mass of fungi and from which fruiting bodies, e.g. Mushrooms, sprout. Quorn is an example of a food made from hyphae.

Hypochlorites: Salts of hypochlorous acid (HClO), such as sodium hypochlorite. Widely used as disinfectants.

Hypocholesterolaemic activity: Ability of a food, nutrient or diet to produce hypocholesterolaemia, a state wherein blood cholesterol level is abnormally low, or to lower high cholesterol levels (as in hypercholesterolaemia) to within the normal range. Reduc- tion of blood cholesterol levels are associated with re- duced risk of cardiovascular diseases. Dietary components possessing hypocholesterolaemic activity include dietary fibre, some plant proteins, some fatty acids, phytosterols and probiotic bacteria. It included as a specific type of hypolipaemic activity. It has alternative spelling hypocholesterolemic activity.

Hypolipaemic activity: Ability of a food, nutrient or diet to reduce the fasting and/or postprandial levels of plasma lipids, including cholesterol and triacylglycerols. Reductions in certain plasma lipid parameters, such as fasting levels of total cholesterol and cholesterol within low density lipoproteins (LDL), and postprandial triacylglycerol concentrations, are associated with reduced risk for cardiovascular diseases. Dietary components demonstrating hypo- lipaemic activity include certain fatty acids, phytos terols and phytosterol-enriched margarines, probiotic bacteria and dietary fibre fractions.

I

Ice cream cones: Thin, slightly sweetened, wafers baked on a waffle iron and curled before cooling to form a cone shape. It is used to hold one or more scoops of ice cream.

Ice cream mixes: Commercial products used in manufacture of ice cream. Contain all the main components of the final product, including milk, cream, sugar, flavourings and emulsifiers.

Ice cream wafers: Thin, slightly sweetened, waffle textured wafers that are usually triangular or rectangular and served as an accompaniment to ice cream or used to make an ice cream sandwich.

Ice cream: Frozen dairy product with creamy, smooth and crystalline consistency. In addition to milk and dairy products such as cream, milk powders, butter and sweetened condensed milk, also con- tains sugar, flavourings and additives such as emulsifiers and stabilizers. The ingredient mix is processed in an ice cream freezer where it is frozen by contact with the refrigerated wall, blades scraping the mixture from the walls while whipping air into the ice cream. The soft-serve ice cream produced can be hard- ened further by placing in a suitable freezing apparatus.

Ice nucleation activity: Promotion of the formation of ice crystals. Agents displaying ice nucleation activity include small particles, such as food particles, and large molecules, such as ice nucleating proteins.

Ice: Frozon or solid form of water, used for numerous food processing applications, including chilling and glazing of foods (e.g. fish). It is used in place of water where intense agitation or prolonged processing might overheat a mixture. The energy necessary to melt ice would heat the same mass of water by 80°C. Small pieces of ice, e.g. ice cubes or crushed ice, may be added to beverages to cool them, while ice crystal characteristics play an important role in determining the quality of frozen foods. Flavoured ice is consumed in the form of ice lollies and water ices. It is also used for cooling.

Icing sugar: It is very finely powdered granulated sugar with no particles larger than 0.1 mm (100 microns) used as an ingredient of fondants and icings that require sweetness and a smooth texture. Anticaking agents, usually starch or tricalcium phosphate, are commonly added to icing sugar.

Icing(s): A paste formed from icing sugar and water with possible additions of lemon juice, egg white, glycerine, colouring and flavouring, used to coat cakes and buns, etc. for decorative purposes. Other ingredients that may be used include butter/margarines, egg whites and colorants.

ICMSF: The International Commission on Microbiological Standards for Foods which sets standards for assigning foods to various classes of health hazard ranging from case 1, no health hazard and low incidence of spoilage, to case 15, severe and direct health hazard possibly influenced by conditions of handling

Idli: A steamed, naturally fermented cake-type product widely consumed as a breakfast food or snack in India. Prepared by fermenting a slurry of ground rice and legumes (usually black gram dhal) and steaming the resulting batters to give products with a soft, sponge-like texture and good digestibility.

Image analysis: Analysis of a sample on the basis of its structure, as determined by non-destructive techniques such as microscopy. Parameters of interest in the image can be both classified and quantified using the human eye or computer programs.

Image processing: Technique that can be used with image analysis in which the image of the sample is processed in some way to make it easier to perform further interpretation. Thus, the image quality is im- proved but no analysis or quantification is performed.

Imbibition: Process of absorption, by soaking up a liquid.

Imitation cheese: Product with the appearance and sensory properties of cheese, but which is differ ent from genuine cheese in composition. It may be based on soybeans rather than milk.

Imitation cream: Product with the appearance and sensory properties of cream, but which differs from genuine cream in composition. It is usually prepared with vegetable proteins and vegetable fats as substitutes for milk-based components. It is also called non-dairy cream.

Imitation dairy products: Substitutes for dairy products, with vegetable-based components (often soy products) usually replacing all or part of the milk constituents. Products have the appearance and sensory properties of dairy products, but differ in com- position. Nutritional properties of the imitation products may not match those of the dairy products they are intended to replace. It is commonly produced types include imitation cheese, imitation milk and imitation cream.

Immobilization: Process by which microbial, plant and animal cells, and macromolecules (e.g. enzymes) are attached to solid surfaces or entrapped within gels. They can then be used in applications such as bioconversions and biotransformations, affinity chromatography and biosensors.

Immobilized cells: Microbial, plant and animal cells that have been attached to solid surfaces or entrapped within gels. It can be used in bioconversions not possible with isolated enzymes and in biosensors. Entrapment is the most commonly used method for immobilization; gels used include agar, alginates, carrageenans, polyacrylamides and polyurethane.

Immobilized enzymes: Enzymes that have been attached to solid surfaces or entrapped within gels. Immobilization methods include covalent attachment or ionic binding to solid carriers or supports (e.g. celluloses, synthetic polymers and DEAE-cellulose), cross-linking with bifunctional reagents, encapsulation (e.g. in liposomes) and entrapment within gels. Immobilized enzymes often offer a number of advantages over free enzymes, such as ease of reuse and in- creased stability.

Immune response: Reaction of the body to foreign substances (antigens). Antibodies produced by lymphocytes in response to the antigens can destroy the antigens directly or label them in a way that makes them susceptible to attack by white blood cells. White blood cells specific to the antigens (T-cells) may also be produced. It is synonymous with immunological response.

Immunoaffinity chromatography: Chromatography technique in which the stationary phase (immunosorbent) is prepared by immobilizing antibodies specific to the analytes of interest onto the surface of a rigid or semi-rigid support. It is used as a clean up or pre-concentration step in an analytical procedure as well as a separation technique.

Immunoassay: Analytical technique in which sub- stances are measured using specific antibodies that binds to the corresponding antigens. Binding is measured by use of antibodies labelled with radioactive isotopes, enzymes (enzyme immunoassay) or fluorescent dyes.

Immunochemical analysis: Analytical techniques in which specific immune reactions are em- ployed in the investigation.

Immunofluorescence: Immunological technique in which antibodies labelled with a fluorescent dye are used to detect antigens in the samples.

Immunogenicity: Extent to which a substance can cause an immune response. It is affected by a number of factors, including nature of the substance, dose and previous exposure of the host.

Immunology: Science concerned with the way in which the body reacts to foreign substances. It includes immunity, components of the immune system and diagnosis of disease.

Impedance: Opposition to the flow of current in an electrical circuit.

Impellers: Devices for driving an item forwards, employed in food processing.

Impingement drying: Drying technique originally used for paper and textiles but more recently applied to foods. Gas jets are arranged in such a way that the gas, e.g. superheated steam or hot air, impinges perpendicularly on the food to be dried. The gas is directed at high velocity, removing moisture from the surface of the food. Processing time is reduced compared with that required for other types of drying.

Imports: Goods or services that are produced abroad but purchased for use in the domestic economy.

Improvers: Additives that improve the quality of the final product. It is used predominantly in the bakery industry. It includes flour improvers which enhance the breadmaking properties of flour.

Induction heating: Heating, e.g. of foods, by production of an electric or magnetic state by the proximity (without contact) of an electrified or magnetized body.

Infant foods: Foods designed to meet the nutritional needs of infants, such as infant formulas and weaning foods. A wide range of processed infant foods is available in industrialized countries, including rusks, pureed ready meals, fruit drinks and cereal-based dishes. Foods are typically fortified with minerals and vitamins, and designed to be low in sugar and salt.

Infant formulas: Liquid foods for infants used as a substitute for human milk. It is usually take the form of modified cow milk products (milk infant formulas), which aim to mimic the composition of human milk. Formulas may also be based on milk from other species, soymilk or other products in order to meet the nutritional needs of infants suffering from intolerance to cow milk.

Infant milk formulas: Preparations for feeding to infants and young children, intended to satisfy their specific nutritional requirements. It may be based on cow milk or soymilk. It is also called human milk substitutes.

Infectivity: Ability of pathogens to become estab- lished within or on the tissues of a host, or the capabil ity of pathogens to be transferred from one organism to another.

Infestation: Condition in which a host is occupied or invaded by parasites, e.g. ticks, lice or mites which may live on the surface of a host, or worms which may live within the organs of a host.

Information processing: Evaluation of data using a computer, to generate usable information.

Infrared (IR) irradiation: Electromagnetic radiation having a wavelength just greater than that of red light but less than that of microwaves, emitted particularly by heated objects. Application of infrared radiation is to extend shelf life of foods.

Infrared: Section of the electromagnetic spectrum of which the radiation has lower energy than the visible spectrum and wavelengths ranging from 750 nm to 1 mm.

Infuser: A small perforated closed container which allows boiling water to come into contact with tea, herbs, spices, etc. thus extracting flavour from the solid.

Infusing: It is process of bringing a liquid into contact with some food, herb or spice so as to transfer aroma, taste and soluble components to the liquid from which the solid material is strained. e.g. brewing of coffee or tea, extraction of vanilla from a pod and of aromatics from a bouquet garni.

Infusions: Extracts produced by soaking a substance, usually of plant origin, e.g. spices, teas or fruits, in a solvent, usually water. Solvent-soluble components, including flavour compounds and aroma compounds, leach out from the material into the solvent.

Inositol phosphates: Antinutritional factors found in foods, especially cereals and legumes, which can compromise the absorption of minerals from the gastrointestinal tract. It may be present in a range of forms, from bis phosphates up to hexaphosphates (also known as phytates). To improve the nutritional value of foods, both exogenous and endogenous phytases can be utilized to hydrolyse the higher inositol phos- phates into lower phosphates, which generally have lower capacities to bind minerals.

Inositol: A water-soluble carbohydrate which is found in fruits and cereals either free or combined. It has a role in fat metabolism and also appears to be essential for the transmission of nerve impulses as the concentration in nerves is much higher than in the blood and is correlated with the speed of response, since both decline with age. It is also a major constituent of human semen. It has no known toxicity. A polyol which is occur widely in foods as the free form, as inositol phosphates or as a component of phosphatidylinositol. It participates in cell signalling as a part of a membrane secondary messenger system and can also act as an antinutritional factor.

Insect foods: Insects that are eaten as foods in many parts of the world, including China, Japan and rural areas of Africa and South America, where they can serve as a valuable and readily available source of proteins and minerals. Types of insect consumed include grasshoppers, crickets, locusts, bees and ants. Most species are roasted, fried or boiled prior to consumption, although a few are eaten live. Insect foods are generally regarded as taboo in the Western World, although some insect products are available as novelty foods.

Insecticides: Chemical substances used to kill insects. It is used primarily to control pests that infest crops or to eliminate potential disease-carrying insects in specific areas. It is classified into several groups, the most important of which are carbamate insecticides, fumigant insecticides, organochlorine insecticides, organophosphorus insecticides and pyrethroid insecticides. Residues persisting in foods and the environment can represent a health hazard.

Instant beverages: Dried beverages formulated and processed in a manner giving rapid solubility in water or other liquids.

Instant coffee: Dried (generally freeze dried) coffee extracts processed to a form which dissolves rapidly in water.

Instant flour: Wheat flour processed to make it easily soluble in hot or cold liquids.

Instant food: Food which has been prepared and dehydrated in such a way that it is immediately useable on adding (boiling) water. Processed foods that have undergone instantization, so that they can be easily and rapidly reconstituted by bringing them into contact with a liquid such as milk or water. Common instant foods include gravy granules, instant noodles, milk powders, instant coffee and tea powders.

Instant noodles: Noodles that have been pre-cooked and reconstitute rapidly when hot water is added to them.

Instant soups: Dried soup mixes that are designed to rehydrate rapidly upon addition of water. It is often prepared by freeze drying. Typically it is sold as convenience snack foods/beverages in single serving sachets.

Instant tea: Dried (generally freeze dried) tea extracts processed to a form which dissolves rapidly in water.

Instantization: Processing of dried foods in a way that facilitates preparation or reconstitution of the final product. Common techniques used in instantization include agglomeration of particles and lecithination.

Insulin: One of the mammalian endocrine hormones. This polypeptide is synthesized in the pancreas in re- sponse to elevated blood glucose levels. Deficiencies in secretion of insulin or physiological responses to insulin occur in type I (insulin-dependent) and type II (non-insulin dependent) diabetes mellitus, respec- tively. Diet can be used to control type II diabetes, and information regarding postprandial blood insulin and glucose responses to foods (their insulinaemic and glycaemic index values) is useful in dietary control of this disease.

Integrated pest management (IPM): Approach to crop pest's control that uses a combination of various physical, chemical and biological pest control tactics in an attempt to reduce reliance on chemical pesticides, and hence minimize harmful residues in crops and pollution of the environment. Pest control tactics em- ployed include biological control, use of conventional plant breeding or genetic engineering to improve crop resistance to pests, use of agricultural practices that lessen the degree of pest damage (e.g. mixed cropping, time of planting), and selective use of insecticides or other chemical agents (e.g. insect growth regulators).

Interesterification The process by which fatty acyl residues are interchanged between triglycerides in a mixture of lipids. Can be catalysed by lipases, and may be used to modify the composition and properties of fats and oils.

Interfacial tension: Attractive force between molecules at an interface.

Interferometry: Analytical technique based on differences in refractive index between the sample under investigation and a standard. Measurements are made on an interferometer, an optical instrument in which a beam of light is split and subsequently reunited after traversing different paths, producing interference.

Intermediate moisture foods (IMFs): Semi-moist foods, which do not require refrigeration and can be eaten without further preparation. Foods are preserved by limiting water activity to a level unable to support microbial growth, e.g. by addition of humectants. Examples of intermediate moisture foods include dried fruits, beef jerky and semi-dried sausages.

Intermittent warming: Warming of commodities, such as fruits and vegetables, to room temperature at intervals during storage to prevent chilling injury symptoms from developing. Chilling injury is a problem in most crops of tropical or subtropical origin. Symptoms of chilling injury, such as pitting, discol- oration, internal breakdown and decay, can result in large postharvest losses during marketing. Intermittent warming may, however, cause undesirable softening, increase decay, and cause condensation to form on the product.

International Organization for Standardization (ISO): A network of the national standards institutes of some 130 countries, with a central office in Geneva, Switzerland, that coordinates the system and publishes finished standards.

Intestines: The portion of the gastrointestinal tract which extends from the lower opening of the stomach to the cloaca or anus. Intestines of slaughtered animals form a part of edible offal; after cleaning they may be used as casings for the production of meat products, e.g. sausages.

Intolerance: Group of diseases in which there is inability to digest a particular dietary constituent properly, often resulting in malabsorption syndromes. Examples include lactose intolerance, resulting from lack of a gastrointestinal tract brush border enzyme, and coeliac disease, in which an immunological response to wheat gluten results in histopathological changes to the intestinal mucosa. Exclusion of the relevant component from the diet can result in elimination of the symptoms of the disease, and also, in cases such as coeliac disease, reversal of intestinal pathology.

Inulin: Polysaccharide composed mainly of fructofuranose residues (fructose in the ring conformation) although it also contains a glucopyranose residue. Inulin occurs naturally in some plants, e.g. Jerusalem artichokes and chicory, where it replaces starch as an energy store. It is a prebiotic and is selectively metabolized by, thus encouraging the growth of, probiotic bacteria in the colon

Invert sugar Syrup: Syrup composed of a mixture of glucose and fructose manufactured from sucrose by acid hydrolysis or action of β-fructofuranosidases (invertases). It uses include as a substrate for manufacture of sorbitol or mannitol, and in sweeteners and humectants.

Invert sugar: Simple sugars (monosaccharides) whose configuration has been changed by heat, enzyme or acid treatment so that they will not crystallize and will prevent other sugars from crystallizing.

Invertase: An enzyme obtained from *Saccharomyces cerevisiae* that is used to prevent granulation in thick sugar syrups and fondants

Iodates: Salts containing an IO_3 anion. It includes potassium iodate oxidizing agents, which are added to wheat dough during breadmaking. Iodates are also added to table salt (NaCl; sodium chloride) and infant formulas for iodine fortification of the diet.

Iodine (I): One of the halogens occurs naturally in the diatomic form I_2, and is a bluish-black solid which sublimes to form a bluish irritant gas. An essential dietary mineral which is accumulated in the thyroid gland and used to synthesize the thyroid hormones, including thyroxine, which are important for normal growth and development. Foods particularly rich in iodine include seaweeds and marine fish. Low dietary intakes of iodine can cause hypothyroidism and associated iodine deficiency diseases such as goitre. Fortification of the diet with iodine in the form of iodates or iodides is common.

iodine A trace element required for health especially of the thyroid gland, now included in iodized salt but available particularly from shell fish, seawater fish and seaweed

Iodine values: Measure of the unsaturation of fats or oils, based on the amount of iodine absorbed in a given time. It is also known as iodine number.

Iodized salt: Ordinary salt (NaCl) fortified with 0.01% potassium iodide in order to overcome any deficiencies of iodine in the diet. It is added to foods in order to prevent iodine deficiency.

Iodometry: Redox analysis technique based on reac- tion with iodine/iodide. Strong reducing agents are determined by titration with iodine while strong oxidizing agents react with iodide to form iodine. Iodine is titrated with a standard solution of thiosulfate, using a starch solution as an indicator.

Ion chromatography: Chromatography technique allowing simultaneous determination of anions and cations in a sample by using a sequence of a cation exchange resin column, a detector, an anion exchange column and another detector.

Ion exchange chromatography: Chromatography technique in which separations are carried out on ion exchange resins. Ions from the sample solution that pass into the exchangers are displaced by varying the pH, concentration or ionic strength of the eluting liquid, usually using a gradient. Separation is based on anion exchange or cation exchange depending on the type of resin used.

Ion exchange: Reversible process in which substitu- tion of ions for others of the same charge occurs. Solu- tion containing ions is passed through a molecular network containing groups that can be ionized. Ions in the solution attach to the network, releasing free or mobile ions from the network. The reaction is classified according to the nature of the substituent groups in the network, i.e. cation exchange or anion exchange. Substances acting as ion exchangers or ion exchange resins include aluminosilicates, cross-linked polymers and celluloses. This process is the basis of separation by ion exchange chromatography.

Ionic strength: Parameter which is a function of the charge and concentration of ions in a solution.

Ionization: Process by which a neutral substance becomes charged, forming ions. The conversion is due to the addition or removal of electrons induced by various means, including heating, chemical reaction, exposure to ionizing radiation or passage of an electric current.

ionizing radiation(IR): Electromagnetic radiation which causes chemical changes in materials through which it passes, in particular ionization of compounds and production of short-lived free radicals which cause the death of living organisms. It is used for the sterilization of food in hermetically sealed packs, for the reduction in amount of spoilage flora on perishable foods, for the elimination of pathogens in foods, for the control of infestation in stored cereals, for the prevention of sprouting of root vegetables in storage and for retardation of the development of picked mushrooms (e.g. opening of the caps). Unfortunately there is no current legislation requiring the labelling of foods so treated.

Ionones: Volatile aroma compounds found particularly in fruits, wines and tea. One of the major ionones, β- ionone, has a violet-like aroma.

Ions: Electrically charged atoms or groups of atoms. Positively charged cations result from the loss of electrons and negatively charged anions from their acquisition.

Ipomeamarone: One of the toxic phytoalexins formed in sweet potatoes as a result of mechanical injury or fungal infection.

IR spectra: Absorption patterns resulting from IR spectroscopy analysis of samples. It is serve to analyse the composition of samples, and identify impurities.

Iron (Fe): It forms salts in either the ferric {iron (II)) or ferrous (iron (III)} oxidation states. One of the essential minerals, iron is required to synthesize ferritin, lactoferrin, haemoglobin, cytochromes and other haemoproteins. Iron deficiency in the diet can lead to anaemia, and non- soluble iron salts such as ferrous sulfate are used for fortification purposes. Good sources of iron include meat and meat products, cereals and green vegetables. Bioavailability of iron in the diet is influ- enced by the presence of other chemicals such as calcium and phytates.

Irradiated foods: Foods subjected to irradiation to improve shelf life and eliminate harmful bacteria, insects and other pests. Types of food that can be successfully irradiated include poultry meat and red meat, fruits, vegetables and cereals. Regulations vary between countries as to which (if any) foods may be irradiated. Irradiated spices are currently the only irradiated foods licensed for sale in the UK.

Irradiation: Application of various forms of radiation. In food processing, this can be exposure of items to low doses of high-frequency energy from gamma rays, X-rays or accelerated electrons with the aim of extending shelf life. These rays contain sufficient energy to break chemical bonds and ionize molecules that lie in their path. The two most common sources of high-energy radiation used in the food industry are cobalt-60 (^{60}Co) and caesium-137 (^{137}Cs). For the same level of energy, gamma rays have a greater penetrating power into foods than high speed electrons. The unit of absorbed dose of radiation by a material is denoted as the gray (Gy), one gray being equal to absorption of one joule of energy by one kilogram of food. In which, food is treat with ionizing radiation in order to kill all microorganisms and insects in fruits, cereals, pulses, dried fruits and the like. It also prevents sprouting of roots and tubers. It does not destroy toxins or viruses and may cause chemical changes in the food. It is often used to preserve sealed packs of food and to increase the shelf life of perishables. Since no laws govern the labelling of irradiated food it is impossible to know whether it has been treated in this way. Some authorities consider it a dangerous practice.

Isoamyl acetate: Ester with banana-like odour. One of the natural aroma compounds found as a result of yeast fermentation in beer, sake and wines, and also occurs naturally in fruits such as apples and bananas. Widely used as an added flavour compound in processed foods. It can be produced in microbial fermentations and also enzyme bioconversions.

Isoamyl alcohol: One of the aliphatic alcohols, with a characteristic odour and pungent taste. Synonyms include isopentanol, methyl butanol and isopentyl alcohol. It is used as an esterification substrate for production of isoamyl esters. It is also identified as one of the aroma compounds present in wines, cider and beer as a result of yeast fermentation.

Isoelectric points: pH at which the net charge on a molecule is zero. At their isoelectric points, proteins will not migrate in an electric field.

Isoenzymes: Multiple forms of enzymes that catalyse the same reaction but which differ in characteristics such as primary structure, kinetics, electrophoretic mobility and immunological properties.

Isoflavones: Subclass of the flavonoids, sharing a basic structure of two benzyl rings joined by a three carbon bridge which may or may not be closed into apyran ring. Isoflavones differ from flavones in that the benzyl B ring is joined at position 3 instead of position 2. These phytochemicals are more restricted in occurrence than other flavonoids, but can be found in several legumes, including soybeans, lentils, peas and mung beans. Soybeans and soy products provide a major dietary source of isoflavones, including daidzein and genistein, which display activity as phytoestrogens.

Isoflavonoids: A subclass of the flavonoids which includes isoflavones.

Isoglucose: Fructose sweetener prepared from starch. Starch is dispersed in water and hydrolysed to produce glucose syrups, and the glucose is then isomerized to fructose via a reaction catalysed by glucose isomerases. When produced from corn starch, isoglucose preparations are known as fructose high corn syrups.

Isohumulones: Components of the hops-derived iso-α -acids fraction in worts and beer. It formed by isomerization of humulones during boiling of worts or preparation of isomerized hop extracts. It is an important bitter compound in beer.

Isomalt: Polyol or dissacharide alcohol produced by reduction of sucrose and composed of a mixture of glucomannitol and glucosorbitol. Isomalt has approximately half the sweetness of sucrose and half the calorific value. It has low solubility in water and is used in the manufacture of hard sugar confectionery, chewing gums, preserves and bakers confec- tionery.

Isomaltooligosaccharides: Oligosaccharides produced from starch or corn powder. It is used in the form of syrups as low calorie sweeteners. It is effective in stimulating the growth of *Bifidobacterium* species and also beneficial in preventing dental caries, improving intestinal function and enhancing immune response in humans. It is potentially useful ingredients for functional foods.

Isomerization: Reaction in which the structure of a molecule is altered so that it is converted into one of its isomers.

Isomers: Series of compounds that have the same mo- lecular formula but which differ in structure (structural isomers) or orientation (stereoisomers).

Isopropyl alcohol: A solvent used for food colours and flavourings and in glass cleaners.

Isothiocyanates: Organic compounds containing a nitrogen-carbon-sulfur unit. It is structural isomers of thiocyanates. Many isothiocyanates are pungent volatile compounds released upon damage to tissues, for example in *Brassica* species. Allyl isothiocyanate, which has antimicrobial activity, contributes to the pungency of mustard, watercress, horseradish and wasabi.

Isotonic drinks: Beverages which are isotonic with normal human body fluids, and contain components such as electrolytes and sugars. It is claimed to enhance performance and recovery from physical exertion.

J

Jaggery or Gur: Unrefined brown coloured sugar produced mainly in India by evaporation of sugar cane juices. Or a crude brown sugar made from the sap of coconut or palmyra palm trees, sold in round cakes or lumps.

Jam: A fruit preserve made by mixing fruit or boiling it with sugar and water and adding extra gelling agent if required, usually pectin. The high sugar concentration inhibits but does not totally prevent proliferation of microorganisms, so jam is usually bottled hot in sterile jars with a seal, or sterilized after bottling.

Jasmine: Natural flavourings with warm, spicy characteristics derived from flowers and leaves of jasmine (*Jasminus* spp.). Predominant flavour compounds and aroma compounds include jasmonates, jasmones, benzyl acetate, indol and eugenol.

Jellied milk: Milk to which is added sugar, flavourings, thickening agents and gelling agents. It is also known as jellified milk.

Jellies: Small, soft sweets, usually fruit flavoured, of gelatinous texture, made in various shapes and often coated with sugar. The singular term, jelly, is used to refer to jam-like products, usually clear, that are made from strained fruits containing pectins which are boiled with sugar. Also refers to soft, semi- transparent foods prepared from gelatin which are sweetened, flavoured, cooled in a mould and eaten as desserts.

Jointing: Cutting of animal carcasses into joints.

Juice extractor: An implement for manually or mechanically extracting juice from fruits. For citrus fruits, it usually consists of a ribbed convex cone, on which half the citrus fruit is turned, above a slotted base. Other fruits are generally liquidized and strained.

Juice: The liquid expressed or released from fruits, vegetables or even meats by any of several methods, squeezing, pressing, liquidizing, freezing and thawing, heating, etc. Juice is usually strained to remove solids

Juiciness: Sensory properties relating to the extent to which products, such as fruits, vegetables and meat are juicy or succulent. Dependent on the amount of cell sap released during fracture. Uncooked cell walls fracture across to reveal the internal structure of the cell, thus releasing sap. In contrast, cooked cells tend to separate along the middle lamella, so cell sap is not released in the same way.

Junk food: Food with plenty of Calories, generally in the form of cheap sugar, fat, starch and recovered protein, highly seasoned and flavoured if savoury, but with few or no vitamins or trace elements, & if the major source of energy will lead to suboptimal health

Jute: Rough fibre made from the inner bark of tropical plants belonging to the genus *Corchorus*, especially *C. olitorius* (in India) and *C. capsularis* (in China). Jute fibre is used to make jute board, a strong flexible cardboard often used to make shipping cartons. Also woven into sacking and used for making wrap- ping paper and twine.

K

Kachauri: A type of pitta bread made into a pocket and stuffed with a spiced bean mixture.

Kalakand: Sweetened dairy product that is popular in India. It is made by evaporating acidified buffalo milk.

Kanji: Traditional Indian beverage made from black carrots. Peculiar to the northern plains of India, black carrots are black on the outside but a rich red colour under the skin. The carrots are parboiled in water with salt and other flavourings such as ground mustard seeds and chilli powder. The mixture is then left to ferment in the sun, resulting in a sour and spiced red drink which is consumed as an accompaniment to meals.

Kava: Beverage made by aqueous extraction of powders prepared from rhizomes of kava kava (*Piper methysticum*). It is used in the south Pacific region as a narcotic/stimulant. It is also used in treatment of anxiety and a range of disorders. The pharmacologically active components are lactones. Non-addictive, but can have adverse effects, such as muscle weakness, drying of the skin, and liver damage, if consumed over a long period of time or in high amounts.

Kawal: Strong-smelling pastes prepared by fermentation of leaves of the legume *Cassia obtusifolia*. It is rich in protein. It is used as meat substitutes in soups and stews.

KCl: It is chemical formula for potassium chloride. One of the chlorides widely used in food processing at varying levels to replace salt (NaCl), for example in brines, in order to reduce Na levels in foods, and specifically to produce low sodium foods and salt substitutes.

Kebabs: Pieces of meat, fish and/or vegetables grilled or roasted on skewers or spits.

Kefir grains: Traditionally used in the culture of milk during manufacture of kefir. An irregularly shaped mass of microbial polysaccharide, mesophilic streptococci, leuconostocs, mesophilic and thermophilic lactobacilli, and yeasts. Yeasts comprise the grain core, with yeasts and rod-shaped bacteria immediately out- side, and rod-shaped bacteria predominantly at the periphery. Due to problems with blowing of packages, kefir grains are not used in industrial manufacture of kefir, starters consisting almost entirely of streptococci, with few or no yeasts and lactobacilli, being used instead. The product then lacks the foaming char- acteristics of traditional kefir.

Kefir: Alcoholic fermented milk product made from mare's milk, now from cows' milk traditionally by addition of kefir grains to milk. The traditional product contains alcohol and CO_2 in addition to lactic acid, making it foaming and viscous. Since this can cause blowing of packs, starters with few or no yeasts and lactobacilli are used in industrial pro- duction of kefir. Commercial kefir tends to contain much lower amounts of alcohol than traditionally prepared products. Kefir is generally more digestible than milk and more easily tolerated by lactose-intolerant individuals. It is marketed with various fat contents. It contains 0.6 to 0.9% lactic acid and 0.2 to 0.8% alcohols. It is used to treat gastro-intestinal problems in Russia.

Kegs: Small barrels, often used for transportation or storage of alcoholic beverages, especially beer. It may be made from wood, but are commonly made from plastics or metal.

Keratin: One of the structural fibrous animal proteins found in vertebrate skin and specialized epidermal structures, including feathers, nails, hair, hooves, horns and quills. Keratin-degrading microorganisms and serine proteinases (keratinases) are of interest for bioremediation of wastes from slaughterhouses and food factories wastes from meat and carcass processing.

Kernel: The edible, usually central, part of a nut or stone of a fruit.

Ketchup: Nowadays the term refers to various thick piquant sauces containing sugar, spices, vinegar, and other ingredients such as tomatoes, mushrooms, nuts or fruits. Tomato ketchups are one of the most well known types of ketchup and are a popular accompaniment for French fries, burgers and many other foods. A commercially it sold in bottles to be used cold as a condiment.

Ketones: Types of carbonyl compounds in which the carbonyl substituent is bound to two carbon atoms. Many ketones are important volatile aroma compounds in foods and beverages.

Ketoses: Nonreducing sugars in which, in the acyclic form, the actual or potential carbonyl group is nonterminal (ketonic). Many of these sugars, which are monosaccharides, have the suffix '-ulose'. Examples include xylulose and arabino-2-hexulose (fructose).

Kettles: Metal or plastic containers with a lid, spout and handle for boiling water. It is also metal containers for heating any liquids. Fish kettles are long pans specially designed for cooking fish.

Kewra essence: An extract of the flowers of screwpine used in India to flavour meat, poultry and rice dishes.

Khamir (Yeast): natural yeast used for making soft bread. It is propagated in sour dough fashion with flour and yoghurt.

khas khas: An aromatic grass, *Vetiveria zizanioides* (Poppy seeds), used as a flavouring in Indian cooking. It is grown in India and the south of the USA.

Kheer: A dessert made from rice cooked in milk until it is completely disintegrated and sugar, raisins, chopped dried apricots, bruised cardamom seeds and cinnamon stick are added and the whole cooked for 1.5 hours. Almond slivers are added towards the end. The mixture is then cooled and thinned with a little rose water until a thick pouring consistency, put in individual bowls, chilled and decorated.

Khichhari: A dish of cooked rice mixed with a variety of ingredients from a few sprouted seeds or lentils to an elaborate mixture of spices, nuts and raisins.

Khoa or Khoya: Heat-concentrated milk product usually prepared from buffalo milk and popular in India. It is used as the base material for a number of Indian sweets, such as burfi, peda and gulabjamans.

Khurchan: Concentrated milk product popular in India. It is prepared by simmering whole milk and adding sugar.

Kidney beans: Type of common beans (*Phaseolus vulgaris*) with kidney-shaped seeds. Red kidney beans form an integral part of the Mexican dish chilli con carne. Due to the presence of antinutritional factors, such as lectins, beans must be well soaked in water and cooked prior to consumption.

Killer yeasts: Yeasts (including brewers yeasts, wine yeasts and sake yeasts) which secrete protein or glycoprotein toxins able to kill sensitive yeast strains. This may be disadvantageous, if desirable yeast strains are killed, or beneficial if wild yeasts or contaminating yeasts are eliminated.

Kilning: Final stage of malting, in which steeped germinated malting barley is heated and dried to a specified moisture content. This halts metabolism and enzyme activity in the malt. Kilning temperature and duration may be selected to give malts with a range of colour and flavour.

Kilns: Furnaces or ovens for burning, baking or drying. An oast is a kiln used to dry products such as hops and malt.

Kimchi: Fermented cabbage (usually Chinese leaves) or other vegetables, flavoured with salt, garlic and chillies. An acquired taste which provides most of the vegetable nutrients in the Korean diet. A Western substitute can be obtained by mixing chopped Chinese leaves which have been salted to remove excess liquid with minced spring onions, garlic, ginger, chillies,

Kinema: Traditional Indian product made by fermentation of cooked soybeans, usually with *Bacillus subtilis*. It is rich source of protein, with a stringy texture and characteristic flavour. It is consumed as meat substitutes, usually in a side dish with cooked rice.

Kjeldahl nitrogen: Total nitrogen in a substance, determined by digesting the sample with sulfuric acid and a catalyst. Kjeldahl nitrogen is used extensively for determination of protein levels in foods. In these cases, the nitrogen measured is converted to the equivalent protein content by use of an appropriate numerical factor.

Kneading: It is a mixing of stiff dough by repeatedly compressing and folding in or over. This may be done manually or mechanically by using a dough hook at slow speed. In the case of bread doughs made from strong flour, kneading is used to develop sheets of gluten in the mixture and to incorporate air. The air acts as nuclei for the formation of bubbles of carbon dioxide during rising and proving; the gluten sheets facilitate their retention in the cooking dough. The network of **gluten** strands stretches and expands during kneading, so enabling dough to retain gas bubbles formed by the actions of the leavening agent.

Knives: Sturdy and well balanced cutting instruments consisting of a blade fixed into a handle, or blades on a machine for cutting, peeling, slicing or spreading. Most knife blades are made of steel or ceramic zirconia, a hard material that doesn't rust, corrode or interact with food. Knife handles are usually made of wood, plastic, horn or metal. Preferably, the end of the blade should extend to the far end of the handle, where it should be anchored by several rivets. Knives are tai- lored for specific applications. For example, a chef's knife has a broad, tapered shape and fine edge, which is ideal for chopping vegetables, while a slicing knife with its long, thin blade cuts cleanly through cooked meat. Knives with serrated edges are good for slicing softer foods such as bread, tomatoes and cakes. The easy-to-handle, pointed, short-bladed paring knife is ideal for peeling and coring fruits.

Koji: Cereals or beans inoculated with *Aspergillus* or other fungi and used as a starter for a wide range of Oriental fermented foods and fermented beverages, including miso, sake and soy sauces. It acts as a supplier of various enzymes, such as lipases, which contribute to the quality and functional properties of the products.

Koko: Thin, fermented porridge made from corn, sorghum or cassava flour, either singly or in mixtures. It is often consumed as infant foods in Ghana and Kenya.

Kokum: Common name for the tropical tree, *Garcinia indica*, fruits of which are used in preparation of a spice. The dark purple fruits are picked when ripe, dried and the peel removed for use in foods, where it adds colour and a sour, slightly astringent flavour. Used especially in curries, vegetable dishes, chutneys and pickles. Fats prepared from kokum seeds have been used in cocoa butter extenders suitable for use in chocolate and sugar confectionery. Kokum is also known by a variety of other names, including cocum, kokam and Goa butter.

Kosher foods: Foods permitted under Jewish biblical law and prepared in accordance with Jewish dietary code. Laws relate not only to the types of foods per- mitted (e.g. pork and rabbit meat products are non kosher) but also to the methods of slaughter/ preparation, and to food combinations (e.g. meat products and dairy products may not be mixed). Kosher foods are perceived by many as having been prepared to high standards of wholesomeness and hy- giene, and are currently attracting a new market of non-Jewish consumers who use kosher certification as an indication of quality.

Koumiss: A thick slightly alcoholic fermented and sour drink/milk originally made from ass or camel's milk now made from fresh mare's milk. It is produced using a 2-stage fermentation in which lactic acid bacteria are added, followed by yeasts on completion of lactic fermentation. In addition to lactic acid, it contains ethanol and CO_2, giving a light effervescence. It is similar to but more alcoholic than kefir and rather like a gassy alcoholic yoghurt. It is used to treat gastrointestinal problems in Russia.

Kulfi: It is oncentrated frozen milk product similar to ice cream popular in India and Pakistan.

Kusum: Seeds from the kernels of the tree *Schleichera trijuga*, oils extracted from which are rich in arachidic acid and may be used for culinary purposes.

L

Labelling: Process of attaching labels to items to make them identifiable, or the information included on the labels. For foods, information may include barcodes, brand names, trademarks, illustrative matter, and compositional and nutritional details.

Labels: Pieces of paper, plastics or fabric which are attached to and provide information about an item. For foods, this information may include branding or the trademarks of a food company, the geographical origin, date marking, compositional details, health claims, nutritional values and warnings relating to specific ingredients, e.g. nuts. The content of informa- tion on food labels is often governed by legislation.

Lacquers: Liquids consisting of resins, cellulose esters, shellac or similar synthetic substances dissolved in a solvent, such as ethyl alcohol. Dry to form shiny, hard, protective or decorative coatings for plastics, wood, metals and other products.

Lactacins: Bacteriocins synthesized by *Lactoba cillus* spp. that are inhibitory only to other lactobacilli. Lactacin A is produced by *L. delbrueckii subsp. lactis* JCM 1106 and 1107. It has a narrow host range and is heat labile. Lactacin B is produced by *L. acidophilus* N_2, and its synthesis is chromosomally linked. This protein forms aggregates of molecular weight 100,000 Da; however, the actual molecular weight of lactacin B is 6000-6500 Da. Lactacin F is produced by *L. acidophilus* 88, and its synthesis is plasmid linked. It has a broader activity range than lactacin B, and forms ag- gregates of molecular weight 180,000 Da; however, the actual molecular weight of lactacin F is 25,000 Da.

Lactalbumin (alfa): One of the major whey proteins, accounting for approximately 20% of total whey proteins in cow milk. It is rich in tryptophan and cystine. It is found in genetic variants A, B and C that differ in amino acids composition and have a bearing on the properties and yield of milk.

Lactalbumins: Albumins present in milk. The main protein is α-lactalbumin.

Lactase: The enzyme that converts lactose into glucose and galactose.

Lactates: Salts of lactic acid is used as buffers i.e. to maintain near constant pH in foods and as firming agents. Sodium (E325), potassium (E326) and calcium lactate (E327) are used in the food industry. Lactates such as sodium lactate are widely used in foods as preservatives, whilst calcium or iron lactates can be used in food fortification. Lactate concentrations are frequently determined in foods as a measure of lactic acid levels.

Lactation number: It is a value defining the number of lactations undergone by an animal. It can affect physicochemical properties and functional properties of milk.

Lactation stage: Measure of the number of weeks of lactation that have passed since parturition. Lactation is generally divided into three stages during which three distinct secretions are produced: colostrum; transient milk; and mature milk. Colostrum is produced for approximately the 1st week, transient milk for the following 2-3 weeks and mature milk is pro- duced thereafter.

Lactation: Physiological process involving secretion of milk from the mammary gland, usually beginning at the end of pregnancy and controlled by the hormones prolactin and oxytocin. At the beginning of lactation, colostrum is produced, mature milk being secreted later. In cows, milk yield as well as composition varies during lactation. Yield increases up to the 2nd month of lactation and decreases thereafter. Milk protein and fat contents are lowest during the 2nd month, then increase. Free fatty acids contents and proportions of stearic acid, oleic acid and linolenic acid in milk fat increase as lactation progresses, while proportions of short- and medium-chain fatty acids and linoleic acid decrease. Lactose content of milk decreases as lactation proceeds. Contents of immunoglobulins, minerals and trace elements, and activities of some enzymes increase towards the end of lactation.

Lactic acid bacteria (LAB): Gram positive bacteria (e.g. *Lactobacillus*, *Lactococcus*, *Leuconostoc*, *Pediococcus* and *Streptococcus* species) that are capable of lactic fermentation of sugar substrates. It is used extensively in the food industry as starters to initiate lactic acid fermentation in the production of fermented dairy products (e.g. yoghurt and cheese), fermented meat products (e.g. salami), and fermented plant products (e.g. sauerkraut and sour- dough).

Lactic acid: An important food acid which is the end product of anaerobic fermentation of sugars by *Lactobacillus* spp. It is added as a preservative to reduce pH or made naturally in the production of yoghurt, sauerkraut, some cheeses, olives, other varieties of pickles and in continental sausages. It is present in sour milk, molasses, fruits, beer and wines. It is produced via lactic fermentation of sugars by lactic acid bacteria, a process that is an important step in manufacture of cheese, yoghurt and other acidic fermented dairy products. It is also used for acidulating worts in brewing and in preservation of meat products, such as salami and pepperoni.

Lactic beverages: A types of beverages which is manufactured by of lactic fermentation.

Lactic butter: Butter made from a cream treated with species of *Lactobacillus* to give it a slightly sour flavour

Lactic fermentation: Process by which certain bacteria, such as lactic acid bacteria, convert sugars entirely, or almost entirely, to lactic acid (homolactic fermentation) or to a mixture of lactic acid and other products (heterolactic fermentation). Lactic acid bacteria produce either L(+)- or D(-)-lactic acid or both, depending on the specificity of the NAD-dependent lactate dehydrogenases present.

Lacticins: Bacteriocins synthesized by *Lactococ- cus* spp. Lacticin 481 is synthesized by *L. lactis* subsp. *lactis* CNRZ 481, and is inhibitory towards strains of *Lactococcus*, *Lactobacillus*, *Leucono toc* and *Clostridium tyrobutyricum*. Lacticin 481 is produced optimally at pH 5.5. The molecule has a relative molecular mass of 1700, is rich in nonpolar amino acids and contains the unusual amino acid lanthionine.

Lactitol: A modified lactose is used in sugar-free confectionery

***Lactobacillus* spp.:** Bacteria used for the production of lactic acid in cheese and other milk products and in sauerkraut production.

Lactobacillus bulgaricus: One of the bacteria used in starter cultures for yoghurt and continental cheeses.

Lactobacillus caseii: A bacterium associated with the ripening of Cheddar cheese.

Lactobacillus helveticus: One of the bacteria in starter cultures for yoghurt and Swiss cheeses.

Lactobacillus plantarum: One of the bacteria included in starter cultures for the fermentation of cucumbers and olives.

Lactobacillus sanfrancisc: Bacteria used for the leavening of sour dough bread.

Lactococcus lactis: A bacteria which can ferment traditional foods made from rice to prevent transmission of infection to weaning children in underdeveloped countries. It has the advantage of producing a safe form of lactic acid, the laevo isomer (L-lactic acid) which cannot harm children. Other lactic fermenters produce both the L and D forms of the acid which are easily tolerated by adults.

Lactoglobulins: One of protein (globulins) found in milk. The main protein is β-lactoglobulin which accounts for approximately 50% of the total content of whey proteins.

lacto-ovo-vegetarian: A person who will not eat poultry, animal flesh or fish, but will eat eggs and milk products.

Lactose intolerance: An inability to absorb and metabolize the galactose component of disaccharide lactose due to lack of lactases (β-galactosidases) in the small intestinal mucosa. Undigested lactose remains in the intestinal contents, and is fermented by bacteria in the colon, resulting in explosive and watery diarrhoea. Treatment is to omit lactose from the diet. It is common in persons of Chinese origin.

Lactose syrups: Syrups consisting predominantly of lactose. It is manufactured from whey by removal of whey proteins and minerals using ultrafiltration and ion exchange chromatography, respectively. It is used as sweeteners in dairy products, infant formulas and sugar confectionery.

Lactose: Disaccharide sugar comprising glucose and galactose residues linked by glycosidic bonds. Predominant sugar in milk which may be recovered from whey by removal of whey proteins and min- erals, followed by crystallization of lactose monohydrate and, to a lesser extent, anhydrous lactose. Has low sweetness, approximately 16-20% that of sucrose, and is used in sweeteners for sugar confectionery and infant formulas. Some individuals suffer from lactose intolerance due to an inability to digest this sugar. It is also known as milk sugar.

Lacto-vegetarian: A person who will not eat animal flesh, fish, poultry or eggs but will eat milk and milk products.

Lactulose: Nutritive sweetener produced by isomerization of lactose which has 1.5 times more than sweetness of lactose.

Ladle: A cooking implement for transferring liquids or skimming liquids, consisting of a hemispherical metal scoop attached at an angle to a long handle hooked at the other end. It is usually sized in fluid measure.

Laminates: Materials made up of several layers of reinforcing fibres produced by placing layer on layer and bonding the sheets together, usually with heat or pressure. Laminates include fibreglass, plywood and reinforced plastics.

Lard: Soft, white solid fat traditionally obtained by rendering or melting the internal fats from swine. It is rich in a number of fatty acids, including *sn*-2 palmitic, stearic, oleic and linoleic acids; contains cholesterol. It has a bland flavour and odour and used for cooking and baking purpose.

Lassi: Sweetened **fermented milk** beverage popular in India. It is prepared by stirring **sugar**, water and **flavourings** into **dahi**, giving a viscous, white and mild to highly acidic drink. Or yoghurt diluted with water or mineral water (1 part yoghurt to 1 to 2 parts water), used as a refreshing drink either with sugar or flavoured with e.g. cumin and mint. It is sometimes made from salted and sweetened buttermilk.

Lauric acid: One of the medium-chain saturated fatty acids. It contains 12 carbon atoms and has melting point of 44°C. It is synonym of odecanoic acid. It is slight odour of bay oil. It occurs as a triacylglycerol component of milk fats and vegetable oils including rapeseed oils and palm oils, and is a component of several cocoa butter substitutes. It is identified as an aroma component in cheese.

Lautering: Separation of worts from insoluble material in brewing mashes by running off the worts through the perforated bottom of lauter tuns, in which the insoluble solids are retained.

Leafy vegetables: Leafy plants, the stems and leaves of which are used as vegetables.

Leavening: The process by which dough is made to rise due to fermentation by yeasts.

Lecithins: Products comprising phospholipids. It composed of phosphate esters of diglycerides (mostly oleic acid, palmitic acid and/or stearic acid) esterified to choline via the phosphate group. Due to the presence of both polar and non-polar moieties, the molecule forms micelles and has uses as food emulsifiers. It is prevalent in soybeans and egg yolks; by-products in manufacture of soybean oils. Lecithin is also called phosphatidylcholine.

lectins A type of protein found mainly in legume seeds, some of which

Lectins: Carbohydrate-binding proteins or glycoproteins, synonyms include phytohaemagglutinins and agglutinins. Lectins are of non-immune origin and agglutinate cells and/or precipitate glycoconjugates. It is found in many plant foods and can have detrimental properties as antinutritional factors and toxins, or possible beneficial properties including antitumour activity. Lectins are widely used analytically as specific binding and separating agents. It is highly poisonous and believed to act as a protection for the seed against insect and animal attack. They attack different human populations differently depending on their blood type. They are inactivated by boiling at 100°C for at least ten minutes. It has been reported that the lower boiling point of water at high altitude can prevent inactivation. It is also called haemagglutinins.

Legume proteins: Proteins formed in legume seeds, a very good dietary source of protein.

Legume sprouts: It is produced by germination of legume seeds, commonly mung beans, alfalfa, lentils, soybeans and black gram. It is rich in protein, vitamins and minerals. Fresh sprouts are crisp and tender, and are often eaten raw. In dishes, they are cooked for a short period only to avoid wilting. Also available canned.

Legume: The general name for plants whose seeds are enclosed in two-sided pods attached along one join. Sometimes the whole pod is eaten, but more often the seeds only, either in the green or dried state, are consumed. When dried, they are known as pulses and are an important source of protein. Well over 10,000 varieties are known, most of which are edible.

Legumes Vegetables: It is of the family Leguminosae (Fabaceae). The seeds or beans are contained in pods. Edible products include dry seeds (beans or pulses), immature green seeds, oilseeds (such as soybeans), green pods, spices, shoots, leaves and sprouts. Rich sources of good quality proteins, but generally low in fat (exceptions include peanuts, soybeans and chick peas). It is also good sources of dietary fibre and some B vitamins. Carotenes, vitamin C and vitamin E can be obtained from immature seeds, pods, leaves and sprouts. Some seeds also contain antinutritional factors or toxins that can cause diseases. These can usually be destroyed by careful processing of the seeds.

Lemon essential oils: Distillates of lemon peel used as flavourings. The active component of lemon oils is citral, a mixture of the terpene aldehydes neral and geranial.

Lemonade: Effervescent or still beverages made from lemon juices, or, more generally, carbonated beverages with a lemon flavour. It may be added to spirits before consumption.

Lemongrass oils: Essential oils produced by steam distillation of fresh lemongrass, comprising approximately 65-75% citral.

Lemons: Yellow citrus fruits (*Citrus limon*) that are extremely rich in vitamin C. Total sugar content is relatively low for a citrus fruit. Its citric acid content of approximately 5% makes it too acidic for eating as a dessert. However, lemon juices are widely used as food and beverage flavourings, and lemon peel is also used in foods.

Lentils: Seeds of the legumes *Lens culinaris* or *L. esculenta*, rich in proteins and carbohydrates. It is used to make dhal, in soups or in snack foods. Flour made from the seeds can be used as an ingredient in cakes and infant foods. Young pods of the plant are eaten as a vegetable.

Leucine: One of the essential amino acids. A common protein constituent and free amino acid is found in many foods. Leucine is also a precursor of several aroma compounds and participates in the Maillard reaction. It is produced industrially by fermentation of *Corynebacterium glutamicum*.

Leuconostoc mesenteroides: One of the bacteria included in starter cultures for the fermentation of cabbage, cucumbers and olives and the production of cottage cheeses.

Leuconostoc: Genus of Gram positive, facultatively anaerobic coccoid lactic acid bacteria of the family Streptococcaceae. Occur in dairy products and in fermenting vegetables and fermented beverages. Species may be used as starters in the production of fermented foods. *Leuconostoc mesenteroides* subspecies *cremoris* strains are used as starter cultures in the production of fermented dairy products (e.g. fermented cream, cheese, kefir, buttermilk).

Leukocytes: White, nucleated blood cells that lack haemoglobin, which are found in blood and lymph. It is formed in lymph nodes and bone marrow. It can produce antibodies and move through the walls of vessels to migrate to the sites of injuries, where they surround and isolate dead tissue, foreign bodies and bacteria. There are two major types: those with granular cytoplasm (granulocytes), which include basophils and neutrophils; and those without granular cytoplasm, such as lymphocytes and monocytes.

Light: Source of illumination that makes objects visible; electromagnetic radiation in the wavelength range 390- 740 nm.

Lignin: Random phenylpropanoid polymer component of plants, where it confers strength, rigidity and resis tance to degradation. Lignin is one of the most abun- dant biopolymers, and a major component of insoluble dietary fibre in plant foods.

Lignocelluloses: Complex of lignin and celluloses found in the cell walls of plants, and a component of dietary fibre in plant foods. Plant-derived wastes such as pomaces and bagasse contain lignocelluloses, and these wastes can be hydrolysed chemically or enzymically to release sugars which can be used as microbial fermentation substrates, for example for ethanol synthesis.

Lima beans: Seeds produced by *Phaseolus lunatus*. Variable in size, shape and colour. Rich in proteins and a good source of vitamin A, vitamin C, some of the vitamin B group, fibre and potassium. As well as dried beans and immature beans (often canned or frozen), pods and leaves are also eaten. Mature seeds can contain toxic hydrocyanic acid, which is destroyed by soaking and boiling in water before consumption. Also known as butter beans, sieva beans and Madagascar beans.

Lime essential oils: Essential oils from limes produced by compression of peel or distillation of mashed lime pulps or juices. It is used as flavourings, particularly in carbonated beverages, such as cola beverages. The predominant flavour compound pre- sent is terpineol which is produced from citral during distillation.

Liming: One of several sugar processes is used for purification of sugar juices. It involves addition of some form of lime, e.g. calcium oxide, milk of lime slurry of calcium hydroxide) or calcium saccharate, to sugar juices and heating. The lime neutralizes organic acids present and forms insoluble lime salts with the impurities. Suspended particles from the sugar cane or sugar beets that remain after filtration associate with the precipitate formed. Forms of liming include cold liming, hot liming and intermittent liming; these differ with respect to the order in which addition of lime and heating are carried out.

Limonene: One of the monoterpenoid aroma compounds, with lemon-like odour. Found in citrus fruits and their products, including citrus juices and citrus essential oils. Also found in dill and caraway seeds.

Limonin: One of the main bitter compounds found in citrus fruits. Limonin and other limonoids are highly oxygenated triterpenoids of interest as anticarcinogenic phytochemicals.

Limonoid glucosides: Limonoids with carbohydrate (glucose) substituents; in contrast to limonoids, the glucosides are generally non-bitter. Over 17 different limonoid glucosides have been isolated from citrus fruits, and limonoids are mainly accumulated as glucoside derivatives in mature citrus fruit tissues. Along with limonoid aglycones, the glucosides show possible anticarcinogenicity.

Limonoids: Highly oxygenated triterpenoids found predominantly in citrus fruits. Over 35 limonoids have been identified in citrus species, and many are bitter compounds. Limonoids possess anticarcinogenic activity and also antifeedant activity against insects and termites.

Linalool: One of the monoterpenoid aroma compounds, with floral/sweet/citrus aroma characteristics. Linalool is found naturally in many foods and beverages, and is also added as a flavour compound to processed foods.

Linalyl acetate: Ester with sweet/floral aroma characteristics. This flavour compound is found in several plant essential oils, including bergamot oils, sage oils and citrus oils.

Linamarin: One of the cyanogenic glycosides, linamarin is found in cassava roots. This toxin has to be removed by processing, generally fermentation, before cassava can be eaten safely.

Linoleic acid: One of the polyunsaturated fatty acids, synonym octadecadienoic acid. It contains 18 carbon atoms and 2 double bonds at positions 9 and 12. Linoleic acid is an essential nutrient in mammals, and is present in many plant and animal foods. However, there is some concern that excessive consumption of linoleic acid is harmful as it has been associated with increases in cancer and coronary heart diseases.

Linolenic acid: One of the polyunsaturated fatty acids, synonym octadecatrienoic acid. It contains 18 carbon atoms and 3 double bonds at positions 9, 12 and 15 (ϒ-linolenic acid) or at positions 6, 9 and 12 (ϒ-linolenic acid). γ-Linoleic acid is an essential nutrient in mammals, and is found in many plant oils, especially linseed oils. ϒ-Linolenic acid is found in several plant oils, particularly in evening primrose oils, and is also found at low levels in animal lipids, including those of human milk. ϒ-Linolenic acid is a precursor for arachidonic acid and the prostaglandins.

Linseed oils: Yellow to amber viscous vegetable oils obtained from flax seeds, *Linum usitatissimum*. It is rich in iodine and α-linolenic acid. Polymerize on exposure to air, resulting in thickening. Used as a food oil. It is also known as flax seed oils.

Lipases: Enzymes that hydrolyse tri-, di- or mono- acylglycerols at a lipid-water interface to form free fatty acids and either di- or mono-glycerides, or free glycerol. The term usually refers to triacylglycerol lipases, which act on triglycerides. It can cleave various natural lipids and oils, such as olive oils, soybean oils, coconut oils, butterfat, and pork and beef fats, and can show positional-, fatty acid- or stereo-specificity. Useful for enhancing of flavour during cheese ripening and, due to their esterification, interesterification and transesterification activities, for production of modified esters and lipids, speciality fats and cocoa butter substitutes. Lipases are also active in organic solvents.

Lipids: It is naturally occurring organic chemicals that are characteristically poorly soluble in water but are soluble in organic solvents. Lipids constitute one of the four main classes of compounds found in living tissues, and also one of the major nutrient types, and as a class include oils, fats, fatty acids, long-chain (or fatty) alcohols, triglycerides, phospholipids, waxes, steroids, terpenoids and some hormones and vitamins.

Lipolysis: Hydrolysis (splitting) of lipids by lipases to yield glycerol and fatty acids.

Lipolytic enzymes: Encompasses lipases, lipoprotein lipases and phospholipases.

Lipopolysaccharides: Complexes formed between polysaccharides and lipids. Lipopolysaccharides are an important component of the outer membrane of Gram negative bacteria and are key determinants of antigenicity and toxicity.

Lipoproteins: Conjugated molecules containing proteins and lipids. The lipid may be a phospholipid, triglyceride or cholesterol, or a mixture of these. Serum lipoprotein and lipoprotein-cholesterol profiles are frequently measured as an indicator of diet and health. Also, oxidation of serum low density lipoprotein (LDL) is implicated in the aetiology of coronary heart diseases. Certain functional food constituents such as flavonoids from green tea and red wines have the ability to inhibit LDL oxidation due to their antioxidative activity.

Liqueur: A sweetened alcoholic extract of various herb-, spice- and fruit-based flavourings, used as a drink but also as a flavouring in many dishes, e.g. cointreau, Kümmel, kirsch

Liqueurs: Alcoholic beverages made from spirits or neutral alcohol with addition of other ingredients such as sugar and flavourings.

Liquid egg whites: Pasteurized egg whites in liquid form. Processing conditions confer a long shelf life and ensure that they are free of *Salmonella* contamination. It may be used in the manufacture of meringues and cakes.

Liquid egg yolks: Pasteurized egg yolks in liquid form. Processing conditions confer a long shelf life and ensure that they are free of *Salmonella* contamination. It may be used in the manufacture of mayon naise and salad dressings.

Liquid egg: Pasteurized egg whites, egg yolks or whole eggs in liquid form. The long shelf life and *Salmonella*-free status of such products make them suitable for use by food manufacturers and caterers.

Liquid membranes: Thin layers of liquid, separating two phases: a process stream and a stripping phase. Impurities, e.g. metal ions, can be extracted almost completely by a carrier that is dissolved in the liquid membrane. On the other side of the membrane, strip- ping takes place. While the carrier is stripped continu- ously, the driving force for the extraction remains high. Types of liquid membranes in use include: bulk liquid membranes; emulsion liquid membranes; thin sheet supported liquid membranes; hollow fibre supported liquid membranes; two module hollow fibre supported liquid membranes; and spiral wound membranes.

Liquid nitrogen: Nitrogen gas (N_2) that has been cooled to a temperature less than or equal to 77.4 K, thus existing in a liquefied state.

Liquid smoke: Oil or water extracts of smoke produced from burning woods, often maple, oak or mesquite. It imparts a smoky flavour to foods.

Liquid whole egg: Pasteurized blend of egg whites and egg yolks in liquid form. Processing conditions confer a long shelf life and ensure that the product is free of *Salmonella* contamination. It may be used in the manufacture of doughnuts, cookies, mayonnaise, salad dressings and egg noodles.

Listeria monocytogenes: A bacterium causing illness which grows in soft ripened cheeses (unpasteurized milk has been wrongly implicated), pâtés, shop-prepared salads, etc. It will grow in these foods at less than 4°C with a doubling time of 18 hours. Samples cultured at 37°C normally have a doubling time of 7.4 hours but after subjecting them to cold shock at 4°C the doubling time drops to 2.5 hours, at the same time refrigeration appears to select for more virulent strains. The incubation period is up to 4 weeks and the resulting illness can range from a general feeling of malaise to meningitis and septicaemia. It is more likely to cause stillbirths and miscarriages, and for this reason pregnant women are recommended to refrain from the food items mentioned unless home cooked. It does not grow in raw milk farmhouse cheeses due to the low pH (less than 5), and better hygienic practices. Subsequent pasteurization cannot be used as a let-out, and there is more rapid transfer of the milk from cow to cheese making.

Listeria: Genus of Gram positive, aerobic rod-shaped or coccoid bacteria. Occur in soil, fresh and salt water, sewage sludge and decaying vegetation. *Listeria monocytogenes*, the causative agent of listeriosis in humans, has been associated with foods such as soft cheese, milk, ice cream, raw vegetables, prepared salads, cakes, fermented sausages, sliced cold meat, and raw and smoked fish.

Listeriosis: Infection in humans caused by *Listeria monocytogenes*. It is usually transmitted by contaminated foods. Pregnant women, babies, the elderly and the immunocompromized are particularly susceptible to infection. Symptoms vary from a mild influenza-like illness with high fever and dizziness to meningitis and meningoencephalitis. In pregnant women, intrauterine or cervical infections may result in spontaneous abortion, stillbirth or premature birth. Gastrointestinal symptoms such as nausea, vomiting and diarrhoea may precede more serious forms of listeriosis or may be the only symptoms exhibited.

Lite beverages: Beverages with a low content of alcohol and/or sugar compared with conventional beverages of the same general type.

Lite foods: Foods that are low (light) in calories, fats, cholesterol, sugar and/or salt.

Litesse: Bulking agent and fat substitute which is only partially metabolized in the body. It derived from polydextrose, with small amounts of sorbitol and citric acid. It is used in production of low calorie foods, including bakery products, dairy products, salad dressings and confectionery products.

loaf pan: A rectangular pan with deep, slightly sloping sides, used to bake bread, some cakes and meat loaves

Loaf volume: Space occupied by bread as it rises during baking. It is often measured in cubic centimetres. It is used as a measure of breadmaking quality of cereals, flour and dough.

Loaf: A standard quantity of bread dough usually baked in a rectangular loaf tin to give a characteristic shape. Or any type of food baked in a loaf tin, e.g. meat loaf, fruit loaf, etc.

Lollipops: Large sugar confectionery products on wooden or plastic sticks.

Long life foods: Foods that have a prolonged shelf life, usually under ambient conditions. Includes ultra high temperature (UHT) treated and sterilized products, such as UHT milk, and shelf stable bakery products.

Lotus roots: Underground stems, or rhizomes, of the lotus plant (*Nelumbo nucifera*), commonly used in Asian cooking. Rich in sodium, the vitamin B group, vitamin C and vitamin E. Eaten as a vegetable and also in sweet dishes. Lotus root flesh is creamy-white, with the texture of raw potatoes. Flavour is similar to that of fresh coconuts. Seeds and leaves of the lotus plant are also consumed.

Low alcohol beer: Beer, the alcohol content of which is lower than that normal for the specific type; legal definitions covering the limit differ between countries. Low alcohol beers are made by two general classes of process: formation of lower than normal amounts of alcohol by interrupted fermentation or restricted fermentation (using immobilized yeasts or low fermentation temperatures); or removal of alcohol from normally-fermented beer (by techniques such as vacuum evaporation or dialysis). Sensory properties of low alcohol beer frequently differ from those of normal beer; defects include a worts-like flavour, and lack of typical beer aroma notes formed during fermentation.

Low alcohol beverages: Beverages, the alcohol content of which is lower than that normal for the bev- erage type; legal definitions of the limit differ between countries. Low alcohol beverages are made by two general classes of process: formation of lower than normal amounts of alcohol (by restricted or interrupted fermentation processes); or removal of most of the alcohol from normally-fermented beverages (generally by evaporation or membrane processes). Low alco- hol beverages commonly have sensory properties which differ, to a greater or lesser extent, from those of normal beverages of the same type.

Low alcohol wines: Wines, the alcohol content of which is lower than that normal for the specific type; legal definitions for limits differ between countries. Low alcohol wines are made by two general classes of process: formation of lower than normal amounts of alcohol (by use of glucose oxidase treated musts, early arrest of fermentation, aerobic fermentation or use of special yeasts); or removal of alcohol from normally-fermented wines (by distillation processes, membrane processes, adsorption or extraction). Low alcohol wines commonly have sensory properties which differ from those of conventional wines of the same type.

Low calorie foods: Any foods that are low in calories, i.e. those that are naturally low in calories such as lettuces, and processed foods that have been manu- factured to give a reduced calorie content for a given reference amount, such as low calorie spreads. Although originally developed for those with specific health or weight problems, low calorie processed foods are now consumed by many consumers who perceive them as a healthy option. Sensory properties of these foods have also improved due to developments of new sugar substitutes and fat substitutes. Many of these foods can also be classed as low fat foods.

Low density polyethylene (LDPE): Polyethylene of low density grade. It is less rigid and with better resistance to impact than **high density polyethylene** (HDPE).

Low fat foods: Foods that are low in fats, either naturally or because they have been formulated to contain a reduced fat content compared with a given reference amount. Some of the most popular foods in this sector are low fat dairy products, low fat spreads and bakery products, many of which contain fat substitutes as a means of reducing fat content while maintaining acceptable sensory properties. Much of the growth in this sector is attributed to consumer perception of these foods as a healthy option. It is also classed as low calorie foods.

Low sodium foods: Foods containing relatively low levels of sodium, and therefore deemed suitable for consumption by those suffering from hypertension and certain other diseases. Reduced sodium levels may be achieved by replacement of NaCl with salt substi- tutes containing potassium or magnesium compounds.

Low sugar confectionery: Confectionery in which sucrose is partially replaced with sweetensers (e.g. polyols). It is suitable for consumption by indi- viduals following a special diet.

Low sugar foods: Foods manufactured in such a way that they are low in sugar, such as low sugar confectionery. It is commonly contain sweeteners and bulking agents as sugar substitutes. Such foods may also provide a reduction in calories (low calorie foods) and are regarded as a healthy option by the consumer. The reduced sugar contents may also be beneficial for dental health.

Low-calorie: With a low energy value. The guidelines (not law) used in the EU require such foods to have less than 40 Kcal per 100 g of food.

Low-density lipoprotein: A specific complex of a lipid (fat) and a protein that transports cholesterol in the blood. High levels appear to increase the risk of heart and vascular disease.

Low-fat: A term used for any food which contains less fat than normally expected or a low-fat substitute for a fatty food. Examples are milk, cheese, yoghurt, substitute butter spreads, etc. Guidelines (not law) in the EU require foods so labelled to have less than 5 g of fat per 100 g of food.

Low-methoxyl pectin: Pectin treated to remove methoxyl groups. It can form a gel without sugar.

Low-starch flour: Flour from which most of the starch has been removed. Used for diabetics and makes a bread rather like an open foam or sponge.

Lozenges: Small flat sweets made from icing sugar, glucose syrups, gum arabic/gelatin and flavourings. Sometimes it is medicated, as in the case of cough drops.

LTLT pasteurization: Low temperature, long time batch pasteurization treatment (also known as the holder method) that is applied to liquid foods, particu- larly milk. A quantity of milk is placed in an open vat, heated to 63°C, held at that temperature for 30 minutes, and then pumped over a plate-type cooler prior to bottling or cartoning. In addition to destroying common pathogens, this heat treatment also inactivates lipases, which might otherwise quickly cause the milk to become rancid.

Luminescence: The emission of light from a substance or organism, and which occurs at temperatures below those required for incandescence. It includes photoluminescence, chemiluminescence, electroluminescence, fluorescence and phosphores- cence.

Lumpiness: Texture term relating to product consistency and the extent to which an item contains lumps. Lumpy products contain inhomogeneities in structure, which can be present as invisible defects. Lumpiness has a negative effect on the spreadability of products such as margarines, and hampers the formation of a smooth surface of the spread film.

Luncheon meat: A cooked meat product prepared from chopped pork, ham and/or beef. Luncheon meat is available canned or sliced, and is sold in vac- uum packaging.

Lupanine: One of the toxic alkaloids present in lupins.

Lupins: Species of *Lupinus*, some of which are used as food. Seeds are rich sources of proteins and oils. High levels of alkaloids make some seeds too bitter for consumption, but contents may be reduced by washing in water. Varieties selected as grain crops are low in alkaloids (sweet lupins). Seeds have been used as coffee substitutes and seed flour has been suggested as a substitute for soy meal.

Lutein: One of the most widespread naturally occurring carotenoids. Found in many foods, and particularly fruits and vegetables.

Luteolin: Member of the flavonoids, found in a range of plant foods, including sage, olives, lettuces, endives and citrus fruits. It has also been found in honeys.

Lycadex: Maltodextrin derived from corn, potatoes, wheat and tapioca. It is used as a fat substitute, texture modifier or bulking agent in processed foods, bakery products, dairy products and salad dressings.

Lycopene: One of the carotenoids, particularly characteristic of tomatoes.

Lyes: Aqueous solutions of alkalies, generally sodium hydroxide or potassium hydroxide, of use in food processing treatments such as peeling or sugar processes.

Lysine: One of the essential dietary amino acids. Present as a free amino acid and protein constituent in a wide range of foods. Cereals such as rice and some wheat varieties contain low lysine levels, and both conventional plant breeding and genetic engineering techniques have been used in attempts to increase lysine contents of these dietary staples.

Lysozyme: An enzyme found in egg white which protects the egg from bacterial contamination by destroying the cells of any invading bacteria.

β-Lactoglobulin (Beta): One of the major whey proteins, accounting for approximately 50% of total whey proteins in cow milk. Small globular protein is rich in methionine. It exists as a dimer at neutral pH, with one free thiol group and two disulfide bridges. Several genetic variants that affect milk properties and yield have been identified in cow milk, but variants A and B are most common. It is often used as a surfactant in food dispersions such as emulsions to stabilize polyphasic systems.

M

Macaroni: Thick hollow tubes of pasta which are usually short and curved.

Macaroons: Small chewy cakes or cookies made from ground almonds/almond paste or coconut, sugar and egg whites. It is ften baked on rice paper.

Maceration: Softening or breaking up of foods by soaking in a liquid, or the soaking of foods (usually fruits) in a liquid in order to absorb the flavour. Spirits or liqueurs are often used as the macerating liquid.

Machine vision: Inspection systems in which samples are examined using a camera, the image from which is analysed by computer using image processing algorithms. Operations which can be performed include defect detection, dimensions measurement, orientation detection, grading, sorting and counting.

Magnesium (Mg)**:** One of the essential mineral nutrients which is widely distributed in plant and animal foods, good sources including fruits, vegetables and dairy products. Standard Western diets generally contain adequate levels of magnesium, so fortification is largely unnecessary. Absorption of dietary magnesium may be affected by other dietary nutrients such as calcium, phosphate and vitamin D, and also by some clinical conditions, including alcoholism and diabetes. Magnesium is an important bone constituent and intracellular inorganic cation acting as an essential cofactor in many enzymic reactions. Magnesium deficiency can cause calcification of soft tissues, electrolyte imbalances, gastrointestinal symptoms and personality changes. If taken in excess, magnesium toxicity symptoms can include nausea, vomiting, hypotension and neurological changes.

Magnetic fields: Regions around a magnet within which the force of magnetism acts. Various applications in the food industry include non-thermal preservation techniques.

Magnetic resonance imaging: Non-destructive analytical technique based on nuclear magnetic resonance which is used widely in the food industry. Applications include assessment of meat quality, determination of components in foods and measurement of thermophysical properties.

Maida: Indian refined white flour made from wheat.

Maillard reaction: Chemical reaction that occurs between reducing sugars and the amino groups of proteins or amino acids present in foods, and is responsible for nonenzymic browning. Maillard reaction products cause a darkening of colour, reduced solubility of proteins, development of bitter flavour, and reduced nutritional availability of certain amino acids, such as lysine. Rate of Maillard reaction is influenced by many factors, including water activity, temperature and pH of foods. It is responsible for the browning of meat, bread, chocolate, coffee and other roasted food. It is also contribute to the colour and flavour of foods such as soy sauces, caramels and toffees, milk chocolate and bread. Its products are important functional components of caramel colorants (produced by heating sucrose in the presence or absence of ammonium ions) that are used in a wide range of foods and beverages.

Malai: Cream used in Indian cooking, similar to clotted cream but made by repeatedly boiling the milk and removing the skin from the top until it is all reduced to a crumbly texture.

Malathion: Non-systemic insecticide and acaricide used for control of biting, chewing and sucking insects in a wide range of crops, including pome fruits and stone fruits, vegetables and rice. It is also used to control pests during storage of cereals. It is classified by WHO as slightly toxic (WHO III) and also known as carbofos.

MALDI-TOF-MS: Commonly it is used abbreviation for matrix-assisted laser desorption/ionization time of flight mass spectroscopy. Technique used to de- termine biomolecular structure of substances such as proteins, sugars and oligonucleotides, including those of food origin. Molecules are embedded in a matrix on a metal surface, desorbed into a gas phase by the force of a laser beam, accelerated by an electric field and fly through a drift tube at high vacuum. They are characterized according to molecular weight, which is indicated by the time taken to pass through the drift tube.

Maleic acid: Carboxylic acid which occurs as a colourless, crystalline solid and is used in making synthetic resins. On heating, water is eliminated, forming maleic anhydride, which can be used in modification of proteins, particularly enzymes, and in preparation of copolymers used in packaging of foods. Maleic acid has antibotulinal activity, making it suitable for use in food preservatives, and is also a substrate for synthesis of D-malic acid by many bacteria. A more modern name for maleic acid is *cis*-butenedioic acid. The more stable *trans* isomer is fumaric acid.

Maleic hydrazide: One of the plant growth regulators. It is used particularly to control sprouting in potatoes and onions during storage.

Malic acid: Aliphatic dicarboxylic acid, an important metabolic intermediate in the glyoxylate and tricarboxylic acid cycles and also commonly accumulated in some fruits and vegetables including apples and grapes. This organic acid is the substrate for malo- lactic fermentation by bacteria which produces lactic acid and carbon dioxide and reduces the overall acidity of the fermented products, generally wines, thereby increasing product quality. It is used as flavouring in soft drinks, dessert mixes and pie fillings.

Malnutrition: It is a condition resulting from inappropriate nutrition. It includes both inadequate and excessive dietary intakes of nutrients and/or calories. Insufficient protein intake causes kwashiorkor in children, and a diet deficient in all nutrients causes marasmus. Lack of vitamins causes a wide variety of deficiency diseases, including scurvy, rickets, beriberi and pellagra. Malnutrition may result from eating disorders, such as anorexia nervosa and bulimia nervosa. Over nutrition can lead to toxicity and obesity.

Malt beverages: Beverages based on malt. It may resemble beer, but do not comply with national regulations for beer.

Malt bread: A yeasted bread dough enriched with malt extract, black treacle and dried vine fruits prior to baking to a soft moist loaf malted milk. A mixture of dried milk powder and ground malted barley reconstituted with milk. It served hot or cold.

Malt extract: A concentrated solution of the soluble sugars extracted from malt. It is used in cakes, bread and puddings in the West and as a constituent of sauces and meat glazes in China.

Malt vinegar: A brown Vinegar produced by partial oxidation of the alcoholic liquor (ale) made from the water extract of barley malt. Starch is hydrolysed during malting and the sugars in the resulting hydrolysate are fermented to produce acetic acid. Malt also imparts flavour to the vinegar. Malt vinegar is often used for pickling and as a condiment. It contains about 5% acetic acid.

Malt: Cereals grains which have been steeped, partially germinated, then kilned to terminate germination. The sprouting liberates enzymes which convert the starch into a variety of sugars and polysaccharides of which maltose is the most important. The malting process includes starch saccharification and partially breakdown of proteins present in the grain to yield fermentable material; enzyme activity is also increased. Malt is used mainly in brewing; small quantities are used in making bakery products. Malt is most commonly made from barley, but other cereals such as wheat and sorghum may also be malted.

Maltase: The enzyme which splits maltose into its constituent glucose units.

Malt-houses: Industrial premises used for **malting** of barley.

Malting barley: Barley (*Hordeum vulgare*) cultivars which have composition and germination properties making them suitable for malting and brewing.

Malting properties: Properties of barley or other cereals which determine suitability for malting and quality of the malt produced. These include germination characteristics, composition, proteins and starch modification properties, and activity of enzymes.

Malting: It is a process of conversion of cereals (especially barley) into malt by controlled steeping, germination and kilning to terminate germination.

Maltitol: Polyol, systematic name 4-*O*-α-glucopyranosyl-D-sorbitol, manufactured by hydrogenation of maltose syrups. It has 0.6-0.9 times the sweetness of sucrose and is used in sweeteners.

Maltose syrups: Syrups in which the predominant sugar present is maltose. It is manufactured by hydrolysis of starch and may contain up to 90% maltose.

Maltose: A disaccharide comprising two molecules of glucose linked by a α-1,4-glycosidic bond which is manufactured by hydrolysis of starch. It has 0.4-0.5 times the sweetness of sucrose and is used in sweeteners and as a fermentation substrate in brewing. It is the principal sugar of malt and so called malt sugar.

Malvidin: One of the anthocyanidins, a pigment commonly found in grapes and wines, sometimes as a glycoside. It is also found in other berries.

Mancozeb: Fungicide with protective action used for control of many fungal diseases (e.g. blights, leaf spot, rusts and downy mildew) in a range of fruits, vegetables and cereals. It is classified by Environmental Protection Agency as not acutely toxic (Environmental Protection Agency IV).

Mango kernels: Edible seeds found within the stone of mangoes. Good source of nutrients for humans in times of food shortages. Fats and oils extracted from the kernels have been used in foods, e.g. as cocoa butter substitutes. Meal prepared from the kernels can be used as a substitute for wheat flour in baking.

Mango pickle: An Indian pickle made from chopped very underripe raw mango, destoned but not peeled mixed with turmeric, ground cinnamon, nigella, fennel seed, soaked fenugreek seed, salt and chilli powder, left for two days then covered with warm oil flavoured with dried chillies and allowed to mature

Mango pulps: Soft mass prepared from the flesh of mangoes. It is used in a range of products including beverages, ice cream, yoghurt, bakery products, jams and jellies.

Mango purees: Smooth creamy preparation made from the flesh of mangoes by sieving or reducing in a blender or liquidizer. It is used as sauces or in prepara- tion of products such as fruit juices, fruit nectars, bakery products, ice cream, yoghurt and jams.

Mannans: Polysaccharides containing a high proportion of mannose. Mannans that also contain glucose or galactose residues are known as glucomannans and galactomannans, respectively. Mannans are produced by plants, e.g. konjac glucomannans, bacteria and fungi, including yeasts. It uses include in thickeners and texturizers.

Mannitol: Polyol consisting of six carbon atoms that occurs naturally in plants, plant exudates and seaweeds. It is manufactured by reduction of mannose or reduction and isomerization of glucose. It has approximately 0.6 times the sweetness of sucrose and used as nutritive sweeteners, anticaking agents, stabilizers and thickeners. The name is derived from manna, the sweet exudate from the ash tree, from which it has been isolated. It is also called manna sugar.

Mannose: Monosaccharide consisting of six carbon atoms (hexoses). It has approximately 0.6 times the sweetness of sucrose.

Manometers: Instruments used for measuring the pressure of liquids or gases.

Manometry: It is the measurement of the pressure or tension of gases or liquids.

Maple syrups: Concentrated sugar solution produced by evaporation of maple saps. Sucrose is the predominant sweet substance, comprising approximately 60% of the syrup by weight; hexoses are also present. Maple syrups also contain flavour compounds, e.g. syringaldehyde, and natural colorants, which provide the characteristic maple syrup flavour and amber colour.

Marbling: Streaks of intramuscular animal fats in meat from mammals. Marbling is one of the factors used to assess quality of meat, particularly beef. For example, good quality beef is marbled with fine strands of fat; this fat bastes the meat as it cooks, thus affecting juiciness and tenderness. Lower quality beef has either no marbling or thicker marbling; it tends to be tougher after cooking.

Margaric acid: Carboxylic acid with 17 carbon atoms, member of the saturated fatty acids, with a melting point of 59-61°C. Synonyms include heptadecanoic acid, margarinic acid and *n*-heptadecylic acid. It occurs as a free fatty acid and lipid component of animal fats and vegetable fats.

Margarines: Water-in-oil emulsions usually composed of approximately 80% animal fats or hydrogenated vegetable fats and 20% water, together with emulsifiers, colorants, vitamin A, vitamin D and flavourings. It is usually solid at room temperature. It is used as spreads, butter substitutes, in baking or as cooking fats. Low fat products may contain as little as 20% fats.

Marinade: A mixture of various tenderizing and flavouring agents, usually including an oil, an acid such as vinegar or lemon juice and a selection from wine, sherry, yoghurt, coconut milk, fruit juices and similar liquids, herbs, spices, seasonings and flavour enhancers such as soya sauce. Used to treat raw meat and fish for periods from an hour upwards to tenderize, flavour and add moisture. Sometimes the strained marinade is used in the resulting dish. The classic marinade contains 1 litre of wine, 0.5 litres of vinegar and 0.2 litres of oil to 200 g of chopped aromatic vegetables plus garlic, parsley stalks, bay leaf, peppercorns and cloves and may be cooked or uncooked. It is seasoned liquids used for marination mainly of meat or fish. Usually contain oils mixed with wines, vinegar or lemon juices, and herbs or spices.

Marination: Soaking of foods in marinades, mixtures of ingredients such as oils, vinegar and herbs, before cooking, in order to add flavour or tenderize. Because most marinades contain acidic ingredients (lemon juices, vinegar or wines), marination should be conducted in glass, ceramic or stainless steel, but not in aluminium, containers.

Market research: The activity of gathering information about customers' needs and preferences. Market research uses surveys, tests and statistical studies to analyse consumer trends and to forecast the quantity and locale of a market favourable to the profitable sale of products or services. The social sciences, for example psychology and sociology, are increasingly utilized to provide clues to people's activities, circumstances, wants, desires and general motivation.

Markets: As well as conveying the offering of goods for sale or promotion of products, this term can also cover the regular gatherings for the purchase and sale of food, livestock and other commodities, the outdoor spaces or large halls where vendors sell their goods, or particular areas of commercial or competitive activity.

Marlins: Any of a number of large, fast swimming marine fish species belonging to the family Istiophoridae. Commercially important species include *Makaira indica* (black marlin), *M. nigricans* (blue marlin) and *Tetrapturus audax* (striped marlin). Marketed fresh or frozen and occasionally smoked; also used in manufacture of fish sausages in Japan.

Marmalade: A jam made from citrus fruit, by boiling the juice with water, shredded peel and a muslin bag containing all the pips, pith and excess peel for several hours to extract pectin. The muslin bag is removed, sugar added and the whole boiled for a short time until it reaches setting point.

Marshmallows: Soft aerated confectionery products made from corn syrups, glucose, gelatin and egg whites. It is originally manufactured from the root sap of the marsh mallow plant (*Althaea officinalis*).

Masala dosa: A South Indian pancake made from a lentil flour and water batter which is allowed to ferment overnight. After frying, it is stuffed with mashed potato and spices.

Mashes: Mixtures of ground malt, optionally with other brewing adjuncts, with hot water. Heated under controlled conditions to solubilize and extract fermentable constituents and other materials of impor- tance for the brewing process and beer quality.

Mashing: Preparation of aqueous extracts of malt (optionally together with brewing adjuncts) by heating them in water under a time/temperature regime which will optimize enzymic solubilization and extraction of carbohydrates, soluble nitrogen compounds and other constituents of importance for fermentation and beer quality. Brewing enzyme preparations may be used to enhance the enzymic solubilization process, especially when non-malted adjuncts are used.

Mass spectroscopy (MS)**:** Spectroscopy technique in which separation is based on atomic and molecular mass. Samples are bombarded with electron beams which fragment the molecules. The fragments are accelerated through magnetic fields and sorted on the basis of charge to mass ratio.

Mass transfer: Movement of matter from one place to another, usually considered with reference to a defined boundary, as in the transfer of water within or from a wet product during drying.

Massecuites: Mixture of crystallized sugar and sugar syrups which is produced during manufacture of sugar. It is centrifuged to separate the sugar crystals (which are dried and stored) from the syrup, which undergoes further crystallization to improve sugar yield.

Mastication: First stage in the digestion of foods, whereby food taken into the mouth is processed into a form suitable for swallowing. During mastication foods are chewed, ground and torn with the teeth, and mixed with saliva. Small food particles result which have a large surface area on which saliva can act. Mastication also releases food flavour and aroma. In conjunction with the action of the tongue, a cohesive food bolus is formed of the correct size to pass through the oesophagus.

Mastitis: Inflammation of the mammary gland caused by pathogenic **microorganisms**. In cows, can cause reductions in milk yield and alterations in the compo- sition of milk from infected quarters.

Mawa: Type of condensed milk made by heating milk until boiling and then stirring continuously over a low heat until it thickens to the consistency of cream cheese. It is used in preparation of Indian desserts and sweetmeats. It is also known as khoya.

Maximum residue limits (MRLs): Maximum concentrations of pesticide residues, resulting from the registered use of agricultural or veterinary pesticides that are recommended to be legally permitted or recognized as ac- ceptable in or on a food, agricultural commodity or animal feed.

Mayonnaise: Salad dressings prepared from vegetable oils, egg yolks, vinegar or other acidifying agent (e.g. lemon juices) and flavourings (e.g. mustard). For manufacture of commercial mayonnaise, oil content must be 65% (by weight). Commonly 70-80% (by weight) oil is used to give a thicker product that has been shown to be more acceptable to consumers.

Mead: Alcoholic beverages made by fermentation of a medium in which honeys are the main source of fermentable sugars.

Meal replacers: Products designed for consumption in place of conventional meals for a specific dietary purpose, e.g. weight management.

Mealiness: Sensory properties relating to the extent to which products (usually fruits such as apples, peaches and nectarines) are perceived as being mealy, i.e. soft, powdery and floury. Mealiness is the result of breakdown of flesh into small pieces that tend to be dry in the mouth; it is related to an increase in the levels of water-soluble pectins and decreases in insoluble pectins during ageing. Thus, when eaten, the cells separate easily without the release of cell sap, and the mouth perceives the outside surfaces of the cells rather than the cleaved cells leaking sap.

Meals: Processed foods eaten at mealtimes and/or designed to be one of the main dishes of the day, e.g. lunches, pub meals, ready meals, school meals.

Measuring spoon: Plastic or metal, usually hemispherical spoons used for measuring small quantities of liquid or powder, graded one quarter, one half or one teaspoon (1.25, 2.5 or 5 ml), dessertspoon (10 ml) and tablespoon (15 ml)

Meat analogues: Simulated foods, comparable in structural and mechanical properties to natural meat. They can be produced from various high protein content, raw materials including beans, fish and grain, and also from protein recovered from offal. Ingredients such as dry spun protein fibres, e.g. casein, may be incorporated into meat analogue mixtures as texture imparting materials.

Meat balls: Meat products prepared from chopped meat, which is formed into balls and then cooked. Ingredients may also include onions, breadcrumbs, eggs and seasonings.

Meat emulsions: Meat products which include sausage emulsions and emulsions used in the preparation of comminuted meat products. They are composed of a continuous phase (protein and water) and a dispersed phase (fat particles). They are prepared from meat, such as mechanically recovered meat and offal, and other ingredients, such as non-meat proteins (e.g. sodium caseinate and soy protein isolates). Enzymes may be added to improve the functional properties of meat and non-meat proteins in the emulsions. Mechanical treatment during comminution has major effects on properties of products prepared from meat emulsions.

Meat extenders: Non-meat ingredients used to improve flavour, texture, appearance and nutritional values of meat emulsions. In general, they cost less per kilogram than meat, and include: dairy products, such as dried skim milk, sodium caseinate, milk coprecipitates, whey and whey products, and other milk derivatives; soy protein isolates and concentrates; oilseeds; cereal products; and pea meal, chick pea meal and textured navy bean protein concentrate.

Meat extract: The water -soluble components of meat extracted by heating with water sometimes under pressure above its normal boiling point, filtering off the debris and reducing the solution to a thick dark syrupy paste. It is used for spreading, making drinks and as flavouring. Meat mince is immersed in boiling water to leach out the water-soluble extracts; meat extract (no. 1 extract) is produced by concentrating these extracts. Exhaustive extraction of meat produces a direct extract, which contains a high concen- tration of gelatin. Meat extracts are rich nutritional sources of the vitamin B group, particularly vitamin B_2, vitamin B_{12} and nicotinic acid.

Meat loaf: Meat products commonly prepared from comminuted meat, such as meat mince, poultry mince or fish mince. Meat loaf may include offal, blood and low value meat, such as mechanically recovered meat. Other ingredients may include binders, onions, tomato puree, garlic, white bread, milk, herbs and seasonings. The ingredients are mixed before cooking, usually in a loaf tin; however, meat loaf may also be prepared in casings. Some meat loaf is prepared with colour contrasts or patterns; preparation of these products tends to involve traditional high-cost labour-intensive methods. Once cold, meat loaf can be cut into firm slices.

Meat loaf: Minced raw meat mixed with flavourings, seasoning and extenders, bound with egg or flour, baked in a tin, demoulded and sliced.

Meat mince: Meat cut up or shredded (minced) into very small pieces by the process of mincing. Quality depends on the part of the animal carcass that the meat originated from; in particular, it varies with fat and connective tissue contents. Also known as ground meat or minced meat.

Meat pastes: Comminuted meat products similar to pates, and of intermediate texture, commonly with a meat content of approximately 70%. The non meat portion consists of rusk and water, or other suitable filler such as soy protein concentrates or sodium caseinate. The product is usually heat sterilized after filling into jars or cans.

Meat patties: Round, flat cakes of comminuted meat. Although they may be prepared from meat mince, they may also be reconstituted, e.g. from mechanically recovered meat. Some may include meat extenders. Varieties include beef patties, chicken patties and turkey patties.

Meat pie: A pie filled with precooked meat, gravy, flavouring and seasoning prior to baking.

Meat pies: Meat products in which chopped meat or meat mince is encased in pastry and baked. Meat pies often contain offal and low value meat, such as mechanically recovered meat. They may be prepared in pie dishes that are lined and sealed with pastry, e.g. steak and kidney pie. Pasties are a type of meat pie prepared in a folded pastry case, e.g. Cornish pasties.

Meat sauces: Any sauces that contain meat as the main ingredient. Meat sauces are usually used as an accompaniment to pasta and rice, for example bolognese sauces or meat curry sauces.

Meat substitutes: Simulated foods used as direct substitutes for meat. They may be included in meat products or may provide vegetarian alternatives to meat. Meat substitutes include textured vegetable proteins (TVP), texturized milk proteins, quorn and tofu. Aroma compounds, stabilizers and colorants may be included. It is also known as meat alternatives.

Meat tenderizer: Some form of proteinase, often from papaya or pineapple, which in a water solution is used to break down the muscle fibres in meat. Care must be taken not to use for too long as the enzymes are not used up and the meat may be reduced to soft slush. The enzymes can be deactivated by heat or alcohol.

Meat thermometer: A metal thermometer with the measuring element in a sharp point and a circular indicating dial, placed with its point in the thick part of a joint so as to indicate its internal temperature. Beef and lamb temperatures are 51°C, 60°C and 70°C for rare, medium and well done, pork and veal are 75°C, chicken, ducks and turkey 80°and goose 85°C.

Meat: Animal tissues which are used as food, including those of domestic mammals, poultry, game birds and game animals. Meat is composed of lean muscles, connective tissues, fats, skin, nerves, blood vessels and water. It can be classified as red or white, based on its colour intensity, which results from the proportion of red and white muscle fibres that it contains. Red fibres have higher myoglobin content than white fibres. Composition of meat differs between species and between retail cuts; it depends greatly on the fat to lean ratio, which determines energy value and concentrations of most nutrients. Water content of meat tends to decrease with increasing fat content. Lean meat includes substantial amounts of high biological value proteins; however, meat is also an important dietary source of fat, high bioavailability inorganic nutrients (including Fe, Zn, Cu and Se) and the vitamin B group.

Mechanical boning: Removal of bones from meat or fish, usually before cooking, using specially designed boning equipment.

Mechanical harvesting: Gathering (harvesting) of crops by mechanical means.

Mechanical properties: In relation to foods, physical properties associated with the reaction of foods to stress. It includes parameters such as hardness, viscosity, elasticity and adhesiveness.

Mechanically recovered meat: Meat recovered from bone using separation machinery. Mechanical recovery increases the efficiency of separation and thereby allows the recovery of extra meat per carcass; it is also less time consuming than hand boning of meat. In many systems, meat and bone are forced against perforated plates or cylinders; the meat passes through, leaving the bone to be removed as waste. Composition of the meat recovered varies between the methods used, but in general consists of comminuted meat, bone marrow, collagen, bone and fat. Bone content is very important and must be minimized. Initial raw materials need to have low bacterial counts; they should be handled at low temperature and treated as promptly as possible. Advanced meat recovery (AMR) systems produce a product which is similar in appearance, texture and composition to meat trimmings and similar hand deboned meat products. Other systems produce a paste or batter-like meat product, or liquid meat extracts. Mechanically recovered meat is widely used in meat products. It is also known as mechanically separated meat or mechanically deboned meat.

Medical foods: Foods specially formulated to be consumed by individuals who suffer from disease or health conditions that require special dietary management, because of distinctive nutritional requirements associated with the conditions.

Melamine: It is toxic, nonflammable, colourless to white crystalline solid and insoluble in organic solvents, very slightly soluble in ethanol and slightly soluble in water. When heated, it sublimes slowly before melting at 347°C. It is used to make melamine resins and to give wet strength to paper. Melamine resins are made by polymerizing melamine with formaldehyde; they are used in various coatings, insulators and as adhesives.

Melanins: High molecular weight pigments with reddish-brown to black colour, formed by the action of oxidases on phenols, as in enzymic browning. It is widely distributed in animals and plants, generally bound to proteins. Although a normal constituent of certain foods and beverages, including black tea, melanins can sometimes produce an undesirable discoloration of foods, such as mushrooms, several fruits and shrimps.

Melanoidins: Pigments with yellow to brown colour and malt-like aroma formed by reactions between reducing sugars and amino acids in foods during heating. Formation of these Maillard reaction products is important during food processing pro- cedures such as baking and roasting.

Melting point: A temperature at which a solid changes into a liquid i.e. the solid and liquid forms exists together in equilibrium. A pure substance at a pressure of 1 atmosphere has a single reproducible melting point. The melting point is a characteristic of a pure substance; the presence of impurities lowers the melting point.

Melting: It is a conversion of solid foods (such as butter or chocolate) into a liquid or semi-liquid state by application of heat.

Membrane bioreactors: Bioreactors in which reaction products are removed through membranes by, for example, ultrafiltration, reverse osmosis and dialysis, thus allowing continuous operation. It can be used in processes such as bioremediation of waste water, purification of drinking water, bioconversions and biotransformations. The membranes can also be used as supports for immobilization of enzymes or cells.

Membranes: Solid matrices used for separation of molecules in processes such as dialysis, filtration and reverse osmosis, as supports for immobilization of cells and enzymes, and in techniques such as blotting and hybridization.

Menthol: One of the monoterpenoid aroma compounds and a secondary alcohol. It is characteristic component of mint oils and widely used in mint flavourings.

Menthone: Member of the monoterpenoid aroma compounds, with a ketone functional group. It is present in mint and mint oils, and used in mint flavourings.

Meringues: Confectionery products made by whipping egg whites to foam, incorporating sugar and drying to a crisp finish. The term may refer to small cakes or shells made of this material which have been decorated or filled, e.g. with whipped cream, ice cream or fruits. Also used as toppings added to flans or pies, as in lemon meringue pies.

Metabisulfites: Disulfurous acids, the disodium salts of which are used as **preservatives** and **antioxidants.**

Metabisulphites: Salts similar to the sulphites and hydrogen sulphites used both as sterilants and food preservatives. Sodium-metasulphites, E223, and potassium-metasulphites, E224, are used in the food industry.

Metabolic disorders: Generic term for diseases caused by an abnormal metabolic process. They can be congenital, due to inherited enzyme abnormality (in- born errors of metabolism), or acquired due to disease of an endocrine organ or failure of a metabolically im- portant organ such as the liver.

Metabolism: The process of converting circulating blood components derived from food or the breakdown of body tissues, principally fats, sugars and amino acids, into new or repaired body tissues and into energy by chemical reaction and conversion in the body's cells. The end products of these reactions are transported by the blood and excreted via the lungs, skin or kidneys.

Metal detectors: Electronic devices that give an audible signal when close to metal; used to detect metal foreign bodiesocontaminants during food processing.

Metalloenzymes: Enzymes that contain a bound metal ion as part of their structure. This ion may be required for enzymic activity, either participating directly in catalysis or stabilizing the active conformations of the proteins.

Methanol (methyl alcohol): One of the alcohols, methanol contains a single carbon atom, and is a light, volatile flammable, poisonous, sweet smelling liquid at room temperature. Widely used as a solvent, antifreeze or fuel. It can occur as a fermentation by-product in alcoholic beverages and vinegar.

Methionine: One of the essential dietary amino acids, this thiol-containing amino acid is a common protein constituent in foods. Also a precursor of several organic sulfur compounds which are important in food flavour.

Methyl bromide: Colourless, poisonous gas, synonym bromomethane. It is used in the fumigation of fruits and vegetables to control pests, but now largely being replaced with other fumigants such as phosphine.

Methylcellulose: It is methyl ester of cellulose and prepared by alkali treatment of celluloses followed by methylation of the alkali cellulose with chloromethane. Due to its ability to absorb water and form viscous colloidal aqueous solutions, methylcellulose is often used as a substitute for gums. Food industry uses include as thickeners, stabilizers, emulsifiers, bulking agents and binders. It is often included as an ingredient of fillings for meat pies or fruit pies.

Metmyoglobin: A brown pigment, formed by oxidation of myoglobin, in which water is bound to the ligand, and the haem group of myoglobin is in the ferric (Fe^{3+}) state. In meat, metmyoglobin produces a brown/grey coloration, which is unattractive to consumers; thus, metmyoglobin formation is a major problem in maintaining a stable display of retail meat. Several approaches may be taken to delay the forma- tion of metmyoglobin in meat, including: production of meat from animals fed on antioxidant supplemented feeds; use of modified atmosphere packaging for meat; and treatment of the meat surface with antioxidants, such as vitamin C.

Microbacterium: Genus of Gram positive, aerobic rod-shaped bacteria of the family Microbacteriaceae. Some species, e.g. *Microbacterium lacticum and M.flavum*, occur in dairy products (e.g. spray dried milk, cheese), presumably due to improper cleaning of dairy equipment, and cause spoilage. *M. thermos- phactum* may cause spoilage of vacuum-packaged meat and meat products.

Microbial biomass: Quantitative estimate of the entire assemblage of microorganisms in a given habitat in terms of mass, volume or energy.

Microbial counts: Numbers of microorganisms in a given sample.

Microbial proteins: Proteins produced by microorganisms.

Microbial rennets: Enzymes sourced from microorganisms, commonly fungi that are used as substitutes for animal rennets in coagulation of milk for cheesemaking.

Microbiological quality: Extent to which a substance (e.g. a food) is contaminated with microorganisms.

Microbiological techniques: Techniques used in microbiology, including those used to detect or quantitate microorganisms in substances such as foods and beverages.

Microbiology: Scientific study of microorganisms and their interactions with other organisms and the environment.

Microbreweries: Small breweries making speciality beer in small quantities (generally under 15,000 barrels annually). Frequently, the products are sold on the premises.

Microcrystalline cellulose: A very finely divided form of cellulose used to add bulk and fibre to slimming food, convenience foods, desserts and the like

Microemulsions: Emulsions having a droplet diameter that is too small to be seen by the naked eye, typically 10-100 nm. Applications include edible films, coatings and delivery systems for nutrients and flavourings.

Microencapsulation: The process in which thin films or polymer coatings are applied to small solid particles, droplets of liquids or gases. It can be used to encapsulate enzymes, microorganisms, flavour com pounds, sweeteners and food ingredients.

It is useful for controlled flavour release and enhancing the stability of sensitive ingredients. Methods for microencapsulation include spray drying, spray chilling and spray cooling, extrusion, air suspension coating, liposome entrapment, co-crystallization, molecular inclu- sion and interfacial polymerization.

Microfiltration: A method of sterile filtration that removes particles of approximately 0.1-10.0 micrometers in size, such as large fat globules, large proteins and suspended particles such as microbial cells. Microfiltration is generally used in the clarification and separation of beer, wines and soft drinks, and in the dairy industry for processing of low heat sterile milk.

Microflora: In a microbiological context, refers to all the **microorganisms** present in a particular habitat. It may refer to all the microscopic plants, **bacteria**, **fungi** and **algae** present in a particular habitat in a broader biological context. Also, it can be used to describe the plants, bacteria, fungi and algae that are present in a particular microhabitat.

Microfluidization: High pressure homogenization technique for deagglomeration and dispersion of uniform submicron particles and creation of stable emulsions and dispersions. Combined forces of shear and impact act upon products to create finer, more uniform dispersions and emulsions than can be produced by any other means.

Micrometers: Instruments used in conjunction with microscopes or telescopes for measuring small distances.

Micronization: Indirect infrared (IR) heating method that relies on heat that is generated externally being applied to the surface of a food mostly by radiation, but also by convection, and, to a lesser extent, conduction. IR heating is mostly used to alter eating quality of foods by changing the surface colour, flavour and aroma. The main commercial application of radiant energy is in drying of low moisture foods and in baking and roasting ovens.

Microorganisms: Microscopic organisms which include algae, bacteria, fungi, protozoa and viruses.

Microscopy: Analysis of samples using microscopes which produce magnified images. It includes basic light microscopy and more complex techniques such as electron microscopy.

Microwave cooker: A small oven fitted fitted with a timer, power variation and sometimes elaborate programmers, which cooks by subjecting the food to very high frequency electromagnetic radiation (microwaves) similar to radio waves, generated in a magnetron and piped to the oven through a wave guide. The microwaves are contained within the cooking chamber by the reflective walls but are absorbed by the food, which should be contained in materials such as ceramics, glass and plastics which do not absorb microwaves and so do not themselves get hot.

Microwave ovens: Ovens, ranging in power from approximately 500 to 900 Watts, that use microwaves to cook, heat or defrost foods. A microwave oven cooks with high-frequency electromagnetic waves that cause food molecules to vibrate, so creating friction that heats and cooks the food. Microwaves penetrate only a few centimeters into the food, so the centre of most products is cooked by heat conduction. Non metal containers (such as glass and ceramics) need to be used in these ovens, as microwaves can pass through them (unlike metal) and therefore cook the food from all angles at once. As the microwaves pass through these non metal containers, they are able to stay relatively cool themselves while the food they contain becomes hot. However, during longer cooking periods, the containers can become very hot due to heat conduction from the hot food. To assist in admini- stration of an even distribution of microwaves, some microwave ovens have turntables; others have revolving antennae for the same purpose. Factors that affect how quickly food cooks in a microwave oven include: temperature of the food when cooking begins; volume of food being cooked at one time; size and shape of the food; amount of fat, sugar and moisture in the food; bone distribution; and food density. Because of the way in which foods are cooked in microwave ovens, browning does not occur in the normal manner. Microwave susceptors are used to promote browning.

Microwave susceptors: Devices used in the form of active packaging that cause browning and crisping of foods that are prepared in microwave ovens. A microwave active metal is lightly deposited on a thermally stable substrate (such as PET) and this sheet is laminated onto a back stock that provides rigidity. Once placed in the microwave, these packages will reach temperatures in excess of 170°C almost instanta- neously. The high temperatures allow the food to cook quickly, and promote the Maillard reaction, thus en- hancing browning characteristics.

Microwave thermometer: A thermometer without any metal parts which can be used inside a microwave cooker.

Microwaveable containers: Containers that may be used safely for microwave cooking or reheating of foods. It must be made of materials that will not cause damage to microwave ovens during operation or allow migration of undesirable components into the foods being heated.

Microwaveable foods: Foods suitable for heating in microwave ovens. Ready meals that can be rapidly reheated in this manner and microwave popcorn are some of the most popular types of microwaveable products. Further applications have been limited by problems such as lack of browning in foods cooked in microwave ovens, and arcing dur- ing microwave cooking of foils packaged foods. However considerable advances have been made in development of microwave susceptors and other devices for promoting browning and crisping during microwave heating and cooking.

Microwaveable packaging: Packs or wrappings that may remain on foods during microwave cooking or reheating without causing damage to microwave ovens or causing contamination of the products with undesirable components.

Microwaves: Very-high-frequency electromagnetic waves with a wavelength in the range 0.001 to 0.3 m, shorter than that of normal radio waves, but longer than those of infrared radiation used to heat up and cook foods. Microwaves are used in microwave ovens for cooking, heating and defrosting of foods. Microwaves transfer their energy to food by causing certain types of molecules to vibrate at their frequency. They are generally absorbed in the first cm of the surface and hence solid food to be cooked should be less than 2 cm thick or well stirred from time to time if contained in a deep container. Microwaves are reflected off metal surfaces but isolated metal within an enclosure will heat up and sometimes cause sparking.

Migraine: Condition characterized by severe, usually unilateral, vascular headache. Sometimesit combined with any of a range of other symptoms such as nausea, vomiting and heightened sensitivity to light or sound. In some cases, attacks are triggered by ingestion of specific foods, food additives or beverages. It is commonly suspected dietary triggers include alcoholic beverages, beverages containing caffeine, cheese, some beans, cured meat and chocolate based products.

Migration: Movement of undesirable compounds (e.g. plasticizers from packaging materials) into foods.

Milk beverage: The drinks in which milk is a major constituent. It include milkshakes, flavoured milk, carbonated milk beverages, milk mixed with fruit or vegetable juices or pulps, and products enriched with specific nutrients, e.g. fibre or calcium. It is alternative term for milk drinks.

Milk bread: A white yeast-raised bread containing about 6% of dried milk by weight of the dry ingredients.

Milk chocolate: A high sugar content (greater than 48%) chocolate confectionery with added milk solids.

Milk chocolate: Type of chocolate made by incorporating milk powders along with sugar, chocolate liquor and cocoa butter. It is more widely eaten than dark chocolate and white chocolate. It is compared with dark chocolate; milk chocolate has a creamier texture and taste, and tends to be softer.

Milk clotting enzymes: Enzymes are used in the clotting or coagulation of milk during cheesemaking. Most commonly, rennets extracted from the stomach of young ruminants have been used traditionally in this process, but other sources of enzyme have been developed in the light of shortages of animal rennets and the increasing popularity of vegetarian products. Alternatives include microbial rennets, produced by a range of microorganisms, and enzymes produced by plants, e.g. cardoons.

Milk clotting: The process in which milk is separated into curd and whey by the action of milk clotting enzymes, e.g. rennets, lactic acid produced by bacteria, or a combination of both. It is used in cheesemaking. During clotting (or coagulation), k-casein, which resides on the surface of casein micelles and confers stability, is removed by the action of the enzymes, causing the destabilized casein to precipitate; acid acts by destroying linkages between components of the micelle. Curd produced by using enzymes generally has higher calcium content than that formed by acids.

Milk fat globule membranes: Membranes surrounding milk fat globules, comprising approximately 60% lipid and 40% protein. Enzymes and trace elements are associated with these mem- branes. When broken down, e.g. during churning of cream in buttermaking, fat globules are released and may coalesce.

Milk fat globules: It is emulsified form in which milk fats exist in milk. It is surrounded by milk fat globule membranes. Fat globules have a diameter of 2-6 μm and a large surface area.

Milk fats: Lipids present in milk mainly in the form of emulsified milk fat globules. It is mainly triglycerides, with small amounts of monoglycerides, diglycerides, cerebrosides and free fatty acids. Milk fat globule membranes also contain phospholipids and sterols. Fat content of milk varies greatly among species and among animal breeds. Fatty acid composi- tion of milk fats is governed by many factors, including feed, lactation stage and fat content of milk. Cow milk fat contains a great number of fatty acids, principal ones including palmitic acid, oleic acid, myristic acid, stearic acid, linolenic acid and linoleic acid.

Milk ice: Product similar to ice cream but generally containing less milk fat.

Milk infant formulas: Preparations for feeding to infants and young children intended to satisfy their specific nutritional requirements. It is made from cow milk with the nutrient composition adjusted to mirror that of human milk. Composition is varied according to the age of the infant to be fed.

Milk powders: Products prepared by drying whole milk to a low moisture content, giving a powder with a long shelf life. It is also called dried milk.

Milk protein concentrates: Preparations made by concentration of milk, usually skim milk, by ultrafiltration, during which milk proteins are separated from other milk constituents, followed by drying. Vary in milk protein content according to manufacturing procedures. Used in a similar way to skim milk powders in a variety of foods, including processed cheese, infant formulas, beverages, bakery products, dairy spreads and diet products.

Milk proteins: Proteins found in milk, comprising casein (approximately 75% of total protein) and the whey proteins, including α-lactalbumin, β- lactoglobulin, serum albumin and immunoglobulins. It is used as ingredients in various foods, such as bakery products, coffee whiteners, nutritional beverages and imitation cheese, to modify functional properties and sensory properties. In some food allergies, cow milk proteins act as allergens.

Milk substitutes: Multipurpose term covering re- placements for mothers' milk, e.g. human milk, used in infant feeding (infant formulas) or young animal feeding, as well as products prepared for use by individuals unable to tolerate milk or not wishing to con- sume it for other reasons, e.g. vegans. Depending on the intended consumers, the latter category may or may not contain dairy components, e.g. whey. Non dairy foods used as the basis of milk substitutes include soybeans and oats.

Milk: Secretion of the mammary gland of mammals. Composition varies among species, and is affected by many factors, including feeds and season. When used without further clarification, the term milk is generally accepted to mean cow milk. Cow milk is sold in vari- ous forms that differ in fat content (whole milk, semi skimmed milk and skim milk). Whole milk con- tains approximately 87% water, 4% fat, 3% protein and 5% lactose. It is rich in calcium (approximately 1.2 g/l), riboflavin (2 mg/l), vitamin B_{12} and io dine. It is a good source of vitamin A and vitamin B_1 and contains folates and other vitamins and minerals. Due to risk of contamination with pathogens and spoilage organisms, milk for drinking is generally sold pasteurized, sterilized or UHT (ultra high temperature) treated, although raw milk is sometimes used to make dairy products such as cheese.

Milkshakes: Beverages made by addition of flavourings, often fruits-based, to milk and agitation by beating or shaking, sometimes with the addition of ice cream.

Mill: An implement used to reduce a solid to a fine powder, as with pepper, salt crystals, coffee, spices, etc.

Millet: Small seeds from any of a number of cereal grasses, including common millet (*Panicum miliaceum*), finger millet (*Eleusine coracana*), foxtail millet (*Setaria* *italica*), pearl millet (*Pennisetum typhoideum*) and teff (*Eragrostis tef*). Good sources of many minerals and with good storage properties. It forms the staple diet of much of the world population, especially in Asia and Africa. Consumed like rice or made into various products such as porridge, gruel and bread.

Milling properties: It is ability of solid materials, such as grain, to be ground into powders.

Milling: Milling is a general trade name which normally means reduction of food grain in to various end products like meal, flow, splitted products etc. The meaning of the term milling varies with the crop. For example, milling of wheat refers to a grinding operation to produce flour, whereas in rice industry, milling refers to overall operations in a rice mill i.e. cleaning, dehusking paddy separation, bran removal and grading of milled rice. Milling also refers to extraction of juice, oil or separation of fibre etc.

Mills: Machinery for grinding solid substances, or buildings equipped with such machinery. It includes equipment used for grinding grain into flour.

Miltone: Vegetable-toned milk product developed in India to overcome problems of milk shortages. It is prepared using a protein isolate from peanuts which is added to cow milk or buffalo milk along with sugar and vitamins. It is suitable for drinking on its own, in tea or coffee, or for processing into yoghurt.

Minced meat: Meat which has been finely divided by being chopped or passed through a mincer. Used in stews, hamburger, pies, etc. Mincing converts tough meat into a more palatable and tender form.

Mincemeat: A mixture of dried vine fruits, chopped mixed peel, apples and suet, with sugar, nutmeg, cinnamon, grated lemon zest, lemon juice and sometimes spirits all passed through a mincer, allowed to mature and used as a filling for mince tarts and other sweets, cakes and puddings

Mincer: A culinary implement used for mincing, consisting of a barrel with a hopper above possibly tapering towards one end, into which fits a well-fitting scroll feeder with a pitch which reduces towards the output end. When the scroll is turned either manually or mechanically it forces food towards the output end slightly crushing it. A knife with three or four blades rotates with the scroll and cuts the crushed food as it is forced through a stationary circular disc containing a number of holes. The holes in the stationary disc may be of various sizes to give varying degrees of size reduction. It is particularly used to cut or shred **meat**, into very small pieces.

Mincing: It is a process to reduce the size of solid or semi-solid foodstuffs (mostly for meat) by a combination of chopping and crushing usually in a mincer or food processor.

Mineral oils: Oils derived from hydrocarbon sources, some of which may be of food grade and may be used as food additives. Other mineral oils are not of food grade and may act as food contaminants.

Mineral waters: Natural spring waters or similar waters, which are produced and bottled under conditions specified under national regulations. In the UK, mineral waters may also be used as a general term for carbonated soft drinks.

Minerals: Solid inorganic elements, including metals and non-metals. It is also compounds occurring naturally in the earth's crust. Minerals are not normally volatilized when their organic matrix is ashed to remove car bonaceous materials. Many minerals are essential nutrients in that they are necessary in the diet of humans or animals to allow completion of the life cycle.

Minimal processing: It is a limited processing of products to a level where they maintain the characteristics of fresh foods. Examples include industrial processes such as washing, sorting, cutting, trimming, slicing and dicing. Because a number of minimal processing technologies result in wounding of plant tissues and subsequent acceleration of deteriorative processes, controlled environmental conditions are critical requirements in the transportation, distribution, storage and retail display of these products. Modified atmosphere packaging has become an integral part of minimal processing.

Minimally processed foods: Foods that are processed using technologies that do not significantly alter their fresh-like attributes, but achieve reliable preservation and control over enzyme activity and microbial growth. Most commonly applied to fruits and vegetables. Examples of such processing for fruits and vegetables include washing, sorting, cutting, trimming, slicing and dicing. Other methods of minimal processing include various high temperature short time (HTST) thermal processes in combination with mini- mal processes such as hermetic packaging and refrigeration. In order for minimally processed foods to have a reasonable shelf life, modified atmosphere packaging has become an integral part of many minimal processing procedures.

Mint oils: Essential oils distilled from mint. The characteristic fresh, cool flavour and aroma of mint oils is due to the presence of the terpenoid, menthol. Mint oils are used as flavourings for sugar confectionery, such as mints and chewing gums, and in beverages, e.g. cordials.

Miraculin Flavour modifiers: Miraculin is a flavourless glycoprotein with a molecular weight of 44 kDa that was isolated originally from miracle fruits, berries of the African shrub *Richardella dulcifica*. Foods and beverages having a sour flavour that are eaten subsequent to consumption of miraculin are per- ceived as being more sweet than they actually are.

Miso: A dark brown salty paste tasting rather like soya sauce. It is made by a two step fermentation first producing koji from grains or beans then fermenting the koji mixed with mashed, well cooked soya beans and salt, stirring at intervals. It is aged for up to 3 years. Ingredients and length of ageing are varied to produce products differing in sensory properties. It is used as flavouring and condiment and a good source of B vitamins.

Mithai: Indian sweetmeats made in various ways but often based on evaporated or condensed milk (khoya) and nuts.

Mix: A mixture of dry ingredients to which water, milk, eggs or a combination of these is added and the result may be eaten as is, left to set, or further cooked, e.g. Bread mix, soup mix, cake mix

Mixed spice: Mixtures of ground spices supplied for flavouring, often proprietary brands formulated for various cakes, biscuits, desserts, etc.

Mixers: Electric machines or hand-operated devices for combining food ingredients by beating, mixing or whipping. There are two basic types of mixer: stationary and portable. Stationary mixers tend to be more powerful and can therefore handle heavier mixing jobs; they are usually equipped with an assortment of attachments including dough hooks, wire whisks, paddle-style beaters, and even citrus juicers, ice crushers, pasta makers, sausage stuffers and meat grinders. Portable mixers are small in size, due in part to their small motors, but are consequently easy to store. The term mix- ers is also applied to beverages, such as soda water, tonic water, cola, lemonade or fruit juice that can be combined with spirits to make beverages such as cocktails.

Mixes: Blends of ingredients in dried form that can be reconsitituted in a liquid, sometimes with the addition of other ingredients, to make the desired product. The reconstituted mixture may require cooking or be ready to eat. Products available in this form include cakes, soups and desserts.

Mixing: Combining food ingredients together by hand or using electric machines (mixers).

Mixographs: Instruments used to investigate the physical properties of dough.

Modified atmosphere packaging: Packaging technique used primarily to extend shelf life of foods. Gas composition within the package is changed by altering levels of oxygen, nitrogen and carbon dioxide. This inhibits microbial growth, controls enzymic and bio-chemical reactions, reduces moisture loss and protects against infestation with pests. Used for packaging a wide range of foods.

Modified starches: Starch that has been modified by chemical reaction or physical means in order to adapt it for a specific application or improve its general appli- cability, i.e. to increase stability. Chemical modifications include crosslinking, acetylation, and phosphorylation, reaction with 1-octenylsuccinic anhydride, hydroxypropylation and oxidation. Physical modifications involve pregelatinzation of starch by drying or heating.

Moisture content: Level of moisture held in a substance, stated as a percentage of the wet or dry weight.

Moisture sorption: Process whereby moisture binds to another substance.

Moisture transfer: Movement of moisture in stored products as a result of moisture and temperature changes.

Molasses sugar: A fine grained unrefined sugar containing a large admixture of molasses so that it is very soft. Used in cakes and puddings.

Molasses: Low purity, thick, brown syrups produced as a by-product of sugar refining, and including the syrups remaining after sucrose crystallization has been exhausted. Uses include as a feedstock for microbial fermentation. Molasses from cane sugar refining are also known as blackstrap molasses. Beet sugar molasses contains too many bitter and astringent beet compounds to be palatable. It is a complex mixture of sugar, invert sugar and other plant derived components and minerals.

Molecular weight: Sum of the atomic weights of the atoms in a molecule, usually measured in Daltons (Da).

Molybdenum (Mo): It is one of the essential metal minerals. It is component of xanthine dehydrogenase and sulfite oxidase in animals and also nitrogen-fixing enzymes in some microorganisms and plants. Dietary requirements for molybdenum are very low and deficiency diseases are rare. It is found in vegetables, cereals, oilseeds, fish and water.

Monosaccharide: A simple sugar consisting of five or six carbon atoms in a ring conformation (furanose and pyranose, respectively). The important monosaccharides in food are glucose, fructose and galactose which in various combinations form most of the sugars, starches and carbohydrates found in food.

Monosodium glutamate (MSG): The white crystalline monohydrate sodium salt of the amino acid (L-glutamic acid) used in flavour enhancers. Most commercial production of this additive is from glutamate produced as a result of fermentation by bacteria, including *Micrococcus glutamicus*. It has a distinctive meaty taste in itself but is used to enhance the flavour of other foods especially in Chinese cooking and in convenience foods.. It is also called, aji-no-moto, taste powder, gourmet powder and vetsin.

Monounsaturated fatty acids: Unsaturated fatty acids containing a single double bond. Examples include oleic acid, palmitoleic acid and erucic acid, significant sources of which include olive oils, fish oils and rapeseed oils, respectively.

Mood: Pattern of behaviour exhibited in relation to a current state of mind. It is usually a relatively short-lived and low-intensity emotional state. It can be affected by diet, and can affect consumer response to foods as well as eating habits.

Moringa: Genus of plants native to tropical Asia. Seeds of some species, especially *Moringa oleifera*, yield high quality oils used as cooking oils as well as lubricants. Other plant parts are also used as foods, fruits and leaves as vegetables and roots as a source of spices.

Moromi: Fermenting mash mixture of rice, koji, yeasts and water that is pressed to separate sake from suspended solids during sake brewing. Moromi is also the mash based on koji (derived from soybeans or cereals), with the addition of brines, which is fermented to make soy sauces.

Moulding: Formation of an object out of a malleable substance. It is also use of containers (moulds), usually distinctively shaped, to form food into a specific shape. Moulds can range in size from small, individual candy size moulds to large pudding moulds and cheese moulds. The foods to be moulded (e.g. a gelatin-based dessert) are poured or packed into the mould and then left until they become firm enough to hold their shape.

Mozzarella cheese: Italian soft cheese made from buffalo milk. A plastic, spun-curd cheese is made by coagulating pasteurized milk at 33°C, cutting the curd, treating it with hot water (200°F) and kneading into a shiny lump. Pieces are then taken off, cooled, salted and marketed soon after. It is used for pizzas, lasagne and in salads. It becomes characteristically stringy when cooked. It contains 55 to 65% water, 18 to 20% fat and 16 to 21% protein.

Mucor: Genus of fungi of the class Zygomycetes. It occurs on vegetable matter, soil and dung. *Mucor racemosus* and *M. mucedo* may be responsible for spoilage of bread and meat, while other species may be parastic to stored grains. *M. hiemalis* is used in the production of sufu, and *M. racemosus* and *M. rouxianus* are used in the production of pozol.

Muesli: A mixture of raw or semi-processed and dry cereals with dried vine fruits, chopped nuts, bran, sugar and chopped dried fruit, used as a breakfast cereal and eaten uncooked. Typical cereals used are wheat, rye, barley and oat flakes as well as others less common.

Muffin cakes: Round cakes which may be leavened with yeasts or baking powders and sweetened with sugar. It may be plain, or flavoured with fruits, e.g. blueberries, dried fruits, nuts, chocolate or savoury ingredients such as cheese.

Muffin: A yeast-raised dough made with strong flour, milk, a little semolina and salt, rolled out and cut in 8 cm rounds, proven and cooked on a griddle or hot plate for about 7 minutes each side until golden brown. or A baking powder raised sweet sponge mix baked in deep patty tins to a soft doughy consistency. Sometimes it is flavoured with fresh or dried fruit, glacé cherries, honey or maple syrup.

Muscles: Tissues composed of bundles of specialized cells which are capable of contraction and relaxation to create body movement. There are >600 muscles in an animal carcass; these vary widely in shape, size and activity. There are three types of muscle, namely: skeletal, cardiac and smooth. The largest part of the musculature consists of skeletal muscles and it is this part of animal carcasses that is generally referred to as meat; organs comprised of cardiac or smooth muscle tend to be classified as offal. Muscle tissue also con- tains structural elements (collagen, reticulin and elastin).

Mushrooms: Fruiting bodies of various species of fungi. Eaten raw or used to add flavour to dishes, soups and sauces. Many species are gathered wild, but care must be taken as some are poisonous. The most commonly cultivated species is *Agaricus bis- porus*; other types of commercial importance include shiitake, straw mushrooms, oyster mushrooms and winter mushrooms. It is rich in phosphorus, magnesium, potassium, selenium and riboflavin, and low in fat.

Mustard butter: A compound butter made with 120 g of French mustard per kg of softened butter. It is shaped into a 2.5 cm roll, refrigerated and cut in 6 mm slices for use as a garnish on grilled meats, mackerel and herrings.

Mustard oil: A deep yellow highly flavoured oil extracted from mustard seeds. Used in Indian cooking and in pickles and chutneys. It is usually diluted with blander oil.

Mutton: Meat from mature sheep, which are over one year old, including meat from ewes, rams, wethers and hoggets. Mutton tends to be cheaper than lamb, but also tends to be tougher, darker in colour, fattier and less delicately flavoured. It is the preferred meat for Muslims. It is also known as sheep meat, sheep muscles, ram meat or ram muscles. The best mutton, now difficult to find, is killed at 18 months and is more expensive than lamb.

Mycobacterium: Genus of Gram positive, aerobic rod-shaped bacteria of the family Mycobacteriaceae. It occurs in dairy products, soil and water. *Mycobacterium paratuberculosis*, which causes Johne's disease in cattle, is suspected of causing Crohns disease in humans who consume contaminated milk.

Mycotoxicosis: Disease of humans and animals resulting from the ingestion of **mycotoxins** in foods or feeds.

Mycotoxins: Toxins, e.g. **aflatoxins** and **ochratoxins**, produced by **fungi**.

Myofibrillar proteins: Salt-soluble proteins, including actins and myosin, which are the predominant type of proteins in muscle and are responsible for contraction, texture and water holding capacity. Degradation of these proteins is important for *post mortem* tenderization of meat.

Myofibrils: Elongated contractile elements contained within skeletal and cardiac muscle fibres. The major myofibrillar proteins are actins, myosin, acto- myosin and tropomyosin. Within myofibrils, thick filaments, consisting almost entirely of myosin, and thin filaments, consisting almost entirely of actin, are aligned parallel to each other; in certain regions these myofilaments overlap and this arrangement produces a characteristic striated appearance in myofibrils. Areas of different densities, which are visible within the light and dark bands of myofibrils, include: the isotropic band

(I-band); the anistotropic band (A-band); and the Z-line, a thin dark line, which bisects the I-band. The basic repeating contractile structural unit of myofibrils is called a sarcomere; it extends from Z-line to Z-line. During muscle contraction, actin and myosin within the thick and thin myofilaments interact to form acto- myosin and this causes shortening of the muscle fibres. *Post mortem,* there is lateral shrinkage of the myofibrils in meat. Fluid is expelled from the spaces be tween filaments and is drained by gravity, forming drip.

Myoglobin: Purplish-red protein pigments found in muscles (meat). Myoglobin has one haem unit and one globin chain. In meat, myoglobin content differs between species and between different muscles. Colour lightness of meat is inversely correlated with myoglobin content. Meat colour is affected by oxidation state of myoglobin; the three major myoglobin derivatives are reduced myoglobin (purplish-red), oxymyoglobin (bright red) and metmyoglobin (brown/grey). Colour changes due to oxygenation of myoglobin are reversible. When meat is cured with nitrite, myoglobin is converted into the bright red pigment nitrosomyoglobin. Thermal denaturation of myoglobin to a brown pigment begins at about 65°C; consequently, the red colour of raw meat changes to brown on cooking.

Myosin: Myofibrillar globulins that are the most abundant proteins in meat and the predominant salt-soluble muscle proteins. During muscle contraction, myosin combines with actins to form actomyosin. Myosin molecules are shaped like elongated rods with thickened regions at one end. Proteolysis of myosin by trypsin results in formation of two myosin fractions, myosin heavy chains and myosin light chains. Myosin is insoluble in water and only slightly soluble in acids; however, it is soluble in salt solutions or alkalies. Myosin gelation is a principal factor in obtain- ing good texture in meat products.

N

Nan (bread): A yeast-raised bread made from white flour and yoghurt, cooked very quickly as a large flat pear shaped bread on the side of a tandoori oven. The shaped dough is moistened so that it sticks to the side of the oven and is hooked off when cooked. It served with Indian meals. Or Flat bread originating from northwest India is made from white flour, leavened with sodium bicarbonate and baked in a tandoor.

Nanofiltration: Form of filtration that uses semi-permeable membranes of pore size 0.001-0.1 μm to separate different fluids or ions, removing materials having molecular weights in the order of 300-1000 Da. Nanofiltration is most commonly used to separate solutions that have a mixture of desirable and undesirable components. An example of this is the concentration of corn syrups. Nanofiltration is capable of removing ions that contribute significantly to osmotic pressure, and this allows separation at pressures that are lower than those needed for reverse osmosis.

Naringenin: Non-bitter flavanone found mainly in citrus fruits, but also in other fruits, e.g. tomatoes.

Naringin: Bitter glycoside present in citrus fruits.

Natamycin: Polyene antibiotic with antifungal activity used topically on skin and mucous membranes for treatment of ringworm in cattle and horses. Absorption through dermal tissues of treated animals is normally negligible. Also used as a fungicide on bulbs such as onions and garlic. It is also known as pimaricin.

Natural flavourings: Flavour compounds, also essential oils, extracts and hydrolysates containing flavour compounds, that are derived from natural sources, such as plants, animal foods and edible yeasts. Usually they have little or no nutritive value but are used solely to impart flavour.

Natural foods: Foods produced by natural farming and using natural processing methods, e.g. without the use of added salt, sugar and preservatives. It is similar to organic foods but not necessarily totally organic.

Natural sweeteners: Sweet-tasting substances that occur in nature. Saccharides, such as sucrose (sugar), D-glucose (dextrose) and fructose (laevulose) are the major natural sweeteners used by the food industry. Other natural sweeteners include sweet-tasting proteins (e.g. thaumatin), terpenoids (e.g. glycyrrhizin), steroidal saponins (e.g. polypodoside A), dihydroisocoumarins (e.g. phyllodulcin) and flavo- noids (e.g. neohesperidin).

Near infrared radiation: Infrared radiation which has a wavelength between 0.7 and 2.5 μm. Near infrared is subdivided into very near infrared (0.7-1 micron) and short wave infrared (1.0-2.5 microns).

Nectar: The sugary liquid extracted from flowers by bees which after ingestion is chemically transformed and regurgitated as honey. or Any thick sweetened fruit juice.

Neomycin: Aminoglycoside antibiotic produced by *Streptomyces fradiae*. It is used for treatment of a range of bacterial infections in cattle, sheep, swine and poultry. It is relatively persistent in muscle tissues, eggs and milk. Parenteral use in food-producing animals is not permitted in the EU.

Neotame: High-intensity, non-nutritive sweetener which is an N-substituted derivative of aspartame. Approximately it is 7,000-10,000 times sweeter than sugar. A free flowing white crystalline powder which is water- soluble and heat-stable, and can be used in cooking as well as in tabletop applications. Although similar to aspartame, neotame is degraded differently in the human digestive system, avoiding problems caused by the presence of phenylalanine for people suffering from phenylketonuria.

Nephelometry: Technique used to determine the size and concentration of cells or particles in a solution by measuring the intensity of scattered light. Light scat- tering depends on the number and properties of the particles in the solution.

Net protein ratio (NPR): Weight gain of a group of animals (e.g. rats) fed a test diet plus the weight loss of a similar group fed a protein free diet, and the total divided by the weight of the protein consumed by the animals on the test diet.

Net protein utilization (NPU): Net protein utilization is an index of the nutritional values of proteins. This quality ratio indicates the amount of dietary protein retained in the body under specific clinical conditions. It changes in body nitrogen levels following consumption of a dietary protein are compared with those following consumption of a protein-free diet for the same duration, and then the dietary nitrogen retained in the body is expressed as a proportion of nitrogen intake.

Neural networks: Systems of computer programs and data structures which are modelled on the human nervous system and brain. Incorporate large numbers of processors operating in parallel, each with an individ- ual sphere of knowledge which has been fed into it along with rules about relationships. Networks can use this information to recognize patterns in large amounts of data. Used in the food industry to model processes and predict the behaviour of foods under specific con- ditions. It is also known as artificial neural networks.

Neurological shellfish poisoning: Food poisoning associated with consumption of shellfish containing neurotoxins produced by the dinoflagellate algae *Pytchodiscus brevis*. Gastrointestinal and neurological symptoms normally occur within 3 to 6 hours of ingestion of contaminated shellfish.

Neurotoxicity: Property of being toxic to nervous tissues.

Neurotoxins: Toxins that act specifically or primarily on nervous tissues (e.g. botulotoxins and saxitoxin).

Neutralization: Process of making something chemi- cally neutral, with a pH of approximately 7.

Niacin: A member of the vitamin B group. Generic descriptor for two compounds in foods which have the biological activity of the vitamin: nicotinic acid (pyridine 3-carboxylic acid) and nicotinamide (the amide of nicotinic acid). The metabolic function of niacin is in the coenzymes nicotinamide adenine dinu- cleotide (NAD) and nicotinamide adenine dinucleotide phosphate (NADP), which operate, often in partnership with thiamin and riboflavin coenzymes, to produce energy within the cells. Niacin is found in animal tissue as nicotinamide and in plant tissues as nicotinic acid; both forms are of equal niacin activity. Rich sources of niacin include livers, kidneys, lean meat, poultry meat, fish, rabbit meat, cornflakes (enriched), nuts and peanut butter. Niacin can withstand reasonable periods of cooking, heating and storage. Canning, drying and freezing result in little destruction of the vitamin. In cereals, niacin is largely present as niacytin, which is not biologically available. Deficiency of niacin leads to pellagra (photosensitive dermatitis), depressive psychosis and intestinal disorders. It is previously known as vitamin PP.

Nicotinamide: It is synonym for niacinamide and nicotinic acid amide. The amide form of nicotinic acid which has niacin activity as a constituent of 2 coenzymes (nicotinamide adenine dinucleotide (NAD) and nicotinamide adenine dinucleotide phosphate (NADP)); these coenzymes act as intermediate hydrogen carriers in a wide variety of oxidation and reduction reactions. Nicotinamide can be formed in the body from the amino acid tryptophan; on average 60 mg of dietary tryptophan is equivalent to 1 mg of preformed niacin.

Nicotinic acid: A member of the vitamin B group found in plant tissues. It contributes, along with nicotinamide found in animal tissues, to niacin activity. Its nhemical name is pyridine 3-carboxylic acid.

Nisin: Polypeptide antibiotics produced by *Lactococcus lactis*. It is used in preservatives for a variety of foods, such as processed cheese, meat and meat products, fish, and canned fruits and vegetables.

Nitrates: Salts of nitric acid found in many animal and plant foods as a result of use of nitrate fertilizers, the nitrification process in the soil, or use of sodium nitrate or potassium nitrate food additives. Health risks are associated with conversion of nitrates into nitrites in the gastrointestinal tract. Contamination of drinking water with nitrates from chemicals used in agriculture is a particular concern.

Nitrites: Salts of nitrous acid which are involved in producing the pleasant pink colour of cured meat. They are formed by reduction of nitrates (saltpetre) by bacterial action in the meat or can be included in the curing liquor. Because of the method of production sal prunella contains a little nitrite as impurity. It can be oxidizing agents or reducing agents. It is authorized as food additives for preservation of meat and cheese. Health risks are associated with formation of nitrosamines from nitrites in the presence of amines.

Nitrogen solubility index: Physicochemical properties defined as extracted nitrogen as a percentage of sample nitrogen content, determined between pH 2 and 12.

Nitrogen: Colourless and odourless gas that constitutes approximately three-quarters of the Earth's atmosphere by volume. The common form is dinitrogen (chemical symbol N_2). It is constituent of proteins, amino acids, and many other groups of chemicals, e.g. amines, alkaloids and purines. The inert gas which forms 79% of air, used to fill sealed packaging so as to displace air and prevent oxidation.

Nomograms: Graphical plots in the form of line charts which may be used to solve particular types of equations. Scales for the variables involved in the formula are presented in a way such that corresponding values for each variable are on a straight line intersecting all scales. Thus, when values for two variables are known, the value of a third can be read from its scale.

Nonenzymic browning: Food browning process promoted by heat treatment, which includes a wide range of reactions, such as the Maillard reaction, caramelization, chemical oxidation of phenols and maderization.

Nonreducing sugars: Sugars that do not have a free carbonyl group (ketone or aldehyde) and therefore are not able to act as reducing agents.

Nonthermal processes: Processing techniques that do not require heat. Usually refers to food pasteuriza- tion and sterilization treatments that do not employ heat during processing. Examples include: high pressure processing (inactivates vegetative microorganisms); ultrasonication (inactivates vegetative bacteria and reduces heat resistance of bacterial spores); high voltage electric pulse treatment (electroporation; inactivates vegetative microorganisms); ionizing radiation treatment (inactivates pathogens); high intensity light pulse treatment (inactivates vegetative bacteria); and high intensity magnetic fields processing (inactivates microorganisms).

Noodles: A mixture of grain or pulse flour or other starch product with egg and/or water and salt formed into long flat ribbons of various thicknesses and cross sections. It is cooked by boiling, stir-frying or deep-frying and may be soft or dried. Typical names are listed under egg, rice, cellophane, soba, chasoba, udon, somen, shirataki, bean thread, efu, miswa, Shanghai and all the Italian pasta types.

Norbixin: A golden yellow food colouring obtained from the seeds of achiote.

Novel foods: Foods prepared using unconventional processes (particularly genetic technology), derived from unconventional sources or offering non- nutritional benefits. Examples include biotechnologically derived foods, designer foods and medical foods.

Nuclear magnetic resonance (NMR): Spectroscopy technique based on the magnetic moment of atomic nuclei. An external magnetic field will partially align the axis of spin of spinning nuclei, but some precession about the magnetic field will occur. The precession depends on the magnetic field applied and the magnetic moment of the nucleus (dependent in turn on the chemical state of the atom), and is specific to the type of nucleus. The precession rate, measured by emission or absorption of applied radiofrequency, is used to give details about the composition of the sample.

Nuclear power: Power generated by nuclear reactors in nuclear power plants or stations. Accidents at nuclear power stations have caused fallout of radiocae- sium, and radioactive contamination of growing foods.

Nucleic acids: Polymers of nucleotides in which the 3-position of one nucleotide sugar is linked to the 5-position of the next by a phosphodiester bond. The two major types are DNA and RNA.

Nucleosides: Compounds of purine or pyrimidine bases with a sugar, usually ribose.

Nucleotides: Compounds of purine or pyrimidine bases with a sugar phosphate.

Nut cracker: An implement used to crack the hard outer shell of nuts using a lever or screw thread action

Nut meal: Ground nuts used to substitute for some of the flour in biscuit and cake mixes and in fillings.

Nut oil: Any oil expressed or extracted from a nut, usually carrying the particular flavour of the nut.

Nutrients: Essential dietary factors, such as vitamins, minerals, amino acids and fatty acids, that are required by the body but cannot be synthesized in the body in adequate amounts to meet requirements, so must be provided by the diet. Nutrient deficiency can cause poor growth, deformity, malfunctioning and ste- rility. A range of characteristic deficiency diseases is recognized in humans.

Nutrition: The study of the food requirements of humans and animals in terms of quantity, quality and the content of protein, fat, carbohydrate, fibre, vitamins, mineral, and trace elements, etc. A major aspect considered is the way by which an organism absorbs and utilizes food components. The study of nutrition involves identification of individual nutrients that are essential for growth and maintenance of the individual, interrela tionships among nutrients within individual organisms, and quantitative requirements of organisms for specific nutrients under various environmental conditions in order to optimize health.

Nutritional labeling: Information appearing on labelling or packaging of foods relating to energy and nutrients in the food. The information which must or may be given, and the format in which it must appear, is governed by law in most countries.

Nutritional needs: The amounts of the various components of food required to be eaten for health and well being.

Nutritional values: Indications of the level to which a food contributes to the overall diet. These values de- pend on the quantity of food ingested and absorbed, and the amount of essential nutrients it contains. Nutritional values can be affected by cultivation condi- tions, handling and storage practices, and processing.

Nutritionist: A person who makes a study of and is an expert in nutrition and nutritional needs. Often works with the kitchens and chefs of hospitals, schools and other institutional food providers.

Nuts: Fruits consisting of an edible kernel within a shell, the thickness and hardness of which is varies among types. Kernels have a high fat content and are often used as the source of nut oils. They are also rich in fibre, vitamin E, folic acid and a range of minerals. Nuts are generally available shelled or unshelled; shelled nuts are sold in many forms including raw, blanched, roasted and flavoured. They are eaten out of hand or used in a variety of sweet and savoury dishes. Commonly consumed nuts include walnuts, pistachio nuts, pine nuts, cashew nuts and almonds. Some foods known as nuts are not true nuts, e.g. Brazil nuts are really seeds and peanuts are legumes.

Nylon: Family of strong, elastic polyamide materials, which vary from moderately flexible to strong, tough and rigid products. It can be shaped when heated into forms such as sheets, bristles and fibres. It is resistant to greases and oils. It is used widely as a packaging material for foods in packs such as pouches and boil-in-bags.

Nystose: Fructooligosaccharide comprising three fructose residues and a glucose residue. It produced by hydrolysis of inulin or from sucrose via the action of fructosyltransferases.

O

Oat bran: Outer layer found under the hull of the oat grain which forms the milling fraction.

Oat fibre: Indigestible material derived from oat grains, consumption of which is reported to reduce serum cholesterol.

Oat flour: Flour ground from oats which their outer layers have been removed. It does not contain much gluten and must be combined with wheat flour to make a bread which rises. It is used as an ingredient in bakery products and snack foods.

Oat germ: The germ of the oat grain separated during the process of making oat flour. It is usually mixed with oat bran to a brown powder which can be used to make a thin porridge or added to other flours for use in bread making.

Oat oils: Oils extracted from oats which is highly unsaturated and containing high levels of linoleic acid.

Oatmeal: Coarsely ground or cracked dehusked oats separated by sieving into various particle sizes, usually fine, medium and coarse, but superfine and pinhead can be found

Oatrim: Fibre-rich fat substitute made from oat flour and oat bran which is used in bakery products, salad dressings, confectionery and dairy products.

Obesity: It is a condition in which the body's weight is beyond the limit of skeletal and physical requirements due to excessive accumulation of fat. Possible causes include overeating, an inappropriate diet, genetic factors and metabolic disorders. Obesity is associated with an increased risk of developing a range of diseases, including adult onset diabetes, cardiovas cular diseases and certain types of cancer. In some cases, obesity can be reversed through measures such as adoption of a low-calorie diet, making changes to lifestyle or consuming functional foods designed for this purpose.

Ochratoxin: A Most toxic of the known ochratoxins.

Ochratoxins: Mycotoxins produced by certain spe- cies of *Penicillium* (e.g. *P. viridicatum*) and *Aspergillus* (e.g. *A. ochraceus*) during growth on foods and feeds (e.g. wheat, rye, barley, oats, corn and peanuts). Nephrotoxic and carcinogenic in humans and animals (e.g. cattle and swine) when ingested in contaminated foods and feeds.

Octadecanoic acid: It is synonym for stearic acid, one of the fatty acids that occur naturally in the form of glycerides in animal fats and vegetable fats.

Octadecenoic acid: It is synonym for oleic acid, an unsaturated fatty acid that occurs as the glyceryl ester in fats and oils. It is one of the major fatty acids in cow milk.

Odour activity values: Ratio of the concentration of aroma compounds present in a product to the odour threshold values.

Odour threshold values: Levels at which perception of increasing concentrations of aroma compounds begins. The concept of odour threshold is useful in defining aroma purity, estimating the necessary amount of starting material, serving as a reference point in describing intensity and aroma quality, and evaluating which components present are important in contributing to a characteristic aroma.

Odour: The property of a substance which causes the sensation of smell. It is caused by molecules from the substance binding to receptors in the nasal cavity which send specific signals to the brain. The six main odour qualities are fruity, flowery, resinous, spicy, foul and burnt. See also flavour. It is also called smell.

Oedema: It is excessive accumulation of tissue fluid in body tissues, leading to swelling. It is popularly known as dropsy. It causes of oedema include heart failure, kidney failure, liver failure and malnutrition. Diuretic drugs can relieve symptoms by causing the patient to pass more urine. Allergic reactions may be accompanied by local oedema.

Off flavor/odour: Taints perceived in the mouth upon tasting a product. Off flavours are negative attributes, and affect the eating quality of foods; they may also indicate that a food is spoiled.

Ohmic heating: Thermal processing of foods using energy produced in the form of heat when a current passes through an electrical resistance. In this form of electric resistance heating, the food itself acts as a conductor between a ground and a charged electrode. The food may be immersed in a conducting liquid. Heating is accomplished according to Ohm's law, where conductivity of the food determines the current that will pass between the ground and electrode. Ohmic heating can be used as a cooking technique, and also for pasteurization and sterilization.

Oil expellers: Equipment is used in extraction of vegetable oils from oilseeds. Oils are pressed from the source material using pressure from an auger, which turns inside a barrel. The barrel is closed except for a single hole through which the extracted oil drains. Expellers remove larger proportions of oil from seeds than can be achieved with hydraulic batch presses. Also known as continuous screw presses.

Oil palms: Palm trees, *Elaeis guineensis*, native to tropical Africa. It yield palm oils from the fleshy endosperm of its seeds and palm kernel oils from the seed kernels.

Oil: Fat, which is liquid at ambient temperatures, comprising glycerol esters of a variety of mainly unsaturated, fatty acids. They are extracted from seeds, nuts, fruit, fish and other cold blooded animals. Most natural oils are contain complex flavouring compounds some of which are essential oils. The common seed, nut and fruit oils are almond, coconut, corn, cottonseed, grapeseed, hazelnut, mustard, olive, palm, peanut, poppy seed, rape seed, safflower, sesame, soya, sunflower and walnut. The animal oils are cod liver, halibut liver, whale and snake. It is lipid-rich, viscous substances derived from animal, vegetable or mineral sources that are liquid at room temperature and insoluble in water.

Oilseed proteins: Proteins derived from oilseeds, which have desirable functional properties and nutritional characteristics and may reduce the risk of certain diseases. It may be used as food supplements.

Oilseeds: Seeds, e.g. sesame seeds, sunflower seeds and soybeans, from which vegetable oils may be extracted. The oilseed cake or meal which remains after oils have been extracted is often used as a livestock feed, since it is rich in protein.

Okara: Fibre-rich by-product remaining after extraction of soymilk from ground soybeans. It is also rich in high quality proteins and used in soups, vegetable dishes, sausages and bakery products.

Okra: It is a common name for *Hibiscus esculentus* and good source of vitamin A and vitamin C. Immature pods are eaten as a vegetable, pickled or used to thicken soups and stews. It is also known as okro, lady's fingers, gumbo and bindi.

Oleic acid: A monounsaturated fatty acid of 18 with carbon atoms which when esterified with glycerol produces a low melting point fat which is a constituent of lard making it soft and of many vegetable oils including olive oil. It is prepared by hydrolysis of animal fats, such as tallow, or vegetable oils, such as olive oils, sun- flower oils and soybean oils.

Oleoresins: It is extracts of oil-soluble components of plant materials, usually spices and produced by direct contact of the spices with highly hydrophobic organic solvents, e.g. hexane. The organic solvents can then be evaporated to concentrate the extract. It is used as flavourings by the food industry. Oleoresins are cheaper to produce than essential oils and easier to use than spices, but do not have the full flavour profile of essential oils.

Oleosins: Alkaline proteins found in the oil bodies of plant seeds.

Oligosaccharides: Compounds comprising between three and ten monosaccharides linked by glycosidic bonds. Synthesis is by limited hydrolysis of polysaccharides or via addition of monosaccha- rides, a process catalysed by glycosyltransferases.

Olive oil: The monounsaturated oil obtained from the ripe fruit of the olive tree. The first cold pressing of the hand-picked, destoned and skinned ripe fruit from a single producer is usually the finest flavoured and most expensive, and is treated with the respect accorded to fine wines. Most olive oil is hot-pressed or solvent-extracted from the whole fruit. Olive oils are classified according to the method of production and the acidity (free oleic acid) of the oil. The Italian grades are *afiorato*, extra virgin *(extra vergine)*, superfine *(sopraffino)*, fine *(fino)*, virgin *(vergine)*, and finally the pure or 100% pure commercially blended oils.

Omega-3 fatty acids: Polyunsaturated fatty acids (ω-3) having double bonds in the ω-3 position; found in oily fish. It may have beneficial effects on health, especially resistance to cardiovascular diseases.

Omega-6 Fatty acids: Polyunsaturated fatty acids having double bonds in the ω-6 position. Found in vegetable oils. It may have beneficial effects for health, especially reducing the risks for cancer, stroke and coronary heart diseases. It include arachidonic acid, linoleic acid and ϒ-linolenic acid.

Omelette pan: A heavy-based frying pan with rounded sloping sides usually kept exclusively for omelettes and never washed after use but cleaned with absorbent paper. Omelette pans are proved before first use by heating salt in them.

Omelette: Whole shelled eggs, lightly stirred to mix the white and the yolk, seasoned, then poured into a hot frying or omelette pan greased with butter. As the egg sets, the unset egg is run underneath so that the whole of the egg mixture cooks rapidly with a slightly brown lower surface. The omelette may be turned over to harden the upper surface or folded in half or thirds with or without a precooked filling or topping and served immediately.

Oolong tea: Type of tea in which the tea leaves have been partially fermented, rather than not fermented (as in green tea) or fully fermented (as in black tea).

Opacity: Degree of obstruction an item produces to the transmission of visible light.

Opalescence: A pearly or milky mineral lustre resembling that of opal, resulting from the characteristic internal play of colours, in turn resulting from the reflection and refraction of light passing through adjacent thin layers of different water content.

Optical density: A measure of light absorption of a translucent medium, equivalent to the logarithm of the opacity. In the food industry, optical density is used in measurements of various parameters, including turbidity, browning and bacterial growth.

Optical properties: Physical properties relating to the appearance of a product, including clarity, colour, reflectance, turbidity and fluorescence.

Optical rotation: It is ability of some compounds to rotate a plane of polarized light due to asymmetry of the molecule. If the plane of light is rotated to the right, the substance is dextrorotatory and is designated by the prefix (+); if the plane of light is rotated to the left, the substance is laevorotatory, and the prefix is (-). A mixture of the two forms is optically inactive and is termed racemic. Sucrose is dextrorotatory but is hy- drolysed to glucose (dextrorotatory) and fructose which is more strongly laevorotatory; therefore, hy- drolysis changes the optical activity from (+) to (-). A mixture of glucose and fructose is termed invert sugar.

Orange essential oils: Essential oils produced by compression of orange peel that are composed predominantly of D-limonene but may also contain other aroma compounds, including octanal, myrcene, linalool, decanal, sinensal, ethyl butyrate and valencene. Composition of the essential oils is dependent on the species of orange from which they are produced.

Organic acids: Organic compounds comprising one or more substituents with the chemical formula - CO(OH). Examples include fatty acids, citric acid and acetic acid. Include carboxylic acids.

Organic compounds: Compounds based on a skeleton of one or more carbon atoms. In their simplest forms, carbon atoms are bound to each other and to hydrogen (e.g. hydrocarbons); these include paraffins and olefins. More complex organic compounds have one or more hydrogen atoms substituted with other elements or groups, e.g. halogens, nitrogen, sulfur, hydroxyl groups, as in organic halogen com- pounds, organic nitrogen compounds, organic sulfur compounds and alcohols, respectively. Carbon atoms may form linear structures and ring structures; a hydrocarbon ring comprising six carbon atoms and six hydrogen atoms is known as a benzene ring and organic compounds containing this structure or derived from it are known as arenes or aromatic compounds.

Organic foods: Foods produced by organic farming methods, i.e. without the use of chemical fertilizers or pesticides, and without any additives. The aim is to provide high quality, healthy food free from chemical residues. In the case of livestock, strict attention is paid to animal welfare, growth promoters are banned and use of veterinary drugs is kept to a minimum. Or- ganic foods are regarded as a healthy, environmentally friendly option by the consumer, but future market growth is uncertain due to problems associated with high prices and provision of consistent quality.

Original gravity: Amount of extract (soluble material) present in worts, as calculated from the amount of non-fermented extract left in the finished beer, to- gether with the amount of extract equivalent to the quantity of ethanol present in the beer.

Orthophosphates, calcium: Calcium salts of orthophosphoric acid used as firming agents, anti-caking agents and raising agents in cake mixes, baking powder and dessert mixes. E341 covers calcium tetrahydrogen diorthophosphate, calcium hydrogen orthophosphate and tricalcium diorthophosphate.

Orthophosphates, sodium and potassium: Sodium and potassium salts of orthophosphoric acid used as buffers, sequestrants and emulsifiers in dessert mixes, non-dairy creamers, processed cheese and the like. E339 covers sodium dihydrogen-, disodium hydrogen- and trisodium orthophosphate, E340 covers the equivalent potassium salts.

Orthophosphoric acid: An inorganic acid used as a buffering agent, and in the form of its salts (orthophosphates) as buffers, sequestrants, anti-caking agents, emulsifiers and firming agents in cake and dessert mixes, creamers and processed cheese.

Oryzanols: Ferulic acid **esters** of terpene **alcohols** commonly prepared from **rice bran oils** but which have also been extracted from **corn oils** and barley oils. It is used predominantly as **antioxidants**.

Oryzenin: Glutelin which is one of the main storage proteins in rice.

Osmolality: Concentration of osmotically active parti- cles in a solution, measured in osmoles of solute/kg of solvent.

Osmolarity: Concentration of osmotically active particles in a solution, measured in osmoles of solute/litre of solution.

Osmoregulation: Regulation of osmotic pressure, especially in the body of a living organism.

Osmosis: Passage of water through a differentially permeable membrane, from a region of low concentra- tion of solutes to one of higher concentration. Osmosis stops if the pressure of the more concentrated solution exceeds that of the less concentrated solution by an amount known as the osmotic pressure between them.

Osmotic drying/dehydration: Water removal preservation technique based on the water and solubility activity gradient across a cell's semi-permeable membrane. It involves immersing high moisture foods in an osmotic solution, usually of sugar or salt. Water flows out of the material, and sugar may flow in, depending on several variables. Osmotic drying with osmotic syrup recycling requires two to three times less energy than convection drying. At relatively low process temperatures (up to 50°C), it improves product colour and flavour retention. Application of osmotic drying in the food industry is restricted, as simultaneous solute transfer into the foods can affect product quality.

Osmotic stress: Stress exerted on an item when under osmotic pressure.

Osteoporosis: Weakening and brittleness of the bones, resulting in them becoming liable to fracture. Generalized osteoporosis occurs most commonly in the elderly, and in women following the menopause; it can also result from long-term steroid therapy, infection or injury. Although there is a net loss of calcium from the body, this is the result of osteoporosis, not its cause, and there is no evidence that calcium supplements affect the progression of the disease. A high calcium intake in early life may be beneficial, since this may result in greater bone density at maturity, but the most important factor is regular exercise to stimulate bone metabolism.

Ovalbumin: The main protein comprising about 70% of egg white protein. It is partially coagulated by fast beating especially at higher temperatures and is responsible for the stability of whipped egg white. It starts to coagulate with heat at around 60°C.

Oven temperature: Oven temperature is set according to four different scales, three quantitative, i.e. Fahrenheit °F, Centigrade °C, or gas mark GM, and one qualitative, i.e. Cool <120°C, slow 120°C, moderate 180°C, hot 220°C and very hot 250°C. As a rule of thumb °F are twice °C. This is exactly so at 144 °C; above this °F are slightly less than double and below slightly more than double.

Oven: An insulated enclosed space, usually rectangular but may be other shapes, heated by any type of fuel including solar radiation, either directly or indirectly, with a door or opening and in which food is baked, roasted or cooked.

Overwrapping: Packaging technique in which several packs or multipacks are wrapped together often with cellophane or other plastics films to form a single unit.

Ovomucins: Sulfated glycoproteins found in egg whites which are responsible for their gel structure. It possesses antiviral activity and act as trypsin inhibitors.

Ovomucoid: Heat resistant glycoproteins found in egg whites. Show activity as trypsin inhibitors.

Oxalates: Salts and esters of oxalic acid. Present at high concentrations in fruits and vegetables, e.g. potatoes, spinach, rhubarb, plums, tea and some nuts, where they are regarded as antinutritional factors. High concentrations of oxalates in urine are associated with formation of renal stones.

Oxalic acid: Organic acid comprising two carboxylic acid groups which has many industrial applications including clarification of fats and oils, and acid hydrolysis of starch to produce sugar syrups. Present as oxalates in fruits and vegetables, e.g. spinach, beets and strawberries, where they are considered to be antinutritional factors due to their involvement in formation of renal stones.

Oxidants: Chemicals that are capable of causing the oxidation of other chemicals, i.e. they donate oxygen or remove electrons.

Oxidation: Addition of oxygen to a compound, for example using oxidizing agents. Also includes reactions in which atoms in the reacting materials lose electrons, frequently together with the removal of hydrogen ions. Oxidation-reduction reactions always occur simultaneously; if one reactant is oxidized, another must be reduced. In cooking, the reaction of a food with the oxygen of the air which usually produces off flavours. Oxidation reactions are often promoted by light and high temperature increases the rate of reaction (approximate doubling for every 10°C rise). Anti-oxidants are used to inhibit the reaction.

Oxidative stability: Extent to which a substance can withstand the stress of oxidation.

Oxidizing agents: Chemicals capable of oxidation which are themselves reduced during the process, i.e. they gain electrons. Important oxidizing agents for the food industry include chlorine dioxide, which is used as an antimicrobial agent in water and sea foods, and iodates, which are used as flour improvers.

Oxygen absorbers: Materials which reduce the oxygen contents of food containers and maintain them at a very low level. This inhibits the growth of microorganisms and insects, and oxidative chemical reactions, increasing the stability and shelf life of the packaged products. It is also referred to as **oxygen scavengers**.

Oxygen: Element with an atomic weight of 16 and an atomic mass number of 8. Most common form of free oxygen is the diatomic species, molecular oxygen (O_2). Oxygen is the most abundant element of the Earth (air is composed of approximately 20% O_2). Essential for respiration in animals and aerobic microorganisms, produced by photosynthesis and is a common substituent of organic compounds, including biopolymers, such as proteins and polysaccharides. Reaction of foods with oxygen (oxidation) is a common cause of food spoilage, e.g. oxidation of fats and oils causes' rancidity, and presence of oxygen may allow growth of aerobic food spoilage microorganisms. At high temperatures it causes oxidation and deterioration of food. The rate of oxidation doubles for every 10°C increase in temperature.

Oxymyoglobin: Bright red pigments which represent the reduced form of myoglobin. In oxymyoglo- bin, oxygen is bound to the ligand, and the haem group of myoglobin is in the ferrous (Fe^{2+}) state. When fresh meat is cut and a new surface is exposed to oxygen, the surface colour changes from dark purple to bright red; this colour change, associated with oxymyoglobin formation is known as bloom.

Oxytetracycline: Tetracycline antibiotic used for treatment and control of a wide range of bacterial infections in cattle, swine, sheep, poultry and fish. It is used particularly to treat intestinal and respiratory diseases in swine and bacterial diseases in farmed fish. It residues normally deplete within 5 days post-treatment in cattle, swine, sheep and poultry, but tend to persist for longer periods in fish.

Oxytocin: Peptide hormone (nine amino acids) synthesized in the posterior pituitary gland. Stimulates uterine smooth muscle to induce uterine contractions and promote labour. Also induces secretion of milk in response to a suckling stimulus.

Ozonation: Application of ozone (O_3), a pungent, toxic form of oxygen with three atoms in its molecule, formed in electrical discharges or by UV light. Ozone is used in the purification of drinking water and in oxidizing agents.

Ozone: It is a form of oxygen comprising three oxygen atoms. It is a strong oxidizing agent with broad spectrum antimicrobial activity. It uses within the food industry include in disinfectants; permitted for use on food surfaces. It is also known as triatomic oxygen.

P

Packaging films: Packaging materials in the form of thin sheets which can be wrapped round a product. Films can be made from synthetic materials, such as plastics, or natural substances, such as whey proteins.

Packaging materials: Substances used to make packs. Packaging for foods is commonly made from a variety of materials, including glass, plastics, rubber, wood and paper, which are formed into a range of container types. The type of material chosen depends on the product to be packaged and the intended use.

Packaging: Enclosure or wrapping of products. Functions include product containment for handling, transportation and use, preservation, optimization of product presentation, hygiene and to facilitate product dispensing and use. The term covers retail (primary), grouped (secondary) and transport (tertiary) forms.

Packinghouses: Establishments in which products are packed.

Packs: Containers of varying shapes and sizes made from paper, plastics, cardboard or other materials that are used to enclose items such as food. The term is also used to describe items or groups of items which are packed in containers or enclosed in packaging materials.

Pakora: Indian **snack foods** consisting of pieces of spiced **meat** and/or **vegetables** enclosed in **batters** containing **besan** or chickpea flour and deep fried. Deep-fried and served hot with chutney. The fritters usually contain vegetables, but can be made with shellfish, fish and cooked rice.

Palatability: Sensory properties relating to the extent to which a food is palatable, i.e. possessing satisfactory properties or composition to be acceptable for consumption. Palatability of many foods can be enhanced by selective processing. For example, cereals may be crushed by milling to produce flour, a process that removes or breaks down the indigestible outer husk of the cereal seeds. The flour can be made more palatable by making it into bread or pasta.

Palatinit: It is low calorie sweetener. Disaccharide sugar alcohol produced from sucrose via transglycosylation and reduction reactions. It composed of a mixture of two isomers, namely 6-*O*-(D-glucopyranosyl)-D- sorbitol and 1-*O*-(D-glucopyranosyl)-D-mannitol. It has sweetness of a similar intensity to sucrose, but unlike sucrose it is resistant to enzymic hydrolysis and is therefore a poor substrate for growth of microorganisms, including those responsible for food spoilage and formation of dental caries. It also potentiates flavours of aromatic foods and masks the metallic after taste produced by some artificial sweeteners. It is also called isomaltitol.

Pale soft exudative (PSE) defect: A condition affecting meat, especially pork. It is often stress related and is associated with accelerated *post mortem* muscle metabolism and a low pH value in meat. There is a linear relationship between myosin denaturation and drip loss or surface lightness within the PSE quality class. Excessive colour variation, poor water binding capacity and de creased water holding capacity occur in PSE meat, making it unsuitable for further processing. It is a major problem for the swine industry. Approximately 60% of the PSE defect in pork is associated with porcine stress syndrome (PSS), a genetic disorder which enhances susceptibility to stress. Certain breeds of swine, such as Pietrain, are highly susceptible to the PSE defect. Halothane sensitivity tests have been used to screen breeding swine for PSS with the aim of preventing further propagation of the PSE defect in breeding herds. However, as about 40% of the PSE defect in pork is associated with stress prior to slaughter, and poor meat handling and storage, removal of the mutant PSS gene from breeding stock will not com- pletely eliminate PSE meat. An inherited condition also underlies much of the PSE defect in turkey meat.

Palm kernel oil (palm butter): A white oil extracted from the kernels of the fruits of the oil palm, *Elaeis guineensis*, mainly used for margarine manufacture.

Palm kernel oils: Oils produced from the kernels of the fruits of oil palms, *Elaeis guineensis*, usually by solvent extraction. It is classed as lauric oil and used in the manufacture of margarines, cooking fats and confectionery.

Palm oil: The oil extracted from the fruits of the oil palm, *Elaeis guineensis*, which originated in Africa but is now grown in Malaysia, Indonesia and other high rainfall tropical areas. About half the African production is used as cooking oil, the remainder is used for soap and industrial purposes and after treatment for margarine. The oil sets to a soft solid after extraction. It contains 40% saturated, 40% monounsaturated and 10% polyunsaturated fat. It is as palm kernel oil. It is rich in caro tenes, which are often removed to give the oil a paler colour. In addition to their use as cooking oils, they are also used in the manufacture of soaps and candles.

Palm olein: Olein isolated from palm oils.

Palmitic acid: A saturated fatty acid containing 16 car bon atoms. Present as glyceride esters in many fats and oils, including palm oils, from which it is commonly obtained. It is found in its glycerol ester form in beef and lamb fat and other hard fats

Palmitoleic acid: Monounsaturated fatty acid com- prising 18 carbon atoms and a double bond between atoms 9 and 10. Systematic name is *cis*-å^{9}- hexadecenoic acid. It is component of fats and oils.

Pan: A metal, glass or ceramic vessel usually deep and round but may be other shapes, fitted with one long handle or two small handles.

Pancake: A thin flat cake cooked on both sides with oil or fat in a pancake or frying pan, made from a thin pouring batter of milk, egg, flour and possibly sugar, normally served with a sweet or savoury filling or sauce.

Pancreatin: A protease produced by various microorganisms and used to convert proteins into emulsifying agents

Paneer: Indian cheese-like product made by acid coagulation of heated buffalo milk. It is white in colour with a spongy body and sweet, mildly acidic and nutty flavour. Used in the preparation of many products, including curries, vegetable dishes and sweets.

Panning: Method used to make coated sugar confectionery. It used to make two types of product, i.e. hard centres, such as nuts or dried fruits, covered with chocolate, or chocolate or similar centres coated with sugar. In both cases, the coatings are applied to the centres while they are tumbled in a pan or drum. Temperature control is used to harden chocolate coatings, while sugar coatings are hardened by moisture reduction.

Pantothenic acid: Member of the vitamin B group. Chemically, pantothenic acid is the β-alanine derivative of pantoic acid, and is required for the synthesis of coenzyme A (involved in the metabolism of fats, carbohydrates and amino acids) and of acyl carrier protein (involved in the synthesis of fatty acids). Dietary deficiency is unknown; it is widely distributed in all living cells, the best sources being livers, kidneys, yeasts, and fresh vegetables. Royal jelly is also a rich source. Approximately 50% of pantothenic acid in grains is lost by milling, up to 50% in fruits and vegetables is lost during canning, freezing, and storage, and from 15 to 30% in meat is lost during cooking or canning. Pantothenic acid is reasonably stable in natural foods during storage, provided that oxidation and high temperatures are avoided.

Papads: Traditional Asian snack foods made from a mixture of black gram meal, salt, oils and spices, which is deep fried or toasted.

Papain: A cysteine endopeptidase from the latex of papayas with broad specificity, but with a preference for a residue bearing a large hydrophobic side-chain at the P2 position. Many other plants contain homologues of papain. It is used for tenderization of meat, stabilization of beer, coagulation of milk in cheesemaking and hydrolysis of fish proteins.

Paper chromatography: Chromatography technique performed on blotting paper or filter paper. Components of the sample are separated as a result of interactions between them, the paper and the solvent or mobile phase. It is largely superseded by thin layer chromatography.

Paper: Material manufactured in thin sheets from wood pulp or other fibrous substances. It is used widely as a medium for writing and printing, as a packaging material, as a wrapping material and as an absorbent.

Paperboard: Thick, stiff cardboard, which is composed of many layers of paper or compressed paper pulp. It is also known as pasteboard.

Paprika spice mix: A pungent blend of flavouring agents used to coat chicken prior to barbecueing, consisting of fresh ginger, fresh garlic, paprika and ground cumin pounded into a paste

Parasites: Organisms which live in (endoparasites) or on (ectoparasites) organisms of another species (host), from which they obtain nutrition and/or protection. It is yypically detrimental to the host.

Parasiticol: Mycotoxin produced by *Aspergillus parasiticus.*

Paratha: Flaky unleavened bread made from a mixture of white and brown flour, water and salt, the dough formed into rounds coated with ghee and fried until crisp on the outside.

Parboiling: Partial cooking of foods by boiling briefly in water before cooking by some other means, such as frying or roasting. Dense foods can be parboiled to allow them to be added at the same time as quick cooking ingredients in meals such as stir fry dishes. This means that all the ingredients will complete cooking at the same time. Also refers to the process of soaking and pressure steaming of rice before milling to gelatinize the starch and infuse some of the nutrients from the bran into the kernel.

Parching: Drying of goods such as grain or starchy vegetables is through application of intense heat.

Parchment paper: Waterproof, grease-resistant, stiff translucent paper treated to resemble parchment. It is produced by passing ordinary paper through a zinc chloride or sulfuric acid solution. Used in sheets or as bags to wrap foods.

Paselli: Maltodextrin derived from corn, potatoes, wheat or tapioca. Used as a fat substitute, texture modifier or bulking agent in many foods such as bakery products, dairy products, processed foods and beverages.

Pasta : A variety of extruded, cut or pressed shapes and sheets of pasta dough made either fresh and soft, or dried, from a basic dough of water and/or eggs, salt and hard durum wheat flour and sometimes oil. The simplest pasta is made from strong flour and eggs only (9:5), kneading up to 10 minutes until small blisters appear. There are innumerable names and shapes but the main types are spaghetti, macaroni, noodles, vermicelli, lasagne, canneloni and ravioli. Most pastas are cooked and served with a sauce.

Pasteurization: Thermal treatment of milk and other liquids (such as beer, wines and fruit juices) safe for consumption by destruction of most of the microorganisms present in them. Certain enzymes that would otherwise decrease shelf life are also inactivated by the process. Pasteurization is achieved by application of moderately high temperatures for a short period of time. Variants of the process include HTST pasteurization and LTLT pasteurization. Nutritional values of treated products are not greatly reduced by application of this process, nor are lactic acid bacteria destroyed. Cold pasteurization may be accomplished by radiation and/or chemical methods. Pasteurization precedes the drying of many liquid food products. Milk may be pasteurized at 62.8°C for 30 minutes or at 72°C for 15 seconds, shelled eggs at 64.4°C for 2 to 5 minutes, etc.

Pasteurized milk: Milk that has been heated to a spe- cific temperature for a specified length of time to kill off microorganisms that could cause spoilage or poisoning. This treatment can be carried out at a high temperature for a short time (HTST pasteurization; 72-80°C for 15 seconds) or at a lower temperature for a long time (LTLT pasteurization; 62-65°C for up to 30 minutes). Since pasteurization destroys phospha- tases but not peroxidases, a phosphatase test is used to test the efficacy of the process.

Pasteurizers: Equipment used in pasteurization of milk and other liquid foods to destroy most of the microorganisms present by application of heat.

Pastiness: Sensory properties relating to the extent to which the consistency of a substance is perceived as being pasty or thick.

Pasting properties: Functional properties is relating to the ability of an item to act in a paste-like manner. Pasting properties of starch, e.g. gelatinization temperature, transparency, viscosity and retrogradation, have an important effect on the cooking and processing of foods.

Pastries: Various types of small and sweet bakery products (cakes) made with paste-like dough, often iced, filled or decorated and made of pastry or various sponge or cake mixtures

Pastry A mixture of flour, fat, possibly egg and sugar, the fat usually dispersed as small solid globules coated with flour and the whole brought together with liquid prior to shaping and baking. There are many types of pastry, including shortcrust, flan, sweet, flaky, puff, rough puff, hot water crust, suet crust, choux and filo.

Pastry board: A square or oblong board preferably marble but usually wood on which pastry is rolled out.

Patatin: One of the major storage proteins of potatoes (molecular weight approximately 40 kDa), accounting for 30-40% of total soluble protein. It exhibits esterase activity.

Patents: Official documents issued by a governmental agency granting an inventor or inventors sole rights to use or sell an invention or process described in a patent application for a defined length of time. The patent application includes a written description of the invention, claims which define the scope of exclusivity, and also drawings and diagrams where appropriate. Many processes, pieces of equipment and materials developed for and used in the food industry are covered by patents. These include genetically modified crops and processes used in their production.

Pathogenesis: Cellular events and reactions which occur during the process of disease development.

Pathogenicity: Quality or degree of being capable of causing disease.

Patulin: Carcinogenic mycotoxin produced by various fungi, especially *Aspergillus* and *Penicillium* spe cies. It occurs in fruit juices produced from fruits contaminated with *P. expansum.*

Payasam: A paste made from various ingredients but always containing coconut milk, eaten as a sweet snack or accompaniment to a meal. Typical ingredients are cooked and mashed pulses, cooked rice, cooked sago, cashew nuts, jaggery or sugar.

Pea meal: Flour produced from yellow or green peas. It has been used to make pasta-like products and snack foods.

Pea protein concentrates: High-protein products made from peas. High nutritional quality and good functional properties make them suitable for many uses in the food industry such as manufacture of edible films and inclusion in infant foods and protein supplements. Flatulence causing and antinutritional factors that can become concentrated in the pea protein products must be removed during processing.

Pea proteins: Proteins found in peas, including legumin, vicilin, convicilin and albumins. Protein fractions extracted from peas are purified to yield pea protein isolates and pea protein concentrates that have high nutritional quality and good functional properties, and are used as ingredients of products such as infant foods and protein supplements. They are also used in the preparation of edible films. Flatulence-causing and antinutritional factors that can become concentrated in the pea protein products must be removed during processing.

Peanut butter: : Paste produced from ground, roasted peanuts together with hydrogenated oils (which prevent separation of the peanut oils from the mass), emulsifiers and salt. Ground peanuts sometimes with added peanut oil, salt or sugar with a buttery consistency. It sold as smooth or crunchy, the latter containing chopped peanuts. It is used as a sandwich spread or to flavour some dishes.

Peanut meal: Flour produced from peanuts and rich in protein and fibre. It is used as a protein supplement in a range of products, including milk beverages, bread and biscuits.

Peanut milk: High-protein beverages based on aqueous extract of peanuts to which sugar may be added.

Peanut proteins: Proteins found in peanuts, the main ones being arachin, conarachin I and conarachin II. It is responsible for the allergenicity of peanuts.

Peanut oil: Oil extracted from peanuts containing 20% saturated, 50% monounsaturated and 30% polyunsaturated fat. It has a rather high melting point (freezing point) and a high smoke point. It is used for deep-frying. It is also called groundnut oil, arachis oil.

Peanuts: Seeds produced by the leguminous plant, *Arachis hypogaea.* Up to six seeds develop in the underground pods which are harvested by hand or me- chanical means. Seeds are rich in proteins, minerals, vitamin E and vitamin B complex. As well as being eaten out of hand, roasted, boiled or raw, peanuts are used in cooking and in products such as confectionery, snack foods, peanut butter and salads. A high-protein meal made from peanuts has been in- corporated into a range of foods as a protein supple- ment. The seeds are also the source of groundnut oils. These contain a high proportion of unsaturated fatty acids; uses include cooking and manufacture of margarines. Peanuts are also known as groundnuts, American groundnuts and monkey nuts.

Pearl barley: Whole barley kernels with the husk and part of the bran layer removed by polishing. It is often added to soups.

Pearling: As well as referring to the formation of pearl shaped items, this term relates to the removal of indigestible hulls, aleurone and germ layers from cereals by abrasion. With respect to barley, three successive pearlings removes all of the hull and most of the bran layer, leaving what is termed pot barley. Two to three additional pearlings, followed by sizing with a grading wheel, produces pearl barley. It is also known as attrition milling and abrasive debranning.

Pectic enzymes: Group of enzymes that catalyse degradation of pectic polymers in the cell walls of plants. These enzymes are involved in the ripening of fruits, and have a number of uses in the processing of fruits and vegetables. The group comprises polygalacturonases, pectinesterases, pectate lyases and pectin lyases.

Pectic substances: Pectins and polysaccharides derived from them, such as polygalacturonic acid, polyglucuronic acid and polyuronides.

Peda: Indian sweet made using khoa as the base material. There are regional variations in its manufacture techniques, with consequent effects on sensory and compositional properties. Generally, khoa and sugar are heated to the desired texture and then divided into portions (usually round balls) that are packed in paper- board boxes lined with greaseproof paper.

Pediocins: Bacteriocins produced by several strains of *Pediococcus* spp. that are bactericidal against a wide range of Gram positive bacteria. Plasmid en- coded pediocin A, synthesized by *P. pentosaceus* (FBB-61 and L-7230), has a wide host range against Gram positive bacteria. Pediocin AcH, synthesized by *P. acidilactici* H, is a plasmid encoded, hydrophobic, inhibitory protein with a molecular weight of 2700 Da. Some

Gram negative bacteria can be made sus- ceptible to pediocin AcH when they are sublethally stressed. Antibacterial activity of pediocin AcH is through destabilization of cytoplasmic membranes. Pediocin PA-1, synthesized by *P. acidilactici* PAC 1.0, is a plasmid encoded protein with a molecular weight of 16,500 Da. Both pediocin AcH and PA-1 are ribosomally synthesized. Bactericidial efficiency of pediocins varies greatly under different conditions; some are due to the physical and chemical properties of the molecules, and some are due to environmental factors.

Pediococcus cerevisiae: Bacterium used as a starter culture for lactic fermentation in meat products, e.g. Salami, and in vegetables, e.g. Cucumbers

Pediococcus halophilus: Bacterium uses for the second-phase acidification of soya sauce.

Pediococcus: Genus of Gram positive, facultatively anaerobic coccoid lactic acid bacteria of the family Lactobacillaceae. *Pediococcus acidilactici* and *P. pentosaceus* are used as starters in the manufacture of fermented meat and vegetable products. *P. inopinatus*, *P. dextranicus* and *P. damnosus* may be responsible for spoilage of beer and wines.

Peel: The outer skin or rind of fruits and vegetables. A source of essential oils that may be used as flavourings, dietary fibre, pectins, vitamins and minerals. Peel from some sources, e.g. citrus peel, is used in foods and beverages, eaten candied or chocolate coated, processed into marmalades or incorporated into garnishes. The term also refers to a spade-like device used for moving loaves of bread or pizzas into or out of ovens.

Peeler: An implement used for peeling, consisting of a handled steel blade in which a longish slot with a raised sharp edge has been formed. This is used to remove a uniformly thick continuous slice from the surface of the object being peeled. Also called potato peeler, vegetable peeler

Peeling: Removal of the outer covering, or peel, from fruits or vegetables using knives or special peelers. It is also commonly removal of the shell from hard boiled eggs.

Pelshenke values: Scores that provide estimates of the potential breadmaking strength of wheat in relation to its gluten quality.

Penetration: Process of entry and permeation into an item. Penetration tests are widely used as a simple way to determine yield stress of a product.

Penetrometers: Instruments used for measuring the firmness of foods, especially fruits, on the basis of the depth of penetration of a probe under a known load.

Penetrometry: Technique for measuring the **firmness** of foods, especially **fruits**, based on the depth of penetration by a probe under a known load.

Penicillin G: Natural penicillin antibiotic produced by *Penicillium chrysogenum*. It is active against Gram positive bacteria. It is used for treatment of bacterial infections in all farm animals, particularly for control of mastitis in dairy cows and for treating infections of the gastrointestinal system and urinary and respiratory tract. Its residues in milk and muscle tissues are rarely detectable beyond 5 days from the final treatment. It is also known as benzylpenicillin.

Penicillins: Group of antibiotics widely used to treat bacterial diseases in animals, and constituting the most important group of antibiotics. Classified in three distinct groups: natural penicillins (including penicillin G); penicillinase-resistant penicillins (including cloxacillin and oxacillin); and broad spectrum penicillins (including amoxicillin and ampicillin).

Penicillium: Genus of fungi of the class Ascomycetes. Some species, e.g. *Pencillium digitatum*, *P. expansum* and *P. implicatum* can cause food spoilage and some are capable of causing food spoilage at refrigeration temperatures. Some species produce mycotoxins,e.g. citrinin, luteoskyrin and patulin. Certain species are used in production of organic acids and antibi- otics, while others are used in cheesemaking, e.g. *P. camemberti* (Brie cheese, Camembert cheese) and *P. roqueforti* (Roquefort cheese, Stilton cheese).

Pentoses: Monosaccharides comprising five carbon atoms. Examples include the aldoses, ribose, arabinose and xylose, and the ketose, xylulose.

Peppermint essential oils: Essential oils distilled from peppermint. The characteristic fresh, minty notes, produced by menthol, are not present in the primary distillate but are formed by further processing or natural ageing of the oils. The oils also contain vari- ous quantities of menthofuran, peroxidation of which produces an undesirable aftertaste, and thus content of this molecule influences quality of peppermint es- sential oils.

Peptides: Compounds formed by two or more amino acids linked via peptide bonds, e.g. dipeptides (two amino acids linked), oligopeptides (several amino acids linked) and polypeptides (many amino acids linked).

Peptones: Protein hydrolysates produced via the action of pepsin. Peptones are formed in the stomach during digestion of proteins.

Performance drinks: Non-alcoholic beverages formulated with ingredients claimed to enhance physical or mental performance.

Permeability: Ability of items such as membranes or other barriers to permit fluids to flow through them. Permeability is an important indicator of membrane functionality, and is expressed as a volume flow of liquid through a unit area of membrane at some defined transmembrane pressure. Permeability of food pack aging materials is important in relation to product shelf life. Modified atmosphere packaging of foods can involve use of films with various gas perme- ability coefficients.

Permeation: Passage of fluids through items such as membranes, food packaging materials or other barriers, or, in chemical terms, the diffusion or penetration of ions, atoms or molecules through a permeable substance. In the food industry, knowledge of the level of permeation of gases through functional barriers such as packaging materials is important in relation to product shelf life.

Permissible levels: Recommended limits for the amounts of particular contaminants (e.g. residues of veterinary drugs, heavy metals) that may be permitted in certain foods.

Peroxidases: It involved in ripening of fruits, enzymic browning and degradation of lignin by white-rot fungi. Peroxidases have a number of industrial applications, including use in time temperature indicators, such as those used for investigating inhibition of microorganisms during the thermal processing of low-acid foods, detection of phenols and production of flavour compounds. In addition, the degree of inactivation of peroxidases can be used as an indicator of the extent of blanching in vegetables.

Peroxidation: Formation of peroxides as a result of the action of oxygen, especially on polyunsaturated fatty acids. Vitamin E prevents lipids peroxidation in cells.

Peroxide values: Measure of the number of milli- moles of peroxide absorbed by 1000 g of oil or fat, used as an indicator of rancidity. As fats decompose, peroxides are formed. Chemically, peroxides are capable of causing the release of I from KI. Therefore, the amount of I released from KI added to a fat is a rancidity test. The more peroxide present, the more I re- leased; hence, the higher the peroxide values.

Peroxides: Compounds containing either the peroxide ion, e.g. sodium peroxide, or covalently bonded dioxygen (R-O-O-R), the simplest being hydrogen peroxide. Organic peroxides may be formed via autoxidation reactions or by direct oxidation, processes involved in the development of rancidity of fats and oils.

Persulfates: Salts of peroxodisulfuric acid which are strong oxidizing agents. Ammonium persulfate and potassium persulfate are permitted food additives. Ammonium persulfate is a bakery additive, uses including bleaching agents for starch and food preservatives. Potassium persulfate has uses in defoaming agents. Alternative names include peroxosulfates and peroxodisulfates.

Pervaporation: Membrane separation technique in which a liquid feed mixture is separated by partial vaporization through a non-porous, selectively permeable membrane. A vapour permeate and a liquid retentate are formed. Partial vaporization is achieved by reducing the pressure on the permeate side of the membrane (vacuum pervaporation) or, less commonly, by sweeping an inert gas over the permeate side (sweep gas pervaporation). Vacuum pervaporation at ambient temperature using hydrophilic membranes is used to dealcoholize wines and beer, whereas hydrophobic membranes are used to concentrate aroma compounds such as alcohols, aldehydes and esters.

Pesticides: Chemical substances used to kill plants, animals or other organisms that interfere with agricultural production or are harmful to humans. Major groups include herbicides (for control of unwanted plants), insecticides (for control of insect pests), fungicides (for control of pathogenic or spoilage fungi) and rodenticides (for control of rats, mice and other rodents). Many are non-specific and may be too toxic to organisms that are not considered pests. Some persist for long periods in the environment and can accumulate in the food chain. Residues in foods may represent a health risk to consumers

Pests: Organisms (typically rodents, insects and pathogens) that are regarded as harmful to humans, animals or plants.

pH meters: Instruments for measuring the **pH** of a solution.

pH: Measure of the degree of acidity or alkalinity of a substance. pH (an abbreviation for potential of hydrogen) is defined as the negative logarithm of the hydrogen ion concentration. The scale ranges from 0 (very strongly acid) to 14 (very strong alkaline). A neutral solution, such as pure water, at 25°C has a pH of 7.

Phaeophytins: Brown pigments produced by removal of magnesium ions from chlorophylls using limited hydrolysis. Present in green vegetables as degradation products of chlorophylls; degradation is accelerated by cooking or processing and thus may cause browning in vegetables or vegetable products.

Phase behavior: Activity of the various components of a mixture; of primary importance for food formulation and processing. For example, examination of the phase behaviour of fat mixtures (palm kernel oils, cocoa butter and anhydrous milk fats) can aid in the understanding of softening and bloom formation in compound coatings. Information regarding the phase behaviour properties of biopolymer systems may be useful in the design of new low fat foods.

Phase: A physical state of matter such as gas, vapour, liquid or solid. Two or more phases are common in many foods, e.g. Milk and cream contain two liquid phases, water and butterfat, whipped eggs contain liquid and gas phases, cooked meringue contains a solid and gas phase and some cake mixtures are very complex mixtures of phases.

Phenolases: It is alternative term for catechol oxidases, laccases and monophenol monooxy- genases.

Phenols: Group of organic compounds comprising at least one benzene ring that is covalently bonded to one or more hydroxyl groups. Phenols have wide distribution and applications, and are available in synthetic or natural forms, e.g. lignin and catechols. Its uses include as disinfectants (cresols), in manufacture of azo dyes and plastics, and as flavourings (vanillin), antioxidants (sesamol and NDGA) and pigments (curcumin). Some phenols, e.g. chlorophenol, are also considered to be toxins.

Phosphates: Salts of phosphoric acid used for flavouring and as stabilizers. The important ones used in the food industry are disodium dihydrogen, tetrasodium, tetrapotassium and trisodium diphosphate all classified as E450(a) and pentasodium triphosphate and pentapotassium triphosphate classified as E450(b).

Phospholipids: Lipids comprising a glycerol or sphingosine backbone esterified to two fatty acids and phosphoric acid or a phosphoric acid ester. Examples include phosphatidic acid, phosphatidyl- serine, phosphatidylinositol, phosphati- dylethanolamine and lecithins.

Phosphoric acid: It is synonym for orthophosphoric acid. Acid produced by reaction of phosphates with sulfuric acid or by oxidation of phosphorus followed by addition of water. Permitted food additive that is used to acidify fruit juice beverages and cola beverages, and as a substrate for phosphates.

Photocolorimetry: Colorimetry technique in which results obtained using a colorimeter are recorded permanently using photography.

Photodensitometry: Technique used to determine the density of a substance by examination of photographic negatives. It is used in combination with chromatographic techniques, such as thin layer chromatography, and gel electrophoresis to quantitate separated components. It is also used widely in medicine, where it is known alternatively as radiographic absorptiometry, to assess bone mineral changes.

Photolysis: Cleavage of one or more covalent bonds in a molecule due to the absorption of energy from light or some other form of electromagnetic radiation (e.g. UV radiation, X-rays).

Photometry: Science of visual radiation and the theory of its measurement. Luminous quantities can be meas- ured by the human eye, while radiant quantities are measured by devices sensitive to electromagnetic energy. Photometric measurements are performed using photometers equipped with photoelectric cells of various types and sensitivities.

Photooxidation: Oxidation reactions initiated by the presence of light.

Phulka: Puffed unleavened Indian bread made from wheat flour and similar to tortillas. It is eaten warm as an accompaniment to curries.

Physical properties: Characteristics of substances that do not involve a chemical change, such as density, electrical properties, mechanical properties and optical properties.

Physicochemical properties: Characteristics of chemical systems determined by application of physical principles, i.e. the physical properties of chemical compounds.

Physics: The study of systems and their interactions with one another, in terms of the interrelationship between matter and energy, without reference to chemical change. Traditionally divided into the study of mechanics, electricity and magnetism, heat and thermo- dynamics, optics and acoustics. More modern aspects include quantum mechanics, relativity, nuclear physics, particle physics, solid-state physics and astrophysics.

Physiological effects: Effects thatproducts or their components have on human physiological processes.

Physiology: Study of the function of biological processes within living organisms. It is broken down into the study of the function of particular organs. The concept of homeostasis, the regulation of the internal environment within certain parameters, is central to this science.

Phytates: Salts or esters of phytic acid containing inositol and phosphates as the base. Especially abundant in the outer layer of cereals, in dried legumes and some nuts as both water-soluble salts (sodium and potassium) and insoluble salts of calcium and magnesium. Phytates may decrease absorption of calcium, zinc and iron from the intestine.

Phytic acid: It is hexaphosphoric acid ester of inositol present mainly in cereal grain, nuts and legumes. It tends to bind iron and zinc in the plants and to carry these trace elements through the human gut without absorption. The amino acids methionine and cysteine present in animal products but not in plant-derived protein promote the release of these trace elements and hence their absorption.

Phytochemicals: Physiologically active chemicals produced by plants. It is used in functional foods and nutraceutical foods.

Phytosterols: Steroid alcohols present in plants, particularly in oils and waxes. They have hypocholesterolaemic activity and are thus used in functional foods, such as specially formulated margarines and spreads. Examples include sitosterol and stigmasterol.

Pickle: Food, usually chopped vegetables and fruits, preserved in a flavoured sauce or solution which prevents the growth of microorganisms. Usually used as a condiment or accompaniment and appreciated for its salty or sharp acidic flavour.

Pickled cheese: Cheese that is ripened in brines. Curd is cut into pieces that are put into containers filled with brine or salty whey and left to ripen for several months. Examples of this type of cheese include Feta cheese, Domiati cheese, Brinza cheese and Kareish cheese. Also known as brine ripened cheese.

Pickled eggs: Products prepared by pickling hard boiled eggs in solutions usually of vinegar mixed with flavourings. As well as eggs from chickens, duck eggs and quail eggs are commonly used.

Pickled onions: Small onions (commonly pearl onions) pickled in vinegar mixture or brines. Used as a condiment or garnish.

Pickles: Foods preserved by pickling in a liquid such as vinegar or brines, usually containing spices to enhance flavour. It can be made from vegetables, fruits, meat, eggs or nuts. Popular pickled foods include sauerkraut, cucumber pickles and chutneys.

Pickling: Preservation of foods in a pickling liquid such as vinegar or brines, often containing spices. Foods commonly preserved in this way include vegetables, fruits, meat, eggs or nuts. Pickles can be of various flavours, and can be sweet, savoury or spicy.

Pigeon peas: Seeds produced by *Cajanus cajan*. It grow in long, twisted pods and are usually greyish in **colour**. Young seeds are eaten as a vegetable, but mature seeds are often dried and split, eaten as **dhal** in India. Green pods may also be used as a vegetable and seeds can be germinated to produce sprouts. May be used in- stead of **soybeans** to make **tempeh**. It is also known as red gram.

Pigging: Cleaning of pipes or ducts in processing equipment, including that in food factories, by forcing a tightly fitting, flexible object, such as a brush, blade or swab (pig), through the pipeline in order to scrape or push out the residual contents.

Pigmentation: Colour that a substance exhibits, due to the presence of pigments.

Pigments: Compounds, usually fine, solid particles that give colour or other properties to a tissue, object or substance. For example, chlorophylls impart a green colour to lettuces and peas, carotenes are responsible for the orange colour of carrots, lycopene gives the red colour to tomatoes, anthocyanins contribute the purple colour of grapes and blueberries, and oxymyoglobin gives the red colour to meat. Pigments are sensitive to chemical and physical effects during food processing, and to chemical change during ripening. Pigments may also be added intentionally to foods in the form of food colorants.

Piperine: Alkaloid and flavour compound is isolated from black pepper, where it is primarily responsible for pungency. It my also be used in flavourings for brandy.

Pizza cheese: A soft spun-curd cheese similar to Mozzarella made from cows' milk using either a starter of *Lactobacillus bulgaricus* and *Streptococcus thermophilus* or citric acid to curdle the milk. It is used particularly for pizzas and contains somewhat less water than real Mozzarella, roughly 47% water, 24% fat and 25% protein.

Pizza dough: Typical yeast-leavened dough made of strong flour and tepid water with 30 to 60 ml of olive oil, 10 to 30 ml of active dried yeast and 10 ml of salt per kg of flour. It proved, knocked back and proved again. Exact proportions of flour and water depend on the flour. This dough will keep for 7 days in the refrigerator if oiled and covered with film.

Pizza fillings: Foods used to top pizzas. It include tomatoes, Mozzarella cheese, salami and sea foods.

Pizza oven The traditional pizza oven was brick with a stone floor and a wood fire to one side radiating heat from the brick roof onto the pizza. Modern ovens mimic these characteristics by having a large hot mass of metal.

Pizza: A popular and cheap meal, which was once used for leftover dough from bread baking. The yeast-raised dough is rolled out into a thin circle, covered with sieved tomatoes, oregano and pieces of Mozzarella cheese together with various toppings such as ham, salami, hard-boiled eggs, tuna, anchovies, olives, etc. This is cooked quickly in the oven on a flat sheet or in a shallow dish until the cheese melts and bubbles.

Pizzaiola: A sauce made from skinned and deseeded tomatoes, garlic, parsley, oregano and seasoning. The word is also used for dishes served with this sauce.

Pizzas: Baked tarts of Italian origin composed of a flat base of yeast dough topped with seasoned tomato sauces, cheese (usually Mozzarella cheese) and other foods such as salami, olives, vegetables and sea foods. Traditionally baked rapidly in wood burning ovens and served hot.

Plain flour: Wheat flour made from any type of soft wheat with between 9 and 10% protein with no added raising agents. It is used mainly for cakes and pastries where lessrise and a finer texture than bread is required. It is not used for puff pastry. It is also called household flour

Plant proteins: Proteins sourced from plant material as opposed to animal products. Include vegetable proteins and cereal proteins. It preferred by some consumers due to health benefits. Quality of plant proteins, especially with respect to amino acids composition, varies according to source, but many plant breeding programmes have aimed to improve protein quality of individual crops. Legumes, par- ticularly soybeans, are especially rich in protein.

Plasmids: Autonomously replicating, extrachromosomal, covalently closed circular molecules of DNA found in bacteria, fungi, algae and plants. In bacteria, they often carry genes conferring antibiotics re- sistance. Usually non-essential for cell survival under non-selective conditions and may integrate into the host genome. Used widely as expression and cloning vectors.

Plasteins: Proteins produced by action of proteinases on protein hydrolysates (peptides). Plastein reactions, i.e. transpeptidation and condensation reactions have been used to improve nutritional quality, sensory properties and/or functional properties of proteins, such as fish proteins, soy proteins and whey proteins.

Plasticity: Extent to which a substance can be deformed as a result of application of a stress. When stress is applied in excess of a certain value (yield point), deformation is permanent. Below a certain stress, the elastic limit, most substances will recover their original shape when the stress is removed. Such substances are said to be elastic.

Plasticizers: Substances, typically organic solvents, which are capable of imparting flexibility to a non-plastic material or improving flexibility of a ceramic mixture. It is added during the manufacturing process to decrease brittleness and to promote plasticity.

Plastics bags: Bags made from plastics. It is ued widely as containers for particulate and solid foods.

Plastics bottles: Bottles made from plastics. It is used widely as containers for liquid foods and beverages.

Plastics films: Packaging films, such as cellulose films and polyethylene films, made from plastics. It is used to wrap or to make containers for foods.

Plastics: Synthetic materials made by polymerization, polycondensation, polyaddition or other similar processes from molecules with a lower molecular weight, or by chemical alteration of natural macromolecules. It can be formed into different shapes while soft, generally when heated, and then set into a slightly elastic or rigid form. Synthetic organic polymers which are used as the basis of plastics are referred to as resins. Early plastics were used to make imitations of other materials, but they are now appreciated widely for their own range of useful thermal, electrical, optical and mechanical properties. Major applications of plastics include their use in containers, packaging materials, construction materials, consumer items, adhesives, pipes, textiles and electronic components. Types of plastics used commonly for packaging of foods include polyethylene, polyvinyl chloride and nylon.

Plate counts: Estimations of the numbers of microorganisms in a sample, by means of culturing a solution of the sample on agar plates and counting the number of microbial colonies that grow.

Pneumatic conveyors: Conveyors containing or operated by air or gas under pressure.

Poaching: To cook food, usually eggs or fish, in a cooking liquor at the simmer, i.e. at around 96-97°C in an open or closed pan on the stove or in the oven. Coddling is a kind of poaching and is as equally successful with fish and small pieces of tender meat as with eggs.

Pod: The elongated seed capsule of leguminous plants which contains a row of seeds and splits lengthways into two halves to release the seeds when ripe.

Polar compounds: Compounds that are ionic or are made up of molecules with a large permanent dipole moment. Commonly used as indicators of oil quality. During repeated heating of frying oils in the presence of oxygen, water and foods, triglycerides are broken down into polar compounds such as free fatty acids, monoglycerides, diglycerides, glycerol and polymers. Such decomposition products have a negative effect on the flavour and nutritional quality of the fried foods. To avoid deterioration in food quality and possible health effects for consumers, there are regulations in force in some countries specifying limits for total polar compound levels in frying oils. Once these values have been reached, the oils are pro- hibited for use in food processing. Polar compound profiles can also be used in detection of adulteration of oils such as virgin olive oils with less expensive types.

Polarimetry: Technique in which the identity and quantity of a substance are determined from its effect on the direction of vibration of polarized light.

Polarization: Restriction of the waves of electromag- netic radiation, including light, to one plane or one direction. This property is not directly perceived by the eye but can be detected, in the case of light, by its behaviour after it has interacted with polarizers. Meas- urement of the degree of polarization of electromag- netic radiation coming from an object reveals valuable information not only about that object but also about any material lying between the object and observer.

Polarography: Electrochemical technique in which current flowing through an electrolysis cell is measured as a function of the potential of the working electrode.

Polioviruses: Viruses of the genus *Enterovirus* within the family Picornaviridae responsible for poliomyelitis in humans. Transmission may be through the faecal-oral route via contaminated food or water.

Polished rice: Rice from which the vitamin B-rich outer coating has been removed during processing. It is prized because of its whiteness.

Polishing: Process in which a surface is made shiny and smooth by rubbing against abrasive materials such as metal, rock or wood. With reference to rice, polishing is the final stage in milling, in which hulled and pearled rice is spun in cones that are lined with leather or sheepskin. The fully processed form is called polished rice.

Polycarbonates: Group of synthetic polyesters in which the carboxyl groups are derived from carbonic acid. It is used to make reusable plastics containers for foods, especially bottles for infant feeding.

Polychlorinated biphenyls (PCB): Toxic chlorinated hydrocarbons with many industrial uses, e.g. as pesticide extenders and plasticizers. It proven toxicity to humans and animals includes adverse clinical effects on the gastrointestinal tract and eyes. Environmental contamination with these compounds and their high stability can allow them to enter the food chain, affecting predominantly animal foods. This has led to decreases in industrial use of these compounds. Preparations include Arochlor, Clophen, Fenclor, Kanechlor, Phenoclor, Pyralene and Santotherm.

Polychlorinated dibenzodioxins: Toxic environmental contaminants produced by municipal waste incinerators and chemical, paper and metallurgical industries. Exposure to these toxic organochlorine compounds can occur via the diet, particularly from consumption of animal foods, due to their accumulation in fats.

Polychlorinated dibenzofurans: Potential toxic contaminants of foods, particularly animal foods, where they accumulate in fats. These organochlorine compounds are produced as a result of incineration of municipal waste and as wastes from various industrial processes.

Polychlorinated dibenzo-*p*-dioxins: One of the polychlorinated dibenzodioxins, a group of toxic chemicals which may be contaminants of foods, particularly animal foods.

Polycyclic aromatic hydrocarbons (PAH): Hydrocarbons comprising two or more ring structures, at least one of which is an aromatic (benzene) ring. It is lipophilic pollutants and potential carcinogens. It is found in foods include benzo[*a*]pyrene and phenanthrene; foods affected include cheese, cooked meat and shellfish. It is also called polynuclear aromatic hydrocarbons.

Polydextrose: A bulking agent used in reduced and low calorie foods.

Polydextrose: Low calorie, highly branched polysaccharide composed of randomly linked D-glucopyranose units (average 12-15 units/molecule). It is manufactured from glucose and sorbitol in the presence of citric acid or phosphoric acid; and used as sugar substitutes and fat substitutes in low calorie foods. It can replace mouthfeel, texture and humectancy of sugar, but does not have a sweet flavour. Derivatives of polydextrose with improved flavour are marketed under the Litesse brand name.

Polydimethylsiloxane (PDMS)**:** Polymers consisting of dimethyl silicon oxide monomer units. Colourless viscous oil that is insoluble in water but soluble in hydrocarbon solvents. Its uses include as antifoaming agents or defoaming agents in beverages, such as wines and fruit juices, anticaking agents in foods, e.g. dried dessert mixes, and as a base for manufacture of chewing gums. This polymer is also used as an extraction fibre and a separation matrix for analysis of food components. It is also known as dimethicone.

Polyesters: Synthetic resins in which ester groups link the polymer units. They are heated to harden them into a shape which they do not lose when heated subsequently at normal cooking temp. it is used to make containers for heating foods in conventional or microwave ovens.

Polyethylene bags: Bags made from polyethylene which are used for packaging or storage of foods.

Polyethylene films: Transparent packaging films made from polyethylene which are commonly used in packaging of foods. Desirable characteristics include their low cost, resistance to low temperature and tough, moisture-proof and heat sealable nature.

Polyethylene glycol: Synthetic polymer which exists as a liquid or waxy solid, depending on the degree of polymerization, and thus molecular weight. It is partially soluble in water. A range of applications includes surfactants, catalysts, flocculants, plasticizers, lubricants, solvents and emulsifiers. Uses in the food industry include modification of enzymes, such as lipases and proteinases, and other proteins to alter their properties, and as a plasticizer in edible films and coatings.

Polyethylene naphthalate: Polyester polymer with characteristics making it suitable for food packaging applications. Compared with polyethyleneterephthalate (PET) it has improved oxygen barrier properties, heat and chemical resistance, and stiffness, but is more expensive. Its physical and mechanical properties make it suitable for manufacturing refillable containers and use in hot fill applications. Polyethylene naphthalate is sometimes blended with PET to make plastics containers for foods and beverages, especially bottles for beer.

Polyethylene: Flexible, tough, but lightweight synthetic resin which is a polymer of ethylene and is formed by pressure treatment of ethylene. Used mainly as a packaging material, especially in bags, films and sheets. Density of the polymer varies according to the polymerization process used. **Low density polyethylene** is used for flexible applications, e.g. **polyethylene films**, while **high density polyethylene** is used to make more rigid structures, such as **barrels** and **bottles**. It is also known as polythene.

Polyethyleneterephthalate (PET): Synthetic resin produced from ethylene glycol and terephthalic acid. It is used in production of polyester fibres, plastics bottles for beverages, and food trays for use in con- ventional and microwave ovens.

Polyglycerol polyricinoleate (PGPR): Highly viscous, strongly lipophilic liquid which is insoluble in water or ethanol, but soluble in fats and oils. Used in the chocolate industry as a viscosity reducing agent, and also as an emulsifier in foods.

Polymerase chain reaction (PCR): Method for amplifying DNA sequences using two oligonucleotide primers that flank the sequence of interest and which are complementary to different strands of the DNA sequence. The method involves repeated cycles (typically 20-30) of denaturation, primer annealing and strand elongation using heat stable polymerases. Each newly synthesized DNA strand serves as the template for a subsequent round of synthesis, resulting in exponential amplification of the sequence of interest. It may also be used to amplify mRNA following reverse transcription to cDNA.

Polymerization: Chemical combination of simple molecules (monomers) to form long chain molecules (polymers) of repeating units. In addition polymerization, the monomers simply add together and no other compound is formed. In condensation polymerization, water, alcohol, or some other small molecule is formed in the reaction.

Polymers: Long chain molecules of repeating units formed by the chemical combination of monomers in a process called polymerization. Natural organic polymers include proteins, DNA and latexes, such as rubber. Diamond, graphite and quartz are examples of inorganic natural polymers. Synthetic polymers include inorganic compounds, such as glass and concrete, but the great majority is plastics. Polymers are formed from monomers under the influence of heat, pressure or the action of a catalyst.

Polyolefins: Polymers, including polyethylene and polypropylene made from olefin monomers. It is used as components of plastics films for packaging of foods.

Polyols: Products formed by hydrogenation (reduction) of the free aldehyde or ketone groups of reducing sugars to produce an alcohol group. Examples include sorbitol, mannitol and maltitol, produced by hydrogenation of glucose, mannose and maltose, respectively. It is also known as sugar alcohols.

Polypeptides: Unbranched chains of 10 to approxi- mately 100 amino acid residues linked via peptide bonds. In contrast to proteins, polypeptides have no secondary or tertiary structure.

Polyphenol oxidase: The enzyme in plants which causes browning of fruits and vegetables on exposure to air by oxidation of phenolic compounds to melanin pigments. Its effects are being eliminated by genetic engineering of plant varieties.

Polyphenols: Compounds containing at least two phenol (hydroxybenzene) groups. Plant polyphenols, including catechin and flavonoids, are present in tea, coffee, fruits, fruit juices and wines and have antioxidative activity. Polyphenols in legumes and cereals are regarded as antinutritional factors, due mainly to the effects of tannins, which reduce protein digestibility.

Polyphosphates: Complex phosphates of sodium and potassium used mainly to retain added water without exudation in frozen chickens, ham, bacon and other similar meat products, and also as stabilizers and emulsifiers.

Polypropylene: Synthetic resin prepared by polymerization of propylene. Used as a packaging material for low fat, low sugar foods during microwave heating, but is unsuitable for use in conventional ovens because of its limited heat stability.

Polysaccharide: Long chains or branched chains of simple sugars which make up starch, dextrins, cellulose and other carbohydrates of natural origin

Polysaccharides: Carbohydrates that are composed of at least 10 monosaccharide residues linked via glycosidic bonds. Starch, celluloses, pectins and carrageenans are all polysaccharides. Polysaccharides have multiple applications in the food industry as thickeners, bulking agents, anticaking agents, gelling agents, and substrates for microbial fermentations and manufacture of sweeteners.

Polysorbate 60: It is produced by reaction of ethylene oxide with monostearic acid esters of sorbitol, with an average of 20 oxyethylene groups per molecule. It is used predominantly as emulsifiers, e.g. in cakes, coffee whiteners and non-dairy whipped toppings, but also as foaming agents, flavourings and dough conditioners. It is also called polyoxyethylene (20)-sorbitan monostearate.

Polystyrene: Synthetic resin made by polymerizing styrene. Produced in two forms, i.e. a hard form and a lightweight foam form called expanded polystyrene. There is concern about health hazards associated with migration of styrene monomers, dimers and trimers from packaging materials into some types of foods.

Polytetrafluoroethylene (PTFE): Tough synthetic resin which is used to coat non-stick cooking utensils.

Polyunsaturated fats: Fats and oils that contain at least two carbon-carbon double and/or triple bonds due to the presence of unsaturated fatty acids. Have lower melting points than saturated fats, and are therefore more likely to be oils at room temperature. It considered more beneficial than saturated fats with respect to their influence on risk of developing cardiovascular diseases.

Polyunsaturated fatty acids (PUFA): Fatty acids that contain two or more carbon-carbon double bonds. It have lower melting points than monounsaturated fatty acids or saturated fatty acids with an identical number of carbon atoms. Hence, lipids containing a high proportion of polyunsaturated fatty acids will be more fluid at room temperature. Examples include linoleic acid, linolenic acid and arachidonic acid, with 2, 3 and 4 double bonds, respectively.

Polyunsaturated: A description of long chains of carbon atoms that occur in fats, oils and fatty acids in which several of the carbon atoms do not have as many hydrogen atoms attached as they could and are therefore connected to neighbouring carbon atoms by double or triple bonds.

Polyurethane: Synthetic polymer made from urethane with a wide range of applications. Polyurethane foams have various uses in the food industry, including immobilization of cells and enzymes, and insulation of brewery tanks and utensils. Polyurethane adhesives are used in manufacture of food packaging.

Polyuronides: Pectic substances present in plant cell walls. Comprise polysaccharides composed of uronic acid monomers.

Polyvinyl acetate: Synthetic resin which is a polymer of vinyl acetate. It is used as a component of gum bases and flavour delivery systems in chewing gums. Also used in high-gloss coatings for foods.

Polyvinyl alcohol (PVA): Synthetic resin produced by polymerization of vinyl acetate and hydrolysis of the resultant polymer. Biodegradable and suitable for use as food packaging.

Polyvinyl chloride (PVC): Tough, chemically resistant, synthetic resin, which is a polymer of vinyl chloride. Low cost material that is moisture-proof but with some oxygen permeability. It is used widely in supermarkets as a wrapping material for meat.

Polyvinylidene chloride (PVdC): Transparent, moisture- proof, thermoplastic polymer which has greater heat stability than polyethylene. It is used to make plastics films, e.g. saran, for packaging foods.

Polyvinylpyrrolidinone: Substance used in the food industry to control haze or colloidal stability in beer by preventing oxidation and polymerization of polyphenols. It is similarly used in clarification of wines.

Pomace The residue after juice or oil has been physically crushed out of fruit. The pomace from oil bearing fruits is often treated by solvent extraction to produce inferior oils.

Poori: Puffed deep fried unleavened Indian bread made from wheat flour. It is eaten warm. Plain poori is eaten as an accompaniment to curries; can also be flavoured to make a sweet or savoury product.

Popcorn: Variety of corn with hard kernels that expand on exposure to heat or microwaves to form large, fluffy white masses. Also refers to the edible mass formed by this process, which is eaten as a snack food, often flavoured with salt or a sweet substance such as toffee.

Popping: Process in which cereals and grains are expanded by heating until the outer skin of the kernels is burst with a sudden sharp, explosive sound. It is used particularly in the manufacture of popcorn.

Pork fat: Fat from the pig is graded according to hardness and has many uses in sausages, terrines, pâtés and other items. Back fat tends to be the hardest and belly fat the softest. Lard is rendered pork fat and is traditionally used in pastry and in the cooking of eastern France. It contains about 49% saturated, 42% monounsaturated and 9% polyunsaturated fat.

Pork: Meat from swine, especially when the meat is uncured. Depending on the size of the animal and the part of the swine carcasses from which the meat is cut, colour of pork varies from pale pink to pinky-red. Raw boar meat and sow meat tend to be a stronger red colour than pork from young swine. On cooking, pork becomes paler and may become almost white in col- our. Pork is characterized by clearly noticeable deposits of subcutaneous fat; this fat is white in colour, medium-firm in texture and of a greasy consistency. Pork is a particularly rich dietary source of thiamin, containing approximately 10 times as much as beef. In some religions, pork is considered as unclean and con- sumption is forbidden; conversely, in other parts of the world, notably in China and the Pacific, and in other Asian cultures, pork is highly regarded. Pork quality is affected by halothane genotype and Rendement Napole (RN) genotype in swine. Quality is often categorized as being: pale, soft and exudative (PSE defect); red- dish-pink, soft and exudative (RSE defect); red, firm and non-exudative (RFN; normal); or dark, firm and dry (DFD defect).

Porridge oats: The breakfast cereal produced by heating either pinhead oatmeal or whole oats with steam as they are passed through rollers to flatten them. The pinhead oatmeal produces the normal porridge oats, whilst the whole oats produce oat flakes which can be used in muesli.

Porridge: A kind of gruel made by boiling porridge oats with water or milk or mixtures of both and salt until the desired consistency is reached. It is eaten as a breakfast dish with sweetening and milk or cream. The name is also used for oatmeal, maize, etc. it is boiled to the same consistency with water.

Potable water: Water of composition and hygienic and sensory quality permitting its use as drinking water.

Potassium bromated: Salt that is used primarily in dough conditioners. Other food industry uses in- clude in bleaching agents and improvers. It is also used in malting of barley for manufacture of alcoholic beverages.

Potassium lactate: White solid which is produced on a commercial scale by neutralization of lactic acid with potassium hydroxide. Applications in foods and beverages include flavour enhancers, flavourings, adjuvants, humectants and pH regulators.

Potato chips: Thin slices of potatoes fried until crisp. Eaten as snack foods or served as a garnish or with dips. It may be flavoured with salt or a variety of other flavourings. It is called potato crisps or crisps in the UK.

Potentiometry: Technique in which detection is achieved by measuring the change in electric potential between two electrodes placed in the sample solution.

Pouches: Small, sealed flexible bags which can be used as containers for foods. Commonly made from plastics or foils, and used to store frozen foods or dried foods.

Pouchong tea: Lightly fermented tea, intermediate between green tea and oolong tea.

Powders: Dried foods in the form of fine particles. Food powders include products, which can be reconstituted (e.g. with milk or water) to form liquid foods, and powdered ingredients such as baking powders and spices.

Pralines: Cooked mixtures of crushed nuts and partly caramelized sugar, often used as a centre for chocolates. It may be ground to a paste for use in pastry or candy fillings.

Prebiotic foods: Foods containing nondigestible ingredients with potentially beneficial health effects for the host based on selective simulation of the growth and/or activity of one or a limited number of bacterial species already resident in the colon. Examples of prebiotic components include inulin and nondigestible fructooligosaccharides. It is similarly, probiotic foods.

Precipitation: Process of forcing a substance in solid form from solution. It is achieved through a variety of means, including addition of an agent to the solution and centrifugation.

Predictive microbiology: Determination of the influence of various chemical, physical and biological factors on microbial growth and survival, typically by means of challenge trials or mathematical models.

Predictive modeling: Use of simplified and gener- alized representations (models) of phenomena to forecast the influence of certain factors on events.

Prepared meals: Convenience foods eaten at mealtimes and/or designed to be one of the main meals of the day. It is similar to prepared dishes.

Preservation: Any substance added to food capable of inhibiting, retarding or arresting the growth of microorganisms or of any deterioration of food caused by microorganisms or capable of masking the evidence of any such deterioration. Traditional preservatives included salt, sugar, saltpetre, acids or alcohol. Numbers of synthetic substances and derivatives of natural substances are also used. To make food suitable for long term storage by preventing growth of microorganisms or enzyme attack using a variety of techniques such as refrigeration, freezing, canning, bottling, drying, freeze drying, curing, brining/salting, pickling, fermenting or preserving with sugar, chemical preservatives or alcohol.

Preservatives: Additives that increase shelf life of foods and beverages. Shelf life is determined by rates of growth of spoilage microorganisms and chemical degradation, usually oxidation, of food components. Preservatives are chemicals that inhibit one or both of these processes. Examples of preservatives include organic acids (e.g. lactic acid, propionic acid, formic acid), benzoic acid derivatives (sodium benzoate, hydroxybenzoic acid esters), sulfur dioxide, nitrites and antioxidants.

Preserves: Term applied to preserved foods, usually referring to preserved **fruits**. **Fruit preserves** are made by cooking fruits with **sugar** and sometimes also **pectins**. Differ from fruit **jams** in that preserves generally contain larger chunks of fruit, while jams are similar to thick **fruit purees**. Preserves are used in a similar manner to jams. Other types of preserves include **vegetable preserves** and **fish preserves**.

Presses: Devices used for applying pressure in order to flatten or shape an item, or to extract natural fluids, e.g. fruit juices from fruits or oils from oilseeds or nuts. For oil extraction, screw presses are commonly used in preference to hydraulic presses because they provide a continuous process, have greater capacity, require less labour and generally remove more oil.

Pressing: Process whereby pressure is applied to an item with presses in order to flatten or shape it, or to extract natural fluids. It is used to produce fruit juices from fruits, and vegetable oils from oilseeds and nuts.

Pressure cooker: A vessel which can be completely sealed with a tight fitting lid, fitted with a pressure regulator and a pressure relief valve so that when heated with water inside, the internal pressure rises to some predetermined value above atmospheric pressure with a consequent increase in temperature above the normal boiling point.

Pressure fryer: A deep-fryer with a sealed lid which holds steam under pressure over the surface of the hot fat, thereby reducing the cooking time. It is similar in action to a pressure cooker.

Pressure: The force per unit area applied to a surface. Pressure is usually measured in pascals (Pa), which is defined as 1 Newton per square metre; it can also be measured in millimetres of mercury (mmHg), or millibars. High pressures may be applied in some food manufacturing processes (high pressure processing) or analytical techniques as a preservation process or to enhance results, respectively.

Pressure-cooking: To cook food in the presence of water or steam in a pressure cooker at a temperature above the normal boiling point of water by allowing the pressure to rise to some predetermined value. Food cooks much quicker than normal under these conditions.

Pressure-temperature relationship: The boiling point of water depends on the pressure exerted on its surface and is 100°C at sea level in an open vessel. This rises to 120.5°C at a pressure of 15 psi (pounds per square inch, equivalent to 1 bar) above atmospheric pressure and decreases by approximately 2.7°C for every 1000 m above sea level.

Pretzels: Small, brittle biscuits made from a stiff dough typically formed into loose knots which are boiled briefly, glazed with eggs and baked. It is often topped with salt crystals.

Pricing: Determination of the amount of money expected or required in payment for something.

Principal component analysis (PCA): Statistical technique by which variables in a data matrix are transformed to make them independent of one another. Covariance values are plotted on axes in multidimensional space. The first principal component, describing the majority of the spread of data, corresponds to the first axis in multidimensional space. Higher order axes show less variation, as the data are less correlated.

Printers: Equipment, such as computer peripherals, which are used for printing text or graphics, e.g. on labels for foods. Print quality and printing speed vary greatly between printers. The major types include line printers, matrix printers, letter quality printers and laser printers.

Printing: Process of generating printed material, such as labels for foods, including text and graphics.

Probiotic bacteria: Bacteria which benefit health by promoting a balanced gastrointestinal microflora (e.g. *Bifidobacterium* and *Lactobacillus* species). It is used in the preparation of microbial cultures for use in foods and animal feeds.

Probiotic foods: Novel foods containing viable probiotic microorganisms (particularly lactic acid bacteria, but also some bifidobacteria and yeasts) that have beneficial effects on the health of the host by improving the microbiological balance of the intestine. Examples include bifidus milk, acidophilus milk and yakult.

Probiotic microorganisms: Microorganisms which benefit health by promoting a balanced gastrointestinal microflora. It is used in the preparation of microbial cultures for use in foods and animal feeds.

Probiotics: A rather vague name sometimes used for functional foods but more often for the various beneficial bacteria which hopefully grow in the colon and crowd out the more hostile ones. Typical are *Lactobacillus acidophilus* and *Bifidum longum*, which can be obtained in capsule form or in various yoghurt-type foods.

Process control: Use of computerized systems for automatic control of continuous industrial processes.

Processed cheese: Product made from one or more hard or semi-hard cheese by milling and heating with water, emulsifying agents such as phosphates or citrates, and other ingredients including milk or whey powder, butter, cream, seasonings and flavourings. The mixture is pasteurized at a high temperature to extend shelf life of the product. The heating used during processing stops any further cheese ripening or flavour development. Soft versions con- taining 50% water are used as processed cheese spreads.

Processed foods: Foods which have been subjected to some degree of processing in order to bring about a desired modification, e.g. enhanced shelf life, physicochemical properties, sensory properties or nutritional quality. Examples include chilled foods, frozen foods, canned foods, ready meals, preserves and dietetic foods. It is also known as prepared foods.

Processing equipment: Machinery used in the processing of foods.

Processing lines: Sequences of processing equipment units that are integrated in order to manufacture a complete product.

Processing: Treatment of a raw material, such as a food, usually by applying a series of actions or steps, to produce a specific end product.

Product liability: A producer's legal responsibility for goods, or the liability of manufacturers and traders for damage or injury caused to purchasers or bystanders by their products.

Product technology: Processing procedures em- ployed during the manufacture of foods.

Production: The action or process of producing or being produced, or the bulk of a commodity produced in a given country or area.

Profitability: The monetary difference between the cost of producing and marketing goods or services and the price subsequently received for those goods or services. Profit is an essential competitive feature of buying and selling in the economic system.

Profiteroles: Small, cream puffs made from baked choux pastry shells, which are filled with whipped cream and topped with chocolate sauces.

Prolamins: Seed globulins that are insoluble in water and soluble in water-ethanol mixtures. Rich in proline and glutamic acid but contain small amounts of lysine, arginine and tryptophan, resulting in their being of poor nutritional value. It affects the texture of wheat dough during proving and baking

Proline: Non-essential amino acid whose structure differs from those of other amino acids in that its side chain is bonded to the N of the amino group as well as the C, making the amino group a secondary amine. It has a strong influence on the secondary struc- ture of proteins and is found more abundantly in collagen than in other proteins.

Pronase: A commercial preparation of proteinases from *Streptomyces griseus* containing at least 4 enzymes, including trypsin and a neutral metalloproteinase. It is used for production of protein hydrolysates, and improving the sensory properties of dry fermented sausages and the functional properties of insoluble gluten.

Proof cabinet: A controlled temperature and humidity enclosure, used for proving baked goods.

Proofers: Equipment assisting in the proofing (or proving) of dough, in which airflow, ambient condi- tions (e.g. air temperature and relative humidity) and handling can all be controlled.

Proofing: Stage of breadmaking in which dough is fermented under controlled conditions. During proofing (or proving), starch is converted by enzymes into sugars that are used as growth substrates by the yeasts employed. The breakdown products are carbon dioxide and alcohol. As carbon dioxide is pro- duced, it is retained in the tiny cells formed in the pro- tein matrix during mixing, causing them to grow and the dough to expand. Other products of yeast activity, mainly acids, are also formed during proofing; they contribute significantly to flavour development. Dough expands by a factor of three or four during proofing, and it is important that the skin remains flexible so that it does not tear as it expands. Yeast is at its most active at 35-40°C, so to minimize proofing time, heat transfer to the dough is necessary, to raise its temperature by 10-15°C.

Propan-1,2-diol: A solvent used for food colours and flavourings.it is also called propylene glycol.

Propanol: Alcohol containing three carbon atoms which is also known as propyl alcohol. Used for extraction of glutenin subunits from wheat and phospholipids from fish oils.

Propanone: Colourless, flammable, volatile ketone used as a solvent and as a raw material for making plastics. It produced commercially by fermentation of corn or molasses, or by controlled oxidation of hydrocarbons. It is also known as acetone.

Proprionates: Salts of proprionic acid used as flour improvers and food preservatives. The sodium- E281, calcium- E282 and potassium proprionates E283 are used.

Proprionic acid: E280, a simple fatty acid which occurs naturally in dairy products, now synthesized for use as a flour improver and preservative.

Propyl gallate: Esters of propanol and gallic acid (3,4,5-trihydroxybenzoic acid) with antioxidative activity. It is soluble in fats and thus used as antioxidants for fats, including margarines and edible oils, and meat products. Propyl gallate exhibits synergistic antioxidative activity with BHT and BHA.

Propylene glycol: Aliphatic alcohol used primarily in emulsifiers. Other uses in foods include as anticaking agents, antioxidants, flavourings and humectants. In food processing, propylene glycol is used in freezing media and solvents. It is synonym for 1,2- propanediol.

Protease: A specialized group of enzymes which attack the peptide links (i.e. Between the amino acids) in proteins causing a variety of changes including complete digestion of proteins to amino acids. Used in brewing, baking, cheese making and for flavour enhancement and meat tenderization.

Protein engineering: Use of genetic techniques to modify and enhance the properties of proteins.

Protein bodies: These are roughly spherical structures consisting of protein encapsulated in a membrane which occur in all seeds including cereals. They cannot be broken mechanically but water causes them to swell and break open, this being the process occurring in germination and also in dough production. See also gluten, glutenin, gliadin, prolamin, lectins

Protein concentrates: Products prepared by extracting proteins from animal and plant materials such as vegetables, fish or whey. Protein content varies among preparations. Used to provide protein fortification and enhance functional properties in a wide range of foods. Some of the most commonly used concentrates in the food industry are fish protein concentrates, soy protein concentrates and whey protein concentrates.

Protein efficiency ratio (PER): Biological method for evaluating protein quality in terms of weight gain per amount of protein consumed by a growing animal. PER is used widely in comparing the nutritional values of proteins in individual foods. It assumes that all protein is used for growth and no allowance is made for maintenance.

Protein hydrolysates: Proteins that have been subjected to hydrolysis by treatment with enzymes, acids or alkalies, so that the protein molecule is broken down into peptides and free amino acids. It is easily digestible and used to reduce antigenicity of foods. Applications include as ingredients of medical foods, infant formulas and hypoallergenic products.

Protein isolates: Products prepared by extracting and purifying proteins from animal and plant materials. It have similar properties to protein concentrates, but typically contain about 90% protein. Examples include soy protein isolates and whey protein isolates.

Protein values: Relative nutritional values of proteins based on amino acids composition, digestibility and availability of the digested products. Also the relative biological value defined in various terms, including the ability of a test protein, fed at various levels of intake, to support nitrogen balance, relative to a standard protein.

Protein: Nitrogenous organic compounds con- sisting of linked amino acids that are distributed widely in plants and animals. The sequence of amino acids in proteins is determined by the base sequence of their encoding genes. They serve many roles, such as enzymes, structural elements and hormones, and are essential nutrients. Long chains of amino acids which arrange themselves in many different shapes, some for use as muscle fibres, others to act as structural building blocks of body tissue and others as the enzymes which mediate most body processes. Of the 20 amino acids required 8 (9 in the case of infants) cannot be synthesized by the body and must be supplied in the diet. These are known as essential amino acids. Proteins in food are broken down into amino acids in the gut, these are absorbed into the blood and reassembled as required or burnt to provide energy.

Proteinases inhibitors: Substances that have the ability to inhibit the proteolytic activity of certain enzymes. Such inhibitors are found throughout the plant kingdom, particularly among legumes. Trypsin inhibitors are found in soybeans, lima beans and mung beans. Chymotrypsin inhibitors are found in cereals and potatoes. Proteinase inhibitors are destroyed by heat.

Proteinases: Enzymes that hydrolyse proteins by cleavage of peptide bonds. Endoproteinases cleave within protein molecules, while exoproteinases attack the ends of protein chains removing amino acids one at a time. They are classified as serine proteinases, thiol proteinases, metalloproteinases or acid proteinases. Some proteinases exhibit a high degree of specificity with respect to the peptide bonds they cleave (e.g. trypsin), while others are much less specific (e.g. papain). These enzymes are used in all areas of food production, including the meat, brewing, cheesemaking and breadmaking industries. It is also known by many other names, including proteases, protea- somes and proteolytic enzymes.

Proteinates: Protein products typically obtained by precipitation from the source material at the isoelectric points, followed by a neutralization step (e.g. to form sodium or calcium proteinates). Some of the most widely used proteinates include caseinates, total milk proteinates and soy proteinates, which have applications as functional ingredients in meat products, dairy products and imitation foods. Mineral proteinates are also used in animal and human nutrition as a readily absorbed form of mineral complex.

Proteoglycans: High molecular weight complexes of proteins and polysaccharides that are major con- stituents of structural tissues such as bones, cartilage and muscles, and are also found on the surface of cells. Glucosaminoglycans, the polysaccharides in proteoglycans, are polymers of acidic disaccharides contain- ing derivatives of glucosamine or galactosamine.

Proteolipids: Complexes of proteins and lipids abundant in brains but also found in a wide variety of tissues in animals and plants. In contrast to lipoproteins, they are insoluble in water. The proteins in proteolipids have high contents of hydrophobic amino acids, while the lipids consist of a mixture of phosphoglycerides, cerebrosides and sulfatides. In contrast, lipoproteins consist of phospholipids, cholesterol and triglycerides.

Proteolysis: Hydrolysis of proteins to smaller peptide fractions and their constituent amino acids, catalysed by alkalies, acids or enzymes (proteinases).

Proton resonance: Phenomenon used in nuclear magnetic resonance and proton magnetic resonance in which protons in a static magnetic field absorb energy from an alternating magnetic field at characteristic frequencies.

Protoplasts: Bacterial and plant cells that lack cell walls. Cell walls can be removed enzymically or by growth in the presence of antibiotics that block synthesis of cell wall peptidoglycans. Protoplasts can con- tinue to metabolize and can revert to normal cells under appropriate conditions, although they cannot divide. Bacterial protoplasts are prepared more easily from Gram positive cells than from Gram negative cells.

Protozoa: Unicellular organisms of the subkingdom Protozoa which lack cell walls. Occur in soil and freshwater, brackish and marine habitats. Some are pathogenic in humans and animals. Transmission is typically via raw meat and faecally-contaminated vegetables, salads and fruits.

Proving: To allow a yeast dough to rise both before and after shaping. Even rising depends on the incorporation and dispersion of the correct amount of air in the dough by the mixing and kneading processes. **or** To heat a new frying pan to a high temperature with oil or salt prior to using it so as to fill in minute imperfections in the surface. This prevents certain mixtures containing eggs or other proteins from sticking to it. Such pans should not be washed in detergents.

Pseudocereals: Plant species that do not belong to the grass family, but produce seeds or fruit that are used in the same way as cereal grain to make flour and bakery products. Include buckwheat, quinoa and amaranth grain.

***Pseudomonas*:** Genus of Gram negative, aerobic, curved or straight rod-shaped bacteria of the family Pseudomonadaceae. Occur in soil, water, salads and meat. Some species (e.g. *Pseudomonas fluores- cens* and *P. fragi*) may cause spoilage of meat, dairy products, cream, butter, eggs and fish. Certain species, e.g. *P. cepacia*, have been reclassified under the new genus *Burkholderia*.

Psychrophiles: Organisms, especially microorganisms that grow best at relatively low temperatures. Their optimum growth temperature is generally accepted as being below 20°C.

Psychrotrophs: Organisms, especially microorganisms, that can grow at relatively low temperatures, but grow optimally within the temperature range of 15 to 20°C.

Pubs: Informal name for public houses, also known as inns. Establishments, found chiefly in the UK, consisting of at least one public room and licensed for the sale and consumption of alcoholic beverages. Most pubs now sell meals, often in a separate restaurant area.

Pudding mixes: Dried instant foods consisting of a mixture of pregelatinized starch and other ingredients used to prepare puddings, typically by adding milk.

Puddings: Sweetened, usually cooked, desserts made from various ingredients, e.g. flour, fruit, milk and eggs. It includes milk puddings and steamed sponges. The term may also refer to savoury dishes topped with or surrounded by suet crust or pastry, such as steak and kidney puddings, or to savoury products in a sausage shape enclosed in casings, e.g. black puddings or white puddings.

Puff pastry fat: A high-melting-point fat which can be used to make puff pastry in warm conditions and which will withstand rough handling. Usually free of water. If it contains water the amount of fat in the puff pastry must be increased to give the correct ratio of pure fat to flour.

Puff pastry: Pastry made from 700 to 1500 interleaved layers of a very short flour dough and fat formed by rolling out a rectangular layer of the short flour dough, coating or sandwiching with fat, folding in 3 (3-fold turn) or folding the ends to the centre and then together like a book (book turn), rotating a quarter turn and repeating this, resting for 20 minutes in the refrigerator between turns. 5 book folds and 6 3-fold turns are required. The fat used must be of the same consistency as the dough. Margarine or pastry fat requires a strong flour, butter a softer flour. The ratio of fat to flour in the dough is roughly 1:8 whilst overall; the fat flour ratio is 1:1.

Puffed rice: Rice grains that are heated under pressure which is then rapidly released, causing the superheated steam in the grain to expand and explode the rice grain. Used in a range of food applications, including snack foods, breakfast cereals and confectionery.

Puffing: Method for expanding foods, particularly cereal grains. Grain is subjected to high pressure and/or temperature, before being ejected into a normal atmospheric pressure, causing the samples to expand sharply. Used mainly in the manufacture of breakfast cereals such as puffed rice and puffed wheat, and for making snack foods and puffed rice cakes.

Puffs: Small cakes or tarts made with a casing of puff pastry filled with jam, custard, whipped cream, etc.

Pulping: Crushing of foods, e.g. fruits and vegetables, into soft, smooth and moist masses (pulps).

Pulps: Preparations of a soft, moist consistency, typically obtained by mashing foods, particularly fruits or vegetables. Used in the manufacture of a wide range of foods and beverages, including fruit juices, yoghurt and pie fillings. Also refers to the solid residue remaining after extraction of juices from fruits and vegetables.

Pulsed electric fields: It is used in food processing and preservation. A high intensity electric field is delivered as a series of pulses of direct current to the food for a very short period of time while the food is held between two electrodes. This process results in formation of pores in, and breakdown of, cell membranes; the consequences of this can be microbial inactivation and increased yield of fruit juices during extraction. The risk of dielectric breakdown of foods limits this type of processing primarily to liquid foods, because uniformity of the applied electrical field would be distorted by air bubbles or suspended solids that usually exist in solid foods.

Pulsed field gel electrophoresis (PFGE): Gel electrophoresis technique in which DNA fragments are separated by subjecting the gel to an electric current alternately from two angles at timed intervals.

Pulses: Edible seeds of leguminous plants, including various beans, peas and lentils. Mature seeds are dry and can be stored. Also refers to the plants pro- ducing these seeds.

Pulverization: Reduction into fine particles (powders or dust), usually by crushing, pounding or grinding.

Pumpkin seed oils: Oils rich in unsaturated fatty acids. Frequently used as salad oils; also used as an ingredient of cider vinegar.

Pungency: Sensory properties relating to the extent to which the aroma or flavour of a product (usu- ally onions, chillies, peppers, ginger and rad- ishes) is acrid or pungent.

Pungent principles: Flavour compounds responsible for pungency of foods such as chillies, on- ions, peppers, ginger and radishes.

Punnets: Small lightweight containers or baskets for vegetables or fruits.

Purees: Smooth, thick preparations made by mashing from foods, particularly cooked fruits and vegetables, which have had any coarse fibre removed by sieving or similar means.

Puri: A deep-fried chapati which puffs and swells as it is cooked. Served hot, often with a hot spicy filling for use as a snack. Also called poori, bhatura, bhatoora

Purification: Removal of contaminants or undesirable components from a substance.

Purines: Heterocyclic organic bases that pair with pyrimidines in DNA and RNA, and whose derivatives are important in metabolism. They include adenine and guanine, as well as many alkaloids, such as caffeine and theophylline.

Purity: Extent to which an item or substance is pure, i.e. free from contaminants and adulterants.

Puroindolines: Lipid-binding cereal proteins (puroindoline-a and puroindoline-b) found in wheat which play a significant role in texture of bread crumb. Genetic variation of puroindoline alleles is associated with kernel hardness in wheat, a property known to affect milling and baking qualities.

Purothionin: Disulfide-rich protein of the wheat endosperm which shows antimicrobial activity.

Putrefaction: Typically anaerobic, microbial decomosition of substances (especially proteinaceous and

Pycnometry: Technique for determining the density of a liquid, using a small bottle of accurately measured volume. Density is determined from the ratio between the weights of a given volume of water and the same volume of sample.

Pyrazines: Nitrogen containing, heterocyclic flavour compounds found in many foods and beverages that can be formed during the Maillard reaction.

Pyrene: Toxic four ringed polycyclic aromatic hydrocarbon that can contaminate foods and beverages.

Pyridoxal: One of the three forms of vitamin B_6, the aldehyde form, the others being pyridoxamine (the amine form) and pyridoxine (the alcohol form). The relative proportion of each of the three forms in foods varies considerably. All are equally biologically active.

Pyridoxamine: One of the three forms of vitamin B_6, the amine form, the others being pyridoxal (the aldehyde form) and pyridoxine (the alcohol form). The relative proportion of each of the three forms in foods varies considerably. All are equally biologically active.

Pyridoxine: One of the three forms of vitamin B_6, the alcohol form, the others being pyridoxal (the alde- hyde form) and pyridoxamine (the amine form). The relative proportion of each of the three forms in foods varies considerably. All are equally biologically active.

Pyrolysis: Decomposition of chemical substances as a result of high temperatures. Sometimes used in analysis of foods by gas chromatography and mass spectroscopy, and as part of some processing techniques to add flavour or colour to products.

Pyrones: Heterocyclic flavour compounds found, for example, in roasted malt and chicory. It can also be produced by microbial fermentation. Certain pyrones act as mycotoxins, while others have been found to exhibit antifungal activity.

Pyrroles: Organic nitrogen compounds that can be formed in foods by the Maillard reaction or by other pathways, and contribute to flavour. Some pyrroles exhibit antimicrobial activity. The pyrrole ring structure is also found in many important biologi cal compounds, such as pigments, chlorophylls and haem.

Pyruvic acid: Intermediate in a wide range of aerobic and anaerobic metabolic pathways. It is produced as the end product of glycolysis and is at the starting point of the Krebs' cycle.

Q

Quality assurance: Planned and systematic actions necessary to provide adequate confidence that goods or services will satisfy given requirements. For the food industry, this is a customer-focused management system, whose aim is to guarantee food safety and con- sistent product quality by application of production, processing and handling standards. Proactive food safety programmes, in particular those based on Hazard Analysis Critical Control Point (HACCP) principles, are the foundation of many food quality as- surance systems.

Quality control: A system of maintaining standards in manufactured products by testing a sample against the specification.

Quantitative descriptive analysis: Comprehensive system used in sensory analysis that covers sample collection, assessor screening, vocabulary development, testing and data analysis. Quantitative descriptive analysis (commonly abbreviated to QDA) uses small numbers of highly trained assessors. Once the training sessions have established satisfactory panel performance, and removal of ambiguities and misun- derstandings, the test samples can be evaluated. This is carried out in replicated sessions using experimental designs that minimize biases. Three major steps are required: development of standardized vocabulary; quantification of selected sensory characteristics; and analysis of results by parametric statistics.

Quercetin/Quercitrin: Flavonol glycoside distributed widely in plants. It is found in many foods and beverages, where it exhibits antioxidative activity.

Quick freezing: To freeze food so that it spends a minimum of time between 0°C and –4°C, the region where ice crystals would, if given the time, grow to such a size as to rupture cell walls releasing their contents on thawing. It is usually done by blasting with liquefied gases. Quick frozen food is usually stored at around – 30°C.

Quinic acid: Organic acid that, together with caffeic acid, is a constituent of chlorogenic acid, an anti- fungal metabolite found in certain higher plants. Quinic acid can interact with proteins, influencing their function and digestibility.

Quinine: Bitter alkaloid isolated from cinchona bark, derivatives of which are used in the treatment of malaria. Also used as a bittering agent in carbonated beverages, especially tonic waters, although high doses are thought to be toxic.

Quinoa: A pseudocereal comprising the high protein dried **fruits** and glutinous **seeds** of the plant *Chenopodium quinoa* or *C. album*, which is native to Chile and Peru. It is used to make **flour** and **bread** and rich source of **iron** and **vitamin B_1**.

Quinoline yellow: E104, a synthetic yellow food colouring.

Quinones: Aromatic dioxo compounds that are usually coloured and are constituents of many natural pigments; intermediate products of enzymic browning. Their derivatives include the K vitamins. They function in aerobic and anaerobic electron transport chains, in photosynthesis, and as carriers of reducing equivalents between dehydrogenases and terminal enzyme complexes.

R

Rabadi: Traditional fermented food of India, prepared by fermentation of a mixture of flour made usually from pearl millet, and buttermilk. Cereal flour may be partially substituted by that prepared from soybeans or other vegetables.

Rabri: Concentrated and sweetened buffalo milk product with a flaky/layered texture. It is popular in India. Traditionally, milk standardized to 6% fat is heated at approximately 90°C with repeated removal of clotted cream (malai), sugar is added to the concentrated milk and finally the clotted cream is added back to the con- centrated sweetened milk. In a commercial method, shredded chhana or paneer is used in place of clot- ted cream. Rabri has a relatively short shelf life.

Racking: Process of drawing off wines or beer from the sediment in the barrel.

Radappertization: The production of food free from spoilage microorganisms using ionizing radiation

Radiation: Energy emitted in the form of electromag- netic waves or subatomic particles.

Radical scavenging activity: Ability to trap organic free radicals formed by the splitting of molecular bonds. This protects cellular membranes from oxidative destruction and ultimately prevents DNA damage caused by the action of the radicals which can lead to carcinogenesis. Substances with high radical scavenging activity include antioxidant vitamins, such as α-tocopherol.

Radicals: Highly reactive molecular species which possess an unpaired electron. It is often formed by the splitting of a covalent bond. It may react with macro- molecules (especially DNA and proteins), causing them damage.

Radioactivity: Emission of ionizing radiation or particles caused by the spontaneous disintegration of atomic nuclei.

Radioelements: Elements that undergo spontaneous disintegration of their nuclei with the emission of subatomic particles (α-particles and β-particles) or electromagnetic rays (X-rays and ϒ-rays).

Radiofrequency: Electromagnetic wave frequency between audio and infrared. Radiofrequency technology is used in a number of food processing applications, including heating, drying, tempering, defrosting and pasteurization.

Radioisotopes: Isotopic forms of elements that are radioactive and undergo radioactive decay, properties that make them useful in various analytical techniques and for studying metabolic pathways.

Radiometry: Technique for measurement of incident radiation using radiometers that can be tuned to specific frequencies.

Radurization: The reduction of the content of spoilage organisms in perishable food using ionizing radiation.

Raffinose: Oligosaccharide composed of 3 sugar residues, i.e. fructose, glucose and galactose. It considered one of the antinutritional factors in legumes due to its tendency to cause flatulence.

Raftiline: Preparation consisting of inulin extracted from chicory roots. It is used as a fat substitute, sugar substitute, dietary fibre and bulking agent. Applications include low fat dairy products, bakery products and processed foods.

Raftilose: Registered trade name of an oligofructose sweetener produced by partial enzymic hydrolysis of chicory inulin.

Ragi: Cereal plant, *Eleusine coracana*, that is an important food grain in India and Africa. It is used in porridge and gruel, and to make beer. It is alternative term for finger millet.

Raising agents: Any chemical mixture which liberates the carbon dioxide on heating so as to form small bubbles. Bakery additives that are used for chemical leavening of cakes. Raising agents, such as baking powders (mixtures of tartaric acid and sodium bicarbonate), produce CO_2 on addition of liquid, such as water or milk. On baking, the gas bubbles expand but are trapped by the protein and starch of the flour, and become set as the liquid in the cake mix evaporates.

Raisins: Dried grapes usually made from Thompson seedless grapes. It is prepared by sun or mechanical drying. It is rich in iron with high sugar content and a range of vitamins and minerals. It is eaten out of hand or used in bakery products and various dishes. Golden raisins are amber in colour due to treatment with sulfur dioxide, and are dried with artificial heat, giving a plumper and moister product that is preferred to common raisins for cooking. Muscat raisins are dark and sweet and used in fruitcakes.

Raita: A combination of chopped vegetables and fruit, usually cucumber, onion and bananas mixed with thick yoghurt and flavoured with cumin, coriander and seasoning, served as an accompaniment to other foods.

Raman spectroscopy: Technique based on measurement of scattering of incident light from a laser upon striking the sample. Raman scattered light is of a different wavelength from the incident light. The difference in energy between the incident light and the Raman scattered light is the energy required to make a molecule vibrate or rotate. A Raman spectrum is built up of the energy difference at different intensities, with clear bands representing functional groups. From this information, it is possible to determine the structure of compounds present in the sample.

Rancidity: Sensory properties relating to the extent to which the flavour of a product containing fats or oils is perceived to be rancid (sour or stale). It caused by oxidation of unsaturated fatty acids in fats and oils, resulting in the characteristic disagreeable flavour and aroma. It occurs slowly and spontaneously, and is accelerated by light, heat and certain minerals. Rancidity in foods may be prevented by proper storage, and/or the addition of antioxidants. Peroxide values are used as a measure of rancidity of oils and fats.

RAPD: Amplification of randomly selected genomic sequences by PCR under low stringency conditions using arbitrary primers. Can be used to determine taxonomic identity, study genetic diversity, generate probes and analyse mixed genome samples. It is abbreviation for randomly amplified polymorphic DNA.

Rapeseed meal: Residue remaining after rapeseed oils have been extracted from rapeseeds. Rich in proteins and minerals, but use in foods is limited due to the presence of antinutritional factors, such as glucosinolates.

Rapeseed oils: Oils extracted from rapeseeds, *Brassica napus* and riich in erucic acid, although varieties producing oils low in erucic acid have been developed. It is rich in monounsaturated fatty acids and low in saturated fatty acids. It is often used as cooking oils and also known as canola oils.

Rasogolla: Sweetened dairy product prepared from chhana. Chhana is mixed with flour and other con- stituents, divided into balls and cooked in sugar syrups.

Raw milk: Milk that has not been heat treated to destroy disease or spoilage causing microorganisms. It is used to make some products, especially cheese, but not usually drunk. Sale of raw milk for drinking is prohibited in many countries. It is also called unpasteurized milk.

Ready meals: Convenience foods prepared industrially to a set meals recipe usually by cook freeze or cook chill processing, and requiring no further preparation by the consumer other than reheating.

Ready to eat foods: Convenience foods that require no further preparation by the consumer, such as fast foods, food bars, ready to eat meals and ready to eat cereals. Similar to ready to serve foods.

Ready to eat meals: Convenience foods in the form of meals that require no further preparation by the consumer. It is similar to ready meals.

Ready to serve foods: Convenience foods requiring no further preparation by the consumer, other than reheating where appropriate. Examples include ready to serve dairy desserts, gravy, salads, soups and beverages. It is similar to ready to eat foods.

Recombination: Process similar to reconstitution, but involving addition of substances other than water which have been removed from the product. Examples include addition of butterfat as well as water to dried skim milk to make recombined milk of the desired fat content.

Recombined foods: Products made in a similar way to reconstituted foods, but with the addition of substances other than water which have been removed from the product in its original form during processing. Examples include recombined milk, made by addition of butterfat, as well as water, to dried skim milk to achieve the desired fat content in the final product.

Recombined milk: Product made by reconstituting dried milk with water and other components such as a fat source (e.g. butter) to give a composition similar to that of milk.

Recommended daily intake (RDI): The amounts of vitamins, minerals and other micronutrients (sometimes fat, fibre, carbohydrate and protein) that the government recommends people take in their food or otherwise every day. They tend to be set so as to avoid ill health rather than to promote optimum health and are not related to body weight, age, physical activity, type of work, state of health, etc. Most can be exceeded with safety. The amount of vitamins and some micro-nutrients that remain in food depends on its age, storage conditions and cooking methods. Amounts of nutrients greater than the requirements of almost all members of the population, determined on the basis of the average requirement plus twice the standard deviation, to allow for individual variation in requirements and thus cover the theoretical needs of 97.5% of the population.

Recommended dietary allowance (RDA): Levels of intake of essential nutrients that, on the basis of scientific knowledge, are judged to be adequate to meet the nutrient needs on all healthy people. Allowance is used to avoid the implication that these are absolute standards and to emphasize that the levels of nutrient intake recommended are based on a consensus of scientific opinion and should be reevaluated periodically as new information becomes available. Dietary allowances are higher than physiological requirements in order to allow for a safety factor, which considers bioavailability of the nutrients and to allow for individual variations.

Reconstituted foods: Foods that have undergone reconstitution before consumption, often by addition of a liquid. Examples include soups and bakery products made from mixes, and fruit juices made from concentrates.

Reconstitution: Restoration of a product to its original state and consistency, often achieved by adding a liquid, usually water. It includes addition of water to concentrates and powders.

Recovery time: The time it takes for a cooking appliance (deep fat fryer, oven, blanching liquor, etc.) To return to the required temperature for another batch of food after a batch of cooked food has been removed from it.

Rectification: One of two general methods, the other being simple distillation, used to separate a substance or a mixture of substances from a solution through vaporization. Distillation usually involves boiling a liquid and condensing the vapour that forms in a still. In simple distillation, all the distillate is removed from the still after collection. In rectification, part of the distillate flows back into the still. This portion comes into contact with the vapour being condensed and enriches it. Rectification can also be undertaken using large towers (fractionating columns). As the mixture to be separated is heated, its vapours rise through these columns. Substances that boil at the lowest temperatures form the first fractions. Their vapours rise highest and are carried off by pipes near the tops of the fractionating columns. Separate pipes carry off different fractions at various levels. Reflux (return) of some distillate to the columns produces the most efficient conditions for this method of distillation. Rectification can be carried out with a continuous feed of liquid. During manufacture of vodka, by-products of distillation, such as methanol, are removed from the distillate by rectification using a continuous still.

Recycling: Reuse of renewable resources in an effort to maximize their value, reduce waste, and reduce environmental disturbance. Food packaging wastes such as paper, glass and plastics are often recycled.

Red cooking: A Chinese technique of cooking meat by first browning it then slow cooking it in a tightly sealed casserole with dark soy sauce, sherry, sugar and spices which turns the meat a dark reddish brown colour. It is also called red braising, red stewing.

Red meat: Meat which is red in appearance in its raw state as opposed to the paler types such as chicken, veal, rabbit, etc.

Red wines: Wines which are red in colour, due to the presence of anthocyanins extracted from the skins of red winemaking grapes. It is a hought to have beneficial effects on health due to the anthocyanins content.

Redox potential: Scale of values, measured as electric potential in volts, indicating the ability of a substance or solution to cause reduction or oxidation reactions under non-standard conditions.

Reducing agents: Chemicals capable of the reduction of other chemicals, i.e. they donate electrons or hydrogen. During this process, the reducing agents themselves undergo oxidation. It is also known as reducing substances.

Reducing sugars: Sugars with free aldehyde or ketone groups available for oxidation to form carboxylic acid groups. Reducing sugars are substrates for Maillard reaction with amino acids. Examples include glucose, maltose, lactose and mannose.

Reduction: Loss of oxygen from a compound, e.g. removal by reducing agents. It also includes reactions in which atoms in the reacting materials gain electrons. Oxidation-reduction reactions always occur simultaneously; if one reactant is oxidized, another must be reduced.

Reference materials: Materials of certified composition that are used as standards in analytical procedures.

Refined oil: Culinary oil which has been treated to remove flavouring and other compounds which oxidize easily, which would reduce its shelf life and would depress the maximum temperature at which it could be used.

Refining: It is a removal of impurities or unwanted elements from a substance. It is often used to describe the processing of sugar and oils.

Reflectance: Optical properties relating to the measure of the proportion of light or other radiation falling on a surface which is then reflected or scattered.

Reflectivity: Optical properties relating to the amount of light or other radiation that can be reflected by an item. Rough surfaces reflect in a multitude of directions, and such reflection is said to be diffuse. Smooth, brightly polished or glossy surfaces reflect clearly and sharply at the same angle to the surface as the angle at which the light or heat contacted the surface. Reflectometers are instruments used for measuring the luster or sharpness of reflection of a finished surface.

Reflectometers: Instruments used to measure the colour or gloss of foods based on their reflectance of light.

Refractive index: Measure of the bending or refraction of a beam of light on entering a denser medium (the ratio between the sine of the angle of incidence of the ray of light and the sine of the angle of refraction). It is a constant for pure substances under standard conditions. For example, refractive index is used as a meas- ure of sugar or total solids in solutions, and in determining the purity of oils.

Refractometry: Measurement of refractive index using one of the several types of refractometer.

Refrigerants: Substances with low vaporization temperatures used to promote the refrigeration conditions necessary for chilling foods and beverages. Examples of refrigerants include Freons, ammonia, ice and solid carbon dioxide.

Refrigerated foods: Chilled foods requiring refrigeration prior to consumption.

Refrigerated storage: Process of keeping objects, usually foods, at a temperature that is significantly lower than that of the surrounding environment in order to extend their shelf life by a few days. Refrigeration or cold storage of foods is a gentle method of preservation, having minimum adverse effects on flavour, texture and nutritional values. Refrigeration keeps spoilage reactions (microbial or enzymic) to a minimum, but does not kill microorganisms or inactivate enzymes, instead slowing down their deteriorative effects. Household refrigerators are usually run at a temperature of 4-7°C. Commercial refrigerators are operated at a slightly lower temperature.

Refrigerated transport: Specially designed transport vehicles, such as lorries, rail cars, aeroplanes or cargo ships, with refrigeration systems on board which are designed to protect frozen and perishable foods from high ambient temperatures. The refrigeration systems also cool the hot air mass in the cargo container, and remove the stored heat from the structure of the cargo body. Product integrity is maintained through avoidance of temperature fluctuation.

Refrigeration: Process by which heat is removed from an enclosed space or from a substance for the purpose of lowering the temperature. Refrigeration is chiefly used to store foods and beverages at low temperatures, thus inhibiting the destructive action of microorganisms. Cooling caused by the rapid expansion of gases (refrigerants) is the primary means of refrigeration.

Refrigerator: A cooled sealed storage box, usually with shelves and a front opening door which maintains food at a preset temperature between 1 and 6°C. It is an appliances or compartments kept artificially cool by the use of refrigerants, and which are used to store foods and beverages. Mechanical refrigerators have four basic elements: an evaporator; a compressor; a condenser; and a refrigerant flow control (expansion valve). A refrigerant circulates among the four elements, changing from liquid to gas and back to liquid. In the evaporator, liquid refrigerant evaporates under reduced pressure, so absorbing latent heat of vaporization and cooling the surroundings. The evaporator is at the lowest temperature in the system and heat flows to it. This heat is used to vaporize the refrigerant. The refrigerant vapour is sucked into a compressor, a pump that increases the pressure and then exhausts it at a higher pressure to the condenser. To complete the cycle, the refrigerant must be condensed back to liquid and in doing this it gives up its latent heat of vaporization to a cooling medium such as water or air.

Reheating: Application of heat to a food that has already been thermally processed but then cooled. Cook chill foods and ready meals often need reheating before consumption.

Rehydrated foods: Products made by reconstitution of dried foods, e.g. dried vegetables, with water.

Rehydration: Process by which the water or moisture removed in making dried foods is replaced, so restoring it to near its original quality.

Relative density: Ratio of the density of a substance to the density of a reference material. For liquids or solids, relative density is the ratio of the density (usually at 20°C) to the density of water (at its temperature of maximum density (4°C). It has synomym for specific gravity (sp. gr.).

Relative humidity: The amount of water vapour in air as a percentage of the total amount the air could hold at the same temperature and pressure. If liquid water is present as fog or steam then the relative humidity will be 100%. At 100% relative humidity no evaporation of sweat or water can take place.

Religious food laws: Some religions have particular food prohibitions in particular; Buddhism, usually vegetarian; Hinduism, complex laws depending on caste, but generally no beef; Jainism, strict vegetarian with no eggs but dairy products allowed; Judaism, no horse or pig, seafood must have fins and scales, unfertilized eggs only, dairy products and meat must not be mixed and animals are examined for blemishes, slaughtered and prepared by licensed persons; Islam, no pork or alcohol, special slaughter of animal and no gold or silver plate; Sikh, Muslim or Jewish slaughter not allowed

Renaturation: Reconstruction of proteins or nucleic acids that have previously been denatured, such that the molecules resume their original function. Some proteins can be renatured by reversing the conditions that brought about denaturation.

Rendering: Process applied on a large scale to production of animal fats such as tallow, lard, bone fat and whale oils. It consists of cutting or chopping the fatty tissue into small pieces that are boiled in open vats or cooked in steam digesters. The fat gradually liberated from the cells floats to the surface of the water, where it is collected by skimming. Membranous matter is separated from the aqueous phase by pressing in hydraulic or screw presses; in this way, additional fat is obtained. Centrifuges may also be employed in rendering. Cells of the fatty tissues are ruptured in special disintegrators under close temperature control. The protein tissue is separated from the liquid phase in a desludging type of centrifuge, following which a second centrifuge separates the fat from the aqueous protein layer. It compared with conventional rendering, centrifugal methods provide a higher yield of better quality fat, and the separated protein has potential as an edible meat product. Usually it is done in a heavy pan over a low heat or in the oven at 150°C.

Rennet substitutes: Enzymes used as alternatives to animal rennets for coagulation of milk during cheesemaking. Developed due to shortages of the animal products and in cases where a vegetarian cheese is desired. It substitutes include microbial rennets and vegetable rennets.

Rennet: A mixture of enzymes extracted from the stomachs of suckling animals such as calves, kids and lambs, but usually from the fourth stomach of the unweaned calf. Is is used to curdle milk as the first stage in cheese making. Vegetarian rennets are now made using microorganisms. Some extracts of plants, especially cardoon, are used for the same purpose. Rennin is the main protease enzyme in rennet. It is used to cause coagulation of milk during cheesemaking. Traditionally extracted from the abomasum of young ruminants, mainly calves (animal rennets, calf rennets), but alternative forms (e.g. microbial rennets, vegetable rennets) are now used due to shortages of this type of preparation. The active enzyme is chymosin, but pepsin is also present.

Rennetability: The ease with which milk is coagulated using rennets.

Residence time distribution (RTD): Distribution of times spent by the various components of a food product through a process vessel. It is a critical factor affecting the sizing of holding tubes for aseptic processing of particulate foods. Also, in design of continuous sterilization equipment for liquid food processing, knowledge of flow characteristics, especially residence time distribution is of prime importance.

Residues: Food contaminants derived from a variety of sources, including agricultural chemicals (e.g. pesticides and fertilizers), veterinary drugs, and environmental pollution and manufacturing processes.

Resin: Solid exudate from plants and trees used to seal wounds in the bark or skin and prevent infection. Some such as asafoetida, mastic and pine resin are used as flavourings.

Resins: Group of organic chemicals, usually polymers, which are solid or semi-solid and have high electrical resistance. It is used as chromatography support materials and for manufacture of plastics, including those used as food packaging materials, e.g. epoxy resins used for coating of food containers.

Resistant starch: Starch which is resistant to digestion in the gastrointestinal tract. Resistance may be conferred by: protection by a physical barrier, such as plant cell walls, e.g. starch in seeds and legumes; the highly crystalline nature of some starch granules, such as those in bananas; retrogradation of starch in cooked foods; and modification of starch. It regarded as a source of dietary fibre.

Resistographs: Instruments similar to farinographs used to study rheological properties of dough, and thus evaluate flour quality.

Resorcinol: Resorcinol and its derivatives are used as preservatives in foods, where they exhibit antioxidative activity and inhibit enzymic browning, and for stabilization of vitamin D and vitamin E. Derivatives of this phenolic compound are also useful in the development of high performance packaging materials.

Respiration: Metabolic process in animals and plants by which organic substances are broken down into simpler products with the release of energy, which is incorporated into ATP and subsequently used for other metabolic processes. In most plants and animals, respiration requires oxygen (aerobic respiration), and carbon dioxide is an end product. Anaerobic respiration is the breakdown of food components such as glucose to yield energy in the form of ATP in the absence of oxygen. Anaerobic respiration in yeasts produces ethanol as a waste product, a process that is the basis of manufacturing alcoholic beverages.

Response surface methodology (RSM): Collection of statistical and mathematical techniques used in developing and optimizing processes, developing new products and improving existing products. It is used particularly where several variables affect the process or properties of the product.

Restaurants: A place where complete meals are sold for consumption on the premises. It means any of a wide variety of commercial catering establishments where foods and beverages are prepared and served. Types of restaurants include fast food establishments, cafeterias, canteens and pub restaurants.

Restructured meat products: Small pieces of meat reformed into steaks, chops and roast-like meat products. They may be difficult to distinguish visually from the real product. Minced, flaked, diced or mechanically recovered meat may be used. Often, massaging and tumbling are used to extract salt-soluble contractile proteins from the meat pieces. The pieces become coated with these proteins, which subsequently act as an adhesive when the pieces are thermally processed and compressed. Cohesion of the meat pieces also involves gelation of connective tissue proteins. Also known as reconstituted meat products.

Resveratrol: Polyphenol found in grapes and wines that exhibits antioxidative activity and is thought to protect against cardiovascular diseases.

Retinal: Aldehyde derivative of vitamin A, originally isolated from animal retina. It is formed in the body by cleavage of β-carotene in the intestines. It is necessary for night vision and also known as vitamin A aldehyde, retinene or retinaldehyde, the last form being the preferred alternative if the name is liable to be confused with the adjective meaning pertaining to the retina.

Retinoic acid: Biologically active acid form of retinols; can partially replace retinols in the rat diet. It promotes growth of bone and soft tissue production. However, has no activity in the visual process or the reproductive system and cannot be stored in the body. Retinoic acid is converted by the rat to an unidentified form that is several times as active as the parent com- pound in conventional vitamin A nutritional assays.

Retinoids: Compounds consisting of four isoprenoid units joined in a head-to-tail manner. Vitamin A is a generic descriptor for retinoids exhibiting qualitatively the biological activity of retinol. While preformed vitamin A occurs only in foods of animal origin, retinoids such as β-carotene are found in both animal foods and plant foods. Retinoids have many activities in the body, including control of cell proliferation, cell differentiation and embryonic development.

Retinols: The alcohol form of vitamin A. Vitamin A exists in two forms: retinols, which predominate in mammals and marine fish; and dehydroretinols, which predominate in freshwater fish. Retinols can be reversibly oxidized. Retinols circulate in the blood as a complex with retinol binding protein and transthy- retin.

Retort pouches: Flexible containers commonly made from aluminium foils and plastic laminates. It can withstand in-package sterilization of the product that they contain. Some have zipper-type closures and are resealable.

Retorting: Thermal process is that a part of the food canning process. Batch retorts, of a still or agitating type, and designed to operate with saturated steam or hot water, are used. By processing under pressure, it is possible to use temperatures of approximately 121°C (250°F), which greatly speeds up the destruction of microorganisms and spores.

Retrogradation: Process in which gelatinized (disordered) **starch** reassociates to form a more ordered structure; under optimal conditions starch may recrystallize. It occurs during cooling of cooked starch.

Reuterin: Broad spectrum antimicrobial substance produced by *Lactobacillus reuteri*. Shows potential for use in natural food preservatives.

Reverse micelles: Aggregates of small molecules such as surfactants which assemble in non-aqueous solutions at levels above the critical micellar concentration. In contrast to normal micelles, hydrophilic components associate in the interior of the aggregates. It is widely used to manipulate localized solvent polarity, for example in enzyme catalysis, to provide a hydro- philic environment for the enzymes used in an otherwise non-aqueous solvent. It is also used for selective extraction from mixed solvent systems.

Reverse osmosis: Membrane process, driven by a pressure gradient, in which a membrane separates the solvent (generally water) from other components of a solution. With reverse osmosis, the membrane pore size is very small (0.0001-0.001 micrometers) allowing only small amounts of very low molecular weight solutes to pass through. Even small dissolved molecules, such as salts, are retained by the membrane. At this molecular level, high pressures are required of the order of 10-50 bar because osmotic forces come into play. The largest commercial food applications of reverse osmosis are concentration of whey produced as a by-product of cheese manufacture and clarification of wines and beer. Reverse osmosis systems are additionally used water processing, such as desalination of sea water. It is also known as hyperfiltration.

Rheological properties: Mechanical properties relating to the flow of materials. In food technology, rheological properties relate to concepts such as elasticity, rigidity, shear, stretch, thixotropy and viscosity.

Rheology: Study of the relation between forces exerted on a material and the ensuing deformation as a func- tion of time. In the food industry, rheology provides a scientific basis for subjective measurements such as mouthfeel, spreadability and pourability.

Rheometers: Devices used for measurement of viscosity.

Riboflavin: It is synonym for vitamin B_2 and vitamin G. A water soluble vitamin which occurs mainly in yeasts, livers, milk, eggs, cheese and pulses; milk and dairy products are probably the most important source in the average diet. It occurs in bound form in plant and animal tissues and is not available unless liberated by cooking. It is resistant to heat, but readily destroyed in the presence of light and alkali. It involved in a wide range of oxidation reactions, of fats, carbohydrates and amino acids. It is a constituent of the coenzymes flavine adenine dinucleotide (FAD) and flavine mononucleotide (FMN). Deficiency impairs cell oxidation and results clinically in a set of symptoms known as riboflavinosis.

Ribose: Pentose sugar that forms, with phosphate, the backbone for ribonucleic acids.

Rice bran oils: Oils with high oxidative stability which are derived from the outer layers of the rice grain removed during manufacture of white rice. It is used widely in Japanese cooking as salad oils and frying oils. It is reported to lower serum cholesterol levels due to high contents of oryzanols.

Rice bran: Outer layers of rice seeds and is used as a source of rice bran oils and protein concentrates and as a fibre ingredient in bakery products.

Rice dough: Rice flour mixed with boiling or cold water to make a dough which is used for wrappers, cakes or dumplings. Boiling water makes a translucent wrapper for won ton.

Rice flakes: Rice grain steamed or softened by partial cooking and flattened through rollers, may be cooked or eaten uncooked in e.g. Muesli. Also called flaked rice

Rice flour: Broken rice grains that are milled and used in brewing and distillation, as well as making puddings and bakery products such as biscuits and cakes. It is mainly starch with very little gluten and used in the same way as corn flour and for noodles, sweets and short pastry.

Rice germ oils: Oils extracted from rice germ, a byproduct of rice milling. Rich in vitamin E. Major fatty acids are linoleic acid and oleic acid. Benefits for human health include protection against cardiovascular diseases, lowering of high cholesterol levels and management of menopausal problems.

Rice noodles: Noodles made with rice flour in very long strands and of varying thicknesses and widths. It is usually folded into a compact bundle. When deep-fried they puff up.

Rice vermicelli: Very thin rice noodles which only require softening in hot water, or they may be deep-fried in the dried state

Rice vinegar: Vinegar made by oxidizing the alcoholic beer or wine made from fermented rice starch. It is milder and with a gentler flavour than other vinegars made from fruits and wines. Chinese rice vinegars are available in white, red and black varieties, while Japanese rice vinegars tend to be almost colourless. It is used in salad dressings, a variety of dishes, including sushi rice and sweet and sour meals, and in pickles.

Rice weevils: Common name for *Sitophilus oryzae*, serious pests of stored grain and seeds and develop inside whole grain kernels with no external evidence of their presence. It may be transported into the domestic environment in infested whole grains or seeds, e.g. popcorn or beans.

Rice wine: An alcoholic liquid made by fermenting a cooked ground rice mash. It has a sherry-like taste and is both used as a drink and as a cooking liquid. Clear and amber-coloured varieties are available. Saccharification is by enzymes of starters containing fungi, rather than by malt enzymes, as in Western alcoholic beverages based on cereals.

Rice: Starchy grains produced mainly by *Oryza sativa* that form a staple food, especially in Asia. Brown rice, produced by removal of the hulls, is regarded as a healthier food than white rice, as vitamin B and fibre contents are reduced by removal of the bran and germ. However, parboiling of rice before milling increases the nutritional quality of white rice. Rice is eaten in many forms, as an accompaniment, or a component of dishes such as paella or risotto. It is also used to make breakfast cereals and infant foods, and as the starting material in manufacture of sake.

Ricin: Highly toxic lectin occurring in the seeds of castor beans. It consists of a toxic A-subunit that inactivates ribosomes, and a B-subunit that binds to carbohydrates and is specific for galactosyl residues.

Ricinoleic acid: Fatty acid found in castor oils and other vegetable oils. Useful as a precursor for microbial production of flavour compounds.

Rigidity: Rheological properties relating to the extent to which products (such as food gels, plant cells and meat fibres) are rigid, i.e. solid, firm and inflexi- ble.

Rigor mortis: Stiffening of muscles, which accompanies the *post mortem* loss of ATP and glycogen in muscle fibres; it develops gradually after slaughter of animals. The physical changes in muscles accompanying development of *rigor mortis* include a loss of extensibility and elasticity, shortening, and an increase in tension and firmness. Stiffening results from the formation of permanent crossbridges between actins and myosin filaments in the muscles. *Rigor mortis* does not last indefinitely, as after a period of ageing or conditioning, the muscles gradually lose their stiffness; resolution of *rigor mortis* results from physical degradation of the muscle structure. In many species, onset and resolution of *rigor mortis* occur more rapidly following electrical stimulation of carcasses.

Rigor: Relates to rigidity or stiffness of muscles, as occurs in *rigor mortis*.

Rigorometers: Instruments used to measure *rigor mortis* development in meat and fish on the basis of muscle tension and length.

Rinsing: Washing an item with clean water to remove impurities.

Ripened cream: Cream that has been ripened naturally or by fermentation with starters. It is used in making butter. It is also called sour cream.

Ripeness: Extent to which crops, such as fruits or vegetables, or cheese are ripe (fully developed and mature), and ready for eating.

Ripening: It is a term used in relation to the maturation of fruits, vegetables or cheese. As ripening proceeds, sensory quality of foods improves. Ripening of fruits and vegetables can involve changes in colour and texture. Flavour development is an important stage during the ripening of cheese.

Risk factors: Circumstances (e.g. personal habits, environmental exposure) that lead to the enhanced likelihood of an event occurring (e.g. disease development, food contamination).

Risks assessment: Estimation of the probability of adverse effects occurring due to exposure to specified **health hazards** or the absence of preventive or beneficial measures.

Risks management: Process of minimizing the probability of adverse effects occurring by developing systems to identify, analyse and prevent hazards.

Roast: A joint of piece of meat cooked in the oven on a trivet without any cover at around 230 to 250°C.

Roasted coffee: Coffee beans which have been roasted to develop characteristic flavour and aroma. The degree of roasting required is dependent on the intended style of coffee beverages to be prepared.

Roasted foods: Foods is cooked by dry heating, usually with added fats, in ovens. Maillard reaction products contribute to the characteristic roasted flavour.

Roasted peanuts: Peanuts that have been roasted by conventional oven cooking in-shell or shelled, or by microwave or oil cooking out of their shells. It is usually seasoned with salt or a variety of other flavourings, including garlic, paprika or chilli.

Roasting: To cook food in the oven by a combination of convected and radiated heat with good air circulation around it and usually with fat. The object is to brown the surface of the food, to make it crisp and tasty and to just cook the interior to the right degree. Generally used for meats and root vegetables.

Robotics: The branch of technology concerned with the design, construction and application of robots used for mechanical operation of procedures.

Rock salt: Unrefined salt as mined, used as a heat transfer medium e.g. in baking potatoes and heating oysters, or as a freezing mixture with ice

Rock sugar: Large brown transparent irregularly-shaped crystals of sugar.

Roller drying: Type of web drying in which the material to be dried makes a sinusoidal path around rollers while heat is supplied externally by blowing air.

Roller mills: Mills that crush or pulverize items by means of rollers that move the material and press it against the sides of a revolving bowl.

Rolling: Flattening of an object by passing a roller over it or by passing it between rollers. During baking, a rolling pin is used to flatten dough into a thin, even layer.

Rolls: Small rounded portions of bread made from yeasts-leavened dough. It may have a soft or crisp crust. Also called bread rolls.

Ropiness: Condition responsible for spoilage in products including beer, wines and bread due to the presence of certain bacteria (ropy bacteria) which form polysaccharides and rope-like threads, adversely affecting viscosity and consistency of the product. In yoghurt manufacture, ropy bacteria are sometimes used as yoghurt starters to produce a product with the desired consistency.

Ropy bacteria: Bacteria which produce ropiness in foods. It includes *Acetobacter* species causing ropiness in beer, *Bacillus* species acting on bread and *Leuconostoc* species responsible for spoilage of wines.

Roquefort cheese: French semi-soft blue cheese made from ewe milk. Traditionally ripened in natural caves under the French village of Roquefort-sur- Soulzon. Interior is creamy and white with blue to green-grey veins. Cheese has a pungent flavour with a metallic tang. Frequently used in dressings and salads.

Rotaviruses: Viruses of the family Reoviridae. It occurs in the faeces of birds and mammals. It is responsible for acute gastroenteritis in humans, especially children. It is transmitted by the faecal-oral route via foods, such as salads and fruits, or contaminated water.

Roti: Flat, unleavened bread prepared with corn flour.

Rots: Fungal or bacterial infections of plant tissues that cause softening, discoloration and disintegration.

Rotting: Natural process in which animal or plant tissues is decay or decompose due to microbial activity.

Roughage: Material derived mainly from plant cell walls which cannot be digested enzymically, but which can be partially broken down by intestinal bacteria to produce volatile fatty acids that can then be used as a source of energy. Roughage consists of soluble fibre which reduces levels of blood cholesterol and increases the viscosity of the intestinal contents, and insoluble fibre (cellulose and cell walls). Foods with a high fibre content include wholemeal cereals and flour, root vegetables, nuts and fruits.

Roughness: Physical properties relating to the extent to which the surface of an item feels rough, i.e. not smooth or glossy.

Roux: A base for thickening of sauces, prepared by heating together flour with fats. Sauces produced from this base by addition of liquid (e.g. milk or stocks) and heating, to thicken the liquid, are known as roux sauces.

Royal jelly: Partially digested honeys and pollen formed in the stomach of worker honeybees for feeding of bee larvae. Thought to possess beneficial health properties, and thus marketed as a health food.

Rusks: Light, sweet crisp or hard biscuits or raised bread which are browned in an oven and often used as a food for young children.

Rust: Reddish- or yellowish-brown flaky coating of iron oxide that is formed on iron or steel by oxidation, especially in the presence of moisture.

Rutin: Disaccharide derivative of quercetin, containing glucose and rhamnose. It is found mainly in cereals and at one time known as vitamin P.

Rye bran: Outer layers of the rye grain. It is used as a source of fibre; displays cholesterol lowering activity and anticarcinogenicity.

Rye bread: Bread made either entirely from rye flour or with a blend of wheat flour and rye flour. When made entirely from rye flour, it is often dark grey in colour and lacks the elasticity of wheat bread.

Rye flour: Flour produced by milling of rye grains. It is available in varying degrees of purity and colour (light, medium or dark).

Rye malt: Fermented mashes made from rye grain, which are used in the manufacture of rye whisky.

S

Saccharification: Process by which oligosaccharides and polysaccharides are degraded to produce smaller sugar units. It involves acid, alkali or enzymic (e.g. cellulases, amylases) hydrolysis of glucosidic bonds. Term is used frequently to describe hydrolysis of wastes, e.g. sugar cane bagasse or other lignocellulosic materials to produce substrates for microbial fermentation.

Saccharimeters: Devices used for measuring degree of rotation produced during transmission of polarized light through a sugar solution. When a standardized saccharimeter is used, this property is a function of the concentration of a sugar solution. In the sugar industry the rotation value (Pol) is used as a measure of sucrose content due to the low concentrations of other sugars.

Saccharin: It is a heterocyclic organic sulfur compound (*o*-benzosulfimide) that has approximately 300-600 times the sweetness of sucrose and is used in artificial sweeteners. It is available as the free acid and as sodium or calcium salts. Like sugar, saccharin salts are white crystalline solids that are highly soluble in water, but unlike sugar they are non-nutritive and impart a bitter metallic aftertaste.

Saccharometer: It is a graduated devices i.e.hydrometer which is directly calibrated with the percentage of sugar in the sugar/water solution. It is also used for determination of the density of sugar solutions, based on the level at which the device floats. It is also known as hydrometers.

***Saccharomyces cerevisiae*:** The most common yeast used for converting sugars into alcohol or water and carbon dioxide, for making bread and as a source of some enzymes, e.g. Invertase.

***Saccharomyces exiguous*:** A yeast is used for the leavening of sour dough bread.

***Saccharomyces inusitatus*:** A yeast used for the leavening of sour dough bread.

***Saccharomyces rouxii*:** A yeast used in the third phase of the production of soya sauce.

***Saccharomyces*:** Genus of yeast fungi of the class Saccharomycetes. It occurs in foods and beverages (e.g. fruit juices, fruits and alcoholic beverages), soil and on human skin. *Saccharomyces cerevisiae* is

Safflower oil: A mild-flavoured, high polyunsaturated oil from the seeds of the safflower, which is a good source of vitamin E. It is not suitable for deep-frying. It is used for margarine manufacture and in salad dressings. It contains about 10% saturated, 15% monounsaturated and 75% polyunsaturated fat.

Safflower oils: Oils extracted from seeds of *Carthamus tinctorius* which are rich in linoleic acid. It is used as cooking oils, in salad dressings and in the manufacture of margarines.

Safflowers: Large orange, red or yellow flowers produced by the thistle-like plant, *Carthamus tinctorius*. It is used as a source of food colorants that may be used as a substitute for saffron dye. The plant also has edible leaves and produces seeds from which safflower oils may be extracted.

Saffron: Dried stigmas from flowers of *Crocus sativus* that are used as yellow colorants and spices. The principal pigments of saffron are the carotenoids crocin and crocetin.

Safrole: Organic compound found in various spices and essential oils that has been shown to be carcinogenic in rats. Safrole and its isomer isosafrole are used as flavourings in foods.

Sago: A starch extracted mainly from the pith of the sago palm, *Metroxylon sagu*, using water.). The wet starch that is washed out from the bark can be eaten cooked, or dried to produce flour. The starch is then dried and granulated into small balls known as pearl sago. Pearl sago is produced by forcing wet starch through sieves and drying; this form is used in puddings. It is used for milk puddings.

Sake yeasts: Yeasts (*Saccharomyces* spp.) used for fermentation of saccharified rice mashes in sake manufacture.

Sake: Rice wines made in Japan by fermentation of rice mashes saccharified with koji starters.

Salad cream: A commercial emulsion sauce made to resemble mayonnaise but deriving its texture more from thickeners than from emulsified oil.

Salad dressing: A sauce usually based on oil and an acid such as vinegar, lemon juice or possibly yoghurt, either a stable or unstable emulsion, seasoned and flavoured with herbs, spices, garlic, etc. Generally it is used to coat very lightly the components of a salad especially leaves, but occasionally to bind the ingredients together. Condiments that are served with, and complement the flavour of, salads. Examples include mayonnaise, French dressing and salad cream.

Salad oils: Refined, bleached and deodorized vegetable oils used in preparation of salad dressings. Oils used in manufacture of commercial salad dressings are also subjected to winterization to prevent clouding upon refrigeration. Clouding is caused by formation of crystals of high m.p. triglycerides and may also be inhibited by addition of anti-clouding agents, namely oxystearin, polyglycerol esters and some emulsifiers.

Salad vegetables: Vegetables eaten raw in salads. Include leafy green vegetables, such as lettuces, chicory and watercress, spring onions and radishes.

Salad: A mixture of raw leaves, vegetables, fruit, warm or cold cooked vegetables, sausages, ham, cheese, fish, shellfish, cereal grains, pasta, etc. Virtually any edible foodstuff may be incorporated in a salad but, save for pure fruit salads, all are dressed with some kind of acid based sauce or dressing and seasoned. It is served as a course or meal in its own right or as an accompaniment to other food. Cold dishes consisting of one or more uncooked salad vegetables, such as tomatoes, cucumbers and lettuces, usually sliced or chopped, and often accompanied by a protein source, such as eggs, fish or meat. Also refers to dishes of vegetables served with dressings, such as potato salads or coleslaw, and to cold dishes of cooked rice or pasta mixed with cooked or raw vegetables or fruits. Fruit salads usually comprise sliced mixed fruits served in fruit juices or sugar syrups.

Salami: Highly seasoned, raw, dried sausages, originally produced in Italy. They are prepared from coarsely comminuted meat. There are two major kinds, namely soft salami, which are semi-dry sausages; and dry salami, which are dried slowly to a hard texture. Most are made from fresh pork and include garlic; however, they may be prepared from beef, turkey meat, veal, or from meat mixtures. The majority are cured during preparation, air dried, uncooked and unsmoked, but some smoked versions are produced. Characteristics of salami are affected by: type and amount of meat used; proportion of lean to fat; how finely, uniformly or coarsely the fat appears among the lean; choice of seasonings; and degree of salting and drying.

Salatrim: Fat substitute comprising short and long chain acid triglyceride molecules produced by interesterification of short chain triacylglycerols with fully hydrogenated vegetable oils. Applications include in confectionery, bakery products and dairy products.

Salinity: It is a measure of the total amount of salt in foods and brines.

***Salmonella*:** Genus of Gram negative, facultatively anaerobic rod-shaped bacteria of the family Enterobacteriaceae. Occur in soil, water, foods (e.g. raw meat, raw sea foods, eggs and dairy products) and the gastrointestinal tract of humans and ani mals (especially poultry and swine). *Salmonella* Typhi is the causative agent of typhoid fever, while *Salmonella* Typhimurium and *Salmonella* Enteritidis are responsible for gastroenteritis. Transmission is via the faecal-oral route by contaminated foods or water.

Salmonellosis: Any infection caused by *Salmonella* species. It is usually manifests itself as food poisoning with severe diarrhoea, nausea, vomiting, fever, head- ache and abdominal cramps.

Salt grinder: A small mill like a pepper mill which is used to reduce large crystals of sea salt or similar to a small enough size to put on food

Salt substitutes: Chemicals used to mimic the flavour and/or applications of salt. Concern regarding effects of salt consumption on blood pressure has lead to a search for salt substitutes that do not have hypertensive effects. Potassium, ammonium and calcium salts have been tested as salt substitutes, but these metal ions have been unsuccessful in replacing sodium, underlining the importance of sodium ions in perception of saltiness. Reductions in salt content of processed foods have been possible due to the addition of salt flavour enhancers such as amino acids, yeast extracts, acetic acid and allyl isothiocya- nate.

Salted fish: Fish products preserved or cured with dry salt or in brines, after which they may or may not be dried. In the UK, the term usually refers only to salted white fish species, such as cod, coalfish, haddock and hake.

Saltine biscuits: Thin cracker biscuits sprinkled with coarse salt crystals before baking.

Saltine crackers: Crackers which are thin and crisp-like and are topped with coarse salt crystals.

Saltiness: Sensory properties relating to the extent to which a product tastes of salt.

Salting: To preserve food by immersing it in brine or covering it with dry salt. This replaces the water in the tissues with salt or a strong solution of salt, in which bacteria and other food degrading organisms cannot grow.

Salts: The general chemical name for the compound formed when an acid reacts with a base (usually an alkali) as e.g. Sodium acetate, which is formed by the reaction of acetic acid with the corrosive alkali sodium hydroxide or with sodium bicarbonate which is itself the salt of a weaker acid. The common example is sodium chloride, formed from the highly corrosive hydrochloric acid and it has appropriated the name salt to itself. NaCl obtained by mining or as residues from evaporation of sea water. Chloride is used extensively in food processing and cooking as a taste item, to extract plant juices in fermented vegetables, e.g. Sauerkraut, to solubilize proteins in meat, to assist emulsification, e.g. Frankfurters, and to act as a preservative in e.g. Salted fish and meat. Several different forms of this mineral are used as condiments; table salt, rock salt and sea salt are all forms marketed for this purpose. Commercial salt often includes other salts, such as calcium chloride or magnesium chloride, as anticaking agents. Salt has multiple uses in the food industry, primarily in flavourings, e.g. salted butter and salted nuts, and in aqueous solutions (brines) as preservatives. Other uses include as dough conditioners and curing agents.

Sampling: It is a collection of samples for analysis. Procedures vary according to type of material and analytical technique to be used.

Sandwiches: Snack foods comprising two or more slices of bread (usually buttered), enclosing sweet or savoury fillings (e.g. meat, fish, cheese, eggs, jams). Variations include open sandwiches and toasted sandwiches. Commercial, pre-packed sandwiches form an important part of the fast foods sector in many countries.

Sanitation: Establishment and maintenance of environmental conditions conducive to the preservation of public health.

Sanitizers: Agents used in disinfection or sterilization.

Sansa oils: Low quality vegetable oils that are chemically extracted from press residues of olives. It is may be used as frying oils.

Sapogenins: The aglycone components of saponins occasionally found free in plants but usually present as glycosides. It is may be triterpenoid or steroid in nature.

Saponification: It is process of hydrolysis of fats into constituent glycerol and fatty acids by boiling with alkalies.

Saponins: Glycosides is found in many plants, consisting of sapogenins and sugars. Thought to have a number of beneficial health effects, such as the ability to lower cholesterol levels.

Saran: Class of thermoplastic resins that are polymers of vinylidene chloride. It is made into transparent films, also called cling films, that are resistant to oils and chemicals and used for wrapping foods. It is originally a US trademark. It is also known as saran wrap.

Sardine oils: Fish oils extracted from the body of *Sardina pilchardus*. Contain variable amounts of eicosapentaenoic acid and docosahexaenoic acid. It is may be used in the manufacture of margarines.

Satiety: State in which the desire or motivation for something no longer exists because the need has been satisfied. In the food sense, satiety relates to the physiological sensation of fullness after consumption of a meal. Satiety can also be sensory-specific, e.g. texture and flavour specific satiety; this may significantly contribute to overall satiety.

Sensory-specific satiety refers to the decrease in the perceived pleasant- ness of a food after it has been eaten to satiety, and the smaller amount of that food, relative to other foods, that is subsequently eaten.

Saturated fat: Hard fats in which all the carbon atoms are attached to the maximum number of other atoms, usually two carbon and two hydrogen atoms. Unsaturated soft fats and oils are converted to harder fats by hydrogenation, a process which can produce the trans fatty acids which some suspect to be harmful for pregnant and nursing mothers. Saturated fats are said to be unhealthy but fashions in health change from time to time. Small amount of natural saturated fats (suet, lard, butter, etc.) are probably beneficial.

Saturated fatty acids: Fatty acids that contain no double bonds. Diets rich in saturated fatty acids are thought to increase the risk of developing coronary heart diseases.

Saturated solution: A solution containing the maximum amount of solid (solute) that can be dissolved in and remain in solution in a liquid solvent at a particular temperature. The amount of solute dissolved depends upon and usually increases with temperature. A supersaturated solution may be obtained by cooling a saturated solution but this is unstable and any shock will cause solid to crystallize out.

Sauce mixes: Powders containing all the ingredients required (e.g. fats, flour, seasonings, stabilizers) to produce sauces upon reconstitution with water. The reconstituted powders are usually thickened by heating to produce sauces of the required consistency.

Sauces: Condiments of a pourable or spoonable consistency that are served as an accompaniment to foods in order to enhance the flavour of the food. Sauces may be sweet or savoury, e.g. apple sauces and cheese sauces, respectively, and may be served as a side dish, poured over the food or used during cooking.

Sauerkraut: Dish made by fermenting shredded cabbages, salt and, optionally, spices and sold fresh or in jars or cans. It is rich in vitamin C and B vitamins and eaten as a side dish, in sandwiches and in casseroles.

Sausage casings: Natural, cellulose or collagen casings which are filled with sausage emulsions in the preparation of sausages. Particular types of sausages are prepared in particular types of casings. For example, sheep intestines are used as casings for chipolatas and frankfurters, swine intestines are used as casings for fresh frying sausages, and cellulose casings are used in the preparation of skinless sausages.

Sausage emulsions: Fillings for sausages prepared from comminuted meat, fat, preservatives, spices, salt and sometimes fillers, such as cereals or dried milk solids. Level of NaCl is controlled in order to improve the binding capacity of sausage emulsions, especially those prepared from non-slaughter warm meat. Additives are often included to help preserve, thicken or colour sausages. Extent of comminution of the raw meat materials differs widely, so that sausage emulsions may include small pieces, chunks, chips or slices of meat. Curing ingredients may be added during comminution or mixing, either in dry form or as a concentrated solution. Most sausage emulsions are packed into sausage casings to produce sausages.

Sausage maker: A hollow long tapered cylinder that fits over the outlet of a mincing machine. The casing with one end tied in a knot is placed over the cylinder like a wrinkled stocking and as it is filled with the mixture extruded from the machine is allowed to slip off the cylinder. The long sausage may then be linked or divided as required.

Sausage meat: A meat mixture similar to that for stuffing sausages, used for pies, rolls, turnovers, meat loaf, etc. In the UK, a revolting paste of ground meat scraps, drinde, MRM, rusk, etc. See also **English sausage**

Sausage roll: Sausage meat or similar meat mixture wrapped in puff or shortcrust pastry to form a small roll, egg-washed and baked in the oven

Sausage: Comminuted, seasoned, usually cylindrical, meat products prepared from sausage emulsions stuffed into sausage casings. Commonly, filled sausage casings are twisted at intervals to form links; these vary in shape and size depending on the type of sausages. Sausage production may also involve curing, smoking, fermentation, shaping and/or cooking. Shape or form of particular types of sausages tends to be dictated by tradition. Countries such as France, Italy and Germany have an extensive range of regional speciality sausages. Most sausages are prepared from pork mince or beef mince, but some are prepared from other meats (e.g. chicken mince or donkey mince) or various types of offal (e.g. livers). They often include low value meat, such as mechanically recovered meat or parts of the carcass that are unattractive to the consumer, e.g. the intestines and feet. The six major types of sausages are: fresh (e.g. fresh pork sausages); cooked (e.g. liver sausages); uncooked smoked (e.g. mettwurst); smoked and cooked (e.g. knackwurst); semi-dry (e.g. semidry salami); and dry (e.g. rohwurst).

Sausagemeat: Fresh **sausages** which are sold in bulk without casings. It is often mixed with other meats, formed into patties or balls, or used as an ingredient in **stuffings**.

Sauteing: Frying of foods quickly in a small amount of hot fat or oil in a skillet or special saute pan over direct heat.

Scalded cheese: Cheese which has been made from curds which have been heated to between 40 and 48°C, soft cheeses at the lower temperature

Scalded curd: Curd for making cheese, which has been heated to a temperature between 40 and 48°C

Scalding: It is to pour boiling water over something or immerse in steam for a few moments so as to cook a thin outer layer without affecting the inner part. It is used to remove skin from tomatoes or to loosen hair on animal skin. It is to heat milk to just below boiling point and then immediately cool. It is used to clean cooking utensils by immersing in boiling waterIt also used to seal in flavour by bringing food in water to a temperature which coagulates surface proteins, usually 80°C.

Scaling: Removal of scales from fish skin, generally using blunt knives or special tools called fish scalers.

Scanning electron microscopy (SEM): Electron microscopy technique in which a focused beam of electrons is used to scan the surfaces of suitably prepared samples. Secondary electrons emitted from the samples are detected and used to create detailed images of the structure of the samples. Advantages over light microscopy include greater magnification (up to 100,000×) and much greater depth of field.

School meals: Meals, particularly lunches, but sometimes also breakfasts and evening meals, provided for school pupils, usually by a foods service. Emphasis is placed on planning healthy menus that appeal to children and adolescents and which provide suitable nutrients for these age groups.

Sea foods: All edible marine and freshwater aquatic organisms; includes fish (finfish), shellfish, aquatic mammals, plants and algae. Generally it is regarded as a healthy component of the human diet. Many sea foods are good sources of high quality proteins, unsaturated fatty acids, vitamins and minerals, and are low in fats and calories.

Sea water: Water from marine environments, charac- terized by a high salinity and complex physicochemical structure; covers nearly 75% of the earth's surface. In some countries, desalination is used to produce potable water from sea water.

Sealing: Process of closing openings in containers in such a way as to prevent leakage of the contents or entry of undesirable elements.

Seaming: Process of joining together the edges of food cans to form a seal.

Seasonings: Blends of spices, flavourings and other additives, such as colorants and sweeten- ers, that are used to enhance flavour, aroma and/or overall appearance of foods. Commercial seasonings may also contain anticaking agents. Seasonings are often created for use with particular types of food, e.g. barbecue seasonings or chicken seasonings.

Seaweeds: Multicellular marine algae which are fixed to marine substrates by root-like holdfasts; occur in intertidal or subtidal environments worldwide. It is subdivided into 4 classes: green (Chlorophyta); brown (Phaeophyta); red (Rhodophyta); and blue-green (Cya- nophyta). Many species are edible, providing an excellent source of vitamins and minerals. Agar, carrageenans and alginates are extracted from some species for use as food additives.

Sedimentation: Settling of matter to the bottom of a liquid by gravitational force so as to separate sus- pended solids from fluids.

Seed: The structure containing an embryo, stored food and a seed coat derived from the fertilized ovule of a flowering plant or tree from which a new plant will arise if it germinates successfully. It is variously known as grains, cereals, pips, nuts, stones, kernels, etc.

Self-raising flour: A wheat flour which incorporates a chemical raising agent, usually sodium bicarbonate plus an acid salt for reaction and completed liberation of all the available carbon dioxide. 30 g of baking powder per kg of plain flour may be substituted (9 teaspoons per kg or 4 teaspoons per lb).

Self-service: A type of service of food where the customers go to a counter where food is laid out and usually help themselves to cold items and sometimes to hot items. The food is then taken to a cash receiving point, if to be paid for, where the cost is computed and paid. Cutlery, napkins, seasonings and condiments are usually collected after payment.

Semi skimmed milk: Milk from which some of the fat has been removed. This low fat product is preferred to whole milk by some health conscious consumers, and is used by processors to make low fat dairy products. Semi skimmed cow milk contains approximately 1.7% fat, compared with approximately 4% in whole milk.

Semolina: A technical term for the large pieces of endosperm from wheat which are made by the fluted rolls used to separate the bran from the semolina. The culinary term for a particular type of semolina made from durum wheat. The particle size varies according to the use to which it is going to be put. Fine semolina is used for pasta, coarser varieties are used to make a boiled milk pudding. It purified granular middlings from durum wheat used principally in the manufacture of pasta and milk puddings.

Senescence: Degeneration of plants due to maturation or ageing. Stress due to disease or attack by insects may induce early senescence.

Sensors: Apparatus used in detection by responding to a specific stimulus.

Sensory analysis: Analytical techniques used to determine the sensory properties of foods. The techniques fall into three main classes: discrimination/difference tests; descriptive tests; and he- donic/affective tests.

Sensory properties: Properties that can be detected by the sense organs. For foods, the term relates to the combination of concepts such as appearance, flavour, texture, astringency and aroma.

Sensory scores: Scores given to particular sensory properties of foods by panellists during sensory analysis.

Sensory thresholds: Term used in sensory analysis relating to the levels at which perception of increasing concentrations of a stimulus, such as aroma compounds or flavour compounds, begins. Classical methods for estimating sensory thresholds include probit, graphic, exact, logistic, Spearman-Karber, moving average and up-and-down methods.

Separation: Action or state of division into distinct elements, using techniques such as centrifugation, filtration, sieving, crystallization and distillation. Separation of food components is fundamental for preparation of ingredients to be used in other processes. Some separation methods are used to sort foods into classes based on size, colour or shape, to clean them by separating contaminating materials, or to selectively remove water by evaporation or drying. Centrifugation is used for separation of immiscible liquids and for separation of solids from liquids. Filtration is used for removal of insoluble solids from a suspension by passing it through a filter medium.

Separators: Equipment that facilitates the division of items or solutions into distinct elements. Examples include centrifuges, filters and sieves.

Sequestrants: Additives that bind to or form com- plexes with other chemicals, reducing their reactivity in order to prevent the occurrence of undesirable reactions. Examples of sequestrants include sodium citrate and EDTA which are used to chelate calcium ions (e.g. used to modulate the strength of gellan gels), and phosphates that bind to and enhance the stability of proteins at low pH.

Serine: Non-essential amino acid required for metabolism of fats and fatty acids, muscle growth and ahealthy immune system. It is abundant in meat and dairy products, wheat gluten, peanuts and soy products.

Service charge: The charge added to a restaurant bill, usually as a percentage of the total to cover the cost of providing waiters. It originated in the times when waiters were paid only this percentage for their labour. This archaic and demeaning practice dies out as society develops.

Sesame oils: Seed oils derived from sesame seeds, which are rich in oleic acid and linoleic acid and have high oxidative stability due to the presence of natural antioxidants. Contain sesamin and sesamolin. Due to their nut-like flavour, the oils are used as seasonings as well as cooking oils. It is also known as gingelly oils or til oils.

Sesame seed meal: Residue remaining when sesame oils are extracted from sesame seeds. It is used as an animal feed, a source of proteins and sometimes as a partial substitute for wheat flour in baking.

Sesamol: Natural phenol antioxidants prepared from sesame oils.

Setting: Firming of foods, usually as a result of cooling, as with gelatin-based dishes, such as jelly.

Shami kabab: Small balls made with minced meat and soaked yellow split peas (4:1), cooked in water with flavourings such as garlic, coriander, cumin, mint, ginger, minced onion or the like until soft and all liquid absorbed. The mixture is then processed to a smooth paste, formed into small flat circular croquettes and deep fried in clarified butter on both sides, taking care not to break them.

Shape: Shape is also important in heat and mass transfer calculations, screening solids to separate foreign materials, grading of fruits and vegetables, and evaluating the quality of food materials. The shape of a food material is usually expressed in terms of its sphericity and aspect ratio.

Shear strength: Measure of the resistance of a material, such as a food, to shear stress and the associated deformation caused by the application of this stress. Peak shear strength is the highest stress sustainable just prior to complete failure of a sample under load; after this, stress cannot be maintained and major strains usually occur by displacement along failure surfaces. For material not previously sheared there is a rapid decline in strength with increasing shear until the residual shear strength is reached. The shear strength of a food will influence the rheological properties and mechanical properties of the food during processing, and also the texture and other sensory properties of the food during consumption.

Shear values: Measures of the forces experienced by a material, such as a food, undergoing shear. It is often determined in meat after cooking as an indication of tenderness.

Shear: Force that one plane exerts on a neighbouring plane per unit area of contact, and which causes a deformation in a direction related to the direction of the applied force. Shear forces are applied during food processing such as mixing and extrusion and will affect the texture of the final product. Shear also occurs during mastication of foods.

Shelling: Removal of husks, shells or pods from foods such as nuts, eggs and peas.

Sherbet: Artificial fruit-flavoured effervescent powders eaten as sweets. When mixed with bicarbonate of soda, tartaric acid, sugar and flavourings, may also be used to make beverages..

Shigellosis: Bacillary dysentery caused by infection with *Shigella* species. It is characterized by abdominal cramps, diarrhoea, fever, vomiting, presence of blood, pus or mucus in stools, and tenesmus. Transmission is via the faecal-oral route by contaminated foods (e.g. salads, vegetables, dairy products and poultry meat) and water.

Shops: Buildings or parts of buildings where goods or services are sold.

Shortbread: Sweetened biscuits prepared with a high ratio of butter or other shortenings to flour.

Shortening: It is a process that results from changes occurring in numerous connected sarcomeres in the myofibrils of muscles. It occurs during muscle contraction in living animals, but also during *rigor mortis*. Degree of sarcomere shortening is influenced by muscle fibre type (e.g. oxidative vs. glycolytic) and *post mortem* ambient temperature. If ambient temperature decreases rapidly during the onset of *rigor mortis*, muscle fibres contract to a greater extent than at higher ambient temp. This physiological occurrence is referred to as cold shortening; severe shortening results in reduced meat tenderness. Electrical stimulation is used to reduce toughness associated with cold shortening in meat. Shortenings may be vegetable fats often used in baking. By dispersing as a film throughout batters, they impart crispness or flakiness to bakery products.

Shredding: Tearing or cutting of items into strips of material (shreds). This can be achieved either by hand or by using a grater or a food processor fitted with a shredding disk.

Shrikhand: Fermented milk product usually prepared from buffalo milk and popular in India. Traditionally, the milk is fermented with a mixed starter culture (*Streptococcus lactis* and *S. lactis* var. *diacetylactis*) and chakka is prepared by draining off whey from the resultant curd. Other ingredients, e.g. sugar, colorants and flavourings, are then added to the chakka.

Shrink packaging: Transparent, clinging thermoplastic films used to enclose a product or package. When heated, the film shrinks to fit closely to the package.

Shrinkage: It is defined as the percent change from the initial apparent volume. Two types of shrinkage are usually observed in food materials. If there is a uniform shrinkage in all dimensions of the material, it is called isotropic shrinkage. The non-uniform shrinkage in different dimensions, on the other hand, is called anisotropic shrinkage. Shrinkage is the decrease in volume of the food during processing such as drying. When moisture is removed from food during drying, there is a pressure imbalance between inside and outside of the food. Shrinkage affects the diffusion coefficient of the material and therefore has an effect on the drying rate.

Shucking: Removal of husks from corn, or shells from shellfish such as oysters and clams.

Sides: A butchers' term for the two halves of animal carcasses, divided along the backbone.

Sieve A fine stainless steel wire, plastic or other stranded fine mesh either stretched in a flat circle on a metal or wooden former or made into the shape of a half sphere or inverted cone. Used for separating large from fine particles, for straining liquids or for reducing soft foods to a purée. It is also know as strainers.

Sieving: Process of straining solids from liquids or separating coarser from finer particles using sieves or strainers. Sieving also incorporates air to make ingredients (such as flour) lighter.

Sifter: A closed cylindrical container with a sieve or perforated top at one end for sprinkling flour, sugar or other powders over food items.

Sifting: Process of passing a dry substance through sifters to remove lumps or large particles. Sifting also incorporates air to make ingredients (such as flour) lighter.

Silica gels: Gels formed from polymers of silicic acid. When dried, they are termed silica xerogels and are used as desiccators or as adsorbents, e.g. for clarification of beer by adsorption of cloud forming proteins.

Silica: Silicon dioxide that occurs in crystalline, crypto- crystalline and amorphous hydrated forms. It is ubiquitous component of the diet with numerous applications in the food industry, such as stabilization of beer, refining of vegetable oils, and immobilization of proteins and enzymes.

Silos: Tall towers or pits which are used for storage. Commonly refers to stores for grain, e.g. on a farm or at a mill, but can also be used for storing other commodities including vegetables and milk. It is also applied to airtight structures in which green crops are compressed and stored as silage for animal feeding.

Simmering: Heating of foods in a liquid, such as water, at a temperature that causes the liquid to bubble gently.

Simulated foods: Processed foods that are modified to simulate another kind of food, e.g. by using textured vegetable proteins and flavourings, to mimic texture and sensory properties of the target food. Some of the most popular simulated foods are meat substitutes (e.g. for use in vegetarian foods), butter substitutes and imitation cream. It is also known as imitation foods, analogues or artificial foods.

Simultaneous saccharification and fermentation: Process which involves enzymic saccharification of cellulosic biomass and simultaneous microbial fermentation of the resulting glucose, e.g. to ethanol. Advantages over the traditional two-stage process include the ability to use lower temperatures, thus reducing operating costs. Although these processes can be performed by mixed cultures of an appropriate enzyme-producing microorganism and a fermentative microorganism, recent research has focused on genetic engineering of strains to enable them to ferment cellulosic substrates directly.

Single cell proteins (SCP): Protein-rich biomass produced by large-scale microbial fermentation using a variety of substrates, such as petroleum fractions or carbohydrates. It is used as a source of proteins for use in foods and animal feeds. There is potential for future commercial exploitation of these proteins with advances in fermentation technology.

Single cream: Cream with a minimum butterfat content of 18%. Not suitable for whipping.

Single market: An association of countries trading with each other without restrictions or tariffs.

Sinigrin: Antinutritional glucosinolate with a bitter taste found in *Brassica* spp.

Skim milk powders: Products prepared by drying skim milk to a low moisture content, giving powders with a long shelf life. It is also called dried skim milk and non-fat dried milk.

Skim milk: Milk from which virtually all the fat has been removed (fat content is less than 0.5%). It is preferred to whole milk or semi skimmed milk by some health-conscious consumers and used by processors to make low fat dairy products. Almost total removal of fat means that skim milk differs greatly from whole milk in mouthfeel and also in appearance, having a bluish tinge.

Skimmed milk: Milk with a butterfat content between 1.5 and 1.8%. It is sold in bottles with a silver and red striped top.

Skinning: Removal of the skin from foods such as poultry and fish before or after cooking.

Slaughter: The killing of animals and poultry for food. In developed and some developing countries, slaughter of animals for meat takes place under closely regulated conditions in slaughterhouses. Legislation often dictates that food animals should be slaughtered without undue stress and suffering, and that bleeding should be as complete as possible; effective stunning is of primary importance in achieving these aims. Both

Judaism and Islam prescribe a ritual protocol for the slaughter of animals for human con- sumption; stunning is not used in kosher or halal slaughter.

Slaughterhouses: Places where the slaughter of animals takes place in a hygienic fashion, and where carcasses are prepared for retail to consumers. The term includes abattoirs and butcheries. Usually, carcasses are examined at slaughterhouses by qualified inspectors and only those carcasses that are free from disease are allowed to leave the premises for retail to consumers.

Slicing: Cutting of thin pieces or slices of food from a larger portion using a sharp implement.

Slow cooker: A thermostatically controlled electrically heated deep metal container into which a lidded earthenware or glass, deep and well fitting pot is placed. Used for slow cooking food at or below simmering temperature for long periods of time.

Slow cooking: To cook at a low temperature in the range 80 to 95°C.

Sludges: Usually thick, soft, wet mud or similar viscous mixtures, or alternatively any undesirable solids settled out from a treatment process.

Smoke concentrates: Concentrated smoke flavourings produced by removal of the liquid base in which wood smoke flavour compounds are dissolved.

Smoke flavourings: Flavourings produced by contact of a liquid, usually water or oils, with smoke produced by burning of wood, e.g. hickory, oak or maple.

Smoke point: The minimum temperature at which oils and fats begin to decompose and produce smoke, usually above 160°C

Smoke: Substance obtained by burning certain types of wood, such as hickory, maple, or ash, that is used in the process of preserving or flavouring of foods, including meat and fish.

Smoked fish: Fish which has been processed by smoking. Whole gutted or ungutted fish or fish fillets are smoked at high or low temperature, sometimes after brining, using smoke produced from various types of wood. The kind of wood used affects the flavour and colour of the product. Cold smoking is usually performed at a temperature not exceeding 30°C. Hot smoking is carried out at a temperature sufficient to cause thermal denaturation of the proteins, usually between 50 and 80°C. Common types of smoked fish include smoked salmon, smoked trout, smoked mackerel, smoked eels, smoked haddock and kippers. Some types are mainly eaten cold, while others, especially smoked haddock and kippers, are usually eaten hot.

Smoked foods: Foods preserved and flavoured by treating with smoke, e.g. kippers, yellow fish and smoked meats. In traditional methods, foods are placed directly in the smoke. Other methods use smoke flavourings or liquid smoke, which are applied to the food. Sensory properties of smoked foods are affected by the type of smoke and smoking method used.

Smoked mackerel: Mackerel which has been cooked by smoking. Fish can be smoked whole and gutted, or as fillets, by hot smoking at a temperature between 50 and 80°C or cold smoking at up to 30°C. Usually eaten cold in salads or made into pates.

Smoked salmon: Salmon which has been smoked by one of two methods, i.e. hot smoking or cold smoking. In hot smoking, the process is conducted at a high temperature (50-80°C) and lasts 6-12 hours, the time depending mainly on the size of the fish and strength of flavour desired. Cold smoking is performed at a much lower temperature (up to 30°C) and may take several weeks to complete. Smoked salmon is usually eaten thinly sliced and cold, often in salads and sandwiches, although it can be used as an ingre- dient in many dishes, hot or cold.

Smoked trout: Trout which has been cooked by smoking. The fish are most commonly hot smoked, whole or filleted, at a temperature of 50 to 80°C. Can be eaten cold in salads, made into pates and mousses, or used as an ingredient of soups and other hot or cold dishes.

Smokehouse: A well-sealed chamber in which food is hung on racks for smoking with provision for the generation of smoke from smouldering wood, for temperature control and for the removal of smoke. Such a chamber may range in size from a small box up to a large room.

Smoking: Curing or preservation especially of meat or fish, by exposure to smoke produced by the burning of certain types of wood, such as hickory, maple or ash. Foods can be cold smoked or hot smoked. Hot smoking partially or totally cooks foods. Usually fish, meat and occasionally cheese are smoked but most of the latter so described is flavoured with a synthetic mix of chemicals which mimic the real thing. Meat and fish will be partially cooked and dried which helps to preserve them and the flavour will be improved.

Smoothies: Thick and smooth textured beverages made by blending fruits with yoghurt, milk, ice, ice cream or frozen yoghurt.

Smoothness: Sensory properties relating to the extent to which a product has a smooth consistency, i.e. is perceived to be uniform and regular.

Snack foods: Sweet or savoury foods eaten to provide light sustenance in a quick and convenient format. It is eaten between or as an alternative to main meals. Popular types include sandwiches, cereal bars and potato crisps. It is also known as snacks.

Soaking.

Soaking: Process by which an item is made thoroughly wet by immersion in a liquid either to rehydrate it or to dissolve out excessive salt used in its preservation, or undesirable constituents e.g. Alkaloids, prussic acid, etc. So as to make it suitable for cooking.

Soapiness: One of the **sensory properties**; relating to the extent to which a product tastes soapy.

Soda water: Water carbonated so that it is effervescent when dispensed.

Sodium acetate: Sodium salt of acetic acid. Anhydrous and trihydrate forms of this salt are both used as food additives. The anhydrous salt is hygroscopic and both forms are highly soluble in water. Uses in foods include as part of pH buffering systems, flavourings and preservatives.

Sodium ascorbate: Sodium salt of ascorbic acid (vitamin C). In addition to being a source of vitamin C for fortification of foods, this salt has food industry uses in antioxidants and preservatives. It is also used in curing of meat.

Sodium bicarbonate: Monosodium salt of carbonic acid prepared by reaction of sodium carbonate with carbon dioxide and water. In aqueous solution, the bicarbonate tends to decompose, releasing CO_2. Due to this property, sodium bicarbonate is used in raising agents for bakery products, as an ingredient of baking powders, and in the manufacture of car- bonated beverages. Baking powders contain sodium bicarbonate and tartaric acid, reaction between which increases CO_2 production. Aqueous bicarbonate solutions are slightly alkaline; solution pH increases with agitation, time and increasing temperature due to loss of CO_2 from the solution. This salt is therefore also used as an alkali and as part of pH buffering systems in foods. It is also known as sodium hydrogen carbonate and baking soda.

Sodium carbonate: A mild alkali used for cleaning purposes. See also E500. Also called washing soda, soda ash

Sodium carboxymethyl cellulose: E466, the sodium salt of a derivative of cellulose used as a thickener and bulking agent

Sodium caseinate A food additive (sodium salt) made from milk protein (casein). It is used in a wide range of foods as a source of protein or to enhance functional properties such as water binding capacity, emulsifying capacity, whitening ability and whipping capacity. It is used to maintain colour in sausages and other processed meats

Sodium caseinate:

Sodium cyclamate: One of the cyclamates, artificial sweeteners with approximately 30 times the sweetness of sucrose. White crystalline solid, highly soluble in water and stable at baking temperatures. It has a pleasant flavour profile and thus is often used in combination with saccharin, a molecule that is sweeter than sodium cyclamate, but which produces a bitter aftertaste. It is lso known as sodium cyclohexyl- sulfamate and sucaryl sodium.

Sodium heptonate: A sequestering agent used in edible oils

Sodium hydrogen diacetate:, a sodium salt of acetic acid used as a preservative and firming agent

Sodium hydroxide: The strongest alkali available for domestic or general industrial use. It should be handled with extreme care with all bare skin and eyes protected. Used for dissolving fats from drains, etc. By converting them to soaps and for removing the bitterness from unripe olives. It is also called caustic soda or lye solution.

Sodium lactate: Sodium salt of lactic acid. Hygroscopic, soluble in water and alcohol and odourless, but with a slight salty flavour. It is used in additives including preservatives, emulsifiers, flavour enhancers, humectants, and as part of pH buffering systems. Also known as 2-hydroxypropanoic acid monosodium salt and lacolin.

Sodium metabisulfite: Disodium salt of disulfurous acid that forms acidic aqueous solutions. It is used predominantly in preservatives, but also in antioxidants, flavourings and bleaching agents. It is also called (di) sodium pyrosulfite and sodium bisulfite.

Sodium nitrate: E251, the sodium salt of nitric acid used for curing and preserving meat. **sodium nitrite:** E250, the sodium salt of nitrous acid which is used in curing salts to preserve meat and maintain a pink colour by its reaction with haemoglobin. It is banned in some countries because of fears that it might induce stomach cancers.

Sodium phosphate: An emulsifier used to assist in the incorporation of water into various processed foods such as sausage, luncheon meats, etc.

Sodium polyphosphate: A chemical used to increase the water uptake of poultry and of other meats and bacon so as to increase their weight. All of this water is lost on cooking and can cause problems e.g. When frying bacon.

Sodium saccharin: A sodium salt of saccharin used in the same way as saccharin

Sodium stearoyl-2-lactate: E481, the sodium salt of a stearic acid ester with the lactate of lactic acid used to stabilize doughs and emulsions and to improve the mixing properties of flour and the whipping and baking properties of dried egg white.

Sodium tripolyphosphate: Phosphate often used to improve the physicochemical properties, and increase the quality and shelf life of meat products.

Sodium: Soft, silvery, highly reactive alkali metal with the chemical symbol Na, most commonly found in the form of salt (NaCl). An essential nutrient in the diet, albeit in moderate quantities; excess intake may result in high blood pressure (hypertension).

Soft cheese: Cheese with a creamy, smooth texture made from milk with a relatively low dry matter content and range of fat contents, skim milk soft cheese having a butterfat content of <2% and full-fat soft cheese containing at least 20% butterfat. The limits of water content are 80% for low-fat (2 to 10% butterfat), 70% for medium-fat (10 to 20% butterfat) and 60% for full-fat (greater than 20% butterfat) soft cheeses. It is can be ripened, e.g. Camembert cheese, or unripened (fresh), e.g. Cottage cheese, cream cheese, fro- mage frais, quarg.

Soft drinks: Non-alcoholic beverages, commonly carbonated beverages, often with fruit or cola flavours.

Soft flour: Flour made from soft wheat consisting of unbroken starch granules, mostly used for cakes, biscuits and general thickening and coating. It is much cheaper than strong flour.

Soft frozen beverages: Frozen beverages which are served in a partially frozen or slush state.

Softeners: Additives that increase softness of foods. Examples include glycerides, which are added to bakery products as crumb softeners and to chewing gums to improve texture. It is also used to describe chelating agents that remove ions, e.g. calcium or magnesium ions, from water.

Softening: It is process whereby products such as fruits and vegetables lose their rigidity and firmness, often during ripening or ageing.

Softness: One of the sensory properties; relating to the extent to which a product is firm in texture.

Solanidine: Alkaloid present in the sprouts and skin of green potatoes (potatoes exposed to light). It is formed by hydrolysis of solanine. It is toxic to humans and not destroyed by cooking.

Solanine: Alkaloid present in all green parts of the potato plant, including potatoes that have been exposed to light. It is inhibitor of cholinesterases. Poisoning causes gastrointestinal and neurological disorders. Solanine is not destroyed by cooking.

Solar driers: Equipment used to carry out solar drying, a process that depends on the sun as the source of energy. There are two types of solar driers: direct and indirect. In direct solar driers, air is heated in the drying chamber, which acts as both the solar collector and the drier. An indirect drier comprises two parts: a solar collector and a separate drying chamber.

Solar drying: Drying method that depends on the sun as the source of energy, but which also involves the use of some sort of structure to collect and enhance the solar heat. Solar drying generates higher air temperatures and lower humidities than those produced by sun drying, resulting in faster product drying rates and lower final moisture contents.

Solar energy: Radiant energy emitted by the sun that is captured and used during solar drying processes or converted into electrical energy.

Solar radiation: Radiation emitted by the sun, made up of an extensive range of wavelengths of the spectrum.

Solid phase extraction: Extraction technique for preparation of samples prior to analysis, developed as an alternative to liquid-liquid extraction. Samples are dissolved in solvent and passed through a bed of adsorbent to effect separation of components of interest. Compounds are eluted with small volumes of solvent.

Solid phase microextraction: Type of solid phase extraction in which samples are adsorbed onto a fused silica fibre coated with a stationary phase. The fibre is then inserted into a GC injector, where the sample is desorbed and analysed.

Solid state fermentation: Fermentation of microorganisms on a solid support of low moisture content under non-septic conditions. Energy requirements are low but the process can yield high product concentrations. In addition, downstream processing is facilitated. A variety of agricultural residues (e.g. wheat straw, rice hulls and corn cobs) have been used as supports for production of enzymes and sec- ondary metabolites.

Solid volume (Vs): It is the volume of the solid material (including water) excluding any interior pores that are filled with air. It can be determined by the gas displacement method in which the gas is capable of penetrating all open pores up to the diameter of the gas molecule.

Solidification: It is a process by which an item becomes hard or solid.

Solids not fat (SNF): The solids content of milk excluding the fats content, i.e. the contents of proteins, lactose and salts. It is used as an index of milk quality. Milk contains on av- erage 8.6% SNF.

Solids: Particles whose shape and volume are fixed and are not affected by the space available to them, and which have a tendency to resist forces that would alter their shape.

Solubility: Extent to which one substance dissolves in another. Normal solubility records the maximum mass of a solid that can be dissolved in a specified mass of water to form a saturated solution, and is measured in kilograms per metre cubed. When solubility is ex- ceeded, excess solid appears as a precipitate. Solubility is temperature-dependent. Generally, for a solid in a liquid, solubility increases with temperature; for a gas, solubility decreases with temperature.

Solubilization: Process by which a substance is made soluble or more soluble, especially in water.

Soluble solids: Particles that can be dissolved in flu- ids, especially water.

Solvents: Liquids that dissolve other substances to form solutions. Polar solvents are compounds such as water and liquid ammonia, which have dipole moments and consequently high dielectric constants. These solvents are capable of dissolving ionic compounds or covalent compounds that ionize. Nonpolar solvents are compounds such as ethoxyethane and benzene, which do not have permanent dipole moments. These do not dissolve ionic compounds but will dissolve nonpolar covalent compounds.

Solvents can be further categorized according to their proton donating and accepting properties. Amphiprotic solvents self ionize and can therefore act both as proton donators and acceptors. A typical example is water. Aprotic solvents neither accept nor donate protons; tetrachlo- romethane (carbon tetrachloride) is an example.

Sonication: Process of disrupting biological materials such as bacteria, plants or foods using high-frequency sound waves. It is used widely in preparation and extraction of samples prior to analysis.

Sorbates: It is salts of sorbic acid. Sorbates, including sodium, potassium and calcium sorbates, are used as preservatives for foods, particularly cheese, and beverages, including wines.

Sorbestrin: Thermally stable preparation formed from fatty acid esters of sorbitol and sorbitol anhydrides. It is used as a substitute for vegetable oils in applications such as frying oils, salad dressings and mayonnaise.

Sorbets: Water ices made from water, sugar and sometimes eggs, flavoured with fruit purees or fruit juices and sometimes alcoholic beverages (e.g. champagne). It is frequently served between courses of meals to act as a refresher.

Sorbic acid: A permitted preservative, E200, for use in baked and fruit products. It is may be obtained from berries of the mountain ash but is now synthesized and is a mould and yeast inhibitor. It exhibit antimicrobial activity. The free acid and it salts (sorbates) are used as food preservatives. Sorbic acid also has uses in acidulants and flavourings. Systematic name is 2,4-hexadienoic acid.

Sorbitol: Sugar alcohol (polyol) produced by reduction of glucose or fructose. It occurs naturally and has approximately 0.5 times the sweetness of sucrose. Digestion of sorbitol yields fructose, making it suitable for use as a sweetener for diabetic foods. It is found in many fruits and berries but now synthesized from glucose for use as a humectant and as a sweetening agent for diabetics. It is also known as glucitol.

Sorting: It is systematic arrangement of items in groups or grades.

Soup mixes: Mixes, usually powdered, that are recombined, typically with water, to form soups.

Soup: A flavoured liquid based on meat and/or vegetable extracts in water, milk or occasionally water only, with added ingredients. The term covers many types of product, including: clear, e.g. consommes; creamy, with all ingredients liquidized and often with cream added, e.g. cream of chicken; or thick, with chunks of ingredients floating in the clear liquid base, e.g. broths. Soups are generally eaten hot, but some types, e.g. vichyssoise, are usually consumed chilled. Other popular types include borshch, bouillabaisse and minestrone. Some types are also available as soup mixes or instant soups.

Sour cream sauce: A chicken or game velouté flavoured with vinegar, sweated chopped shallots, white wine and soured cream.

Sour cream: Commercial product made by fermentation of homogenized pasteurized cream with lactic acid bacteria. Used in cooking and as a component of dips. Also known as soured cream and ripened cream.

Sour milk: Milk that has become rancid due to breakdown of fats or a fermented milk. The latter is produced by fermentation of milk (of various species) by lactic acid bacteria (starters). During fermentation, lactose is converted into lactic acid, aroma compounds are formed and milk proteins are partly decomposed to peptides and free amino acids, improving digestibility of the milk. Milk which has naturally curdled or been deliberately exposed to some acid-producing organism and thus curdled at around 30 to 35°C. UHT milk should not be soured. Useful in baking where bicarbonate of soda is used as double the amount of carbon dioxide will be liberated.

Sourdough bread: Bread which uses in place of yeast, dough from a previous batch which has been allowed to ferment with a little warm water and sugar overnight or longer.

Sourdough: Dough which has either been fermented by microorganisms naturally present in flour and/or other ingredients, or by added microbial cultures, e.g. lactic acid bacteria. Fermentation of the dough produces organic acids, which impart a desirable sour flavour to the dough. It is used to make sourdough bread.

Sourness: One of the sensory properties; relating to the extent to which a product tastes sour, i.e. tart, bitter or sharp.

Sous vide foods: Vacuum-sealed pouches of chilled foods preserved by the sous vide process. Foods preserved in this manner undergo minimal heat proc- essing and thus have improved shelf life compared with non-vacuum cook chill methods. Improved eating quality benefits have also been reported. Foods com- monly processed in this way include fruits in syrups and some meals used in catering.

Sous vide meals: Individual meal portions preserved in vacuum-sealed pouches by the **sous vide** process.

Sous vide: Food processing and packaging technique in which fresh ingredients are combined into specific dishes or meals, vacuum packaged in individual portion pouches, cooked under vacuum and then chilled.

Southern blotting: Method for detecting specific DNA fragments. DNA is digested with restriction endonucleases, separated by gel electrophoresis, dena- tured and transferred to a chemically reactive matrix (e.g. nitrocellulose or nylon), on which the DNA frag- ments bind covalently in a pattern identical to that on the original gel. After blotting, target molecules are detected through the use of labelled complementary single-stranded DNA or RNA molecules.

Soy 11S globulins: One of the two major types of soy proteins (the other group are the 7S globulins) that together make up 70% of the total storage proteins in soybeans.

Soy 7S globulins: One of the two major types of soy proteins (the other group are the 11S globulins) that together make up 70% of the total storage proteins in soybeans. Trimeric glycoproteins is comprising α α' and β subunits, which together form conglycinin. It is responsible for softness and adhesion properties of soy products.

Soy beverages: Beverages derived predominantly from soybeans or their products. It is includes soymilk.

Soy cheese: Creamy product made from soymilk. It is used as a replacement for cheese or sour cream.

Soy curd: Product made from soymilk, by coagulation, draining and pressing in a manner similar to that used in cheesemaking. It is rich in protein. Available packaged in water, vacuum packaged or frozen. It is used in a variety of dishes, including soups, casseroles and sauces. It is also called tofu.

Soy globulins: It is storage proteins of soybeans. Main soy proteins are glycinin and β-conglycinin.

Soy glycinin: One of the main soy proteins. An 11S storage protein that, along with β-conglycinin (7S globulin), makes up approximately 70% of storage proteins in soybeans.

Soy ice cream: Frozen dessert made from soymilk and used as a substitute for conventional ice cream.

Soy infant formulas: Products made by mixing soy protein isolates with fats and carbohydrates to give a composition similar to that of human milk. It is used mainly to feed infants who are allergic to cow milk or suffer from lactose intolerance.

Soy lecithins: Lecithins extracted from soybeans and used as emulsifiers in foods. It is also called soy lecithins.

Soy meal: Flour made by grinding roasted, dehulled soybeans. Full fat soy meal is made from soybeans that still contain oil. Defatted soy meal is made using soybeans from which soybean oils have been extracted. Good source of soy proteins, iron, calcium, B vitamins and fibre. It is used in baking and in thickeners for sauces.

Soy pastes: Fermented products prepared from cooked soybeans. It includes Japanese miso and Ko rean doenjang. It is used mainly as seasonings.

Soy protein concentrates: Protein concentrates made by extracting sugars from defatted soy flakes, leaving proteins and fibre. It is used to make meat substitutes and in a variety of products, such as cereal products, bakery products, beverages and gravy.

Soy proteins: Storage proteins found in soybeans. Nutritional and health-promoting properties, combined with functional properties make them useful and widely-used ingredients in food processing.

Soy purees: Preparations made by mashing or blending cooked soybeans. It is used in infant foods and beverages.

Soy sauces: Sauces produced by fermentation of a soybean mash prepared by grinding soybeans with water. A fungus, often *Aspergillus oryzae*, is added to the soybean mash to initiate fermentation. Duration of the fermentation process and addition of other in- gredients influences sensory properties of the sauces. Soy sauces fermented for a shorter period have a less rich flavour than those fermented for a longer period. Addition of molasses produces richer, darker soy sauces, while inclusion of wheat in the fermentation produces lighter products.

Soy yoghurt: Creamy product made from soymilk. It is used as a substitute for cream cheese or sour cream.

Soya bean: The seeds of an erect bushy plant, *Glycine max*, of the pea family, originating in China but now a major crop grown worldwide in frost-free, warm summer climates and containing about 20% oil and 35-40 % protein on a dry weight basis. The pods contain between 2 and 4 seeds which may be green, brown, yellow or black. It is used as a source of bean sprouts, oil, flour, milk, protein curds, vegetable protein and meat analogues and sauce as well as being an important food pulse in its whole state. The beans, especially the oil, are rich in compounds which behave like weak oestrogens and dilute the effect of the body's own oestrogens, thus reducing the risk of breast cancer, but evidence has been reported that raw soya flour fed to rats produced pancreatic cancer. It is rich in high quality soy proteins, unsaturated soybean oils, B vitamins and minerals. Numerous soy products are made from the seeds, including soymilk, cheese-like products (tofu, tempeh, miso) and meat substitutes (soy meal, soy protein concentrates, soy protein isolates).

Soya flour: Flour produced from soya beans with a high fat and protein content, used together with wheat flour in baking and in ice creams and other manufactured foods

Soya oil: Oil extracted from from seeds of *Glycine max* (soybeans) used as cooking oil, for salad dressings and for the manufacture of margarine containing only about 10% of saturated fat. It contains palmitic acid, oleic acid, linoleic acid and linolenic acid. It is used as salad oils or cooking oils, as well as in margarines and shortenings. By-products obtained during processing include lecithins, tocopherols and phytosterols. It said to have protective effects against breast cancer. It is also called soybean oil.

Soya sauce: A highly flavoured liquid which is obtained by long fermentation of soya beans and various cereal grains. It is produced in 3 stages. In the first, polished rice is fermented with *Aspergillus oryzae*. This is then mashed with boiled soya beans and crushed roasted grains of wheat, incubated at 30°C for 3 days, diluted with brine then fermented with *Pediococcus halophilus* to reduce the ph. In the third stage the mixture is fermented slowly for between 1 and 3 years with *Saccharomyces rouxii*. Both light and dark varieties are produced with varying amounts of salt added.

Soybean sprouts: Legume sprouts are produced by germination of soybeans and rich in protein, vita- mins and minerals. It is widely used in Asian dishes such as egg rolls and stir fried meals. It is also used in soups, casseroles, sauces, bakery products, and raw in salads. Dried sprouts can be eaten as snack foods or used as a substitute for nuts in bakery products or dishes.

Soymilk: Product prepared by cooking dehulled, ground soybeans in water and filtering of the solid matter (okara). It is rich in B vitamins, protein and iron. It is used similarly to milk as a beverage, as the basis of soy products such as soy yoghurt, soy ice cream and soy cheese, and in cooking and baking. Available in regular, low-fat and flavoured forms or as a powder.

Space flight foods: Meals designed for consumption in the confined microgravity environment encountered on space flight programmes. Originally bite-sized cubes or squeezed from a tube, space flight foods have now evolved into more appetizing meals that can in corporate frozen, refrigerated and ambient foods. A typical meal tray could include a foil beverage pouch, and individual servings of lightweight easily rehydrated foods, intermediate moisture foods and thermostabilized, aseptic fill, natural form foods. Early research into providing assurance against microbial contamination in space led to development of the HACCP concept.

Spaghetti: Thin solid pasta made by extruding a pasta paste though circular holes about 2 to 3 mm in diameter, either cooked immediately or cut into lengths (up to 40 cm) and dried for future use.

Sparkling wines: Wines which contain sufficient dissolved carbon dioxide to result in effervescence when the bottle is opened. The high carbon dioxide content may be achieved by secondary fermentation (in the bottle or in a tank) or by carbonation.

Spearmint: Common name for *Mentha spicata*, the leaves of which are used as spices. It has a sweet, minty (fresh and cool) flavour due predominantly to the flavour compound

L-carvone. Essential oils distilled from spearmint are also used as flavourings, particularly for chewing gums.

Species identification: Recognition of the animal source of products containing meat, fish or milk. It is used to detect adulteration or establish authenticity. Methods used to identify the species of origin include electrophoresis, isoelectric focusing and genetic techniques. It can also refer to determination of microbial species.

Specific conductivity: It is a measure of the ability of a substance to transport an electric charge. It is inversely proportional to the electrical resistance of a substance and dependent upon its dimensions (length/crosssection) if the substance is a solid. For a solution, specific conductivity is measured between electrodes spaced 1 cm apart with a cross section of 1 cm^2. For dilute solutions, specific conductivity is proportional to electrolyte concentration.

Specific gravity: It is ratio of the density of a substance to the density of a reference material. For a liquid or solid, specific gravity is the ratio of its density (usually at 20°C) to the density of water {at its temperature of maximum density (4°C)}. It has no units and substances with a specific gravity greater than 1 will sink in water, whilst if less than 1 they will float. Sugar solutions and brines have a specific gravity greater than 1, oils and fats less than. It is synonymous with relative density.

Specific heat: Heat capacity of a substance per unit mass. The amount of energy required to raise the temperature of unit mass of an object by a unit increment in temperature (measured in Joules per Kelvin per kilogram).

Specific rotation: Optical properties relating to the rotation that a beam of light of a given wavelength undergoes, relative to its plane of polarization, as it passes through a solution of a given density, path length, concentration and temperature.

Spectra: Pattern of properties arranged in order of increasing or decreasing magnitude. In analytical ap- plications, the property measured varies according to the analytical technique being employed. In mass spectroscopy, a mass spectrum with a range of masses is produced. An emission spectrum represents the range of radiations emitted when a substance is heated, bombarded by electrons or ions, or absorbs photons. An absorption spectrum shows the energies absorbed from a continuous spectrum of radiation by an absorbing medium. Spectra produced by an unknown substance can be compared with those of a standard to give information about the composition of the sample.

Spectrofluorometry: Spectroscopy technique in which the intensity of fluorescence of a sample is measured as a function of wavelength. A pair of monochromators is used, one of which selects the excitation wavelength and the other the emission wavelength.

Spectroscopy: Series of techniques in which absorption or emission of radiant energy of various wave-lengths is used to measure chemical concentrations or structures. It includes atomic emission, atomic absorption, IR and mass spectroscopy.

Sphericity: Sphericity is an important parameter used in fluid flow and heat and mass transfer calculations. Sphericity or shape factor can be defined in different ways. According to the most commonly used definition, sphericity is the ratio of volume of solid to the volume of a sphere that has a diameter equal to the major diameter of the object so that it can circumscribe the solid sample. For a spherical particle of diameter Dp, sphericity is equal to 1.

Spices: Aromatic plants or parts of plants, e.g. roots, leaves or seeds, in various forms (native, dried, ground, whole) used primarily for their **flavour** rather than for any nutritional benefit.

Spinning: Texturization process usually applied to protein isolates. For example, biodegradable films can be prepared by spinning soy protein isolates in a coagulating buffer, and wet spinning methods can be used to produce edible protein fibres from a variety of materials, such as soy proteins, casein and blood plasma proteins. The term can also be used to describe the process used in the manufacture of cotton candy. Chocolate can be spun moulded.

Splitting: It is process of breaking any substance by forcefully into parts. Like the cutting of animal carcasses into left and right sides using a saw during processing. Also relates to undesirable processes, such as the damage that can occur to fruits (such as tomatoes, cherries and grapes) when their peel splits upon absorption of excess water, and fruit splitting, a physiological disorder of peel development in citrus fruits. Water stage fruit split is an erratic and complex problem often causing major crop losses to susceptible cultivars of pecan nuts. In the beverage industry, corks placed in wine bottles can be susceptible to splitting. Problems are also associated with premalting (splitting) of malting barley, which is thought to be caused by alternating periods of sunny and rainy weather during ripening of the grain. Egg shells can split during boiling; this results from excess internal pressure in the egg, due to the egg contents having a higher coefficient of thermal expansion than the shell. Canned kidney beans are liable to split during storage, and sausage casings can split during cooking.

Spoilage bacteria: Bacteria typically involved in the spoilage of foods.

Spoilage yeasts: Yeasts typically involved in the spoilage of foods.

Spoilage: Deterioration of a food by chemical, physical or microbial means.

Sponge cakes: Light, porous cakes made using self-raising flour, sugar, beaten eggs and flavourings. Butter or oils may be added, although many sponge cakes contain no shortenings.

Sponge Dough: It is used in breadmaking which contains a proportion of the flour, all of the yeasts, yeast foods, malt and sufficient water to make a stiff dough. Fats may also be added, together with a proportion of salt; this controls fermentation which takes place over 3-5 hours.

Spores: Usually unicellular, dormant reproductive or resting bodies produced by microorganisms under conditions of environmental stress (e.g. extremes of temperature and dehydration). It is resistant to unfavourable environmental conditions, and capable of germinating and developing into vegetative cells when environmental conditions are favourable, without fusion with another cell.

Sports drinks: Soft drinks formulated to enhance or maintain the performance of sports people, or to improve their recovery after a sporting event or training session. Generally contain ingredients such as sugars and electrolytes.

Sports foods: Products formulated to contain precise levels of nutrients and other ingredients intended to enhance sports performance in athletes.

Sporulation: It is a process by which spores develop in microorganisms.

Spray driers: Equipment for manufacture of dried foods from liquids, such as production of dried milk from liquid milk, by spray drying. Liquids are sprayed as a fine mist into a hot-air chamber, where they dehydrate; solids fall to the bottom of the chamber as dry powders.

Spray drying: Process for manufacture of dried foods from liquids. The liquid food is generally pre- concentrated by evaporation to reduce the water content. The concentrate is then introduced as a fine spray or mist into a tower or chamber with heated air. As the small droplets make intimate contact with the heated air, they flash off their moisture, become small particles, drop to the bottom of the tower and are re-moved. The advantages of spray drying over other types of drying include the need for only a low heat and short time, which leads to better quality product.

Spread: A semi-solid savoury paste spread on bread for sandwiches or on toast for canapés, also any soft food item which can easily be spread.

Spreadability: Texture term relating to the ease with which a product can be spread.

Spring roll: A thin rectangle of pastry made with eggs instead of water, wrapped securely around a cylinder of filling and deep-fried until crisp and golden. The filling consists of various mixtures of cooked chopped meat, poultry, shellfish or vegetables.

Springiness: One of the sensory properties; relating to the extent to which a product springs back quickly when squeezed, bent, pressed or stretched.

Sprouting: It it alternavie term of germination, meaning the process whereby seeds or spores begin to grow. Also describes the production of sprouts in potatoes and other tubers during storage. Sprouting can be controlled by storing susceptible vegetables in the dark and at low temperatures, or by the use of sprouting inhibitors. In this, to keep seeds, beans or grains in warm damp conditions after soaking in water so that they germinate and begin to grow and generally convert starch and protein into sugars and plant tissue. Many seeds become more palatable and digestible after this process.

Srikhand: Fermented milk product usually prepared from buffalo milk and popular in India. Traditionally, milk is fermented with a mixed starter culture (*Streptococcus lactis* and *S. lactis* var. *diacetylactis*) and chakka is prepared by draining off whey from the resultant curd. Other ingredients, e.g. sugar, colorants, flavourings, are then added to the chakka.

Stabilization: Process of making or becoming stable. Stabilizers such as agar, alginates, car- rageenans and gums are used for the stabilization of foods.

Stabilizers: Additives included in food formulations to prevent separation of ingredients and thus improve appearance and shelf life. Common uses include stabilization of oil and water components in emulsions, e.g. in salad dressings, of air incorporation into foams, e.g. in whipped cream, and of proteins in beer, precipitation of proteins producing cloudiness. Examples of stabilizers include gums and hydrocolloids.

Stachyose: Non-reducing tetrasaccharide found in legumes and other plants, hydrolysis of which gives two molecules of galactose, and one each of glucose and fructose.

Staining: Marking or discoloration with something that is not easily removed, such as penetrative dyes, pigments or chemicals.

Stainless steel: Type of steel which contains chromium. It is resistant to tarnishing and rusting. It is widely used in equipment and utensils for the food industry.

Staling: It is a process by which foods cease to be fresh or pleasant to eat. For example, bread becomes dry and hardened when stale, due to changes in the structure of starch.

Standardization: Process by which substances and procedures are made uniform. In the dairy industry, the term refers to adjustment of the fat content of milk to a given level. Milk from different batches is blended to the desired fat content. It is used especially to ensure the uniform quality of cheese milk.

Standards: Something used as a measure, norm or model in comparative evaluations; a benchmark or specification.

Stanol esters: Fatty acid esters of plant stanols (phytostanols). Commonly used in enrichment of foods such as spreads, yoghurt and food bars to produce products which may have a cholesterol lowering action. It reduces levels of total and low density lipoprotein cholesterol in blood by inhibiting absorp- tion of cholesterol in the intestine.

Stanols: Hydrogenation products of sterols which occur naturally in plants (phytosterols). It is less abundant than the corresponding plant sterols. Like plant sterols, stanols reduce levels of total and low density lipoprotein cholesterol in blood by inhibiting absorption of cholesterol in the intestine. Stanol esters are commonly used in enrichment of foods such as spreads, yoghurt and food bars to produce products which may have a cholesterol lowering action.

***Staphylococcus aureus*:** A food poisoning bacterium whose toxins are not destroyed by cooking. It is found in meat pies and meat products containing salt and in confectionery. The incubation period is 2 to 6 hours and the duration of the illness 6 to 24 hours. The symptoms are nausea, vomiting, diarrhoea, and abdominal pain without fever. There may be collapse and dehydration in severe cases.

Staphylococcus carnosus: Bacteria used as a starter culture for lactic fermentation in meat products such as salami and in vegetables such as cucumbers

Staphylococcus: Genus of Gram positive, facultatively anaerobic coccoid bacteria of the family Micrococcaceae. Occur on the skin and mucous membranes of humans and animals. *Staphylococcus aureus* may be responsible for food poisoning due to consumption of contaminated foods (e.g. meat and meat products, eggs, salads, bakery products and dairy products). *S. carnosus* is used as a starter culture in the manufacture of fermented sausages.

Staple crops: Food plants which provide the major source of energy and/or protein for most of the world's population. The five most important are rice, wheat, corn (maize), cassava (energy only) and beans.

Starch granules: Native structure of starch, comprising discrete aggregates of amylose and amylopectin. Arrangement of the starch polymers is highly organized and some crystalline regions are present due to strong interactions between amylopectin chains. Gran- ules also contain minor amounts of protein, lipid, ash and moisture. Starch granule size and composition vary between plant species and varieties.

Starch hydrolysates: Sugar syrups is produced by hydrolysis of starch slurries. Starch hydrolysis is commonly achieved by the action of acids, e.g. hydro- chloric acid, or amylases; degree of hydrolysis determines the saccharide composition of the syrups. Dextrose equivalent of a hydrolysate is a measure of the degree of hydrolysis relative to the dextrose (D-glucose) content, i.e. 100% dextrose equivalent denotes full hydrolysis. Glucose syrups, maltose syrups and maltodextrins are starch hydrolysates and are substrates for other starch-based sweeteners, such as fructose high corn syrups and crystalline sugars.

Starch syrups: Sugar syrups produced from starch by hydrolysis with acids or amylases. Sugar composition of the syrups is dependent on the degree of hydrolysis, which is measured in terms of the dextrose content of the syrup (the dextrose equivalent value).

Starch: Polysaccharide that is the main energy store of plants. It composed of molecules of amyloses and amylopectins. Amount of each polymer, which varies between plant species, influences the functional properties of starch, such as gel forming ability of starch pastes. In addition to its role in cereal flour or meal used as a base for breadmaking, and manu- facture of other bakery products and pasta, starch has many applications in foods, including as thickeners, anticaking agents, and coatings and binding agents. Starch is often chemically or physically modified in order to improve its applicability for food processing, e.g. to increase thermal stability or alter the texture. It is an odourless and tasteless polysaccharide which is the principal carbohydrate store in plant tissues, tubers and seeds and is an important constituent of the human, herbivore and omnivore diet.

Starter: A bacterial or fungal culture used to initiate fermentation, and thus flavour development, in fermented foods, e.g. *Lactobacillus* in milk to make yoghurt, *Rhizopus oligosporus* in tempeh or the lactic cultures used to replace the microorganisms and enzymes lost during milk pasteurization preparatory to making cheese

Statistical analysis: It is a group of mathematical techniques by which analytical results can be examined on the basis of probability theory.

Steaks: Thick slices of high-quality meat taken from the hindquarters of animal carcasses. They are usually cooked by grilling or frying.

Steal table: A stainless steel table or counter with openings to take food containers heated by steam or hot water. It is used to keep food hot prior to service.

Steam: Hot vapour into which water is converted when heated. It condenses in the air into a mist of miniature water droplets. It is used as a source of energy or in cooking of foods.

Steamed bread: Bread is prepared by baking dough in ovens which are heated to a constant temperature using closed pipes through which steam is passed.

Steaming: Cooking of foods by heating in steam produced from boiling water. The food to be steamed can be placed in steaming apparatus over boiling or simmering water in a covered pan. Steaming has advantages over boiling in terms of retention of flavour, colour, shape, texture and nutrients content of foods.

Stearic acid: A saturated fatty acid which contains 18 carbon atoms. Found abundantly in animals (in hard beef and lamb fats) and plants. Even though consumption of saturated fatty acids has been linked with an increased the risk of coronary heart diseases, data suggest that stearic acid may be neutral with respect to effects on serum cholesterol levels. It is responsible when esterified with glycerine for the hardness of the fat.

Stearin Triglycerides: It is present in both animal fats and vegetable fats; found particularly in solid fats, such as tallow and cocoa butter. It may also be synthesized by esterification of stearic acid with glycerol. Its uses include as emulsifiers and surface-finishing agents for chocolate and sugar confectionery. Also known as tristearin, glyceryl tristearate and octade- canoic acid 1,2,3-propanetriyl ester.

Stearoyl lactylates: Salts of the stearoyl lactylate anion prepared by reaction of stearic acid with lactic acid. The nature of the cation in the salt influences the functional properties of the lactylate, e.g. the sodium salt is soluble in water whereas calcium stearoyl lactylate is not. Its uses include as emulsifiers, dough conditioners and stabilizers.

Steel: Strong, hard grey or bluish-grey alloy made from iron with carbon and usually other elements. It is used widely as a structural and fabricating material. Also refers to a rod of roughened steel which is used for sharpening knives.

Steeping: Soaking of ingredients such as tea leaves, herbs and spices in water or other liquid until the flavour is infused into the liquid. The liquid used is usually hot. Also refers to soaking of barley or other cereals as part of the malting process, and during which imbibition occurs prior to germination.

Steers: Castrated, adult male cattle, which are widely used for beef production. Compared with bulls, steers are easier to handle and their carcasses are less af fected by stress related conditions, such as the DFD defect. However, steers grow more slowly, convert feed less efficiently and achieve lower carcass weights than bulls. Steer meat tends to be lighter in colour than bull beef.

Stellar: Fat substitute based on starches such as potato starch or tapioca. It is used as a bulking agent or texture modifier in salad dressings, condiments, sauces, bakery products and dairy products.

Sterilization: Destruction of all microorganisms and spores in or on a material, such as food, by various means, including the application of chemicals, heat, radiation or filtration. Conventional sterilization involves in-container sterilization, usually at temperatures between 115 and 120°C for 20-30 minutes. Commercial sterilization does not always meet this definition, because some harmless, heat resistant bacteria may still be present. The criterion for food sterility is a process which will ensure no surviving botu- lism bacteria or their spores. The common guideline is to use a multiple of 12 for the D-value (121°C) of *Clostridium botulinum*, or its equivalent.

Sterilized cream: Tinned cream sterilized at about 120°C. It has a slight caramel flavour and cannot be whipped.

Sterilized milk: Milk heated to 120°C for a few minutes or to just over 100°C for 20 to 30 minutes and aseptically sealed in a narrow necked bottle with a crown cork. This process kills all pathogenic and most other microorganisms and spores and the milk will keep for 1 to 2 years. The milk is off white in colour and has a characteristic caramel, boiled milk taste which is not to everyone's liking.

Steroids: Complex polycyclic lipids with a hydrocarbon nucleus. It includes sterols, bile acids, various hormones and saponins.

Sterols: Steroid **alcohols** found widely in animals and plants which have an aliphatic hydrocarbon side chain of 8-10 C atoms at the 17-β position and a hydroxyl group at the 3-β position.

***Stevia rebaudiana**:* Plants native to South America, the leaves of which have a sweet flavour. Analyse have revealed the presence of at least eight sweet com- pounds in the leaves, the most widely used of which is stevioside.

Stevioside: Terpenoid obtained from the leaves of *Stevia rebaudiana* that has 200-300 times the sweetness of sucrose and is stable at baking temperatures. However, it can also have an undesirable bitter and liquorice-like aftertaste. Stevioside sweeteners are commercially available, although they have not been approved as food additives in all countries, e.g. the USA. It is also known as steviosin.

Stew: A mixture of meat, usually tough and cut into small pieces, cooked slowly with vegetables, cooking liquor and flavourings on top of the heat source until tender, a process which can take many hours.

Stewing: Cooking foods slowly and for a long period of time in a small amount of liquid in a closed dish or pan to make a stew. Stews usually contain meat, vegetables and a thick soup-like broth. Stewing not only tenderizes tough pieces of meat but also allows the flavour of the ingredient components to blend.

Stickiness: One of the rheological properties; relating to the extent to which an item is cohesive or adhesive. This term also relates to the extent to which a food adheres to the palate during mastication.

Stiffness: One of the rheological properties; relating to the extent to which an item is stiff, i.e. firm and rigid. When stress is applied to a material, strain is produced in the direction of the stress; stiffness is the ratio of the stress divided by the strain.

Stir frying: **Cooking** method in which food is cut into small pieces and fried over a very high heat in a pan with a large surface area, e.g. a wok, with constant stirring. Very small amounts of oil or fat are used. It is associated particularly with Asian dishes.

Stirring: Manual or automated processing action involving circular movements of a utensil (e.g. a spoon) within a food mixture. Allows ingredients to become well mixed together and where required, distributes heat throughout the mixture.

Stock cube: A mixture of salt, MSG, vegetable and/or meat extracts and various flavourings pressed into soft crumbly cubes or blocks each individually wrapped. One stock cube will usually make 0.4 litres (0.75 pints) of stock. The salt and fat content and their flavour make them unsuitable for the finest cooking.

Stock syrup: A basic sugar solution with many uses containing 0.3 kg of granulated sugar and 0.1 kg of glucose per litre of water. Often 0.4 kg of granulated sugar with no glucose is used, but it may tend to cause crystallization.

Stock: Water flavoured with extracts from herbs, spices, vegetables and/or bones by long simmering. It used as a base for soups and accompaniments such as gravy and sauces. It is available commercially as liquid products or in dried form.

Stoneground flour: Flour which has been ground between a stationary and a rotating stone as opposed to rollers which are used to grind most flour nowadays. The slight frictional heating modifies the flavour of the flour.

Stoppers: Plugs for sealing holes, particularly for sealing the necks of **bottles**.

Storage life: The time for which a stored item remains usable.

Storage proteins: Proteins that accumulate within commodities such as cereals, seeds and legumes, and serve as nitrogen sources essential for germination. They usually occur in an aggregated state within mem- brane surrounded vesicles (e.g. protein bodies and aleurone grains), and are often built from a number of different polypeptide chains. Storage proteins have no enzymic activity. In cereals, storage proteins are de- posited in the endosperm; in legumes, they are deposited in the cotyledon. Seed storage proteins are synthe- sized in large quantities over a limited period of time. In dicots, these proteins are deposited in the embryo as well as in the endosperm of the developing seed. Storage proteins are deficient in several essential amino acids, and because they account for the majority of the protein, seed proteins generally have limited nutritional value. Storage proteins of different cereals have distinct structural characteristics that are responsible for their unique functional properties. For example, it is the composition of seed storage proteins (gliadins and glutenin) that dictates flour and dough properties.

Storage temperatures: Recommended storage temperatures for food are precooked, –3°C for up to 5 days; chilled, less than 5°C; eggs in shell, –10 to-16°C for up to 1 month; bakery produce, –18 to – 40°C, blanched vegetables, less than –18°C; apples and pears, –1 to +4°C.

Storage: Maintenance of commodities, for example fresh or processed foods, under controlled conditions for extended durations while maintaining quality. Undesirable quality changes that may occur during storage include changes in nutrient levels or colour, development of off flavour or loss of texture. Most foods benefit from storage at a constant, low temperature (cold storage or frozen storage) where the rates of most degradative reactions decrease and quality losses are minimized. However, some products, e.g. canned foods or dried foods, are processed in such a way that they may be kept at ambient temperature with no loss in quality. Careful control of atmospheric gases, such as oxygen, carbon dioxide and ethylene (controlled atmosphere storage), is important in extending the storage life of many products, such as fruits and vegetables.

Store: A designated room, cupboard, refrigerator or freezer where foods are kept in their purchased or partially processed condition prior to use.

Stores: Places, such as rooms or warehouses, where items such as foods are kept under controlled conditions for extended durations, for future use or sale.

Stoves: Devices for **cooking** or **heating** of foods and operated by burning fuel or using electricity.

Strain 0157:H7: A strain of the gut bacteria, *Escherichia coli*, found in hamburger-type minced beef which has picked up toxin-producing genes from *Shigella*. These toxins can destroy gut and kidney cells, leading to diarrhoea and kidney failure. The incubation period is 3 to 4 days. Common in the USA, and has occurred in Europe, Africa, Australia and Japan. The organism is killed by a temperature of 70°C for 2 minutes.

Strainers: Devices for straining liquids, semi liquids or dry ingredients to separate out any undesirable solid matter. These utensils have a perforated or mesh bottom, and are usually made from stainless steel, plastic or aluminium. Available in a variety of sizes, shapes and mesh densities.

Street foods: Fast foods sold by street vendors, particularly in developing countries. Often associated with high microbiological risk due to lack of hygienic food preparation and holding areas.

***Streptococcus*:** Genus of Gram positive, anaerobic coccoid lactic acid bacteria of the family Strepto- coccaceae. Occur on the skin, mucous membranes and in the gastrointestinal tract of humans and animals. *Streptococcus salivarius* subsp. *thermophilus* is used in starters for yoghurt and cheese (e.g. Emmental cheese and Parmesan cheese) manufacture. *S. agalactiae* and *S. uberis* may be responsible for mastitis in cattle.

S. pyogenes is the causative agent of strep throat and scarlet fever, which can be transmitted via contaminated foods (e.g. milk, dairy products, eggs and salads). Other species (*S. faecalis*, *S. faecium*, *S. durans*, *S. avium* and *S. bovis*) are responsible for diarrhoeal disease via ingestion of contaminated foods (e.g. meat products, milk and cheese).

Stress proteins: Proteins which are synthesized by an organism in response to environmental stress, e.g. heat shock, exposure to toxic substances, exposure to UV radiation or viral infection. Examples include the heat shock proteins. Produced to protect the organism from destructive consequences of the stress conditions encountered, but also play a role in normal cell physiology. Appear to act as molecular chaperones, assisting in the folding/refolding of other proteins. Prevent stress-induced protein aggregation by binding to surfaces exposed as a result of destabiliza- tion of protein structure. It may also be involved in repair of damaged proteins.

Stress relaxation: One of the rheological properties; relating to the process of stress decay, i.e. the stress response that is apparent after subjecting a material to a certain strain.

Stress resistance: Ability of an organism to withstand environmental stress.

Stress: Any unusual events or conditions which bring about physiological or behavioural changes in animals. In addition to fear and physical trauma, it includes en- vironmental factors such as cold, heat, humidity, light, sound and wind. The term stress also describes the re- sults of such events or conditions. Stress often occurs when animals are faced with unfamiliar, threatening or harmful situations. Transport to markets or abattoirs and poor pre-slaughter management of animals are widely recognized as causes of animal stress. Animal stress is not only an animal welfare issue, but is also associated with various defects in meat including the DFD defect and the PSE defect. Susceptibility to stress differs greatly between species, breeds, genders and individual animals.

Stretch: One of the rheological properties; relating to the ability of an item to be drawn out in length (extended).

Stretching: Making something that is soft or elastic longer or wider without tearing or breaking. An integral part of the manufacture of some cheeses, e.g. Mozzarella cheese, where the curd is stretched during processing.

Stroke: Sudden attack of weakness often affecting just one side of the body. Brain tissue is damaged due to blockage of a blood vessel as a result of thrombosis, atherosclerosis or haemorrhage. Severity of the stroke depends on the region of the brain affected and the extent of damage. Hypertension and hypercholes- terolaemia are major risk factors.

Strong flour: Flour of any type made from a hard wheat, usually containing between 11 and 12% protein which is mostly gluten. It is used mainly for bread and puff and choux pastry.

Structural genes: Genes that encode substances such as enzymes, structural proteins and RNA molecules, rather than genes that serves regulatory purposes.

Structured lipids: Lipids that have been modified to change the position and/or the composition of their constituent fatty acids. Typically triacylglycerols containing mixtures of medium and long chain fatty acids.

Stuffings: Savoury mixtures of chopped and seasoned ingredients which are either used to stuff poultry or other meat joints prior to roasting, or served as a meat accompaniment.

Stunning: Methods used to immobilize animals and birds before slaughter. It includes electrical stunning, captive bolt (projectile) stunning and CO_2 immobilization. Stunning is carried out immediately before bleeding; it aims to render the animal unconscious without stopping the action of the heart, which aids the bleeding procedure. Although stunning procedures involve some stress, they decrease stress responses when compared with bleeding without immobilization; consequently, stunning influences the properties and composition of meat. Overall effectiveness of stunning depends on the design and careful operation of the equipment used.

Styrene: Unsaturated liquid hydrocarbon, which is a by-product of petroleum manufacture. It is polymerized to make resins and plastics that are used as packaging materials for foods. There is concern about health hazards associated with migration of styrene monomers, dimers and trimers from packaging materials into some types of foods.

Suberin: Aromatic polymer similar to lignin, to which are aliphatic components such as ω-hydroxy acids, dicarboxylic acids and long chain alcohols are attached. It is found in waxes and in the cell walls of plants, and also deposited at wound sites.

Subtilisin: A protease obtained from various Bacillus strains of bacteria, used for flavour production from soya bean and milk protein (casein)

Succinylation: Introduction of succinyl groups into a compound or substance. It is usually achieved by reaction with succinic anhydride. Such modification is used to alter the physicochemical properties, functional properties or nutritional quality of substances such as proteins and starch. Succinylation has also been used to modify the properties of enzymes such as papain.

Succulence: One of the sensory properties; relating to the extent to which a product (e.g. meat or fish) is succulent or juicy. Degree of succulence of a food can be measured using a succulometer, in which samples are compressed to squeeze out juices, the volume of which can then be recorded.

Sucralose: Non-nutritive, artificial sweeteners produced by chlorination of sucrose; three hydroxyl groups on the sugar are substituted by chlorine atoms. Sucralose has 400-800 times the sweetness of sucrose, a flavour profile similar to that of sucrose and no aftertaste. Sucralose is stable at baking, pasteurization and extrusion temperatures. It is less reactive than sucrose and thus interacts less with components of foods or beverages to which it has been added.

Sucrose: Disaccharide comprising a molecule of glucose and a molecule of fructose. Sucrose occurs naturally and is extracted commercially from sugar cane and sugar beets to yield the crystalline sweetener marketed as sugar. Sweetness of sucrose is the milestone by which sweetness of all other sugars and/or sweeteners is compared.

Suet: Hard, white fatty tissue surrounding the kidneys of cattle and sheep. It is used in baking, frying and in the manufacture of tallow.

Sugar cooking: The process of modifying sugar (pure granulated sucrose) by dissolving it in a little water then heating it without stirring to various temperatures to produce various non-crystalline consistencies until finally there is caramelization which changes its flavour and chemical composition. The stages are: small thread (103°C), large thread (107°C), soft ball (115°C), hard ball (121°C), hard crack (157°C), and caramel (182°C). With practice these stages can be judged by the behaviour of the sugar between the fingers or when dropped into cold water or on a cold plate. In the

USA the stages recognized are thread (110°C), soft ball (112°C), firm ball (118°C), soft crack (132°C), hard crack (149°C) and caramel (154 to 170°C). The differences may be due to differences in the sugar. In France and Europe generally, the density of the sugar water mixture in °Be is often used but is rather inconvenient.

Sugar alcohols: Products formed when aldehyde or ketone groups of sugars are hydrogenated (reduced) to alcohol groups. Examples include sorbitol, mannitol and lactitol, produced by hydrogenation of glucose, mannose and lactose, respectively. It is also known as polyols.

Sugar batter biscuit method: A method of making biscuits, e.g. Langues de chat, by creaming butter and sugar, mixing in eggs to form a stable emulsion, then blending in the dry ingredients

Sugar beet: A variety of beet, *Beta vulgaris*, whose roots when mature contain a high concentration of sucrose which is extracted using water and crystallized from the solution sugar cane. The thick bamboo-like canes of a tropical grass, *Saccharum officinarum*, about 2 m long by 4 to 5 cm diameter Sometimes sold in pieces as a sweet for children to chew, but usually chopped and crushed to a give a liquid from which sucrose is crystallized

Sugar cane bagasse: Cane sugar processing waste that is composed of unextracted sugar and the remains of the sugar cane after milling. It is used as a fuel source, in feeds, as a substrate for microbial fermentation and for paper and board manufacture.

Sugar confectionery: Collective term for foods which have sugar as a principal component, e.g. chocolate, candy, fudges, jelly confectionery, sweets and toffees.

Sugar juices: Sugar containing solutions obtained by crushing sugar cane, or by hot water extraction of sugar beet cossettes. Sugar is crystallized from the juices following removal of impurities.

Sugar pans: Vessels, usually made of metal, e.g. steel plate, in which evaporation of sugar juices and crystallization of sugar are performed.

Sugar processes: Processes involved in the manu- facture of sugar, such as carbonatation, liming, evaporation and crystallization.

Sugar refineries: Factories where raw cane sugar is purified to produce granulated sugar.

Sugar substitutes: Chemicals used to mimic the flavour and applications of sucrose, e.g. sweeteners.

Sugar syrups: Concentrated aqueous solutions of sugars. Include syrups of individual sugars, such as glucose syrups and fructose syrups, and syrups extracted from specified sources, e.g. corn syrups and maple syrups.

Sugar The sweet-tasting water-soluble monosaccharides and disaccharides crystallized from plant juices (sucrose, fructose) or obtained, usually as syrups, by breaking down starch (glucose, maltose). It is alextracted from either sugar cane or sugar beets, purified to at least 98% purity.

Sugar: Commercial name for crystalline sucrose

Sulfitation: Use of salts of sulfurous acid, mainly sulfites, for applications including inhibition of bacterial growth, prevention of spoilage or oxidation, and control of browning in foods. Sulfites, which may be added as preservatives to packaged and processed foods, can cause a severe allergic response in certain individuals.

Sulfites: Inorganic salts of sulfurous acid that are used as food preservatives since they exhibit antimicrobial activity and antioxidative activity, and prevent enzymic browning. However, they are potentially cytotoxic and mutagenic, and may be allergenic to hypersensitive individuals. Hence, their use is regulated strictly.

Sulfur dioxide: Gas that is used in preservatives and bleaching agents, e.g. for beet sugar, and in stabilizers for vitamin C. Degrades vitamin B_1, and thus it is not recommended for use in foods rich in this vitamin.

Sulfur: Non-metallic element with the chemical symbol Essential in that it is a component of cysteine, methionine, vitamin B_1 and biotin. However, there appears to be no requirement for S in any other form.

Sulphites: Salts of sulphurous acid which is formed when sulphur dioxide is dissolved in excess water, used as food preservative. Common in the food industry are sodium sulphite E221 and calcium sulphite E226.

Sulphur dioxide: E220, a pungent and irritating gas which dissolves in water to form sulphurous acid or its salts the sulphites, hydrogen sulphites and metabisulphites, all of which liberate the gas in solution. It is one of the most common preservatives used in foods.

Summer sausages: Spicy, semi-dry fermented sausages, which are cooked and dried after fermentation. Commonly prepared from pork and/or beef, but may also be prepared from meat mixtures including chicken meat or turkey meat. Natural pigments, e.g. betalaines, may be used to simulate a cured meat colour in summer sausages. High quality is achieved by use of frozen concentrated lactic acid bacteria starters and control of lean to fat ratios in the meat. Varieties include landjaeger and thuringer.

Sunlight: Light emitted from the sun. it is used in sun drying and solar drying of foods.

Sunset Yellow: Orange monoazo dye used in artificial colorants. It is soluble in water or glycerol but only slightly soluble in ethanol. It has a reddish-yellow hue in concentrated solution that becomes yellow on dilution; the dye is colour stable at extrusion temperatures, pH 3-8 and in the presence of organic acids and alkalis commonly used in food processing, such as citric acid and sodium bicarbonate. Sunset Yellow is often blended with tartrazine and used in colorants for low fat spreads. Also used, in combination with other colorants, to colour a range of products, including bakery products, beverages (e.g. cola beverages), sugar confectionery and ice cream. It is also known as FD&C Yellow No. 6 and Food Yellow 3 and CI 15985.

Supercritical CO_2 extraction: Extraction process that uses supercritical carbon dioxide (CO_2) as the selective solvent. The polarity of CO_2 limits its use to extractions of relatively apolar or moderately polar solutes. Thus, a small amount of a polar organic solvent (e.g. methanol, acetonitrile, water), called a modifier or entrainer, is usually added to the supercritical fluid for extraction of more polar compounds. CO_2 is frequently used as the extraction solvent in supercritical fluid extraction because it is in a supercritical state at a relatively low temperature (31°C) and pressure (73 atmospheres), making it a suitable choice from an instrumental point of view. Extraction using supercritical CO_2 also avoids the use of dangerous or toxic organic solvents and the gas is easily removed by reducing the pressure.

Supercritical fluid chromatography (SCFC): Chromatography technique that uses a supercritical fluid as the mobile phase. It developed for analysis of substances not separated effectively using liquid chromatography or gas chromatography, including triglycerides and fatty acids.

Supercritical fluid extraction: Extraction process that uses a supercritical fluid as the selective solvent. A supercritical fluid is the mobile phase of a substance intermediate between a liquid and a vapour, maintained at a temperature greater than its critical point. Carbon dioxide (CO_2) becomes a supercritical fluid when held above its critical temperature and pressure. Instrumentation for supercritical fluid extraction con sists of a solvent supply, a pump, a cooler to cool the pump head, an extraction cell that is mounted in a ceramic heater tube, a heater controller to monitor the temperature of the extraction cell, a restrictor connected to the outlet of the cell, a restrictor heater, and a collection vial. The possibility of varying the solvent strength of the supercritical fluid by alteration of pressure makes supercritical fluid extraction extremely versatile in its applications.

Supercritical HPLC: HPLC technique in which a supercritical fluid is used as the mobile phase.

Supermarkets: Large self-service shops selling foods and household goods.

Surface active agents: Substances such as surfactants that reduce surface tension by interaction with non-mixing substances at phase boundaries.

Surface active properties: Functional properties relating to the ability of a compound to reduce the surface tension of a liquid, thereby increasing wettability or blending ability. Surfactants used as additives in the food industry have surface active properties.

Surface tension: Force on the surface of a liquid that makes it behave as if the surface has an elastic membrane. It caused by forces between the molecules of the liquid: molecules at the surface experience forces from below, whereas those in the interior are acted on by intermolecular forces from all sides. The surface tension of water is very strong, due to intermolecular hydrogen bonding. Surface tension causes a meniscus to form, liquids to rise up capillary tubes, paper to absorb water, and droplets and bubbles to form. It is measured in Newtons per metre.

Surfactants: Substances that concentrate at phase boundaries and reduce surface tension. It contains hydrophilic and hydrophobic regions which align at inter faces to promote mixing of phases. Above a particular concentration, the critical micellar concentration, surfactants form micelles which encapsulate one phase within the other. It is used to produce oil/water emulsions and for encapsulation of lipid soluble flavourings in processed foods. Emulsifiers, such as fatty acid esters, are surfactants as are sodium dodecyl sulfate (SDS) and Tween.

Surimi: Fish products comprising refined, stabilized, frozen fish mince. Refining and stabilization are achieved by washing repeatedly in fresh water to remove soluble protein, straining, pressing to restore water content to natural levels (approximately 80%), followed by incorporation of sugar, sorbitol and polyphosphates. It is used to make products such as kamaboko, fish sausages and sea food analogues such as imitation crabmeat.

Suspension cultures: Cell cultures maintained in liquid media, which grow in suspension rather than at- tached to a surface within the culture vessel. It can include cultures of plants, animals and microorganisms. It may be used in the manufacture of fermented foods and beverages (e.g. dairy products, beer and wines), but also for more specialized fermentation products, including some food ingredients.

Sweet corn: A variety of maize, *Zea mays saccharata*, with a higher proportion of sugars in the maturing kernels. Usually boiled or roasted and eaten directly from the cob after being buttered and salted. The kernels are often sold separately either canned or frozen. Unscrupulous suppliers will often sell maize cobs, *Zea mays*, which are less sweet and more chewy.

Sweet cream: Cream in which no acidity has developed and used to make sweet cream butter, the cream being ripened by warming only, with no addition of butter starters.

Sweeteners: Additives with a sweet flavour that are added to foods as sugar substitutes. Grouped ac- cording to the nutritional value of the sweetener into: nutritive sweeteners that may be metabolized and/or incorporated into the glycolytic pathway in cells to produce energy, e.g. starch-derived sweeteners; fruit-derived sweeteners, e.g. honeys, lactose and maple syrups; and non-nutritive or non-carbohydrate based sweeteners. Sweeteners may also be classified as natu- ral (existing in nature), e.g. carbohydrate-derived sweeteners, stevioside, thaumatin, glycyrrhizin, or artificial (produced by organic synthesis and not present in nature), e.g. sucralose, aspartame, ace- sulfame K, cyclamates, saccharin.

Sweetmeats: Any sweetened delicacy, especially sweets or, less commonly, cakes.

Sweetness: One of the sensory properties; relating to the extent to which a product tastes sweet. Sweet- ness of artificial sweeteners is often expressed in relation to that of sugar (sucrose).

Sweets: Small shaped pieces of confectionery, which are usually made with sugar or chocolate.

Swelling: An increase in the volume of a gel or solid associated with uptake of a liquid or gas.

Swells: Canned foods where the can bulges because of the gas produced by internal bacterial action.

Swiss cheese: A pale yellow cheese with large holes and a slightly nutty flavour that is made in Switzerland, e.g. Emmental cheese and Gruyere cheese. Also a US term for any hard cheese that contains relatively large bubbles of air.

Syneresis: Contraction of a substance, usually a gel, when allowed to stand and the resulting exudation of liquid from the gel. Control of syneresis is a key step for increasing curd yield and improving cheese quality. It is also important for yoghurt quality. Syneresis depends on a combination of specific and nonspecific interactions at the protein level, many of which also occur during curd formation.

Syrup: A concentrated solution of sugar (sucrose or sucrose with a third its weight of glucose) in water used as an ingredient in many sweet dishes and drinks. Concentrations can be measured in °Beaumé, kg of sugar per litre of solution or kg of sugar per litre of water, the last named being the most convenient. In these last units, a light syrup has between 0.1 and 0.4 kg, a medium syrup between 0.4 and 0.8 kg and a heavy syrup between 0.8 and 3.0 kg of sugar per litre of water. Stock syrup has 0.4 kg of sugar per litre of water.

T

Table jellies: Fruit flavoured sweetened desserts set with gelatin or similar gelling agents.

Table salt: A fine-grained, free-flowing salt suitable for being dispensed at the table from a salt shaker. It is pure sodium chloride with the addition of anti-caking agents, a function which was performed before additives by a few grains of dry rice.

Tagatose: Ketose monosaccharide comprising six carbon atoms (hexoses); an isomer of galactose. It has sweetness similar to that of sucrose but no calorific value, making it suitable as a low-calorie sweetener and bulking agent. Formed by bacterial fermentation using galactitol as substrate or produced from lactose via isomerization of galactose.

Taints: Sensory properties relating to the perception of off flavour or off odour in a product. Taints in foods can be related to, for example, warmed over flavour in ready meals or boar taint in pork products.

Tallow: Solid animal fats normally derived from cattle or sheep tissue, containing high levels of saturated fatty acids and monounsaturated fatty acids (triglycerides of stearic acid, palmitic acid and oleic acid). It is white, flavourless, odourless and solid at room temperature. It is usually prepared by heating suet under pressure in closed vessels. It is used for frying and in shortenings.

Tamper evident closures: Closures designed to ensure that any unauthorized interference is evident.

Tamper evident packaging: Packaging designed to ensure that any unauthorized interference is evident.

Tandoor: An unglazed barrel-shaped, deep clay pot with a top opening, used as an oven in India. It is heated by charcoal in the base or gas. The food, e.g. Nan bread, is either stuck on the inside or put on long skewers which are either rested over the opening or stood with their points at the base propped up against the rim.

Tandoori spice mix: A selection from cinnamon bark, nutmeg, dried ginger, dried red chillies, paprika, cardamom, dried garlic, cumin and coriander seeds, ground to a fine powder, the colour adjusted to a bright orange with food colourings and used for marinating and coating meat prior to tandoori cooking

Tandoori: A method of cooking on a greased spit in a clay oven. The oven is often a deep pot and the spits stand vertical from the bottom resting on the rim. It gives a crisp dry finish.

Tanks: Large storage chambers or containers, particularly for gases or liquids.

Tannic acid: Polyphenol which displays antimutagenicity, anticarcinogenicity and antioxidative activity. It is used as a flavouring food additive, a clarifying agent and a refining agen in beer, wine and cider and other natural brewed drinks, but may inhibit the absorption of dietary iron.

Tannins: A mixture of strong astringent and complex polyhydroxybenzoic acid derivatives found in plants, particularly tea leaves, red grape skins and the bark of trees, with the ability to coagulate proteins. It is antinutritional factors inhibiting the bioavailability of vitamins and minerals, and may be carcinogenic. However, it also possesses antimicrobial activity, antioxidative activity and antitumour activity.

Tarka: A style of finishing in Indian cookery involving the fierce searing of precooked food in ghee and seasonings to flavour and coat the surface of the food.

Taro: A tropical and subtropical plant, *Colocasia esculenta*, brought from Southeast Asia to the Caribbean as a staple food for slaves. The tubers are boiled, baked or roasted. Taro is a good source of potassium and fibre. Leaves contain carotenes and are rich in vitamin C. The corm is eaten cooked; if not well enough cooked, irritation of the mouth results due to oxalate crystals.

Tartaric acid: Organic acid present in fruits and isolated from potassium tartrate films produced as a byproduct in winemaking. Tartaric acid, as well as sodium and calcium tartrates, have many uses as food additives, including as flavourings (acidulants) imparting a fruity flavour, humectants, antioxidants, sequestrants and as part of a pH buffering system. Tartaric acid is also a substrate for production of the raising agent, cream of tartar (potassium hydrogen tartrate) which is an ingredient of baking powders. Systematic name is 2, 3-dihydroxybutanedioic acid.

Tartrates: It is salts of tartaric acid used as stablizers. Crystallization of tartrates in wines is a problem, since the wines are then generally considered unacceptable by consumers.

Tartrazine: Synthetic bright yellow pyrozole dye used in artificial colorants for foods and beverages. In aqueous solution, tartrazine shows high stability when exposed to acids and alkalis, moderate stability to light and heat (stable at extrusion and baking tempera- tures) and poor stability in the presence of ascorbic acid. It is synonymous with FD & C Yellow 5 and CI 19140. It is suspected of causing hyperactivity and allergies in some children.

Tarts: Open pastry cases made with shortcrust pastry, which are frequently baked blind (or empty) and then filled with sweet fillings such as fruits, jams or custards, or sometimes savoury mixtures, e.g. cheese or vegetables.

Taste panels: Groups of individuals, untrained or trained, used to sample products and assess their flavour, with a view to providing an insight into consumer preferences. Taste panels are used in research, product development and for purposes of evaluating new and competitive products, and are not restricted to evaluating flavour. Texture, colour and many other quality factors can be measured meaningfully.

Taste: Sensation produced by stimulation of the taste buds on the tongue. The tongue can distinguish five separate tastes (sweet, salt, sour, bitter and savoury/umami). It is often used as an alternative term for flavour.

Tawa: A circular, slightly dished steel or iron griddle with a handle, used for making Indian breads or, when placed below another dish, to reduce the heat especially on a difficult to control heat source.

Tea bags: Tea packaged in small portion-size permeable bags for easy preparation of tea beverages.

Tea beverages: Hot or cold beverages prepared from tea leaves or infusions.

Tea cake: A flat, slightly sweetened yeast-raised round bread bun with currants and mixed peel about 10 cm in diameter. Usually it is split, toasted and buttered.

Tea granules: Instant tea products comprising granules of dry tea extracts which are reconstituted into tea beverages on addition of water.

Tea seed oils: Vegetable oils extracted from the seeds of tea species such as *Thea sasangua* or *Camel- lia oleifera*. It is used as salad oils and cooking oils.

Tea smoking: A method of smoking meat or fish using a domestic steamer. The base is lined with foil and equal volumes of brown sugar, uncooked rice and an aromatic tea (e.g. Lapsang souchong) are put on the foil; the food is placed on the steaming rack with a tight fitting lid. It is then placed over a high heat for 2 minutes to generate smoke, on a low heat for 10 minutes and then let stand.

Tea tree oils: Essential oils distilled from leaves of *Melaleuca alternifolia*, a tree native to Australia and certain parts of Asia. Major constituents of the oils are terpinen-4-ol, 1,8-cineole and ϒ-terpinene. The oils have a warm, spicy flavour. Tea tree oils exhibit antimicrobial activity and are used as an antiseptic. Although more commonly used for their therapeutic properties, tea tree oils are also used as food flavour ings, including as a substitute for nutmeg.

Tea: Hot or cold beverages made by infusion of dry, prepared leaves of *Camellia sinensis* in water. The main types are black tea, in manufacture of which the fresh tea leaves have undergone fermentation before drying, and green tea, in which the fresh tea leaves have not undergone this fermentation. Oolong tea and pouchong tea have undergone partial fermentation, and are intermediate in character between green and black teas.

Tempe/Tempeh: A firm substance like cheese made by soaking and boiling soya beans, inoculating them with a fungus *Rhizopus oligosporus*, packing them into thin slabs wrapped in polythene or banana leaves pierced with holes and leaving to ferment. It is an easily digestible and well-flavoured source of protein and is usually shallow-fried. It is used as meat extenders or meat substitutes. It cooked in a variety of ways or added to dishes such as sauces, soups and casseroles. Some types of tempeh are made from other materials, e.g. bongkrek is made by fermentation of presscake of coconuts or coconut milk residue.

Temper: Measure of the degree of crystallization of cocoa butter in chocolate and the type of crystals present.

Temperature abuse indicators: Devices used to give an indication of whether products have been exposed to inappropriate temperatures that could cause damage during transport, distribution or storage. For example, indicators can be used to show whether frozen foods have been thawed during handling or storage; thawing during distribution can potentially affect quality and safety. Indicator devices often produce a visible, irreversible colour change to show when temperature abuse has occurred. Microbial indicators may also be used to detect exposure to temperature abuse, especially in animal carcasses. For example, poultry products that have been maintained at the correct tem- perature will have fairly constant counts of coliforms, while those that have been warmed will have higher counts.

Temperature probe: A small pointed rod about 3 to 4 mm diameter which measures the internal temperature of food in which it is inserted. It is used to determine the degree of roasting of meats and for checking that microwaved food is heated through.

Temperature: A measure of the intensity of heat such that heat always flows from a high intensity to a low one. It is measured in cooking on the Fahrenheit scale or Celsius scale using thermometers.

Tempering: It is to bring to a desired consistency, texture, temperature or other physical condition by blending, mixing, kneading or standing and as in to temper a pancake batter. It is a stabilization of chocolate by application of a melting and cooling process. Chocolate is tempered to stabilize the cocoa butter, a fat that can form crystals and cause bloom in the finished product. The classic tempering method includes the following stages: melting of the chocolate; working two-thirds of the melted chocolate on a marble slab with a metal spatula until it becomes thick; transferring the thickened chocolate back into the remaining melted

Tenderization: Mechanical or chemical processes by which meat can be made easier to cut or chew, so improving its tenderness. Mechanical methods break down tough fibres in the meat, usually through pounding. Pounders can be made of metal or wood, and can be a variety of shapes and sizes. Chemical methods that can also be applied to soften meat fibres include application of long, slow cooking, marination in acidic marinades and use of commercial meat tenderizers. Most meat tenderizers are composed primarily of papain, an enzyme extracted from papayas; they can also contain salt, sugar (usually glucose) and anticaking agents (usually calcium stearate).

Tenderizing: To break down the connective tissue of meat, octopus tentacles and the like by beating with a spiked mallet, by chemical treatment with acids, proteases or marinades, by hanging meat to allow natural enzymes to break down the fibres and by the correct application of heat at the right temperature

Tenderness: Sensory properties related to the extent to which a product, such as meat, is tender, i.e. soft, palatable and chewable. Tenderness can be measured using tenderometers.

Tenderometers: Instruments used to measure tenderness or the stage of maturity of produce, particularly peas, on the basis of the force required to cause shearing.

Tensile strength: Measure of the resistance that a material produces to a pulling stress (tensile stress); measured in Newtons per square metre.

Tensiometry: Measurement of surface tension.

Terpenes: Unsaturated hydrocarbons consisting of isoprene units found in many higher plants and essential oils. Typically, volatile compounds with pleasant odours used as flavourings. Terpenes are major components of citrus essential oils but, since they are not responsible for the characteristic flavour and readily oxidize and polymerize to produce unpleasant flavours, they are generally removed by distillation or solvent extraction. It is found to inhibit food spoilage yeasts.

Terpenoids: Volatile compounds found in plants and essential oils which are important for flavour. Certain terpenoids exhibit antioxidative activity, anticarcinogenicity and antimutagenicity.

Terpineol: Monocyclic monoterpene alcohol used in flavourings. It is found naturally in essential oils, citrus juices and wines, and can be produced by microbial transformation of limonene.

Terpinyl acetate: Flavour compound with antifungal activity that is found in essential oils.

Texture profile analysis: Analysis of the texture of a food in terms of mechanical properties, geometrical characteristics, and fat and moisture contents, at specific points during the mastication process.

Texture: Sensory properties relating to the feel of a surface or product, or the impression created by a surface structure or the general physical appearance of a surface. A major factor is affecting the mouthfeel and quality of a food.

Textured vegetable proteins (TVP): Plant protein products that are shaped and textured to form particles, or shaped pieces, such as chunks and strips, usually by spinning or extrusion technology. Typically formulated with added colorants and flavourings, and used as meat substitutes. Soy proteins are most commonly used, although other proteins, such as wheat gluten, can also be used.

Texturization: Process by which sensory properties of a substance are altered, e.g. to produce a particular feel, appearance or consistency.

Texturizers: Additives that improve the texture of foods. Examples include gums, hydrocolloids and polydextrose, used as fat substitutes to add body to low fat foods and calcium chloride, which is added to canned fruits and vegetables to maintain firmness of the product.

Texturizing agents: Substances which act as texturizers, improving the texture of foods.

Texturometers: Devices used to measure texture properties of foods, by analysis of physical attributes such as hardness, cohesiveness and crush resistance.

Thaumatin: Non-nutritive natural sweeteners iso- lated from fruits of *Thaumatococcus danielli*, a plant native to West Africa. The sweet flavour of *T. danielli* fruits is attributed to two proteins of approximately 22 kDa, designated thaumatin I and II. Both thaumatin proteins are approximately 1000-2000 times as sweet as sucrose (weight for weight). Commercial thauma- tin preparations are complexed with aluminium to improve their stability. Thaumatin is soluble in water and alcohols and is synergistic with acesulfame K and saccharin. Aqueous solutions of the sweetener have high thermal stability and are stable over the pH range 2-10. However, factors which influence thaumatin structure, e.g. reducing agents, affect its sweetness. Although used as a sweetener, thaumatin has a liquorice-like after taste. It is commonly used in flavour enhancers, e.g. in chewing gums. Synonymous with katemfe and sold under the trade name Talin.

Thawing: It is transition of an item from a frozen to an unfrozen state.

Theaflavins: Flavonoids which contribute significantly to the colour and flavour of black tea, and are used as markers of quality. It possessess antitumour activity and antioxidative activity.

Theanine: Amino acid found in tea. As well as improving the flavour of tea, theanine has a relaxing effect, improves learning ability and lowers blood pressure. It has also been found to help prevent D- galactosamine-induced liver injury in rats.

Thearubigins: Flavonoid **pigments** found in **tea** which contribute to the **flavour**, depth of **colour** and **body**.

Theobromine: Purine alkaloid similar to caffeine that is found in cocoa, chocolate, soft drinks and tea. It acts as a stimulant and may be toxic.

Theophylline: Purine alkaloid that contributes to the flavour of and is used as a marker of quality in tea, coffee, soft drinks and chocolate. It acts as a stimulant.

Thermal properties: Properties that influence the heating rate and response to heating of a material.

Thermal capacity: Thermophysical properties relating to the extent to which a material can retain heat.

Thermal conductivity: Thermophysical properties relating to the rate of conduction of heat through a material, measured in Joules per second per metre per Kelvin.

Thermal diffusivity: Thermophysical properties relating to the extent to which an item diffuses or spreads heat throughout its mass.

Thermal expansion: Increase in size (e.g. length, volume, surface area) of a body in response to heating. For liquids, expansivity observed directly is called the apparent expansivity, as the container holding the liquid will have expanded also with the rise in temperature. Absolute expansivity is the apparent expan- sivity plus the volume expansivity of the container.

Thermal processes: Processes involving heating that are used to produce desirable changes in products, such as protein coagulation, starch swelling, textural softening and formation of aroma compounds. Undesirable changes can also occur with application of thermal processes, such as losses of vitamins and minerals, and loss of fresh appearance, flavour and texture. Examples of thermal processes used in the food industry are: HTST processing; LTLT processing; electric heating; ohmic heating; microwave heating; and blanching.

Thermal processing: Application of heating methods to the processing of foods. Techniques in the category include: HTST processing; LTLT processing; electric heating; ohmic heating; microwave heating; and blanching.

Thermal stability: Thermophysical properties relating to the ability of materials to maintain stability when subjected to various temperatures of applied heat. If food ingredients or additives are heat stable, it is possible for them to be used successfully in products which have to be thermally processed. It is synonymous with heat stability.

Thermistors: Semiconductors used for measuring temperature on the basis that their electrical resistance decreases with increasing temperature.

Thermization: Heat treatment of foods at a temperature lower than that used for pasteurization, with an upper limit of about 65°C for 20s. Thermization is less severe for the product and associated microorgan isms than pasteurization.

Thermoanaerobacter: Genus of Gram positive, anaerobic rod-shaped thermophilic bacteria. Some spe cies are used in the production of thermostable proteinases.

Thermocouples: Devices for measuring or sensing a temperature difference, consisting of two wires of different metals connected at two points, between which a voltage is developed in proportion to any temperature difference.

Thermoluminescence: Luminescence produced by heating a solid substance. Caused by emission of pho- tons of light by free electrons and holes trapped in the solid.

Thermometer: Instruments for measuring and indicating temperature, typically consisting of a graduated glass tube containing mercury or alcohol which expands when heated and contracts when the temperature falls. Thermometers are tailored for different purposes. For example, specific instruments are available for use during the manufacture of sugar confectionery or cooking of meat (to ascertain that the meat has reached the desired degree of doneness), and also for temperature monitoring in freezers, refrigerators and ovens. It is essential for cooking especially for determining when meat is cooked, for oven, freezer, refrigerator, fat and sugar temperatures and for ensuring adequate defrosting and cooking in microwave ovens thermos flask A double-walled vessel made from silvered glass or stainless steel in which the space between the inner and outer wall is a vacuum, mounted in a protective case and fitted with a closure. The vacuum reduces transfer of heat by convection and the silvering reduces transfer by radiation, thus the contents may be kept hot or cold for a considerable period of time (heat is eventually transferred by conduction). Other types of insulation, e.g. Foamed polystyrene or polyurethane, are used for the same purpose especially in larger rectangular containers.

Thermophiles: Organisms, especially microorganisms that grow best at relatively high temperatures. Their optimum growth temperature is generally accepted as being above 50°C. Thermophilic bacteria are also known as thermophiles.

Thermophysical properties: Properties that influence the heating rate and response to heating of a material. Examples of thermophysical properties are thermal conductivity (the ability of a material to conduct heat) and specific heat (the ability of a material to store heat).

Thermostats: Devices that automatically regulate temperature to a specified value or range, or activate devices at a set temperature.

Thiamin: It is synonym for vitamin B_1 and vitamin F. It is member of the water soluble vitamin B group and active in the form thiamin pyrophosphate, a coenzyme for decarboxylation reactions in carbohydrate metabolism. It helps to maintain normal nervous system activity and regulates muscle tone of the gastrointestinal tract. Severe deficiency is clinically recognized as beriberi. Thiamin is found in unrefined cereals, beans, meat (especially livers, kidneys, hearts and pork), yeasts, potatoes, peas and nuts. Cooking losses can be as much as 50%.

Thickeners: Additives or substances, usually carbohydrates or proteins which may have been chemically modified that increase the viscosity of liquid foods. Unlike gelling agents, do not promote the formation of gels. Gums and starch are important thickeners in the food industry.

Thickening: It is a process of making or becoming thicker and usually more viscous by a variety of techniques including reduction, addition of cooked starches, gums, proteins such as white of egg or yolk or gelatine, by emulsifying a discontinuous phase into the liquid or by adding finely ground solids. For example, sauces are thickened using corn starch.

Thickness: As well as relating to consistency and viscosity, this term relates to measurement of the depth of a substance such as backfat on animal carcasses.

Thin layer chromatography (TLC): Chromatography technique in which sample components are separated as the sample travels, under the influence of a solvent, up an inert plate coated with a sorbant.

Thinning: In plant cultivation, removal of young plants to allow remaining plants more room to grow, or re- moval of selected fruits from a plant so that the other fruits can increase in size.

Thiobarbituric acid (TBA) values: Values (commonly abbreviated to TBA values) used for assessing lipid oxidation in foods and other biological systems, using thiobarbituric acid (TBA). Two molecules of TBA react with one molecule of malonaldehyde to produce a red pigment; the amount of pigment produced is measured spectrophotometrically. Extent of lipid oxidation, reported as the TBA value, is expressed as milligrams of malonaldehyde equivalents per kilogram of sample, or as micromoles of malonaldehyde equivalents per gram of sample. The TBA test may be performed di- rectly on the sample, its extracts or distillate.

Thixotropy: Property of a material that enables it to stiffen in a relatively short time on standing, while, upon agitation or manipulation, it can change to a very soft consistency or to a fluid of high viscosity, the process being completely reversible.

Thyme: Common name for plants native to Mediterranean countries of the genus *Thymus*, leaves and flowering tops of which are used as spices. The most commonly used variety is *T. vulgaris*; other spice varieties include *T. citriodorus* (lemon thyme), *T. zygis* and *T. serpyllum* (wild thyme). The predominant flavour compounds of thyme are thymol and carvacrol. Thyme extracts and essential oils are used as flavourings in the food industry.

Thymol: Phenolic derivative of cymene that is isomeric with carvacrol. It is present in essential oils, and exhibits antioxidative activity and antimicrobial activity.

Time intensity: Sensory analysis techniques are used to measure the intensity of a specific food attribute as a function of time. It is usually used to investigate the temporal behaviour of flavour compounds, such as sweet and bitter molecules, and the release of volatile compounds from foods. Such techniques are impor- tant in the reformulation of foods that results in struc- tural modification.

Time temperature indicators (TTI): Quality control devices designed to monitor and register accumulated temperature exposure of foods over time. It is used to alert the distributor or consumer to conditions which may render a particular food hazardous. Usually fixed to the product at the point of distribution and read by the receiving establishment. Time temperature indicators have been used on food rations employed in the armed services; as such rations may be subjected to high temperatures during transit and may also be stored and used in high-heat locations. On rations, each time temperature indicator consists of an outer reference ring and an inner circle. The inner circle darkens with time, and darkens more quickly as the temperature increases; therefore, the darker the circle, the less fresh the food. TTIs are classified according to working principle, type of response, origin, and application and location in food, and can be biological (microbiological and enzymic), chemical or physical systems.

Tin plate: Iron or sheet steel which is coated with the chemical element tin. It is used to make containers and cans for food storage and preservation.

Tin: Silvery-white metal, with the chemical symbol Sn. It is also refers to various metal containers used for food storage or preparation. Examples include lidded air-tight storage containers made of tin plate or aluminium, open-topped metal containers used for baking food, e.g. cakes, and sealed containers made from tin plate or aluminium used for preserving foods. In the UK, the term is often used as being synonymous with the term cans.

Titratable acidity: It is a measure of the total acidity in a sample, both as free hydrogen ions and as hydrogen ions still bound to undissociated acids. Determined by addition of a standardized base to the sample until a predetermined endpoint is reached. The endpoint may be assessed by a change in the colour of an indicator at a particular pH. This test can be used to determine milk quality and to monitor the progress of fermentation in cheese and fermented milk.

Titration: Technique in which reagent solution is added to the analyte until the reaction is complete. It is commonly based on oxidation-reduction or acid-base reactions, complex formation or precipitation. The end point of the reaction may be measured by a range of methods, including spectroscopy, change in colour of an indicator or changes in voltage or current passing between a pair of electrodes in the reaction solution.

Titrimetry: A method for determining the amount of an analyte by reacting it with a reagent solution until the supply of analyte is exhausted.

Toast: Sliced bread which has been placed near a fire or grill so that it becomes brown and crisp.

Toaster: An implement consisting of a spring loaded carrier for a slice of bread moving in a vertical slot with electrically heated elements on either side. The bread is placed in the carrier which when depressed, automatically switches on the elements and a timer which after a preset time releases the catch which holds the carrier down thus releasing the toast and switching off the elements.

Toasting: Cooking or browning of a food, e.g. bread, almonds or other nuts, by exposure to radiant heat.

Tocopherols: Members of the vitamin E group that are fat soluble and have antioxidative activity. In chemical terms, tocopherols are terpenoids. Four isomers exist that have vitamin E activity - α-, β-, ϒ- and k-tocopherols, the most important of which is α-tocopherol. Tocopherols are found in wheat germ oils, butter, egg yolks and leafy vegetables, and are important in the stabilization of cell membranes by protecting them from the damaging effects of oxygen free radicals, which are produced by various disease processes and toxic substances.

Toddy: Type of palm wine made in Southeast Asia by fermentation of sap of coconut palms (*Cocos nucifera*) or other palm species.

Toffees: Hard sugar confectionery products made from boiling together butter or vegetable oils, milk and sugar. Similar to caramels, although the temperature used to boil the ingredients is higher than that used for caramels.

Tofu: Soy curd product with a texture similar to that of compressed Cottage cheese and made like cheese by coagulation of soymilk and draining of the curd. It is a good source of proteins and B vitamins and available in firm, soft and silken forms that have different uses. Firm tofu is cubed and cooked or added to a variety of dishes. Other forms are used as substitutes for sour cream or yoghurt.

Tolerance: Maximum level of a given, potentially harmful, substance (e.g. mycotoxins, heavy metals, pesticides) permitted in foods or beverages.

Tomatine: Glycoalkaloid saponin present in high concentrations in green tomatoes. It is toxic to many fungi and bacteria.

Tomato concentrates: Products made by concentration of tomato pulps by processes such as reverse osmosis, evaporation and ultrafiltration. Uses include as flavour enhancers or in the manu- facture of tomato juices.

Tomato pastes: Rich concentrates produced from tomatoes by cooking, straining and reducing. Used as the base for sauces and soups. It is available commercially in cans, jars and tubes.

Tomato pulps: The soft, succulent parts of tomatoes or preparations made from them by mashing and concentration. Used in the preparation of many cooked dishes.

Tomato purees: Smooth, thick liquids produced from tomatoes by cooking and straining. It is used as the base for soups and sauces. Available commercially in jars, cans and tubes.

Tomato sauces: Condiments produced from tomatoes, seasonings and other additives. Tomato based sauces are used as toppings for pizzas and pasta dishes and in many other dishes, such as stews and casseroles.

Tomato seed oils: Vegetable oils extracted from tomato seeds produced as a by-product in canning of tomatoes. It is high in unsaturated fatty acids and used as cooking oils.

Tongues: A part of edible offal, often sourced from calves, lambs, oxen and pigs. Tenderness, flavour and texture vary with species and age of the source animal. Tongues may be sold fresh or brined; brining produces a pink colour and intensifies flavour. They are eaten hot or cold after boiling, skinning and slicing, or are used to produce meat products, such as brawn.

Tonic waters: Carbonated soft drinks containing bitter compounds such as quinine.

Top fermenting yeasts: Brewers yeasts which are non-flocculent and remain at the top of the beer during fermentation. It is commonly used for ale and other British style types of beer.

Toppings: Sweet or savoury food items such as sauces, pizza fillings or icings, used to garnish/top other foods.

Tortilla chips: Popular salted snack foods. Typically it is prepared by cutting extruded corn masa into chips, baking and frying. It is eaten in the same way as potato crisps or as an accompaniment to dips. Also available flavoured with a variety of flavourings.

Tortillas: Round, thin unleavened pancakes originating from Mexico which are traditionally made with corn flour and baked on a hot surface.

Torula yeast: Highly nutritious yeasts (*Candida utilis*) grown on media such as ethanol and sulfite liquor wastes. It is rich source of proteins and vitamins (especially B vitamins) and used as an animal feed supplement and a food additive.

Total quality management (TQM): Management philosophy geared towards continuous improvement of product quality to meet, exceed and anticipate customer requirements.

Total solids (TS): Total amount of **solids** in a product.

Total soluble solids (TSS): Total amount of **soluble solids** in a product.

Toughness: Sensory properties relating to the extent to which a product such as meat is hard to chew or cut due to its innate resistance, hardness and leathery texture. In a physical sense, toughness is defined as the energy required to propagate a fracture by a given crack area, generally derived from the area un- der a force-extension curve.

Toxicity: Quality or degree of being poisonous.

Toxicology: Scientific study of the nature, effects and detection of toxins, and the treatment of conditions caused by them.

Toxins: Poisonous substances, especially those that are produced by one living organism, and are poisonous to other living organisms.

Trace elements: Elements that are essential nutrients but are required only in minute amounts (mg or micrograms/day) by humans. Examples are chromium, copper, manganese and zinc.

Traceability: The ease with which origin or developmental history of something can be found by investigation.

Trade agreements: Treaties designed to facilitate trade between two nations or a group of nations. In the absence of trade agreements, many nations impose special taxes (tariffs) and take other actions to discourage importation of foreign goods. Trade agreements usually seek to reduce or eliminate such barriers.

Trademarks: Words or symbols established by use or legally registered as representing a product or com- pany. The term 'trade name' may sometimes be used to refer to a name that has the status of a trademark.

***trans* Fatty acids:** Fatty acids produced during the hydrogenation of fats and oils, which are found in foods such as vegetable shortenings, margarines and partially hydrogenated vegetable oils. It have several adverse effects on health, such as increased risk of coronary heart diseases, increased levels of cholesterol and low density lipoproteins, and reduced levels of high density lipoproteins.

Transgenes: Foreign genes introduced into the genomes of transgenic organisms early in development. Transgenes are present in both somatic and germ cells, and are inherited by offspring in a Mendelian fashion.

Transgenic animals: Genetically engineered animals or their offspring that contain genetic material from at least one unrelated organism inserted into their genomes.

Transgenic plants: Genetically engineered plants or their offspring that contain genetic material from at least one unrelated organism inserted into their genomes.

Translucency: Optical properties relating to the extent to which an object diffuses light passing through it, so that objects cannot be seen clearly.

Transmission electron microscopy (TEM): Electron microscopy technique in which the image forming rays are passed through or transmitted by the sample.

Transparency: Optical properties relating to the extent to which an item allows light to pass through it so that bodies can be clearly seen.

Triacylglycerols: Lipids composed of glycerol esterifed at all three of its constituent carbon atoms with one or more fatty acids. Triglycerides are com- ponents of natural fats and oils and have multiple uses in the food industry, including as emulsifiers, coatings and encapsulating agents. It is synonymous with triglycerides.

Trimming: Making an item neat by cutting away irregular or unwanted parts. In the food industry, usually it is applied to removal of fats from meat.

Triolein: Triglyceride of glycerol esterified with three molecules of oleic acid (9-octadecenoic acid). A natural component of fats and oils, triolein is used in the food industry in stabilizers and in solvents for flavourings and fat-soluble vitamins. It is also called glyceryl trioleate.

Trucks: Large motor vehicles designed to transport heavy loads. It is used in a wide range of applications, including transport of livestock to slaughterhouses, carriage of grain and other raw materials to processing facilities and transfer of processed foods from factories to retail premises. The term also describes vehicles used for carrying freight on a railway. Forklift trucks are vehicles with power operated horizontal prongs that can be raised and lowered and are used for transporting goods, especially those stacked on pallets, in warehouses and factories.

Trussing: It is a process of tying up the wings and legs of poultry carcasses in preparation for cooking. Skewers, thread, string or pins may be used. It helps the food to maintain a compact shape during cooking.

Trypsin inhibitors: Proteins found in a range of foods, including soybeans, peanuts, peas, lentils, and raw egg whites, which inhibit the activity of trypsin. It is denatured, and hence inactivated, by heating.

Tryptophan: Essential amino acid important in the synthesis of haemoglobin, plasma proteins and nicotinic acid.

Tuberculosis: Infectious disease most commonly caused by the bacillus *Mycobacterium tuberculosis* which is characterized by the formation of nodular lesions (tubercules) in the tissues. Tuberculosis is associated with poor living conditions, such as nutritional deficiency and inadequate housing. Transmission of tuberculosis is by inhalation of infected droplets. Treatment is by long-term administration of antibiotics.

Tubers: Swollen and fleshy underground stems of plants, usually high in starch. Include potatoes.

Tumbling: As well as being a process by which surface irregularities are removed from an item by rotating it in a tumbling barrel, this term also refers to a process by which the quality of meat can be improved. The mechanical action of tumbling alters the structure of mus- cle proteins. Tumbling can also be used to increase the rate of uptake of marinades by meat pieces.

Tumours: It is a growths in the body caused by the abnormal proliferation of cells. Some food components are thought to possess antitumour activity. Tumours may be benign (i.e. grow at one site only) or malignant (i.e. they destroy the tissue in which they arise and spread to other parts of the body). Benign tumours, which are covered by a capsule, are usually harmless but may become very large, exerting pressure on neighbouring tissues and producing severe effects. In malignant tumours, which are not enclosed by a capsule, cell division is rapid; cells show partial or complete loss of function and bear little resemblance to the tissue cells from which they originated. Malignant tumours cause extensive damage.

Turbidimetry: Measurement of turbidity of a solution, usually using a turbidimeter, an instrument that records the loss of intensity of a light beam passed through a solution containing suspended particles.

Turbidity: Optical properties relating to the extent to which a solution is turbid, i.e. cloudy or hazy. Turbidity in solutions is caused by the presence of finely suspended matter.

Turmeric: Common name for a plant native to Asia, *Curcuma longa*, the dried ground rhizomes of which are used as spices. Turmeric is deep yellow in colour due to the presence of curcumin, desmethoxycurcumin and bisdesmethoxycurcumin. Used in natural colorants, particularly in mustard, pickles and other spicy condiments, curry seasonings, and fats and oils. The predominant flavour compound of turmeric is turmerone. The majority of commercially available turmeric is cultivated in India, leading to the alternative name, Indian saffron. It is also known as CI natural yellow 3 and CI 75300. Extracts and essential oils of *C. longa* rhizomes are also used as colorants and flavourings.

Tyramine: Biogenic amine formed by microbial decarboxylation of tyrosine. It may be formed in foods such as ripened cheese, chocolate, wines and fermented foods. Consumption of contaminated foods can cause increased blood pressure and migraine.

Tyrosine: Non-essential amino acid which can be synthesized from phenylalanine in humans. It is an important precursor of adrenaline, noradrenaline, thyroxine and melanins. Tyrosine isomers can also be formed by ϒ- irradiation of phenylalanine and their detection can therefore be used as an indicator of irradiation of foods.

U

UHT milk: Milk heated by UHT treatment to prolong shelf life.it is also known as long life milk.

Ultra High Temperature (UHT) treatment: A heat treatment (direct or indirect) used to sterilize foods prior to packaging in which to kill all microorganisms that would otherwise spoil the product. Following UHT treatment, foods are filled into pre-sterilized containers in a sterile atmosphere. Food products processed by UHT treatment include liquid products (e.g. milk, some fruit juices, cream, yoghurt, wines, salad dressings), foods with discrete particles (e.g. infant foods, tomato products, some fruit juices and vegetable juices, soups), and foods containing larger particles (e.g. stews).

Ultracentrifugation: Centrifugation in centrifuges which have the ability to develop centrifugal fields of up to 100,000 times that of the gravitational field. Ultracentrifugation is generally used for analytical purposes, such as the determination of physico-chemical properties of food polysaccharides using sedimentation analysis.

Ultrafiltration: Selective membrane separation process, driven by a pressure gradient, in which suspended solids, colloids, emulsified solids such as fat-protein complexes, and dissolved macromolecules with molecular weight in the range 10,000-100,000 Da are retained by the membranes. Molecules that do not pass through the membranes constitute the retentate. Lower molecular weight dissolved materials that pass through the membrane under a driving force of relatively low hydrostatic pressure (1-10 bar) are the permeate. Ultrafiltration is generally used in the concentration and fractionation of large molecules from materials such as cheese whey and milk.

Ultra-heat treated: (A liquid) which has been rapidly heated to 132°C, held at this temperature for 1 to 2 seconds (up to 6 minutes for some products), then rapidly cooled, generally in a stainless steel plate heat exchanger through which the liquid flows as a very thin film. This treatment sterilizes the liquid without too much change in its flavour. However the process does not necessarily destroy enzymes which may induce rancidity after several months.

Ultrapasteurization: It is a process of heating foods, especially milk and liquid egg products, at a high temperature for a short time, sufficient to kill any pathogens present. It is used to extend the shelf life of the product without greatly affecting its nutritional properties. A typical process for ultrapasteurization of milk would involve heating at 280°F for at least 2 seconds. Ultrapasteurized products are aseptically packaged and stored under refrigeration.

Ultrasonics: The science and application of ultrasonic waves that have a frequency above those that are audible, generally defined as above 20,000 hertz.

Ultrasound: Sound or other vibrations having an ultra-sonic frequency. Generally, ultrasound is classified as any acoustic wave above the normal range of human hearing, i.e. above 20,000 hertz, but, in practice, the term usually refers to a much higher frequency used for a specific application.

Ultraviolet (UV) radiation: Electromagnetic radiation hav- ing a wavelength just shorter than that of violet light but longer than that of X-rays.

Unbleached flour: Flour which has not been treated with a bleaching agent to whiten the colour

Uncooked curd: Curd for cheese making which has not been heated to more than 40°C

Unleavened bread: Bread made from flour, water and salt without using a raising agent or incorporating air

Unleavened: Without any natural or chemical raising agent

Unpolished rice: Rice from which the bran and most of the germ has been removed. See also brown rice

Unsaponifiable matter: Substances present in fats and oils which are not glycerides and which are resistant to saponification with strong alkalies. Content varies among different types of oils and fats, and can thus be used as a source of information for their characterization and authentication.

Unsaturated fats: Fats, found at high levels in vegetable oils, that contain one or more carbon-carbon double or triple bonds. Thought to lower plasma cholesterol levels and reduce the risk of cardiovascular diseases when used to replace saturated fats in the diet.

Unsaturated fatty acids: Fatty acids containing one or more carbon-carbon double bond. Those that contain one double bond are termed monounsaturated fatty acids and include oleic acid, while those that contain two or more double bonds are termed polyunsaturated fatty acids and include linoleic acid. It is found at high levels in vegetable oils and fish oils, and thought to lower plasma cholesterol levels and reduce the risk of coronary heart diseases.

Unsaturation: State in which an organic compound contains double or triple bonds and thus shows in- creased capacity for reaction relative to saturated compounds. Used especially with respect to fats and oils. The degree of unsaturation refers to the number of double and triple bonds within the compound. This is expressed in terms of iodine values, determined by the weight of iodine absorbed by the substance under investigation. With respect to fats and oils, degree of unsaturation is important for their characteristics and health considerations, unsaturated forms having benefits with respect to blood cholesterol levels and risk of cardiovascular diseases development.

Use-by-date: Date mark used on packaged food which is intended to be consumed within 6 weeks. Under UK regulations this date may be changed by the manufacturer after the first date has expired.

Uthappam: A thicker variety of the South Indian dosa, with items such as vegetables, onions, tomatoes, chillies and coriander added to the batter before cooking rather than used as a stuffing afterwards

UV spectroscopy: Spectroscopy in which samples are identified on the basis of absorption of light of ultraviolet wavelength.

V

Vacuum cooling: It is a technique based on liquid evaporation which produces a rapid cooling effect in products containing free water. Suitable only where removal of the free water will not cause structural damage and where there is no barrier, e.g. a thick wax cuticle, to water loss. Subjecting suitable products to vacuum pressure allows part of the water contained in them to boil out at relatively low temperatures. Used successfully in reducing postharvest deterioration in fruits and vegetables, thus prolonging shelf life, during processing of some products, including liquid foods and bakery products, and rapid cooling of cooked meat, fish products and ready meals.

Vacuum drying: Removal of liquid from a solid mate rial while in a vacuum system, to lower the temperature at which **evaporation** takes place and thus prevent heat damage to the material. It may be used with frozen food.

Vacuum evaporation: Concentration technique in which the use of high temperatures is avoided by subjecting the substance to a vacuum, causing it to boil at a lower temperature. The process is performed in a chamber surrounded by a water jacket through which water is circulated to control temperature. Particularly useful for products where heat-induced protein de- naturation should be avoided, e.g. liquid egg whites and skim milk.

Vacuum flask thermostat: An electrically or mechanically operated instrument which monitors the temperature in an enclosure and if it deviates from a preset value increases or reduces the energy input to or abstracts more or less energy from the space to maintain the preset temperature. Commonly used in gas or electric ovens, refrigerators and freezers, deep fat fryers, slow cookers and the like.

Vacuum pack: A long-life pack in which food is sealed under vacuum in polythene or other clear plastic pouches. Also known as sous-vide when the contents are cooked or partially cooked food or meals for use in restaurants.

Vacuum packaging: Packaging process in which some or all of the air is removed from flexible or rigid containers before sealing. This form of packaging is used to preserve flavour, inhibit bacterial growth and prolong the shelf life of food.

Vacuum pans: Sealed devices that control the crystallization of solids from liquids by lowering the pressure within the sealed container. Vacuum pans are widely used for crystallization during the manufacture of sugar.

Vacuum: A space entirely devoid of matter or from which the air has been completely removed. In practical terms, a vacuum is an enclosed region of space in which the pressure has been reduced (below normal atmospheric pressure) sufficiently so that processes occurring within the region are unaffected by the re- sidual matter.

Valine: Essential amino acid important for growth. Good sources include soy meal, brown rice, Cottage cheese, fish, meat, nuts and legumes.

Valves: Mechanical devices, either manual or automatic, for controlling the passage of fluids through pipes or ducts.

Vanaspati: Grainy hydrogenated vegetable oils used as an alternative to ghee in India and Pakistan. Similar to margarines and often fortified with vitamin A and vitamin D.

Vanilla: Natural flavourings produced by curing of fully grown but unripe beans (pods) of *Vanilla planifolia* or *V. tahitensis*. Curing causes hydrolysis of glucovanilla to produce glucose and the flavour compound, vanillin. Glucose is then involved in nonenzymic browning via the Maillard reaction with bean proteins. Major vanilla producing countries are Mexico, Madagascar and Tahiti, each country producing vanilla with a distinctive flavour profile. Although vanillin is the main flavour component of vanilla it comprises only about 3% of the total flavour compounds and aroma compounds. Thus composition of minor flavour and aroma compounds is an important determinant of flavour. The flavour is mellow and the aroma fragrant and unmistakeable. It is used to flavour desserts and chocolate. Synthetic vanillin is widely available and is used in most manufactured goods. It has a heavier odour and a disagreeable aftertaste.

Vanillic acid: Phenolic compound produced as an intermediate in bioconversions of ferulic acid to vanillin. Also found as a pollutant in olive oil mills effluents.

Vanillin: Substituted phenol that is the main flavour compound of vanilla. Synthetic vanillin is also manufactured for use in flavourings. It is used as a cheaper alternative to vanilla in a wide range of foods, such as ice cream, bakery products, sugar confection- ery and beverages.

Vaporization: Process by which moisture or another substance is diffused or suspended in the air, becoming converted into vapour. Examples include the rapid change of water into steam, especially in boilers.

Vat: A large tub used for large-scale processing of cheese, pickles ot to hold or store liquids, and also used for fermenting and ageing wine and cheesemaking.

Vegan diet: Strict vegetarian diet which contains no animal foods of any kind.

Vegan foods: Vegetarian foods suitable for a vegan diet, i.e. excluding meat, eggs, milk, butter, cheese and all other animal foods.

Vegetable burgers: Patties made from mashed or chopped vegetables, sometimes also containing cereal or nut ingredients, eaten as an alternative to meat based burgers such as beefburgers. Commonly used ingredients include beans, mushrooms, on- ions and carrots. Spices and condiments are added to produce the desired flavour. Health benefits compared with meat-based burgers include low fat and sodium contents, little or no cholesterol content and increased dietary fibre levels. It is also known as veggie burgers.

Vegetable extract: A dark brown paste made from hydrolysed vegetable protein and vegetable flavourings, possibly mixed with yeast extract. It is used as a flavouring agent especially by vegetarians.

Vegetable fats: Lipid-rich vegetable products that is solid at room temperature. It may be produced by hydrogenation of vegetable oils. It is used in cooking and as food ingredients. Include cocoa butter, sal fats, shea nut butter and vanaspati.

Vegetable oils: Lipid-rich vegetable products that is liquid at room temperature. It is extracted from plant material including seeds, fruit or nuts. It is often contain phytosterols. It is used widely as cooking oils and salad oils and as flavourings. It include cottonseed oils, olive oils, sunflower oils, soybean oils and essential oils.

Vegetable pickles: Vegetables preserved in liquids such as brines or vinegar and eaten as an accompaniment to a meal. Examples include pickled onions and cucumber pickles.

Vegetable preserves: Vegetables that have been preserved by immersing in brines, vinegar or oils.

Vegetable proteins: Proteins sourced from vegetable tissue. It is preferred by some consumers due to health benefits. Quality of vegetable proteins, especially with respect to amino acids composition, varies according to source, but many plant breeding programmes have aimed to improve protein quality of individual crops. Legumes, particularly soybeans, are especially rich in protein. Textured vegetable proteins, usually derived from soybeans, are used as meat substitutes and meat extenders.

Vegetable pulps: Preparations made from vegetables by mashing the cooked flesh. It is used as ingredients in various dishes, such as soups, sauces and casseroles.

Vegetable purees: Vegetables that have been mashed, usually after cooking, to a smooth, thick consistency by various means, such as forcing through sieves or blending in food processors. Used as garnishes, side dishes or ingredients in dishes such as sauces and soups, or beverages.

Vegetable rennets: Enzymes sourced from plant materials that are used as substitutes for animal rennets in coagulation of milk for cheesemaking. It includes enzymes extracted from flowers of cardoons or curdle thistle (*Cynara cardunculus*).

Vegetable salads: Dishes prepared from a mixture of vegetables, raw or cooked, sometimes served in sauces or dressings.

Vegetable soup: A mirepoix of mixed vegetables sweated in butter, flour added and cooked out without colour, white stock, a bouquet garni and sliced potatoes added, simmered and skimmed for 1 hour, bouquet garni removed, the remainder liquidized, seasoned, consistency adjusted and served accompanied by croûtons.

Vegetables: Plants cultivated for an edible part, e.g. root, tuber, leaf or flower buds (as in broccoli and cauliflowers), or the edible parts of such plants.

Vegetarian diet: Diet based on plant foods, and which excludes meat and fish, and, in some cases, other animal foods. Lacto-ovo vegetarians consume dairy products and eggs, while those following a vegan diet consume no animal products at all. Vegetarianism is adopted for a variety of reasons, including ethical and religious beliefs as well as for nutritional/health benefits. The positive health effects reported for the diet have been attributed to relatively low contents of fats and cholesterol and the high contents of some vitamins and minerals. Inclusion of supplements in the diet may be necessary to prevent the risk of deficiency in vitamin B_{12} and some minerals, such as iron, zinc and iodine.

Vegetarian foods: Meat-free foods suitable for inclusion in a vegetarian diet. It include pasta, soy products, vegetable burgers and simulated meat substitutes. Much of the recent growth in the vegetarian food market has been fuelled by non-vegetarians who are keen to cut down on meat consumption and who perceive vegetarian foods as a healthy option.

Vegetarian: A person who avoids eating the flesh of once living creatures although some will eat animal products such as eggs, milk and milk products and fish.

Vending machines: Machines that dispense articles such as packaged foods or beverages, usually when a coin or token is inserted.

Vermicelli: A very fine thin pasta usually bundled up like a bird's nest and used in soups.

Veterinary inspection: Governmental surveillance of food producing animals to ensure a clean, wholesome, disease-free meat supply that is without adulteration. There are approximately 70 diseases that animals can transmit to man; for this reason, inspections are made by veterinarians at places of animal slaughter and at meat processing facilities.

Vicine: Antinutritional glycoside present in faba beans that can cause favism (haemolytic anaemia), thus limiting the nutritional value of these beans.

Video image analysis: Computer-aided technique in which photographic images of a sample are analysed to give information about particle structure and dispersion.

Vinegar: Fermented condiment that is essentially a solution of 4-6% acetic acid. The word is derived from the French, meaning sour wine, as vinegar was originally produced as an unwanted by-product of winemaking. Several types of vinegar with characteristic flavour profiles are produced by fermentation of various substrates, including apples, cider, grape musts, wines and malt. Vinegar fermentation is a 2- stage process. The initial alcoholic fermentation of sugars in the chosen substrate is carried out by *Saccharomyces* spp., while the acetic fermentation of the alcohol produced to acetic acid is carried out by acetic acid bacteria in the presence of O_2. Due to the acidic nature of vinegar, it is also used in acidifying agents and preservatives.

Vinyl chloride: Flammable, possibly carcinogenic, gas which is polymerized to make polyvinyl chloride and also used as a propellant in aerosols. It is synonym chloroethene.

Vinylidene chloride: Colourless liquid which is polymerized to make the thermoplastic material polyvinylidene chloride (PVDC). It is synonym 1,1- dichloroethene.

Viomellein: Mycotoxin produced by species of *Aspergillus* and *Penicillium*. It may be synthesized in stored cereals contaminated with these fungi.

Virulence: Capacity of pathogens to cause disease. Generally it is indicated by the severity of infection in the host.

Viruses: Non-cellular microorganisms that consist of a core of RNA or DNA enclosed in a protein coat and, in some forms, a protective outer membrane. It can live and reproduce only in susceptible living microbial, plant, human and animal host cells. Causative agents of many important diseases of humans, animals and plants.

Viscoelasticity: Rheological properties relating to the reaction of a product to a stress or strain, consisting partly of a viscous element and partly of an elastic one.

Viscometers: Instruments for measuring the viscosity of liquids. It is also called viscosimeters.

Viscometry: Measurement of viscosity of a liquid, usually performed with viscometers.

Viscosity: Measure of the ease with which a fluid can flow when subjected to shear stress, measured in Newton seconds per square metre or Pascal seconds. Low viscosity, e.g. that of a gas, allows flow through a fine tube to be quite rapid, whereas high viscosity (as with thick oils) makes motion sluggish. Viscosity arises from the intermolecular forces in a fluid (internal friction); the stronger these forces, the greater the viscosity. With a rise in temperature, attraction be- tween the molecules is reduced, enabling them to move more freely. Thus, golden syrup is more viscous than cream, which in turn has a higher viscosity than water.

Vision systems: Systems of visual feedback based on various devices, such as video cameras, photo cells, or other apparatus, allowing a robot to recognize objects or measure their characteristics. Vision systems are widely employed in quality control processes in the food industry.

Vital gluten: Wheat protein complex separated from starch in a wheat flour dough and dried. It is used to improve strength of bread dough.

Vitamin B_5: A water-soluble vitamin (pantothenic acid) destroyed by boiling, found in all animal and plant tissues, especially poultry, liver, fish, eggs, potatoes and whole grains. It is converted to co-enzyme A in the body and as such is involved in all metabolic processes. There is no known toxicity. Lack of the vitamin can cause headaches, fatigue, impaired motor coordination, muscle cramps and gastrointestinal disturbance. Severe deficiency caused the burning feet syndrome observed in prisoners of war in East Asia during World War II.

Vitamin B12: A cobalt-containing water-soluble vitamin (cobalamin) which together with folic acid has a vital role in metabolic processes and in the formation of red blood cells. It is responsible for the general feeling of well-being in healthy individuals. It is normally found only in animal products, particularly ox kidney and liver and oily fish. Vegetarians and vegans should take supplements which are produced by a fermentation process. Lack of vitamin B12 which may be due to its absence in the diet or poor absorption in the gut can cause a form of anaemia (Addison's pernicious anaemia). It is normally prescribed together with folic acid.

Vitamin A: A long-chain fat-soluble alcohol which exist in various forms of which the most active is retinol which is found in animal tissues. Precursors of the vitamin are widely distributed in vegetables as carotenes which are transformed in the intestinal wall into Vitamin A. It is concerned with the integrity of epithelial tissues (skin and mucous membranes) and of the retina, especially for low light conditions, and is also needed for the correct functioning of many body cells. The two vitamin A forms are: retinols, which predominate in mammals and marine fish; and dehydroretinols, which predominate in freshwater fish. Vitamin A is present in yellow and green leafy plants as provitamin A, of which there are several forms. The most important ones in human nutrition are the carote- noids, α- and β-carotene and cryptoxanthin. These are converted to the active vitamin in the intestinal wall and liver. Richest sources of preformed retinols are fish liver oils, egg yolks and fortified milk. Biologically active carotenoids are found in dark green leafy vegetables and yellow fruits and vegetables, such as squashes and carrots. In humans, common signs of vitamin A deficiency are poor growth, lowered resistance to infection, night blindness and rough scaly skin. Severe deficiency leads to kera- tomalacia and xerophthalmia.

Vitamin antagonists: Antinutritional factors which are present in some natural foods and do not function as vitamins, even though they are chemically related to them. As a result, they cause vitamin deficiencies where the body is unable to distinguish them from true vitamins, and incorporates them into essential body compounds.

Vitamin B complex: The whole complement of B group vitamins.

Vitamin B group: Group of water soluble vitamins generally found together in nature and basically related in function, although unrelated chemically. These include vitamin B_1 (thiamin), vitamin B_2 (riboflavin) the vitamin B_6 group (pyridoxine, pyridoxal and pyridoxamine), the vitamin B_{12} group (the cobalamins), nicotinic acid (niacin), folic acid (pteroylglutamic acid), pantothenic acid and biotin.

Vitamin B_1: A water-soluble vitamin which must be taken daily. It maintains normal carbohydrate metabolism and nervous system function and is found in high concentration in yeast and the outer layers and germ of cereals. Other major sources are beef, pork and pulses. It has no known toxicity but colours the urine a bright yellow. Lack of vitamin B1 causes beri-beri, the first deficiency disease to be recognized which led to the discovery of vitamins. Also called aneurin, thiamine

Vitamin B_{12}: It knowas as cyanocobalamin and member of the vitamin B group, found in foods of animal origin such as livers, fish and eggs. Vitamin B_{12} is the coenzyme for methionine synthase (EC 2.1.1.13), an enzyme important for the metabolism of folic acid, and methylmalonyl coenzyme A mutase (EC 5.4.99.2). Absorption of this vitamin requires the presence of an intrinsic factor. Failure of absorption, rather than dietary deficiency, is the major cause of pernicious anaemia.

Vitamin B_{13}: It has synonym as orotic acid and an intermediate in the biosynthesis of pyrimidines, and growth factor for some microorganisms.

Vitamin B_2: A water-soluble vitamin (riboflavin), essential for metabolic processes and for cell maintenance and repair. It is stored in the liver, kidneys and heart muscles. It is widely distributed in all leafy vegetables, in eggs, milk and the flesh of warm blooded animals. Lack of vitamin B2 causes soreness of the lips, mouth, tongue and eyelids. It has no known toxicity.

Vitamin B_3: A water-soluble vitamin which occurs in three forms, niacin, nicotinic acid and nicotinamide. The first two can cause skin flushing and should only be taken by diabetics and persons with peptic ulcers under strict medical supervision. Both niacin and nicotinamide forms are essential for bodily health. They are widely distributed in foodstuffs. Meat, fish, wholemeal flour and peanuts are major sources. Maize contains the vitamin in a non-absorbable form and it is for this reason that it is treated with lime which makes the vitamin available as well as improving the taste. Lack of the vitamin causes pellagra summarized as diarrhoea, dermatitis and dementia. It can also cause severe lesions when the skin is exposed to light.

Vitamin B_6: A water-soluble vitamin which consists of three compounds, pyridoxal, pyridoxol (also called pyridoxine) and pyridoxamine, each with a different function. They are found in low concentration in all animal and plant tissues especially fish, eggs and wholemeal flour. They are involved in protein, fat and carbohydrate metabolism and in brain function. Lack of the vitamin can cause lesions around the eyes, nose and mouth, peripheral neuritis and, in infants, convulsions. It is not recommended that supplements exceed 300 mg per day. All are equally biologically active.

Vitamin C: A water-soluble vitamin (ascorbic acid) which is synthesized in the bodies of most animals except humans. It obtained from vegetables and fruit. It is essential for growth of bones and teeth, maintenance of blood vessel walls and subcutaneous tissues, and wound healing; dietary deficiency results in scurvy, the effectiveness of the immune system and is thought by some to play a role in the prevention of cancer. Vitamin C deficiency results in scurvy, a disease still found in the poor and the old. Large amounts (up to 8 g daily) are recommended by some doctors and more famously by Linus Pauling, the chemistry Nobel prize winner. It is an an antioxidant nutrient present in a wide range of foods. It is used in food additives, with applications in food antioxidants and bakery additives. It has no known toxicity

Vitamin D: A fat-soluble vitamin (calciferol) whose main function is regulating calcium and phosphate metabolism (bone formation and repair). It is not normally present in nature but its precursors or provitamins, vitamin D2 and D3 are found in milk, cheese, eggs, butter, margarine (fortified) and especially oily sea fish. These provitamins are converted in the body to vitamin D by the action of sunlight. Lack of vitamin D causes rickets (deformation of the bones) in children and liability to fracture in adults. The vitamin is toxic in excess, 30,000 IU for adults and 2,000 IU for children. It is group of several related sterols. The most important members are vitamin D_2 (ergocalciferol or calciferol) and vitamin D_3 (cholecalciferol). The former is synthesized by irradiation of the plant provitamin ergosterol, and the latter is produced from the provitamin 7-dehydrocholesterol (found underneath the skin) on exposure to UV light from the sun. Vitamin D is also considered to be a prohormone. Fish liver oils and foods fortified with vitamin D are the major dietary sources; smaller amounts are found in livers, egg yolks, sardine and salmon. Severe deficiency in children results in rickets; deficiency in adults leads to osteomalacia.

Vitamin D2: A precursor of vitamin D. it is also called ergocalciferol. It is synthesized by **irradiation** of the plant provitamin **ergosterol**.

Vitamin D3: A precursor of vitamin D. and also called cholecalciferol and one of the group of sterols which constitute vitamin D. Fat soluble vitamin necessary for formation of the skeleton and for mineral homeostasis. It is produced on exposure to UV light from the sun from the provitamin 7- dehydrocholesterol, which is found in human skin.

Vitamin E: A fat-soluble vitamin (tocopherol) found in small quantities in soya beans, other seeds, butter, margarine (fortified), vegetables, whole grains, eggs and liver. It exists in several forms indicated by a Greek prefix, the alpha form being the most potent. It is a very powerful antioxidant and is vital for normal procreation and for cell processes. There is some evidence that it has anti-cancer activity. Natural sources are the best and though it has no known toxicity it may affect some medical conditions and supplements should only be taken on medical advice. Deficiency diseases in otherwise normal humans have not been reported.

Vitamin K: Group of fat-soluble vitamins essential for production of prothrombin and several other proteins involved in the blood clotting system, and the bone protein osteocalcin. Deficiency causes impaired blood coagulation and haemorrhage; vitamin K is sometimes called the antihaemorrhagic vitamin. Two groups of compounds have vitamin K activity: phylloquinone, found in all green plants; and a variety of menaquinones synthesized by intestinal bacteria. Dietary deficiency is unknown, except when associated with general malabsorption diseases.

Vitamin K: This vitamin relates to a group of chemical quinones, some fat-soluble and others water-soluble, which are synthesized by human gut microorganisms and are found in abundance in brassicas and spinach and in moderate concentration in tomatoes and pig's liver. Lack of the vitamin causes haemorrhages especially in new born infants. It is usually only administered under medical supervision.

Vitamin K_1: It is synonym for phylloquinone. Fat-soluble vitamin found in all green plants. Especially abundant in alfalfa and green leafy vegetables. Essential for production of prothrombin, and several other proteins involved in the blood clotting system, and the bone protein osteocalcin. Deficiency causes impaired blood coagulation and haemorrhage. Two groups of com- pounds have vitamin K activity: phylloquinones; and a variety of menaquinones synthesized by intestinal bacteria.

Vitamin K_3: It is synonym for menadione. Synthetic compound with vitamin K activity, used in prevention and treatment of hypoprothrombinaemia, secondary to factors that limit absorption or synthesis of vitamin K. Two to three times more potent than naturally occurring vitamin K.

Vitamin P: Group of plant bioflavonoids, including rutin, naringin, hesperidin, eriodictin and citrin, which affect the strength of capillaries in the body. Bioflavonoids are found as natural pigments in vegetables, fruits and cereals. In addition to their effect on capillary fragility, it is claimed that bioflavonoids function as follows: they are active antioxidative compounds in foods; they possess a metal-chelating capacity; they have a synergistic effect on ascorbic acid; they possess bacteriostatic and/or antibiotic activity; and they possess anticarcinogenic activity.

Viticulture: Cultivation of vines for production of winemaking grapes or table grapes.

Vitrification: Phenomenon whereby a substance is cooled rapidly to a low temperature such that the water it contains forms a glass-like solid without undergoing crystallization. The temperature at which the transition into a glassy solid occurs is the glass transition temp. Glass formation can result in stabilization of non-equilibrium systems, including most foods. In the glassy state, physicochemical deterioration is inhibited, effectively preserving the system. Vitrification temperature can be used as an indicator of food safety and storage stability.

Vodka: Spirits originating in Russia and northeast Europe, made from grain or potatoes. Generally it is rectified to have neutral flavour and aroma, but some types contain added flavourings.

Volatile compounds: Compounds that are readily vaporized. It often has a characteristic aroma and is therefore often flavour compounds and aroma compounds.

Volatile fatty acids: Fatty acids that, apart from being present in some foods, are produced by bacteria in the human intestine and the rumen of cattle from undigested starch and dietary fibre. To some extent, they can be absorbed and used as a source of energy. Volatile fatty acids formed in the colon may show anticarcinogenicity.

Voltammetry: Electrochemical technique in which the relationship between voltage and current flowing between electrodes in a reaction solution is measured. It utilizes a working electrode, where the reaction occurs, an auxiliary electrode for current flow and a ref- erence electrode that is used to measure the potential of the working electrode.

Volumetric analysis: Titration technique based on measurement of the volume of reagent required to react completely with the analyte.

W

Wafer: A thin, crisp, unsweetened biscuit with a papery texture made by cooking a batter between hot plates, served with ice cream and sometimes sandwiched together in several layers with a sweet or savoury cream filling to form a wafer biscuit

Waffles: Light, crisp, golden brown, indented raised cakes leavened with baking powders or yeasts and typically baked in a special waffle iron, which cooks both sides simultaneously. It is often consumed as a breakfast food, acompanied by butter or maple syrups or cold with whipped cream or ice cream.

Walnut oils: Relatively expensive oils extracted from walnuts. The distinctive nutty flavour and aroma make them popular for use in salad dressings, drizzling on to cooked foods and in cooking. Sometimes used as an alternative to olive oils. To prevent development of rancidity, walnut oils are best stored in a cool, dry location, out of direct sunlight.

Walnuts: Nuts produced by trees of the genus *Juglans*, the most economically important species being *J. regia* (common or Persian walnuts), *J. nigra* (black walnuts) and *J. cinerea* (butternuts or white walnuts). Ripe nuts are rich in vitamin E and B group vitamins, while younger fruits also contain vitamin C. Used as dessert nuts, and as ingredients in confectionery, bakery products and ice cream. Oils extracted from the nuts contain a high proportion of unsaturated fatty acids and have a range of food uses.

Warehouses: Large buildings in which raw materials or manufactured goods are stored.

Warmed over flavor: It is a characteristic off flavour primarily associated with cooked meat and poultry meat in chilled ready meals and other cook chill foods. In cooked meat and poultry held at chilled storage temperatures, this stale, oxidized flavour becomes apparent within a short time (48 hours), particularly if the product is stored under air. Modified atmosphere packaging under low oxygen levels helps to delay the onset of oxidative warmed over flavour.

Warming: The process by which an item is heated slightly to the point of being warm.

Washed curd: Milk curds which have been separated from the whey then steeped in cold water one or more times. This lowers the acid content of the curds and gives a coarser cheese.

Washed-rind cheese: A surface-ripened cheese which relies on bacteria on the surface to develop the flavour. They are frequently washed to discourage mould growth and to encourage the growth of bacteria and usually have an orange-red rind and a pungent smell although the paste tends to be sweet to the taste. They are normally small to give a high surface to volume ratio.

Waste water: Unusable, discarded water (effluents) resulting from processing procedures. In the food in- dustry, waste water is commonly produced by breweries, dairies, distilleries, olive oil mills and palm oil mills. It must be disposed of safely, often after treatment, to minimize pollution.

Wastes: Unusable, unwanted or discarded materials. In the food industry, wastes can result from application of processing procedures, and consist of solids such as pomaces, feathers and sludges. By recycling, some materials in wastes can be reclaimed for further use.

Water activity (a_w): A measure of the effective water content (from the point of view of a microorganism) of a food, not necessarily related to the actual water content (% by weight). It is defined as the ratio of the pressure exerted by water vapour in equilibrium with the food to the pressure exerted by water vapour in equilibrium with pure water at the same temperature as the food. (Both increase with temperature but the ratio remains roughly constant.) A food containing no water has a water activity of zero and pure water has a value of 1.0. Freezing reduces water activity by turning liquid water into ice. In foods, it represents water not bound to food molecules; the level of unbound water has marked effects on the chemical, microbiological and enzymic stability of foods.

Water binding capacity: Extent to which a substance can bind water.

Water biscuit: A thin crisp plain biscuit made of flour, salt and water only, similar to a cream cracker and usually eaten with cheese

Water concentrations: The highest water concentrations (% by weight) at which microbial spoilage will not occur are approximately 13 to 15% for wheat flour, 14 to 20% for dehydrated vegetables, 10 to 11% for dehydrated whole egg and 15% for fat-free meat.

Water holding capacity: Extent to which a substance can hold and retain water. It is related to the solubility of the sample.

Water ices: Frozen sugar confectionery made from water and sugar and flavoured with fruit juices, fruit purees or other fruit flavourings. It is used to make some types of ice lollies.

Water vapour: Water that is in its gaseous state, especially when below its boiling point.

Water: Colourless, odourless and tasteless liquid with the chemical formula H_2O, which is essential for plant and animal survival. It is widely drunk as a beverage, usually after some form of disinfection. It is used in the food and beverage industries in many ways, including as an ingredient, in the form of process water, and in cooling and heating systems.

Wateriness: One of the sensory properties; relating to the extent to which a product is watery, i.e. runny and wet.

Wax coatings: Wax-based materials used to coat and preserve the quality of fruits and some types of cheese.

Waxes: White translucent materials including bees-wax, but also a wide variety of similar viscous substances, such as carnauba wax. It is used as coatings for foods or to make candles and polishes.

Weaning foods: Infant foods used during the transition from consuming solely human milk or infant formulas to introduction of a mixed diet. Types of weaning food differ widely between cultures, but initial weaning foods are frequently based on cereals, of a puree-like consistency, and are introduced individually in order to detect allergies to particular foods.

Weaning: It is a process of gradually replacing mother's milk or milk substitute with other types of food in the diet of an infant or other young mammal. For infants, weaning foods are initially of a puree-like consistency and are often based on cereals, but other textures and types of food are introduced as the process proceeds.

Weevils: Common name for various **insects** of the family Curculionidae. It is also known as snout **beetles**. It is often highly destructive pests of crops and stored cereal grains, e.g. the **alfalfa** weevil (*Hypera postica*), the **grain** weevil (*Sitophilus granarius*) and the **rice** weevil (*S. oryzae*). Larvae of some species can be destructive to **fruits**, **nuts** and **grain**.

Weighing machines Devices, also called scales, used to determine the weight of an object. The simplest weighing mechanism is the equal-arm balance, which consists of a bar with a pan hanging from each end and a support (fulcrum) at the centre of the bar. Precision balances used in scientific laboratories can measure the weight of small amounts of material down to the nearest 1 millionth of a gram. Such weighing machines are enclosed in glass or plastic to prevent wind drafts and temperature variations from affecting the measurements. Electronic scales, which use electricity to measure loads, are faster and generally more accurate than their mechanical counterparts; in addition, they can be incorporated into computer systems, which makes them more useful and efficient than mechanical scales.

Weighing: It is a process of determining the weight of an object of food items or sample.

Wet milling: It is a process for separation of a substance into its constituent parts by a combination of chemical and mechanical means. It is used mainly in processing of corn, but can also be applied to other cereals such as sorghum, wheat and rice. Cereals are steeped in water with or without sulfur dioxide to soften the kernels before removal of the germ and separation of the other components. The main product is starch, which can be further processed in the case of corn to manufacture sweeteners or ethanol. Other products include fibre, gluten and oils, such as corn fibre oils.

Wettability: One of the physical properties; relating to the ability of a solid to absorb a liquid, such as water, as it spreads over the surface of the solid.

Wheat bran: Protective outer layer of the wheat grain which is removed from commercial flour by bolting or sifting. It is added to foods such as breakfast cereals or bread as a source of fibre.

Wheat bread: Bread made from wheat flour. White wheat breads are made from finely sifted wheat flour, while whole wheat bread is prepared by incorporating the fibre-rich outer layers of the wheat grain.

Wheat breadmaking: It is a process by which bread is made from wheat.

Wheat classification: Wheats are classified either as hard vitreous, hard mealy, soft vitreous or soft mealy.

Wheat composition: Protein 8 to 15% and fat around 2%. It is the protein type and not the percentage composition which is the underlying cause of hardness in wheat.

Wheat dough: Unbaked, thick, plastic mixture of wheat flour and a liquid, such as water or milk. It may contain yeasts or baking powders as leavening agents. It used predominantly to make bread; dough used to make other products, e.g. pizzas, biscuits, noodles, may vary in composition from bread dough.

Wheat flakes: Partially boiled cracked wheat crushed between rollers then dried, lightly toasted and used in muesli and as a breakfast cereal

Wheat flour: Product resulting from grinding wheat grains. Wholemeal flours are obtained by grinding whole wheat grains, while white flour is produced by separating wheat germ and wheat bran from the endosperm. It is used to prepare a range of bakery products such as bread, cakes and biscuits.

Wheat germ oil: The oil extracted from wheat germ used as a health supplement and high in vitamin E, which is destroyed by heating wheat hardness A measure of the ease of grinding of wheat into flour determined either by standard milling tests (time to grind to a particular size in standard apparatus), by biting the wheat seed (miller's test) or by measuring the force necessary to penetrate the wheat with a sharp point (micro penetration test). It is rich in linoleic acid & tocopherols; also contain α-linolenic acid.

Wheat germ: It is vitamin and lipid-rich embryo (sprouting portion) of the wheat grain. Milling of grain to produce white wheat flour results in separation of the germ, which may then be used to enrich bread and breakfast cereals. It constitutes about 2% of the total weight of the grain, also used in dietary supplements.

Wheat gluten: Complex formed when wheat proteins are mixed with water. It consists of glutenin and gliadins. Gluten forms an elastic network during kneading of dough, which is important for the texture of the bread. Gluten content of wheat varies among varieties.

Wheat malt: It is germinated wheat grains used in brewing and distillation.

Wheat starch dough: A dough suitable for food wrappers made from 2 parts wheat starch, 1 part tapioca starch and a little salt, briskly mixed with 4 parts of boiling water and a little oil, kept warm whilst resting then kneaded to a silky soft dough. Small chestnut-sized pieces can be rolled out or pressed out with the fingers or back of a knife.

Wheat starch: A gluten free wheat flour consisting mainly of starch. It is used as a thickener and is mixed with tapioca starch (2:1) to make a boiling water dough suitable, after kneading & rolling or flattening, for wrapping small parcels of food such as dim sum.

Wheat: Grain of cereal grasses belonging to the genus *Triticum* (particularly *T. aestivum*, and *T. durum*) which contains gluten, a protein complex important for the breadmaking properties of this grain. It contains roughly 85% endosperm, 13% bran and 2% wheat germ. It is used to make many food products, including pasta and breakfast cereals; wheat flour is used widely to make bakery products such as biscuits, cakes and bread.

Whey beverages: Drinks, sometimes sports drinks or nutritional beverages for specific population groups, based on whey. It can be alcoholic or non-alcoholic.

Whey cheese: Cheese prepared by concentrating whey and coagulating the proteins with heat and acids. The resulting curd is strained and possibly pressed. Milk or cream may be added to increase fat content or improve cheese flavour. Ricotta cheese is a well-known whey cheese.

Whey concentrates: It is concentrates prepared from whey and used in a variety of foods to supplement their nutritional value. Its Uses include preparation of sports foods and sports drinks, and dietetic products.

Whey cream: Any cream or fat still remaining in whey after the curds have been separated in cheese making

Whey protein concentrates: A products prepared from whey by separation of whey proteins using precipitation or ultrafiltration. Precipitation at a high temperature and low pH followed by centrifugation produces a concentrate of denatured, insoluble whey proteins. Ultrafiltration followed by vacuum evaporation and spray drying produces a concentrate of non- denatured, soluble proteins. Concentrates varying in composition can be made by controlling manufacturing conditions. Uses include adjustment of protein contents of various products, including infant formulas, dietetic products and protein-enriched foods for specific groups of people, e.g. athletes. Foaming properties of whey protein concentrates make them suitable for use in aerated foods and as replacements for egg whites.

Whey proteins: Milk proteins that remain in whey after manufacture of cheese. Sometimes it is called serum proteins. It consists of albumins (α-lactalbumin and serum albumin) and globulins (mainly β- lactoglobulin).

Whey: The translucent liquid which is formed when coagulated milk separates into a semi-solid portion (curds) and a liquid portion (whey). It contains most of the lactose of the milk and a small amount of protein and fat. Whey is sometimes used in making whey cheese, but is produced in large amounts as a waste, disposal of which poses problems for the dairy industry. Although mainly used in animal feeds, whey can be utilized as an ingredient in some foods and as a fermentation substrate. It is also called serum or lactoserum.

Whipped butter: Softened butter whipped to incorporated air so as to make it easy to spread.

Whipped cream: Cream containing between 35 and 40% butterfat which is whipped to incorporate air and to begin linking up the fat globules to make it a semi-solid. Cream in which the volume has been increased (overrun) by 90-100% by whipping in air. If whipping is carried on for too long the fat will become the continuous phase and it will turn to butter. For this reason, food processors must be carefully watched when being used to whip cream.

Whipping capacity: The extent to which a food can be whipped.

Whipping properties: It is a functional properties relating to the ability of a food to be whipped, increasing the volume by incorporation of air.

Whipping: Beating of ingredients, particularly cream and egg whites, during which air is incorporated into them, increasing their volume and creating a froth.

Whiskey: Alternative spelling of whisky. Spirits made by distillation of fermented mashes made from saccharified cereals, using raw materials, distillation conditions and ageing periods as specified by national regulations for the specific whiskey type.

White pepper: Common name for *Piper nigrum*, fruit of which are ground to produce spices. It is compared with black pepper, which is produced from fully grown, but unripe, fruit of *P. nigrum*, white pepper has a more delicate flavour. The major flavour compound of white pepper is piperine.

White sugar: Purified crystalline sugar containing approximately 1% moisture. It is dried to produce granulated sugar.

Whiteners: Substances used to whiten or bleach foods such as flour or fish. It may be used as substitutes for fresh milk in beverages including coffee (coffee whiteners), tea or cocoa, or in sauces. It is available as liquids or powders. These are prepared from milk proteins or non-dairy proteins (e.g. soy proteins) and fats, blended with other ingredients such as sugar, emulsifiers, stabilizers, buffers, flavour ings and colorants.

Whiteness: One of the optical properties; relating to the extent to which an item is white, i.e. snowy and milky in appearance.

Whole milk: Milk from which none of the fat has been removed. Fat content of milk varies according to species, being approximately 4% in cow milk. Milk is also available in other forms from which some (semi skimmed milk) or almost all (skim milk) of the fat has been removed. These other forms are preferred by some consumers wishing to limit their intake of fats.

Wholegrain foods: Foods made from whole, unrefined grains or wholegrain ingredients. Wholegrains contain the entire edible parts of a grain kernel, i.e. the germ, endosperm and bran, and are rich in many nutrients which are generally lost during refining. In addition, wholegrains are low in fat and cholesterol. Wholegrain foods include wholemeal bakery products and pasta, some breakfast cereals and brown rice. Consumption of wholegrain foods has been associated with a number of health benefits including reduced risks of developing certain cancers and heart disease.

Wholemeal: Flour or bread made from the entire cereal grain with none of the bran or germ removed.

Wine distillates: Intermediate products or finished spirits made by distillation of wines.

Wine gums: Sugar confectionery products with a chewy texture made with sucrose, glucose and either gum arabic or gelatin. It is often fruit-flavoured. Similar to fruit gums and to fruit jellies, although the latter are softer due to a higher moisture content.

Wine vinegar: Vinegar produced by acetic fermentation of wines, e.g. red wines, white wines or sherry. Wine vinegar has a wine-like flavour and is used more as a flavouring than as a condiment, e.g. as an ingredient of salad dressings. The normal concentration is 3 to 4 percent of acetic acid.

Wine yeasts: Yeasts used for fermentation of grape musts to produce wines. It may be spontaneously occurring yeasts, or pure yeasts cultures. It is mainly *Saccharomyces* spp., although other genera of yeasts may play a role in the early stages of fermentation.

Winemaking: It is a process of manufacture of wines. The basic process comprises crushing grapes, alcoholic fermentation of the grape juices and ageing of the wines. Many additional processes may be applied, including maceration, clarification, chaptalization, filtration, fining and, in the case of sparkling winemaking, secondary fermentation.

Wines: Alcoholic beverages manufactured by alcoholic fermentation of fruit musts or fruit juices. Generally refers to beverages produced from grapes (*Vitis* spp., mainly *V. vinifera*). Fruit wines are made from other fruit musts or juices. The term wines may also be used to refer to rice wines (made from saccharified rice mashes), and palm wines (made from palm sap).

Winnowers: Devices for blowing air through grain in order to remove the chaff. Winnowing is also used to separate the shell and some of the germ from cocoa beans during manufacture of chocolate.

Winterization: It is a process of removal of traces of waxes and higher melting glycerides, or stearin, from fats. Waxes are generally removed by rapid chilling and filtration. Separation of stearin usually requires very slow cooling in order to form crystals that are large enough to be removed by filtration or centrifugation. Cottonseed oils and groundnut oils are winterized to produce salad oils that remain liquid at low temperatures. Tallow and other animal fats are winterized for simultaneous production of hard fats and oleo oil. It is also known as destearination.

Withering: It is a process whereby plant material or foods become dry and shrivelled. Controlled withering can be undertaken either chemically or physically (including techniques such as freeze withering, solar wither- ing and warm air withering). Withering is commonly the first stage in the processing of teas. In some regions, wines are made from grapes which have been partially dried by withering in the sun before pressing.

Wood smoke: Smoke produced from the burning of wood. The type of wood used (e.g. oak, hickory, mesquite) influences the properties of the smoke and governs its application. It is used in flavourings and/or preservatives. Foods which are commonly processed using smoke include fish and meat. Smoke flavourings may be added to barbecue sauces or marinades.

Woolliness: Extent to which products, usually fruits, have a woolly texture, i.e. are dry and spongy. Woolliness is an adverse sensory property and physiological disorder, involving lack of juiciness, internal browning and inability to ripen, without variation in tissue moisture. It is associated with an imbalance in pectolytic enzymic activity during storage. Onset of woolliness can be quantified instrumentally and is characterized as a lack of crispness, low hardness values and low juiciness.

World Health Organization (WHO): The World Health Organization (WHO) is a specialized agency of the United Nations (UN) that helps countries to improve their health services and coordinates international action against diseases.

World Trade Organization (WTO): The World Trade Organization (WTO) is an international body based in Geneva, Switzerland, that promotes and enforces the provisions of trade laws and regulations. The WTO has the authority to administer and police new and existing free trade agreements, to oversee world trade practices, and to settle trade disputes among member states. The WTO was established in 1994 when the members of the General Agreement on Trade and Tariffs (GATT), a treaty and international trade organization, signed a new trade pact. The WTO was created to replace GATT, and began operation on 1 January 1995. The WTO has a significantly broader scope than GATT, expanding the GATT agreement to include trade in services and protections for intellectual property. 128 nations were contracting parties to the new GATT pact at the end of 1994, and became members of the WTO. By early 2000, the WTO had 136 members, and about 30 other countries had applied for membership. The WTO is controlled by a general council made up of member states' ambassadors who also serve on various subsidiary and specialist committees. The ministerial conference, which meets every two years and appoints the WTO's director-general, oversees the General Council.

Worts: Clarified extracts prepared from mashes based on malt, sometimes with addition of brewing adjuncts, and subsequently fermented to form beer. Worts are generally boiled with hops to extract hop bitter compounds.

Wrapping: Packaging, e.g. paper or soft material, used to cover or protect a food, particularly during retail and after selection by the consumer.

X

Xanthan gums: Gums produced by the bacterium *Xanthomonas campestris*. These gums are exo-polysaccharides composed of repeating pentasaccharide units comprising a cellulose backbone and trisaccharide side chains of D-mannose and D-glucuronic acid residues. The gums also contain variable quantities of pyruvic acid. It used widely in the food industry as thickeners due to their ability to produce highly viscous, highly stable aqueous solutions. Other uses include as emulsifiers, stabilizers and binding agents, and to provide body, e.g. in low fat foods.

Xanthophylls: Group of neutral yellow or brown carotenoid pigments that are oxygenated derivatives of carotenes and distributed widely in plants. It is useful as food colorants.

X-ray fluorescence spectroscopy: Spectroscopy technique in which the sample is irradiated with X-rays, causing emission of a characteristic X-ray photon and fluorescence, which is measured using a spectrophotometer.

X-rays: Penetrating electromagnetic radiation of very short wavelength, able to pass through many materials. X-rays are produced by bombarding a target, usually made of tungsten, with high-speed electrons. The shorter the wavelength of the X-ray, the greater is its energy and its penetrating power. Longer wavelengths, near the UV-ray band of the electromagnetic spectrum, are known as soft X-rays. The shorter wavelengths, closer to and overlapping the gamma-ray range, are called hard X-rays. A mixture of many different wavelengths is known as white X-rays, as opposed to monochromatic X-rays, which represent only a single wavelength. X-rays are used in the food industry for a wide range of analytical purposes, including detection of foreign

Xylitol: Naturally occurring polyol comprising 5 carbon atoms which has equivalent sweetness to sucrose. It is mnufactured by hydrogenation of xylose. It is used in sweeteners, especially for low sugar confectionery, since it is non-cariogenic.

Y

Yakult: Yakult Original is a fermented milk drink that contains over 650 crores of beneficial bacteria called *Lactobacillus casei* Shirota (SHIROTA strain). It has a unique sweet and tangy taste and is enjoyed by people of all ages. It is one of the probiotic foods that is drunk to help maintain the health of the gastrointestinal tract. It has benefits like (i) Improves Digestion: The Unique bacteria in Yakult Original helps to balance the gut microbiota, aiding digestion and promoting regular bowel movements. (ii) Boosts Immunity: Regular consumption can enhance the body's natural defences by stimulating the immune system. (iii) Supports Gut Health: By increasing the number of good bacteria in the gut, Yakult Original helps maintain a healthy digestive system.

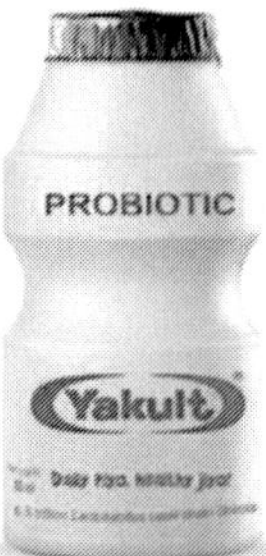

Yeast biomass: Quantitative estimate of the total population of yeasts present in a given habitat, in terms of mass, volume or energy.

Yeast extracts: Water soluble fraction of autolysed yeasts. During autolysis, yeast enzymes hydrolyse cytoplasmic proteins and carbohydrates. Insoluble cell wall material (cellulose) is removed, e.g. by centrifugation, to leave a clear extract of water soluble cellular material that is rich in amino acids and other nutrients. Yeast extracts are used as flavourings, as a source of nutrients for microbial fermentation, and as a source of B group vitamins for fortification of foods.

Yeast: A Unicellular microorganism (fungi) which reproduces by budding from the parent microorganism. The most important in cooking is *Saccharomyces cerevisiae* which has a multitude of variants used principally for converting sugars to alcohol or water and carbon dioxide as in beer, wine, and bread production. Yeasts can also excrete some enzymes which break down polysaccharides into simple sugars. Yeasts work best around 30 to 35°C and are killed above 60°C. It is capable of fermenting carbohydrates into alcohol and carbon dioxide. Some are responsible for food spoilage, while others are economically important as agents in breadmaking, brewing and winemaking, and in the production of single cell proteins, B vitamins and other fermentation products.

Yoghurt starters: Microbial cultures inoculated into milk to produce acidity by fermentation during manufacture of yoghurt. Commercial starter preparations generally contain *Lactobacillus bulgaricus* and *Streptococcus thermophilus*.

Yoghurt: A fermented product made from any milk treated with a culture of *Lactobacillus bulgaricus* and possibly *Streptococcus thermophilus* at a temperature of 37 to 44°C. The fermentation is stopped by cooling to below 5°C after 4 to 6 hours when the liquid will have developed a lactic acid flavour and will be more or less thick, possibly even a gel. Different cultures of the microorganisms and the different milks lead to country-specific textures and flavours. The raw natural yoghurt so obtained may be further pasteurized, sweetened, flavoured, thickened with gums or starches, have fruit added or be treated in a variety of other ways to satisfy Western tastes. It is made from whole milk, semi skimmed milk or skim milk, in a range of thicknesses, stirred or set, and in plain or flavoured varieties. Flavoured yoghurt is mixed with sugar and flavourings or fruits. It is also made into frozen yoghurt, a product resembling soft serve ice cream. Commercially, yoghurt is made using yoghurt starters (generally *Lactobacillus bulgaricus* and *Streptococcus thermophilus*). Other bacteria beneficial to gastrointestinal health, e.g. *L. acidophilus* and *Bifidobacterium bifidum*, may also be added. Pasteurization destroys the bacteria in yoghurt; unpasteurized product is known as live yoghurt. Yoghurt is rich in calcium and iodine and a source of protein and B vitamins. Many spelling variants for yoghurt are used in various parts of the world, including yogurt, yoghourt and yogourt.

Z

Zein: Prolamin which accounts for approximately half of the total storage proteins in corn. Contains minimal concentrations of lysine and tryptophan, but is rich in leucine.

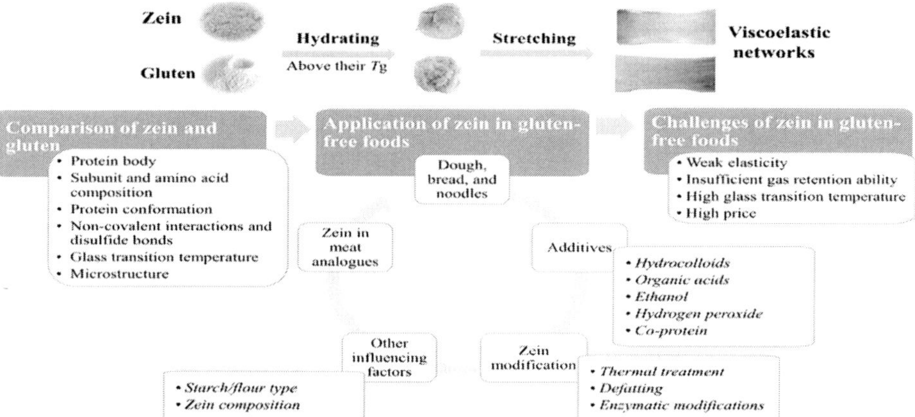

Zeolites: Crystalline, hydrated alkali-aluminium silicates. Useful as catalysts for production of invert sugar from sucrose, downstream processing of flavour compounds, detoxification of contaminated foods and feeds, and as molecular sieves.

Zero energy cool chamber (ZECC): Zero energy cool chambers is an on-farm rural oriented storage structure which operates on the principle of evaporative cooling and has been constructed using locally available raw materials such as bricks, sand, bamboo, rice straw, vetiver grass, jute cloth etc. The chamber has been constructed above the ground and comprises of a double-walled structure made up of bricks. The cavity of the double wall is filled with riverbed sand. The upper part of the chamber was covered with vetiver grass mat on a bamboo frame. Zero energy cool chamber (ZECC) is a double wall structure having space between the walls which is filled with porous water absorbing materials. These pads are kept constantly wet by applying water. When unsaturated air passes through wet pad, transfer of mass and heat takes place and the energy for the evaporation process comes from the air stream. Losses can be minimized by using best post-harvest handling techniques during storage, transportation and distribution to market. There are various technologies available to create and maintain optimal temperature, relative humidity and atmospheric composition for harvested fruits and vegetables during storage.

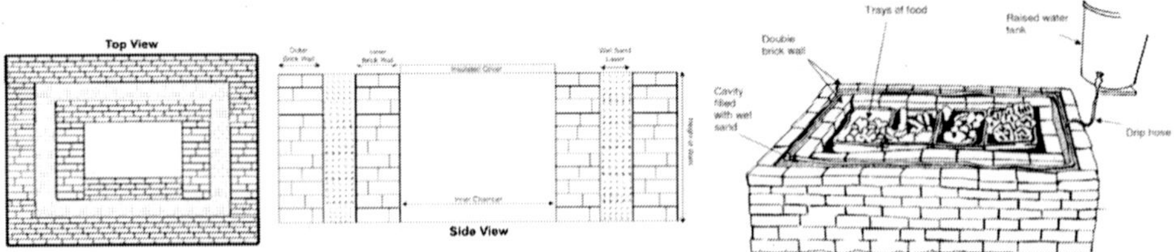

Zymase: Various enzymes which together induce the alcoholic fermentation of carbohydrates usually obtained from living yeast but may be added apart from the yeast.

Zyme: Yeast, the origin of the word enzyme, as the first enzymes were extracted from yeast.